Modern Projects and Experiments in Organic Chemistry

Miniscale and Standard Taper Microscale

Modern Projects and Experiments in Organic Chemistry

Miniscale and Standard Taper Microscale

JERRY R. MOHRIG
Carleton College

CHRISTINA NORING HAMMOND
Vassar College

PAUL F. SCHATZ
University of Wisconsin, Madison

TERENCE C. MORRILL
Rochester Institute of Technology

W. H. Freeman and Company
New York

Acquisitions Editor: Jessica Fiorillo/Yolanda Cossio
Development Editor: Robert Jordan
Marketing Manager: Mark Santee
Project Editor: Penelope Hull
Assistant Editor: Guy Copes
Media and Supplements Editor: Charlie Van Wagner
Text designer: Marsha Cohen
Cover designer: Michael Jung
Cover photo: Juniper Pierce
Illustrator: Fine Line Illustrations
Illustration Coordinator: Bill Page
Production Coordinator: Paul W. Rohloff
Compositor: Progressive Information Technologies
Printer and Binder: R R Donnelley & Sons Company

Library of Congress Control Number: 2002111343

ISBN-13: 978-0-7167-9779-1
ISBN-10: 0-7167-9779-8

Printed in the United States of America
Fifth printing

W. H. Freeman and Company
41 Madison Avenue, New York, NY 10010
Houndmills, Basingstoke RG21 6XS, England

www.whfreeman.com

Contents

Preface

Organic chemistry is an experimental science, and students learn its process in the laboratory. The primary goal in teaching laboratories should be to involve students in the process of organic chemistry. Students should be taken beyond the "cookbook" approach of verifying what the lab manual says to address questions whose answers come from the students' experimental data. Drawing reasonable conclusions from experimental results is at the heart of the process of science. Synthesizing organic compounds and answering questions that concern their reactions require the mastery of good laboratory technique and build scientific skills.

The Challenge

The weaknesses and relative ineffectiveness of organic "cookbook" experiments are well known, but the quandary has been to find a practical way to remedy the situation. It's usually not possible to ask sophomore and junior organic chemistry students to design experiments from scratch. Probably the best inquiry-driven organic laboratory pedagogy of the twentieth century was qualitative organic analysis. However, modern organic chemists no longer use chemical classification tests and the synthesis of solid derivatives to determine the identities of organic compounds. These methods are so dated that most lab programs have cut back drastically on their use.

What Is Different About This Book?

Modern Projects and Experiments in Organic Chemistry is a significant revision of the projects and experiments of *Experimental Organic Chemistry—A Balanced Approach: Macroscale and Microscale*. Many of the projects and experiments in this manual provide the opportunity for students to test ideas and address experimental questions in an atmosphere where they can succeed, making the organic chemistry laboratory rewarding and enjoyable for both students and their teachers. In some cases, we have recast standard experiments from a verification format to a question format, in which students answer questions by interpreting their experimental results. In other cases, we present new multiweek projects designed especially for inquiry-driven laboratory experiences. Every experiment and project has been repeatedly class tested. They use NMR and IR spectroscopy and analysis of reactions by TLC and GC extensively. Many IR spectra and 200- or 300-MHz FT NMR spectra of starting materials and products are on a CD-ROM that is available to users of *Modern Projects and Experiments in Organic Chemistry*.

Unlike other organic chemistry lab texts, *Modern Projects and Experiments in Organic Chemistry* has multiweek projects designed both for introductory first-semester lab work and for the more sophisticated second-semester laboratory. The projects are almost evenly divided between first- and second-semester course material. In fact, it is possible to plan an entire laboratory program around the project approach. Using inquiry-driven multiweek projects has several advantages. They provide a flexible, active-learning environment that allows students to become much more engaged with their laboratory work. They are readily adaptable to different lab session lengths, and they lower lab costs. Projects also engender effective teamwork opportunities, an important aspect in the process of modern science. Seven of the projects, as well as seven experiments, are designed for teamwork.

Our approach in *Modern Projects and Experiments in Organic Chemistry* is to provide experimental directions in the context of asking students to solve problems by the analysis of their experimental data. We state a purpose or propose a question at the start of every experiment or multiweek project. Nine of the 28 experiments and 13 of the 15 projects are question driven. Others deal with the synthesis of organic compounds and with learning basic laboratory techniques in the context of running organic reactions and purifying their products. A prelaboratory assignment at the beginning of every inquiry-based experiment and project provides context for the question being addressed and allows students to experience more effective learning in their laboratory work; it also gives instructors a built-in focus for their prelab discussions. The projects and experiments are carefully designed to teach students how to evaluate their experimental data and to reward their efforts with success.

Techniques in Organic Chemistry

The most effective way to use *Modern Projects and Experiments in Organic Chemistry* is in coordination with *Techniques in Organic Chemistry* by the same authors. *Techniques in Organic Chemistry* teaches the most important techniques used in organic chemistry, and specific references to sections of it appear throughout *Modern Projects and Experiments in Organic Chemistry*.

Modern instrumental methods play a crucial role in supporting the investigative experiments that provide active-learning opportunities. It has been said that you can't teach twenty-first-century science with nineteenth-century technology. Modern spectroscopic (NMR and IR) and chromatographic (GC and TLC) techniques are particularly important to understanding the experimental process of modern organic chemistry. With its focus on the interpretation of spectra and how they can be used to answer questions posed in the laboratory, *Techniques in Organic Chemistry* provides the practical information that students need to use spectroscopy effectively.

- *Techniques in Organic Chemistry: Miniscale, Standard Taper Microscale, and Williamson Microscale* (ISBN 0-7167-6638-8)

Computational Chemistry Experiments

An additional aspect of *Modern Projects and Experiments in Organic Chemistry* is the presence of seven experiments and projects with computational chemistry options, which have been designed to provide theoretical insights on students' experimental work. Experiment 7 presents the basics of computational chemistry as background that students can use in subsequent experiments.

Flexibility

A wide variety of experiments and projects are present in *Modern Projects and Experiments in Organic Chemistry*, permitting a great deal of flexibility in planning a laboratory program. The mix of microscale and miniscale experiments allows students to start with miniscale work to gain confidence before progressing to microscale. Many experiments include both miniscale and microscale procedures. Our miniscale experiments use the smallest reagent quantities that lead to successful results. The scale is tailored to fit the needs of the reaction and the methods used to characterize the product. Our microscale experiments use 150–250 mg of starting material, which provides significant savings in chemical and waste-disposal costs yet ensures that students almost always have the satisfaction of a successful outcome.

Modern Projects and Experiments in Organic Chemistry is published in two complete versions to suit whichever type of glassware is available in the organic chemistry laboratory. The miniscale procedures are designed for use with either 19/22 or 14/20 standard taper glassware. Users of 14/10 standard taper microscale glassware and users of the Williamson/Kontes microscale glassware will both find a compatible version to directly meet their needs. Both versions feature the same experiments and multiweek projects.

- *Modern Projects and Experiments in Organic Chemistry: Miniscale and Standard Taper Microscale* (ISBN 0-7167-9779-8)
- *Modern Projects and Experiments in Organic Chemistry: Miniscale and Williamson Microscale* (ISBN 0-7167-3921-6)

Safety and Waste Disposal

In any laboratory experience, safety and waste disposal are important issues. As in *Experimental Organic Chemistry—Macroscale and Microscale*, the issues have been considered in the development of every experimental procedure. Each procedure includes information on the safe handling of chemicals and equipment. Cleanup instructions on how to handle reaction by-products and waste materials are provided at the end of every procedure.

Custom Publishing

Because of the significant cost of chemistry textbooks, we envision that many professors will want to take advantage of W. H. Freeman and Company's custom publishing option for *Modern Projects and Experiments in Organic Chemistry*. To make the option effective, we have written each project and experiment as a stand-alone entity. Instructors can choose only the experiments and projects they wish to use in their lab classes and incorporate their own additional material so that the manual is perfectly organized to suit their course.

Visit http://custompub.whfreeman.com to learn more.

CD-ROM

An optional CD-ROM featuring video clips of many organic chemistry laboratory techniques plus many IR and 200- or 300-MHz FT NMR spectra of starting materials and products is available shrink-wrapped with either version of *Modern Projects and Experiments in Organic Chemistry* for a slight additional cost. To order the shrink-wrapped package, please use one of the following ISBNs.

- *Modern Projects and Experiments in Organic Chemistry: Miniscale and Standard Taper Microscale with CD* (ISBN 0-7167-5744-3)
- *Modern Projects and Experiments in Organic Chemistry: Miniscale and Williamson Microscale with CD* (ISBN 0-7167-5748-6)

Instructor's Manual

We have given special consideration to the *Instructor's Manual,* which will be available on the W. H. Freeman and Company Web site at www.whfreeman.com/mohrig. To receive an access code for this special password-protected Web site, instructors should contact their W. H. Freeman representative or send an e-mail message to chemistrymktg@whfreeman.com. It includes approximate times required for completion of the projects and experiments, the amounts of reagents, equipment, and supplies needed, good ways to dispense reagents, notes to the instructor about our experiences with the experiment or project, and answers to all questions.

ACKNOWLEDGMENTS

We have benefited greatly from the thoughtful insights and helpful suggestions of the following reviewers:

David Alberg, Carleton College
Christopher J. Cramer, University of Minnesota, Minneapolis
Christopher M. Hadad, Ohio State University
Scott T. Handy, SUNY, Binghamton
Gretchen E. Hofmeister, Gustavus Adolphus College
David K. Johnson, State University of New York–College at Geneseo
Scott B. Lewis, James Madison University
Rita Majerle, South Dakota State University
T. Andrew Mobley, Grinnell College
Michael W. Pelter, Purdue University, Calumet
Robert P. Pinnell, Scripps, Pitzer and Claremont McKenna Colleges
Nancy I. Totah, Syracuse University
Thottumkara K. Vinod, Western Illinois University
Susan E. Walden, University of Oklahoma
Jane E. Wissinger, University of Minnesota, Minneapolis

We thank our students and colleagues at Carleton College, Vassar College, and the University of Wisconsin, Madison, who have participated in class testing the projects and experiments developed for this new edition.

We wish to thank Jessica Fiorillo and Yolanda Cossio, our editors at W. H. Freeman and Company, for their vision and direction of this revision, Robert Jordan, Development Editor, for his helpful discussions in preparing the manuscript, Marsha Cohen, Designer, for the elegant book and cover design, Penelope Hull, Project Editor, for her masterly orchestration of the production stages, and Mark Santee, Chemistry Marketing Manager, for his enthusiastic direction of the marketing. JRM and CNH express heartfelt thanks for the patience and support of our spouses, the late Jean Mohrig and Bill Hammond, during the years that we have worked on this project.

PART 1

Experiments

EXPERIMENT

1 EXTRACTION OF CAFFEINE FROM TEA

PURPOSE: To learn several basic techniques of organic chemistry in the context of extracting caffeine from tea.

In Experiment 1 you will extract caffeine from tea leaves and in the process learn how to do the laboratory techniques of extraction, filtration, and evaporation of a solvent, as well as drying methods.

Pure caffeine is a white substance that melts at 236°C. It makes up as much as 5% of the weight of tea leaves. The biological action of caffeine includes cardiac and respiratory stimulation, and it has a diuretic effect as well. By structure, caffeine is closely related to the heterocyclic bases guanine and adenine found in deoxyribonucleic acids (DNA).

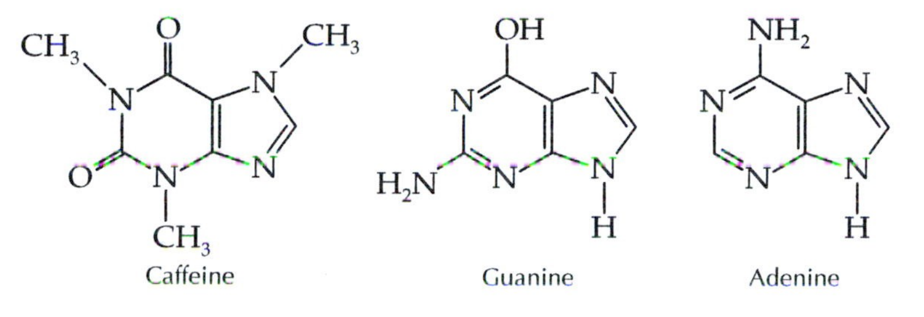

Caffeine Guanine Adenine

A number of plants contain caffeine, and its use as a stimulant predates written history. The origins of tea and coffee are lost in legend. Caffeine is a natural constituent of kola nuts and cocoa beans. Cola soft drinks may contain 14–25 mg of caffeine per 100 mL (3.6 oz), and a sweet chocolate bar weighing 20 g (0.7 oz) may contain 15 mg of caffeine. Whereas most tea has 3–5% caffeine by weight, coffee beans contain only about 2% caffeine. Yet a cup of coffee can have about 3.5 times as much caffeine as does a cup of tea. How can this be? Coffee is usually ground extremely fine or is heated in its brewing; tea leaves are simply steeped in hot water for a few minutes. Furthermore, less tea than coffee is used to brew one cup of beverage. A cup of tea contains about 25 mg of caffeine. Flavinoid pigments and chlorophylls also contribute to the color of a tea solution.

Drinking decaffeinated coffee has become much more popular since better methods of extracting caffeine from coffee beans have become available. One of the latest methods takes advantage of the weakly basic nature of caffeine. It uses carbon dioxide under high pressure on wet, raw coffee beans under conditions where CO_2 has the characteristics of both a gas and a liquid. An acid-base reaction between the acidic CO_2 and basic caffeine extracts the caffeine from the coffee beans.

Obtaining pure caffeine from tea requires a method for separating caffeine from the other substances found in tea leaves. Cellulose, the primary leaf component, poses no problem, because it is virtually insoluble in water. However, a large class of weakly acidic molecules called tannins also dissolve in the hot water. Tannins are colored compounds having phenolic groups that make them acidic. If calcium carbonate, a base, is added to the water, solid calcium salts of these acidic tannins precipitate from the aqueous tea solution. After filtering out the insoluble tea leaves and calcium salts, the caffeine can then be separated from the alkaline tea solution by the process of liquid/liquid extraction [see Technique 8] using dichloromethane, an organic solvent in which caffeine readily dissolves. Although caffeine is soluble in water (2.2 g/100 mL at 20°C), it is far more soluble in dichloromethane (10.2 g/100 mL at 20°C). Therefore, this extraction takes advantage of a distribution coefficient (K) of 4.6 [see Technique 8.1 for a discussion of distribution coefficients]. The calcium salts of the tannins remain dissolved in the aqueous solution. Although chlorophylls have some solubility in dichloromethane, the other pigments in tea do not. Thus, the dichloromethane extraction of a basic tea solution yields a solution of nearly pure caffeine, which has a slight green color from the chlorophyll impurity.

After the extraction procedure, the organic solution of dichloromethane and caffeine is dried with an anhydrous inorganic salt [see Technique 8.7]. Crude caffeine is recovered as a solid residue by evaporation of the dichloromethane.

MINISCALE PROCEDURE

Techniques Vacuum Filtration Apparatus: Technique 9, Figure 9.4
Extraction: Techniques 8.2, 8.3, and 8.4
Drying Organic Liquids: Techniques 8.7 and 8.8
Boiling Stones: Technique 6.1
Solvent Removal: Technique 8.9

SAFETY INFORMATION

Dichloromethane is toxic, an irritant, absorbed through the skin, and harmful if swallowed or inhaled. Use it in a well-ventilated hood. Wear gloves and wash your hands thoroughly after handling it.

Solid **caffeine** is toxic and an irritant. Avoid contact with skin, eyes, and clothing.

Before you begin your laboratory work, be sure to read Technique 8, Extraction and Drying Agents. It tells you how to do an extraction and explains why it works.

Place 9–10 g of tea leaves in a 400-mL weighed (tared) beaker; record the mass of the tea leaves (Note 1). If you use tea bags, four bags should contain about 10 g of tea; remove the tea leaves from the bags and place the tea in the beaker. Add 4.8 g of calcium carbonate and pour 125 mL of water over the tea. Boil the mixture gently on a hot plate for 15 min, stirring every minute or two with a stirring rod.

Let the tea mixture cool to about 55°C, then filter it, using vacuum filtration apparatus [see Technique 9, Figure 9.4], through S&S No. 410 or Whatman No. 54 filter paper (Note 2). Pour the tea mixture into the Buchner funnel in two portions. If the filter paper clogs while the first portion is filtering, replace it with a fresh piece before filtering the remainder of the tea mixture.

Cool the filtered solution to 15–20°C in an ice-water bath. Set a 125-mL separatory funnel in a support ring, and pour the cooled tea solution into

The amount of caffeine transferred from the aqueous layer to the dichloromethane layer depends on the amount of contact between the two phases; swirl the funnel fast enough to produce a good mixing of the phases without forming an appreciable amount of emulsion.

the separatory funnel (be sure the stopcock is closed). Add 15 mL of dichloromethane to the funnel. Stopper the separatory funnel, hold the stopper firmly in place with your index finger, and invert the funnel. Immediately open the stopcock to vent the vapors. Rotate the inverted funnel for 2–3 min so that the two layers swirl together many times, opening the stopcock frequently to vent the funnel (Note 3).

Allow the layers to separate and then drain the dichloromethane layer into a 50-mL Erlenmeyer flask. If a small emulsion layer is present at the interface between the organic and aqueous phases, add it to the Erlenmeyer flask. Cork the Erlenmeyer flask to prevent evaporation of the dichloromethane. Add 15 mL of fresh dichloromethane to the separatory funnel (still containing the tea solution) and repeat the extraction process. Again, allow the layers to separate and drain the dichloromethane layer, including any emulsion layer, into the Erlenmeyer flask containing the dichloromethane solution from the first extraction. Pour the tea solution out of the top of the separatory funnel into a beaker.

This process is called **washing** *the organic layer.*

Rinse the separatory funnel with water before pouring the combined dichloromethane solutions back into the funnel; add approximately 20 mL of water. Stopper, invert, and rock the funnel gently to mix the two layers. Some emulsion layer may be present at this point (Notes 4 and 5). Drain the lower dichloromethane layer slowly into a clean, dry 50-mL Erlenmeyer flask.

Add anhydrous magnesium sulfate to the dichloromethane solution [see Technique 8.7]. Cork the flask and allow the mixture to stand for at least 10 min, swirling the flask occasionally.

Weigh (tare) a dry 50-mL Erlenmeyer flask on a balance that measures to 0.001 g. Place a fluted filter paper in a dry conical funnel and filter the drying agent from the dichloromethane solution [see Technique 8.8], collecting the filtrate in the tared 50-mL Erlenmeyer flask. Rinse the magnesium sulfate remaining in the flask with approximately 2 mL of dichloromethane and also pour this rinse through the funnel. Add a boiling stick or boiling stone to the flask containing the dichloromethane solution so that it boils without bumping.

Evaporate the dichloromethane in a hood on a steam bath or a water bath heated on a hot plate. Alternatively, the dichloromethane may be removed by evaporation, using a stream of nitrogen or air in a hood or with a rotary evaporator [see Technique 8.9]. Continue the evaporation until a dry greenish residue of crude caffeine forms on the bottom of the flask. Weigh the flask and determine the mass of crude caffeine. Calculate the percent recovery. Cork the flask and store it in your laboratory drawer for purification and analysis in Experiment 2.

CLEANUP: Place the tea leaves in the nonhazardous solid waste container. Wash the tea solution remaining from the initial extractions and the water remaining in the 125-mL separatory funnel down the sink. Allow the flask containing the magnesium sulfate drying agent to dry in a hood before putting the spent drying agent in the container for inorganic waste.

Notes About Miniscale Procedure

1. *The 10 g of tea leaves used in this procedure should contain at least 300 mg of caffeine. You should be able to recover 10–30% of this amount.*
2. *If the mixture is filtered when it is too hot, messy bubbling occurs in the filtrate and some solution may be lost. Yet if it is filtered when it is too cool, the gelatinous material that separates on cooling clogs the pores of the filter paper. Fast, nonretentive filter papers such as Schleicher and Schuell (S&S) No. 410 and Whatman No. 54 work well.*
3. *Although you will often want to shake a separatory funnel vigorously to bring about efficient mixing of the two layers, in this extraction a frustrating emulsion (a milky looking mixture that does not separate easily into the two phases) results from such action. Swirling the two layers together rather than shaking should prevent or minimize the formation of an emulsion.*
4. *If only a thin layer of emulsion exists at the interface between the aqueous phase and the dichloromethane solution, push a small piece of glass wool to the bottom of the dichloromethane layer with a large stirring rod. The glass wool will break the membranes of the emulsion.*
5. *If the amount of emulsion is more than a thin film, set up a vacuum filtration using a 4.5-cm Buchner funnel containing a silicone-treated filter paper such as Whatman 1PS. Drain the lower layer and any emulsion layer directly from the separatory funnel into the Buchner funnel with the vacuum source turned on. Pour the remaining aqueous layer out of the separatory funnel, before carefully pouring the filtrate into the separatory funnel (be sure to use a conical funnel in the top of the separatory funnel). Then slowly drain the lower organic phase into a clean, dry 50-mL Erlenmeyer flask.*

MICROSCALE PROCEDURE

Techniques Vacuum Filtration Apparatus: Technique 9, Figure 9.4
Using Pasteur Pipets: Technique 8.5
Microscale Extraction: Technique 8.6b
Drying Agents: Techniques 8.7 and 8.8
Solvent Removal: Technique 8.9

SAFETY INFORMATION

Dichloromethane is toxic, an irritant, absorbed through the skin, and harmful if swallowed or inhaled. Use it in a well-ventilated hood. Wear gloves and wash your hands thoroughly after handling it.

Solid **caffeine** is toxic and an irritant. Avoid contact with skin, eyes, and clothing.

Before you begin your laboratory work, be sure to read Technique 8, Extraction and Drying Agents. It tells you how to do an extraction and explains why it works.

Pour the contents of two tea bags into a tared (weighed) 150-mL beaker; weigh and record the mass of the tea. Add 50 mL of water and 2.4 g of calcium carbonate to the beaker containing the tea. Boil the mixture gently on a hot plate for 10 min.

Let the tea mixture cool to about 55°C. Using a 5-cm Buchner funnel and a 125-mL filter flask, filter the mixture using vacuum filtration [Technique 9, Figure 9.4] through S&S No. 410 or Whatman No. 54 filter paper.

Pour 12 mL of water into a clean 150-mL beaker and mark the outside of the beaker at the 12-mL level. Discard the water before pouring the tea solution into the beaker. Boil the tea solution on a hot plate until the liquid level reaches the 12-mL mark on the beaker. Cool the solution briefly on the bench top.

Your yield of caffeine is directly related to how thoroughly the organic and aqueous phases are mixed with each other.

Transfer the tea solution to a 15-mL screw-capped centrifuge tube, then cool the tube in a water bath containing a few ice chips for 2–3 min. Add 2 mL of dichloromethane to the centrifuge tube, cap the tube tightly, and shake the mixture vigorously. Spin the tube containing the mixture in a centrifuge for approximately 1 min to separate the phases and disperse the emulsion (milky layer).

SAFETY PRECAUTION

Balance the centrifuge by placing another centrifuge tube of equal mass in the hole opposite the one containing the sample tube.

Prepare a Pasteur filter pipet or a Pasteur pipet fitted with a syringe [see Technique 8.5]. Insert the pipet to the bottom of the centrifuge tube. Draw the lower organic layer into the pipet until the interface between the two layers is exactly at the bottom of the centrifuge tube [see Technique 8.6b]. Transfer the pipet to a second 15-mL centrifuge tube and expel the organic phase into the tube. Cap the second tube to prevent evaporation of dichloromethane. The tea solution remains in the first centrifuge tube.

Repeat the extraction process two more times, using a new 2-mL portion of dichloromethane each time. After each extraction, transfer the lower dichloromethane layer to the second centrifuge tube that contains the dichloromethane/caffeine solution from the previous extraction.

Add anhydrous magnesium sulfate to the combined dichloromethane/caffeine solution [see Technique 8.7]. Cap the tube and allow the mixture to stand for a minimum of 10 min, swirling the tube occasionally.

Prepare a microfunnel from a Pasteur pipet and cotton as shown in Technique 8.8, Figure 8.17b. Be sure that the cotton is packed tightly at the top of the tip of the pipet. Clamp the microfunnel so that the tip is inserted about halfway into a 25-mL Erlenmeyer flask whose mass is known to the nearest 0.001 g. Transfer the dried dichloromethane/caffeine solution to the microfunnel, using a clean Pasteur filter pipet. Rinse the magnesium sulfate remaining in the centrifuge tube with about 0.5 mL of dichloromethane and transfer this rinse to the microfunnel.

Evaporate the dichloromethane on a steam bath in a hood. Alternatively, the dichloromethane may be removed by evaporation using a stream of nitrogen or air in a hood. Continue the evaporation until a dry greenish residue of crude caffeine forms on the bottom of the flask. Weigh the flask and determine the mass of crude caffeine. Calculate the percent recovery. Cork the flask and store it in your laboratory drawer for purification and analysis in Experiment 2.

CLEANUP: Place the tea leaves in the container for nonhazardous solid waste. Wash the tea solution remaining from the initial extractions and the water remaining in the centrifuge tube down the sink. Allow the centrifuge tube containing the magnesium sulfate drying agent to lie on its side in a hood until the residual dichloromethane evaporates before disposing of the spent drying agent in the container for inorganic waste.

Questions

1. Why is the tea boiled with water in this experiment?
2. Why is the aqueous tea solution cooled to 15–20°C before the dichloromethane is added?
3. Why does the addition of salt (NaCl) to the aqueous layer sometimes help to break up an emulsion that forms in an extraction?
4. The distribution coefficient for caffeine in dichloromethane and water is 4.6. Assume that your 100-mL tea solution contained 0.30 g of caffeine. If you had extracted with only one 15-mL portion of dichloromethane, how much caffeine would have been left in the water solution? How much would be left in the water after the second 15-mL dichloromethane extraction? How much caffeine would be left in the water solution if only one extraction with 30 mL of dichloromethane were performed?
5. Why is less caffeine actually isolated than is suggested by the calculation in Question 4?

EXPERIMENT

2

PURIFICATION AND THIN-LAYER CHROMATOGRAPHIC ANALYSIS OF CAFFEINE

PURPOSE: To continue the study of the caffeine that you isolated in Experiment 1.

> In Experiment 2 you will use sublimation to separate chlorophyll from caffeine and thin-layer chromatography to assess the purity of the caffeine.

The extraction of caffeine from tea leaves yields a remarkably pure product, considering the complex composition of tea. However, the isolated caffeine is not completely pure. The greenish color immediately suggests the presence of an impurity, because caffeine is white. The major impurity is the chlorophyll present in the leaves.

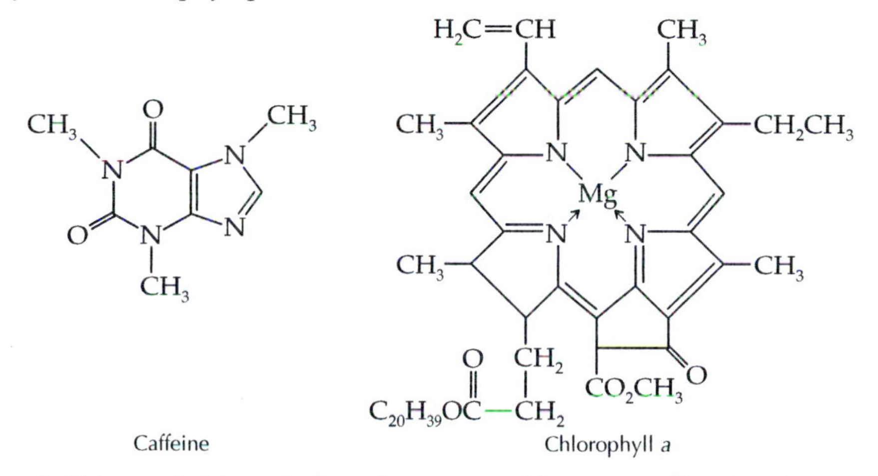

Caffeine

Chlorophyll *a*

Caffeine and chlorophyll can be separated by recrystallization, but in this experiment, you will use a simpler technique, vacuum sublimation [see Technique 12]. When a substance sublimes, it goes directly from the solid to the gas phase without ever being a liquid. Sublimation works well for purifying an organic compound if three conditions are satisfied: (1) the compound vaporizes without melting or decomposing; (2) the vapor recondenses to the solid; (3) the impurities present do not sublime.

After completing the sublimation, you will use thin-layer chromatography (TLC) to assess the purity of your caffeine sample.

MICROSCALE PROCEDURE

Techniques Sand Bath for Heating: Technique 6.2
Sublimation: Technique 12
Thin-Layer Chromatography: Technique 15

Sublimation of Crude Caffeine

Begin heating a sand bath to 170–180°C. Place about 30 mg of your crude caffeine from Experiment 1 in the bottom of a 25-mL filter flask or a side-arm test tube. Fit the filter adapter on the inner test tube or 15-mL centrifuge tube and assemble the sublimation apparatus as shown in Figure 2.1. Half fill the inner test tube or centrifuge tube with cold tap water (ice is not necessary for this sublimation).

SAFETY PRECAUTION

> The inner test tube or centrifuge tube must fit tightly in the filter adapter, or the difference in pressure between the atmosphere and the vacuum inside the apparatus may push the tube forcibly against the bottom of the filter flask or side-arm test tube, shattering both.

Clamp the flask securely in the sand bath and attach the side arm to a water aspirator or the vacuum line with a guard flask (filter flask or trap bottle) between them. Use thick-walled tubing. Turn on the vacuum source.

Adjust the level of the filter flask so that the sand covers a few millimeters of the flask wall. Be sure that the temperature of the sand bath does not rise above 180–185°C, because a higher temperature will cause the crude caffeine to char or melt. As you heat the sample, you will notice white caffeine migrating first to the surface of the crude caffeine, then to the walls of the flask and the outside of the cold inner tube.

When the residue in the bottom of the flask has become dark green and no more caffeine appears to be collecting on the inner tube and the flask walls, the sublimation is complete. Carefully raise the sublimation apparatus from the sand bath and slowly let air back into the system by

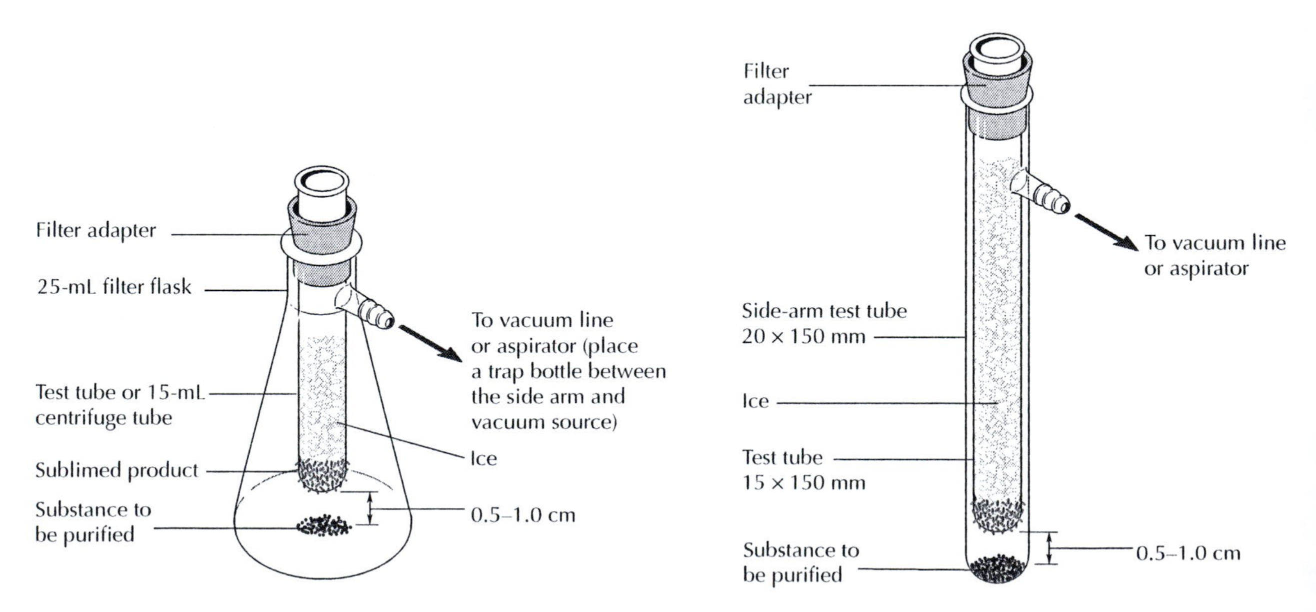

FIGURE 2.1 Microscale sublimation apparatus.

gently removing the rubber tubing from the water aspirator nipple before turning off the aspirator. If you are using a vacuum line, turn off the vacuum and slowly remove the rubber tubing from the side arm. Carefully remove the inner test tube from the flask. Use a spatula to scrape the pure caffeine onto a piece of tared (known mass) weighing paper, and weigh your purified product. Then store the caffeine in a small test tube. The sublimed caffeine should be much lighter in color than the crude material.

Thin-Layer Chromatography

SAFETY INFORMATION

> **Dichloromethane** is toxic, an irritant, absorbed through the skin, and harmful if inhaled. Avoid inhalation or contact with skin, eyes and clothing. Wear nitrile gloves and dispense it only in a hood.
>
> **Ethyl acetate** is toxic, an irritant, and very flammable. Avoid inhalation or skin contact. Keep the developing chamber closed except when inserting or removing a TLC plate.
>
> Never look directly into an **ultraviolet lamp.** Radiation in this region of the spectrum can cause eye damage.

Read Technique 15, Thin-Layer Chromatography, before undertaking this part of the experiment.

Plastic-backed or aluminum-backed precoated silica gel TLC plates with fluorescent indicator and 2.5 × 6.7 cm dimensions work well for the chromatographic separation of caffeine.*

Weigh about 8 mg of your remaining crude caffeine and dissolve it in 0.50 mL of dichloromethane in a small test tube. Cork the test tube when the solution is not being used.

Dissolve 6–8 mg of your sublimed caffeine in 0.50 mL of dichloromethane in a second test tube. Label the two tubes. Again, cork this solution when it is not being used.

Ethyl acetate, a solvent of intermediate polarity, will separate caffeine from the highly polar chlorophyll molecules.

Prepare a developing chamber using a wide-mouthed, capped bottle. A piece of 7- or 9-cm filter paper makes a good wick to ensure solvent saturation in the developing chamber's atmosphere. Add 4–5 mL of ethyl acetate to the developing chamber, cap the chamber, and shake it vigorously. After the filter paper and the atmosphere have been saturated, the solvent should be 2–3 mm deep in the bottom of the chamber. If you have a greater depth, remove some of the solvent with a Pasteur pipet and save it in a corked test tube. If you do not have enough solvent, add another 1–2 mL of ethyl acetate to the developing chamber.

*TLC plates are available from many suppliers. In developing the experiment, we used Eastman Chromatogram Silica Gel sheets with fluorescent indicator.

Prepare several micropipets according to the instructions in Technique 15.3 or obtain micropipets from the supply in the laboratory.

Spotting technique: Just touch the micropipet against the TLC plate and immediately lift it. Longer contact with the adsorbent leads to a diffuse spot, not one containing a more concentrated sample.

Make a mark about 2 mm long with a pencil at the long edge of a 2.5 × 6.7 cm silica gel TLC plate approximately 1 cm from the bottom edge, as shown in Technique 15, Figure 15.4. Dip a clean micropipet into the known solution of caffeine available in the laboratory and apply one spot of the solution at about one-third the width of the TLC plate. With another micropipet, apply one spot of your crude caffeine solution at about two-thirds the width of the TLC plate. Avoid holding the micropipet against the TLC plate for a prolonged time; this practice only produces a larger diffuse spot, not a spot containing a more concentrated sample.

Do not disturb the developing chamber during the chromatography.

Develop the chromatogram by placing the TLC plate in the ethyl acetate developing chamber with a pair of tweezers. Keep the chamber capped at all times. Allow the solvent to rise up the plate until it is about 1.0 cm from the top.

Remove the plate from the chamber with a pair of tweezers and *immediately mark the solvent front* with a pencil line before the solvent begins to evaporate. After marking the solvent front position, let the solvent evaporate in a hood for 2–3 min.

Visualize the results of your chromatogram by shining ultraviolet radiation (254 nm) on the plate in a darkened room or dark box. You should notice dark spots wherever the caffeine is present on the plate. Outline each spot with a pencil while the plate is under the ultraviolet lamp. When you have the TLC plate in ordinary light, you can see the small, pale green spot of chlorophyll; circle it with a pencil as well. Alternatively, use iodine vapor to visualize the plate (Note 1).

Calculate the R_f values for caffeine and chlorophyll under your experimental conditions [see Technique 15.1 and Figure 15.2]. Be sure to record in your notebook all the data needed to reproduce your R_f values; draw diagrams of your chromatograms. Also estimate the size of the spots and note whether there was any tailing (Note 2).

Prepare a second TLC plate with spots of the known caffeine solution and your sublimed (purified) caffeine solution. If a chlorophyll solution is available, prepare another TLC plate with spots of chlorophyll and your crude caffeine. Repeat the development and visualization procedures and again calculate the R_f values. Replenish the ethyl acetate in the developing chamber if the level falls below 2 mm.

CLEANUP: Pour any remaining ethyl acetate into the container for flammable or organic waste. Pour the dichloromethane solutions of crude and pure caffeine into the container for halogenated organic waste. Place the solvent-saturated filter paper and the chromatograms in a waste container for solids in a hood.

Notes About TLC Procedure

1. *An iodine chamber may be used to visualize the spots on the developed chromatogram [see Technique 15.6]. The spots must be outlined with a pencil immediately after removing the plate from the iodine chamber. The iodine visualization method shows a small amount of an impurity having a higher R_f value than caffeine.*

2. *If your spots turn out to be very large and show tailing, you can spot less sample on a different plate by diluting your caffeine solution with a few more drops of dichloromethane and by using a micropipet with a smaller-diameter tip. If the spots are very small or very faint, you can spot more sample by applying repetitive spots to the same position; let the solvent evaporate before overspotting the first spot.*

OPTIONAL EXPERIMENT Evaluating TLC Parameters

TEAMWORK: Work in teams of two students in examining the effects of changes in conditions on the TLC analysis of caffeine. You can decide how to divide the tasks equitably.

SAFETY INFORMATION

Ethanol is flammable.

Hexane is extremely flammable and an irritant. Wear gloves when handling it.

You will probably use TLC analysis a number of times during your study of organic chemistry, so you might want to see how variable conditions change your results. Calculate the R_f value for caffeine (and chlorophyll) using one of the following experimental conditions (or more, if you have time). Remember that you can apply two or more separate samples on each plate and do not forget to mark the starting point and solvent front on each TLC plate.

1. Use repetitive spotting to apply three times as much of the crude caffeine sample to a TLC plate as you used in the experiment. In a separate application, try six times as much. Use ethyl acetate to develop the plate. Calculate the R_f value.

For optional experiments 2–5, the solvent in the developing bottle needs to be changed. Remove the previous solvent and the paper wick from the bottle and dry the jar. Put a new filter paper wick and 3–4 mL of the next solvent in the developing jar. Shake the capped bottle to ensure saturation of the paper.

2. Use a more polar solvent to develop your chromatogram. Make up a solution containing 1 part ethanol to 9 parts ethyl acetate by volume.

Do the rest of the chromatography in the usual way on your crude caffeine. Calculate the R_f value.

3. Try the thin-layer chromatography of crude caffeine using a developing solvent of 1:1 (v/v) solution of ethanol/ethyl acetate. Calculate the R_f value.
4. Try the TLC of crude caffeine using hexane as the developing solvent. Calculate the R_f value.
5. Try the TLC of crude caffeine using 0.5% acetic acid in ethyl acetate as the solvent. Calculate the R_f value. Compare the R_f value to that of caffeine developed in ethyl acetate.

CLEANUP: Pour your dichloromethane solutions of crude and purified caffeine into the container for halogenated organic waste. Pour any remaining developing solvents—ethyl acetate, ethyl acetate/ethanol, or hexane—into the container for flammable or organic waste. Place the solvent-saturated filter paper and the chromatograms in a waste container for solids in a hood.

Questions

1. Why does caffeine have a larger R_f value than chlorophyll?
2. Why can there be no breaks in the thin-layer surface of a TLC plate?
3. Two compounds have the same R_f (0.87) under identical conditions. Does this show that they have identical structures? Explain.
4. Explain the observed effect of the presence of acetic acid in the developing solvent on the R_f value of caffeine in optional experiment 5.

EXPERIMENT

3 SYNTHESIS OF ETHANOL BY FERMENTATION OF SUCROSE

PURPOSE: To use the biochemical process of fermentation to produce ethanol.

In Experiment 3 you will use the enzymes of yeast to catalyze the synthesis of ethanol from sucrose. Then you will purify the ethanol by fractional distillation and assess its purity by density measurements.

Sucrose, common table sugar, is a disaccharide of molecular formula $C_{12}H_{22}O_{11}$. Some of the many enzymes in yeast catalyze its hydrolysis into the monosaccharides fructose and glucose, which after conversion to their phosphates undergo glycolysis, leading to ethanol and carbon dioxide by a series of reactions. The fermentation solution also contains a mixture of potassium phosphate, calcium phosphate, magnesium sulfate, and ammonium tartrate. Over 100 years ago, Pasteur discovered that these salts increase yeast growth and promote the fermentation process. In addition to ethanol, the fermentation produces small amounts of other compounds, such as acetaldehyde, 1-propanol, 2-propanol (isopropyl alcohol), 2-methyl-1-butanol, and 3-methyl-1-butanol, all of which contribute to the odor of biosynthesized ethanol.

$$\underset{\substack{\text{Sucrose} \\ \text{MW 342.3}}}{C_{12}H_{22}O_{11}} \xrightarrow[\text{invertase}]{H_2O} \underset{\text{Glucose}}{C_6H_{12}O_6} + \underset{\text{Fructose}}{C_6H_{12}O_6} \xrightarrow{\text{zymase}} \underset{\substack{\text{Ethanol} \\ \text{MW 46.1}}}{4CH_3CH_2OH} + 4CO_2$$

The ethanol formed in the reaction serves to inhibit the fermentation process by killing the yeast cells, so fermentation stops when the alcohol content of the solution approaches 12% by volume. Death of the yeast cells accounts for the approximately 12% alcohol content of most wines. Fractional distillation must be used to obtain more concentrated ethanol solutions [see Technique 11.4]. However, fractional distillation still does not yield 100% pure ethanol, because the azeotropic mixture of 95% ethanol and 5% water (by weight) boils at 78.1°C, whereas 100% ethanol boils at 78.4°C, [see Technique 11.5].

All lab operations following the fermentation can be completed within one laboratory period. The action of the yeast on sucrose, however, does not occur quickly, so the reactants need to be mixed together a week before the ethanol is to be isolated.

MINISCALE PROCEDURE

Techniques Boiling Points: Technique 11.1
Simple Distillation: Technique 11.3
Fractional Distillation: Technique 11.4
Azeotropic Distillation: Technique 11.5

SAFETY INFORMATION

Celite is a lung irritant. Avoid breathing the dust when handling this material.

Fermentation

Note to instructor: Prepare Pasteur's salt solution by dissolving 2.0 g of potassium phosphate, 0.20 g of calcium phosphate, 0.20 g of magnesium sulfate, and 10.0 g of ammonium tartrate in 860 mL water.

Oxygen in the fermenting mixture causes oxidation of ethanol to acetic acid (vinegar).

Place 40 g of sucrose in a 500-mL round-bottomed flask. Add 200 mL of water and 3.0 g of dry yeast. Stir until the sugar dissolves and no yeast granules remain. Then add 35 mL of Pasteur's salt solution and stir to mix. Close the flask with a one-hole rubber stopper fitted with a piece of bent glass tubing, as shown in Figure 3.1. Fill an 18 × 150 mm test tube halfway with saturated aqueous calcium hydroxide (limewater) and submerge the other end of the glass tubing about 1 cm below the surface of the solution. Limewater serves to exclude atmospheric oxygen from the fermentation mixture and simultaneously prevents a pressure increase from the CO_2 formed by the reaction. Store the mixture at room temperature for 1 week to allow complete fermentation. (Bubbling ceases when the reaction is complete.)

When the reaction is complete, add 10 g of Celite filter aid* to the fermentation mixture and stir to wet the Celite. Set up a vacuum filtration apparatus with a 500-mL filter flask [Technique 9.5, see Figure 9.4]. Wet the filter paper with water and turn on the vacuum source. Pour the reaction mixture slowly onto the filter paper. Rinse the flask with a few milliliters of water and also pour this rinse over the filter cake. Save the filtrate for the next step.

*Celite is a trade name for diatomaceous earth, a powdered inert material made from the shells of diatoms. The tiny particles of yeast cell debris clog the pores of filter paper. The Celite catches the cell debris before it reaches the filter paper, thereby allowing rapid filtration of the solution.

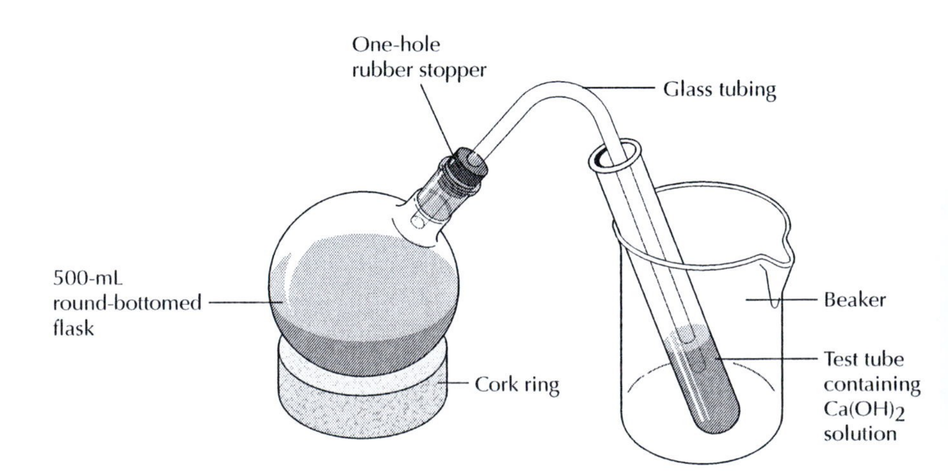

FIGURE 3.1
Apparatus for fermentation of sucrose.

Simple Distillation

Wash the 500-mL round-bottomed flask and pour the filtrate into it. Add two boiling stones. Assemble the apparatus for simple distillation, as shown in Technique 11.3, Figure 11.6, substituting a 100-mL graduated cylinder for the receiving flask. Heat the ethanol solution to boiling and adjust the rate of heating so that the distillate flows into the receiving flask at a rate of about 1 drop per second. Record the temperature as each 2-mL portion of distillate is collected. Turn off the heat and remove the heating mantle from the round-bottomed flask after 50 mL of distillate have been collected.

Tare (weigh) a dry, corked 10-mL Erlenmeyer flask on a balance that weighs to 0.001 g. Carefully pipet 10.0 mL of the distillate into the tared flask and then weigh the flask again. Use this information to determine the density of your ethanol solution. From the density value, you can then determine the ethanol content using Table 3.1.

Fractional Distillation

Pour all the distillate (including the 10.0 mL used for the density determination) into a 100-mL round-bottomed flask, add two boiling stones, and assemble the fractional distillation apparatus shown in Technique 11.4, Figure 11.15, substituting a 10-mL graduated cylinder for the receiving flask. The fractionating column is packed with stainless steel sponge from scouring pads available in any supermarket.

Heat the solution to boiling, then moderate the rate of heating so that the vapor (visible by noting the position of the condensate ring) in the fractionating column ascends slowly, thereby giving the vapor and condensate time to equilibrate. During the course of the distillation, gradually increase the rate of heating so that vapor continues to reach the condenser. A distillation rate of 1 drop every 2–3 s is usually satisfactory.

TABLE 3.1 Density, percentage by weight, and percentage by volume of ethanol in H_2O at 20°C

Density	% by weight	% by volume	Density	% by weight	% by volume
0.9893	5.0	6.2	0.8557	75.0	81.3
0.9819	10.0	12.4	0.8436	80.0	85.5
0.9752	15.0	18.5	0.8310	85.0	89.5
0.9687	20.0	24.5	0.8180	90.0	93.3
0.9617	25.0	30.4	0.8153	91.0	94.0
0.9539	30.0	36.2	0.8125	92.0	94.7
0.9450	35.0	41.8	0.8098	93.0	95.4
0.9352	40.0	47.3	0.8070	94.0	96.1
0.9248	45.0	52.7	0.8042	95.0	96.8
0.9139	50.0	57.8	0.8013	96.0	97.5
0.9027	55.0	62.8	0.7984	97.0	98.1
0.8911	60.0	67.7	0.7954	98.0	98.8
0.8795	65.0	72.4	0.7923	99.0	99.4
0.8676	70.0	76.9	0.7893	100.0	100.0

You must not allow the distillation to stop until you have collected the amount of distillate desired. A smooth, continuous progression through the fractions described in the next paragraph, with ever-increasing temperature changes, indicates that the procedure is being done correctly. Record the temperature as each milliliter of distillate is collected.

Label two 50-mL Erlenmeyer flasks "fraction 1" and "fraction 2." Collect fraction 1 from 77–80°C, fraction 2 from 80–96°C, and fraction 3 above 96°C. If the fractionating column is working well and the rate of distillation is satisfactory, the observed boiling point should remain steady while the ethanol/water azeotrope distills (fraction 1). When 10 mL of distillate have been collected, quickly pour the distillate into the flask labeled "fraction 1." Continue adding 10-mL portions to this flask until the temperature rises above 80°C. At this point, quickly add whatever amount is in the cylinder to fraction 1 and begin collecting fraction 2. If the column is efficient, fraction 2 will be a small volume collected during a rapid rise in temperature to 95°C or slightly higher. When the temperature begins to stabilize again, pour the distillate into the flask labeled "fraction 2." Continue the distillation until 4 mL of fraction 3 have been collected.

Weigh a dry, corked 10-mL Erlenmeyer flask, pipet 10.0 mL of fraction 1 into the flask, and weigh it again. Calculate the density of fraction 1 and determine the % ethanol by weight and volume from Table 3.1. From the total volume of fraction 1, its density, and the weight % of ethanol, calculate the percent yield of ethanol from the fermentation reaction. Submit your product (all of fraction 1) to your instructor in a container labeled with your name, the name of the product, the boiling-point range, the volume, the density, and the date.

CLEANUP: Place the Celite in the nonhazardous solid waste container. The residues in the boiling flasks from both the simple and fractional distillations may be washed down the sink. Fractions 2 and 3 may also be discarded down the sink.

Treatment of Data

Plot the data for both the simple distillation and the fractional distillation on the same set of coordinates, using the volume of distillate as the abscissa (x-axis) and the temperature as the ordinate (y-axis).

Questions

1. After the fermentation, what is the precipitate in the $Ca(OH)_2$ solution? Write the balanced equation for the reaction that produces it.
2. Why is it impossible for 100% pure ethanol to be obtained from fractional distillation of the fermentation mixture?
3. What is the composition of fraction 3?
4. Does your graph of the simple and fractional distillations show any difference in the shapes of the two curves? What does this difference, if any, indicate about the efficiency of fractional distillation relative to that of simple distillation for separating the ethanol/water azeotrope from water?

EXPERIMENT

4

SYNTHESIS OF SALICYLIC ACID

PURPOSE: To convert oil of wintergreen into salicylic acid by a base-catalyzed hydrolysis reaction.

In Experiment 4 you will carry out a reaction at an elevated temperature, purify the solid product by recrystallization, and evaluate its purity by the melting point.

Salicylic acid is a white, crystalline compound, commonly used for the removal of warts from the skin. Derivatives of salicylic acid are familiar and important compounds. Among them are methyl salicylate (oil of wintergreen) and acetylsalicylic acid (aspirin). Methyl salicylate has a fragrant, minty smell that has made it a favorite flavoring in candies. It is a major constituent of oil of wintergreen, making up over 90% of the oil from the wintergreen plant. While you explore the chemistry of these compounds, you will be learning techniques for purifying and analyzing organic solids.

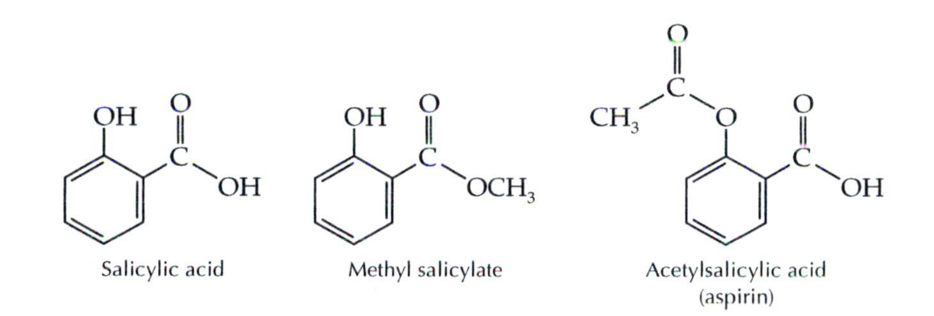

Methyl salicylate has two functional groups that are important in this experiment; one is the ester group and the other is the phenol group.

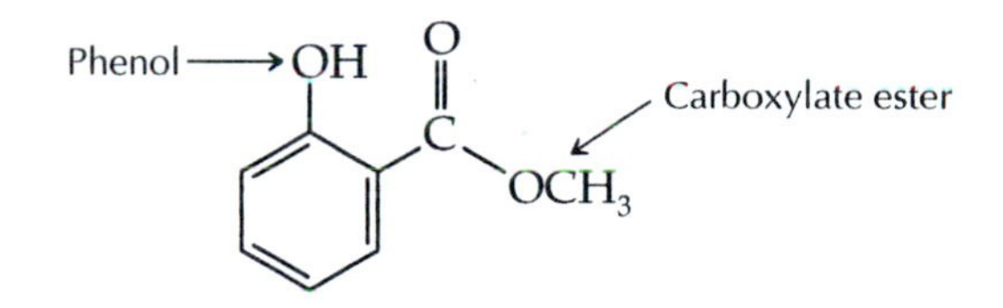

The phenol group of methyl salicylate is a weak acid that rapidly reacts with NaOH, a strong base, to form a sodium salt:

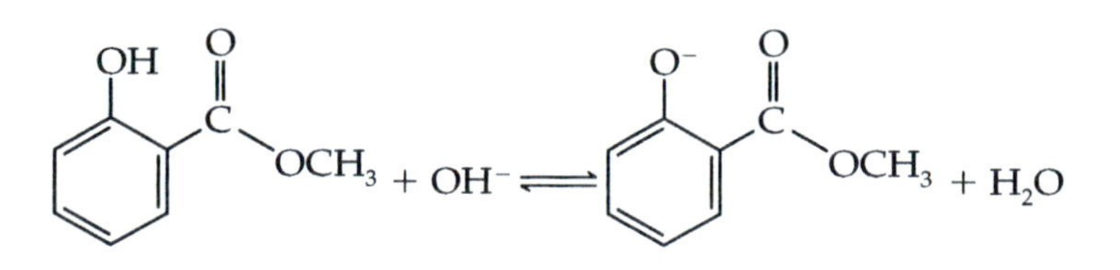

The ester group of methyl salicylate can be converted into a carboxylic acid by base-promoted ***hydrolysis.*** Hydrolysis is the splitting apart of the ester group by a molecule of water, and this reaction is catalyzed effectively by NaOH. Hydrolysis of an ester yields a molecule of an alcohol (in this case methanol) and a molecule of the carboxylic acid (here, salicylic acid). The carboxylic acid rapidly reacts in the strongly basic reaction mixture to form the carboxylate anion. After the reaction, the carboxylic acid is recovered by acidifying the reaction mixture.

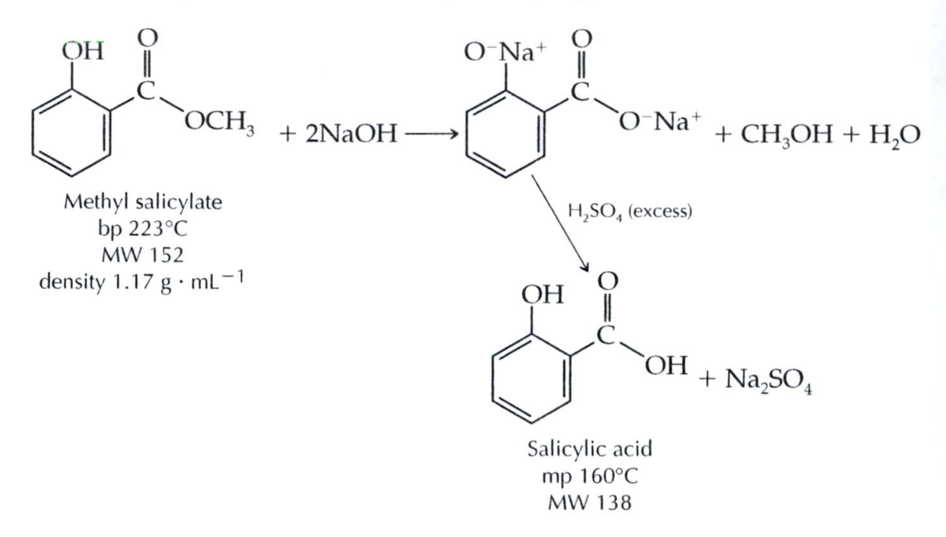

As with most chemical reactions, the hydrolysis of esters goes faster at high temperature. Even using a concentrated sodium hydroxide solution for catalysis, the hydrolysis would take a long time at room temperature. Therefore, the reaction mixture is heated at the boiling point of the solution, a process called ***refluxing.*** Boiling causes the solvent vapor to move up the reflux condenser, where the water-cooled surface condenses the vapor back to a liquid. The condensed liquid then drops back into the reaction flask.

Although hydrolysis of an ester involves a functional group that you will study much later in organic chemistry, you can think of the reaction in a simple way and not be too far afield. Reactions of two chemical species are often dominated by charge complementarity; that is, plus attracts minus and minus attracts plus. Even if a molecule is not ionic, polar bonds lead to partially positive and negative regions within a molecule. The hydroxide anion reacts at the partially positive carbon atom of the carbonyl (C=O) group of the ester. The result is a displacement of the alkoxide anion ($R'O^-$), which is a strong base and quickly forms the alcohol product by removal of the acidic proton from the carboxylic acid. This produces the corresponding carboxylate anion, causing the overall hydrolysis reaction to be irreversible:

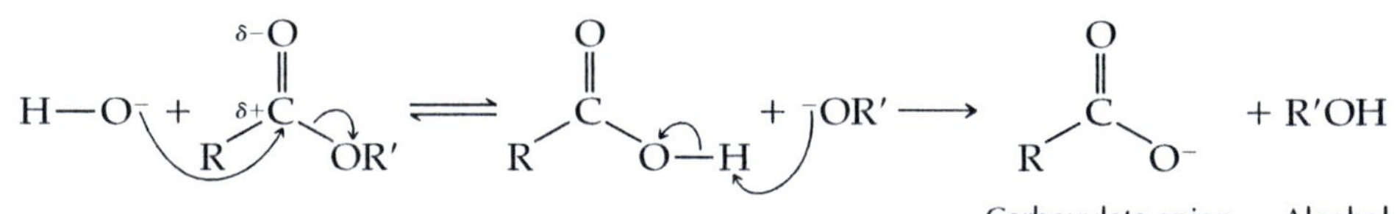

Addition of a strong acid after hydrolysis leads to protonation of the carboxylate salt, and the carboxylic acid is produced. In this experiment, the salicylic acid is insoluble in cold water and can be collected by vacuum filtration. The methanol that is also formed is soluble in water and is not recovered.

MINISCALE PROCEDURE

Techniques Miniscale Reflux: Technique 7.1
Boiling Stones: Technique 6.1
Heating Methods: Technique 6.2
Miniscale Recrystallization: Technique 9.5
Melting Points: Technique 10.3

SAFETY INFORMATION

Sodium hydroxide is corrosive and causes burns. Wear gloves and avoid contact with skin, eyes, and clothing. Notify the instructor if any solid NaOH is spilled.

Sulfuric acid solutions are corrosive and cause burns. Avoid contact with skin, eyes, and clothing.

Methyl salicylate is toxic and an irritant. Wear gloves and avoid contact with skin, eyes, and clothing.

Solid NaOH is hygroscopic and rapidly absorbs water from the atmosphere. Keep the reagent bottle tightly closed when not in use.

Clamp a 100-mL round-bottomed flask to a ring stand or support lattice. Place 4.6 g of sodium hydroxide and 25 mL of water in the flask; stir the mixture until the solid dissolves. Add 2.0 mL (2.3 g) of methyl salicylate. A white solid will quickly form (see Question 3). Attach a water-cooled reflux condenser to the round-bottomed flask [see Technique 7.1].

Always grease ground glass joints when refluxing a basic solution.

Add one or two boiling stones to the reaction mixture to prevent bumping of the solution when it is heated and place a heating mantle or a sand bath under the flask. Heat the reaction mixture under reflux for 15 min. The solid that forms initially will dissolve as the mixture is warmed.

After the reflux period, remove the heating mantle and let the mixture cool to room temperature. Placing the flask in a beaker of tap water speeds the cooling process.

Carefully add 3 M sulfuric acid solution in approximately 3-mL increments until a heavy white precipitate of salicylic acid forms and remains when the mixture is well stirred. You will need approximately 15–20 mL of the sulfuric acid solution.

After you have added just enough 3 M sulfuric acid to give a heavy white precipitate, add an additional 2 mL of acid to ensure complete precipitation of the salicylic acid. Cool the mixture in an ice-water bath to about 5°C. Collect the precipitated crude product by vacuum filtration, using a Buchner funnel [see Technique 9.5, Figure 9.4].

Recrystallization of Salicylic Acid

Read Technique 9, Recrystallization, carefully before doing this part of the experiment.

Recrystallize the crude salicylic acid by the following procedure. Heat about 60 mL of water in a 125-mL Erlenmeyer flask on a hot plate until the water almost boils.

SAFETY PRECAUTION

Use a pair of flask tongs to hold any hot flask.

Remember that the correct amount of solvent is just a little over the minimum *that will dissolve the crystals when the recrystallization solution is boiling.*

Place the crude salicylic acid in another 125-mL Erlenmeyer flask and add a boiling stone or wooden boiling stick. Carefully pour approximately 20 mL of hot water over the salicylic acid and heat the flask containing the salicylic acid to boiling. Continue adding approximately 5-mL portions of hot water until the solid has completely dissolved, allowing a little time after each addition for the dissolution process to occur. When dissolution is complete, add 8–10 mL of excess hot solvent. Estimate the total amount of solvent you used and record this in your notebook.

Set the flask on the bench top until extensive crystallization has occurred throughout the solution and it has cooled nearly to room temperature. Then place the flask in an ice-water bath for 10 min. Collect the crystals, using vacuum filtration. When the filtration is complete, disconnect the vacuum source and pour a few milliliters of ice-cold water over the crystals to dissolve any impurities coating the crystals from residual crystallization solution. Turn on the vacuum source again to remove the solvent.

Pull air through the crystals for a few minutes to facilitate drying. Water does not evaporate very quickly, so the crystals should be kept open to the atmosphere on a piece of filter paper or on a watch glass at least overnight to complete the drying process. You can also dry salicylic acid more quickly by heating it in an oven at 90–100°C for 10–15 min.

Weigh your recrystallized salicylic acid when it is thoroughly dry. Calculate your percent yield. Your instructor may ask you to obtain an IR spectrum of your salicylic acid [see Technique 18.4]. If so, compare your spectrum with that shown in Figure 4.1.

Melting Point

Read Technique 10, Melting Points and Melting Ranges, before doing this part of the experiment.

Carefully follow the directions in Technique 10.3 for determining the melting point of your dried salicylic acid. This compound can sublime with prolonged heating; nevertheless, using a sealed capillary tube should not be necessary. Reference tables give the melting point of salicylic acid as 159 or 160°C. Be sure to record the melting range that you observe; you will probably find a melting range of 2–3 degrees.

Submit your product, properly labeled, to your instructor, unless you are instructed to use it for Experiment 5.

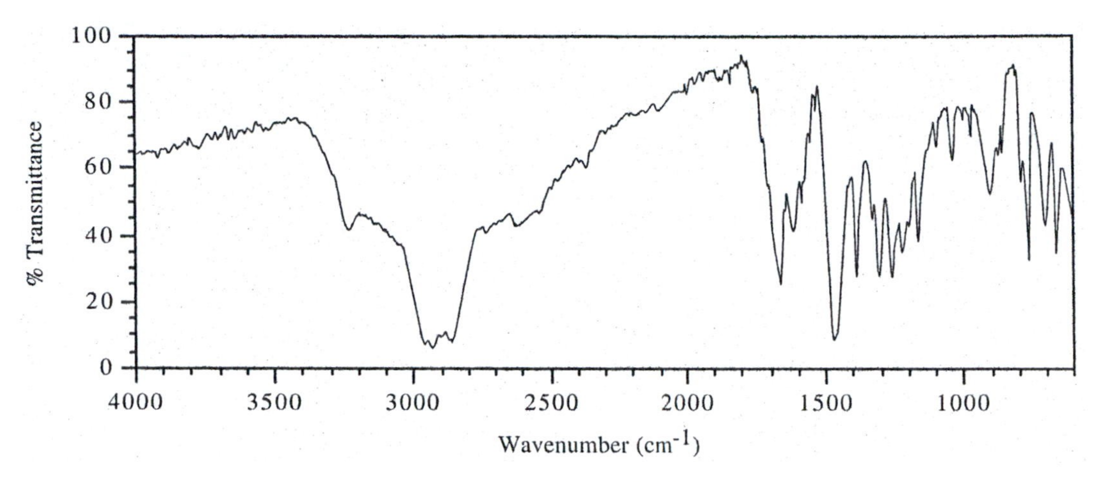

FIGURE 4.1 IR spectrum of salicylic acid (Nujol mull).

CLEANUP: Neutralize any excess acid in the filtrate from the reaction mixture with sodium carbonate before washing the solution down the sink or placing it in the container for aqueous inorganic waste. Dilute the filtrate from the recrystallization with water and wash it down the sink.

MICROSCALE PROCEDURE

Techniques Microscale Reflux: Technique 7.1
Boiling Stones: Technique 6.1
Heating Methods: Technique 6.2
Microscale Recrystallization: Technique 9.7a
Melting Points: Technique 10.3

Solid NaOH is hygroscopic and rapidly absorbs water from the atmosphere. Keep the reagent bottle tightly closed when not in use.

SAFETY INFORMATION

Sodium hydroxide is corrosive and causes burns. Wear gloves and avoid contact with skin, eyes, and clothing. Notify the instructor if any solid NaOH is spilled.

Sulfuric acid solutions are corrosive and cause burns. Avoid contact with skin, eyes, and clothing.

Methyl salicylate is toxic and an irritant. Wear gloves and avoid contact with skin, eyes, and clothing.

Usually the ground glass joints of microscale glassware are not greased, but the solution of a strong base, such as NaOH, can cause the joint between the condenser and flask to "freeze" together if it is not greased.

Pour 3.5 mL of water into a 10-mL round-bottomed flask. Add 0.48 g of sodium hydroxide to the flask. Swirl the flask gently until the solid dissolves. Add 0.20 mL (230 mg) of methyl salicylate measured with a graduated pipet to the NaOH solution. A white solid will quickly form (see Question 3). Add a boiling stone. Attach a water-cooled reflux condenser [see Technique 7.1, Figure 7.1b] after you have applied a light coating of grease to the lower ground glass joint of the condenser. Gently tighten the screw cap to hold the joint together.

Place the reflux apparatus in the flask depression of the aluminum block or in a sand bath. Turn on the water to the condenser. Begin heating the block or sand bath to 120–130°C. Heat the mixture at reflux for 15 min. The white solid that formed initially will dissolve as the mixture is warmed. After the reflux period, lift the apparatus out of the aluminum block or sand bath and cool the mixture to room temperature. Placing the flask in a small beaker of tap water speeds the cooling process.

Remove the condenser and then set the flask in a 50-mL beaker. Carefully add 3 M sulfuric acid solution to the reaction mixture in approximately 0.5-mL increments until a heavy white precipitate of salicylic acid forms and *remains* when the mixture is well stirred; then add another 0.5-mL increment to ensure complete precipitation of the salicylic acid. You will need to add 2–3 mL of acid. Cool the flask or beaker containing the salicylic acid in a 100-mL beaker containing an ice-water mixture. Collect the product by vacuum filtration, using a Hirsch funnel, as shown in Technique 9.7a, Figure 9.7.

Recrystallization of Salicylic Acid

Read Technique 9, Recrystallization, before doing this part of the experiment.

Remember that the correct amount of solvent is just a little over the minimum *that will dissolve the crystals when the recrystallization solution is boiling.*

Transfer the crude product from the Hirsch funnel to a weighing paper, using a spatula. Roll the paper into a funnel or crease it, and transfer the salicylic acid crystals to a 10-mL Erlenmeyer flask. Add 2 mL of water to the flask containing the crystals, put a boiling stick or stone in the flask, and heat the mixture to boiling on a hot plate. Add water in 0.5-mL increments until the solid dissolves in the boiling solvent; then add an additional 0.5 mL of water. Estimate the total amount of solvent you used and record this datum in your notebook.

SAFETY PRECAUTION

Use a pair of flask tongs to hold any hot flask.

Set the flask on the bench top until extensive crystallization has occurred throughout the solution and it has cooled nearly to room temperature. Then place the flask in an ice-water bath for 5 min. Collect the crystals using vacuum filtration and a Hirsch funnel. When the filtration is complete, disconnect the vacuum source and pour approximately 1 mL of ice-cold water over the crystals to dissolve any impurities coating the crystals from residual crystallization solution. Turn on the vacuum again to remove the solvent.

Pull air through the crystals for a few minutes to facilitate drying. Water does not evaporate very quickly, so the crystals should be kept open to the atmosphere on a piece of filter paper or on a watch glass at least overnight to complete the drying process. You can dry salicylic acid more quickly by heating it in an oven at 90–110°C for 10–15 min.

Weigh your recrystallized salicylic acid when it is thoroughly dry. Calculate your percent yield. Your instructor may ask you to obtain an IR spectrum of your salicylic acid [see Technique 18.4]. If so, compare your spectrum with that shown in Figure 4.1.

Melting Point

Determine the melting point of your salicylic acid as described in the Miniscale Procedure.

CLEANUP: Neutralize any excess acid in the filtrate from the reaction mixture with sodium carbonate before washing the solution down the sink or placing it in the container for aqueous inorganic waste. Dilute the filtrate from the recrystallization with water and wash it down the sink or pour it undiluted into the container for flammable organic waste.

Questions

1. Describe the factors that could cause you to obtain an incorrect melting point.
2. Suppose the material that you are recrystallizing fails to precipitate out of the cold solvent. What would you do to recover the material from the solution?
3. What is the white solid that quickly forms when the methyl salicylate is added to the sodium hydroxide solution?
4. Suppose that you obtain 1.0 g of salicylic acid from the hydrolysis of 1.3 mL of methyl salicylate. Calculate your percent yield.
5. (a) Use Table 4.1 and the total estimated volume of solvent that you used for the recrystallization to calculate the amount of salicylic acid that would remain dissolved in the cold recrystallization solvent if you cooled it to 10°C before filtration.

   (b) By how much would this loss lower your percent yield?

TABLE 4.1 Solubility of salicylic acid in water

Temperature, °C	Amount of salicylic acid per 100 mL water, g
0	0.10
10	0.13
25	0.23
50	0.63
75	1.8
90	3.7

EXPERIMENT

5

ANALGESICS AND SYNTHESIS OF ASPIRIN

PURPOSE: To make aspirin from salicylic acid and to analyze common analgesics.

In Experiment 5.1 you will synthesize acetylsalicylic acid, purify it by recrystallization, and determine its purity by a melting-point determination. In Experiment 5.2 you have the opportunity to carry out the thin-layer chromatographic analysis of over-the-counter analgesics.

Even though extracts of willow leaves and bark have been used for centuries for their pain-relieving (analgesic), fever-reducing (antipyretic), and anti-inflammatory properties, only in the late 1800s was the active ingredient of willow and poplar bark discovered to be salicylic acid. This substance, it was found, could be synthesized cheaply and in large amounts, but its use had severe limitations because of its acidic properties. Membranes lining the stomach and passages leading to it are irritated by salicylic acid. Its side effects were often worse than the original discomfort. A breakthrough came in 1893 when the acetyl derivative of salicylic acid was synthesized; it proved to have the same kind of medicinal properties without the high degree of irritation to mucous membranes.

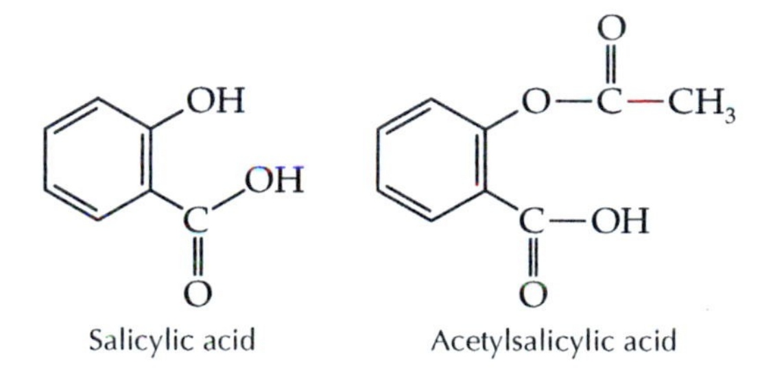

Salicylic acid Acetylsalicylic acid

With the advent of a number of new analgesics in the last 30 years, aspirin is no longer the first painkiller that most people turn to. Aspirin, like acetaminophen and ibuprofen, inhibits the activity of the cyclooxygenase enzyme that catalyzes the formation of prostaglandins, which are involved in the causation of pain and inflammation in the body. Aspirin has been cited as a contributing factor to Reye's syndrome, which can lead to death and is especially dangerous in children. Small amounts of aspirin are effective in reducing heart attacks and strokes by hindering the formation of blood clots.

In Experiment 5.1, aspirin is prepared by acetylating salicylic acid in a process called esterification. The acetyl group comes from acetic anhydride, and the reaction is catalyzed by phosphoric acid:

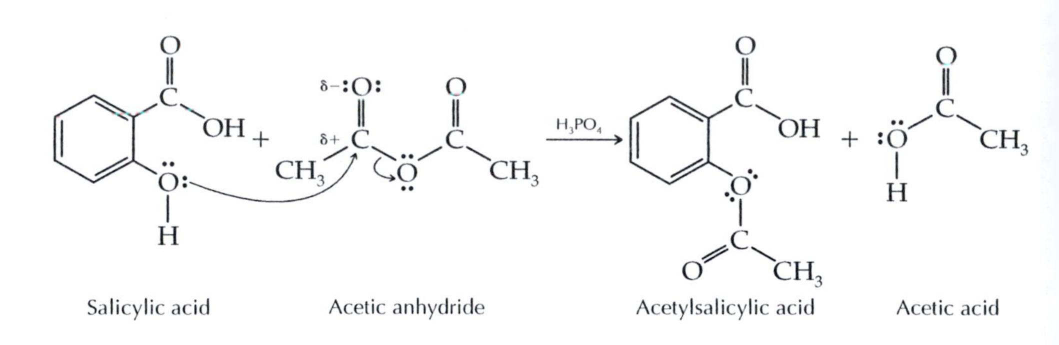

Experiment 5.2 is the thin-layer chromatographic analysis of widely used analgesics. TLC is hard to beat for fast qualitative drug analysis, say if there was the possibility of a drug overdose. The identifications could also be based on melting points or spectroscopic characteristics, but these can be difficult to use with mixtures, and many medicinal preparations contain mixtures.

Acetaminophen (Tylenol), ibuprofen (Advil and Motrin), and aspirin are widely used analgesics. Naproxen (Aleve) is a newer analgesic that is often administered in the form of its sodium salt.

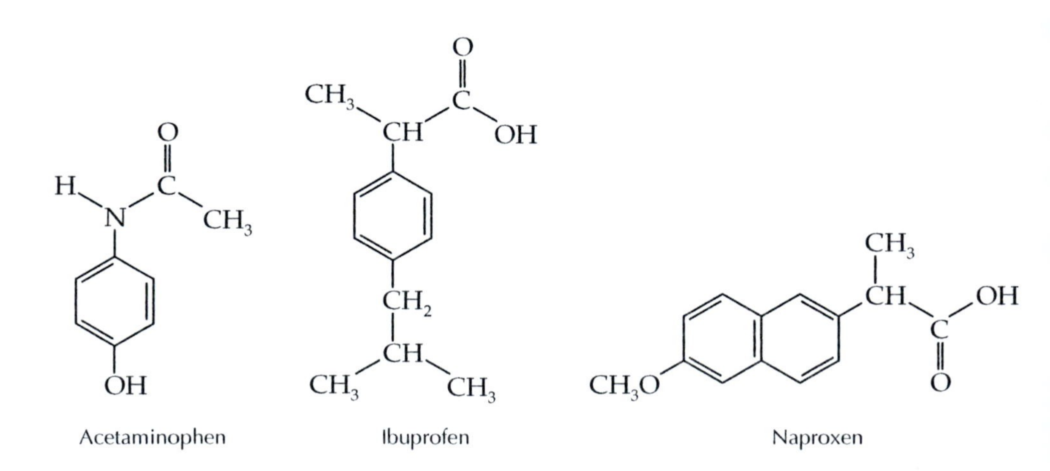

Some analgesics contain mixtures of medicinal agents, for example acetaminophen and aspirin. In addition, caffeine may be included as a stimulant. Sometimes an antihistamine is included; for example, Tylenol PM contains diphenhydramine hydrochloride as a sleep aid.

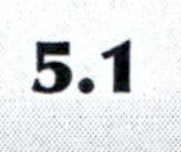

5.1 Synthesis of Acetylsalicylic Acid

PURPOSE: To synthesize aspirin from salicylic acid, purify it, and determine its purity by a melting-point determination.

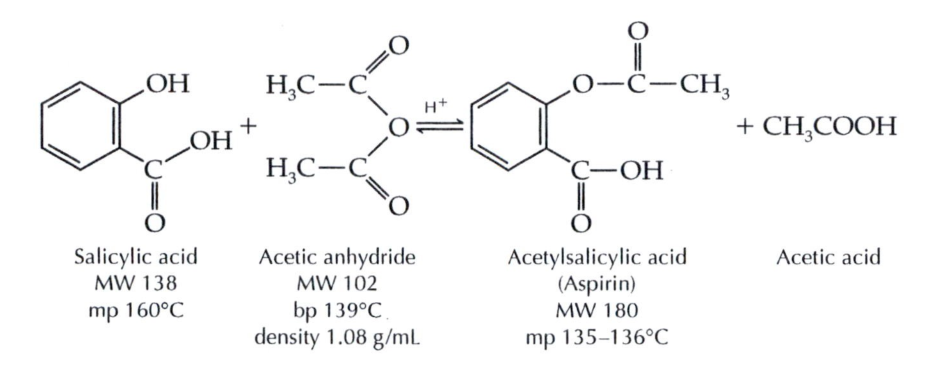

MINISCALE PROCEDURE

Techniques Mixed Solvent Recrystallization: Technique 9.2
Recrystallization: Technique 9.5
Melting Points: Technique 10.3

SAFETY INFORMATION

Salicylic acid is toxic and an irritant to skin, eyes, mucous membranes, and the upper respiratory tract. Avoid breathing the dust. Avoid contact with skin, eyes, and clothing.

Acetic anhydride is toxic, corrosive, and a lachrymator (it causes tears). Wear gloves, use it in a hood, and avoid contact with skin, eyes, and clothing.

Concentrated **phosphoric acid** (85%) is corrosive and causes burns. Avoid contact with skin, eyes, and clothing.

Weigh 1.0 g of the salicylic acid that you synthesized in Experiment 4 or from the supply available in the lab; your salicylic acid from Experiment 4 must be thoroughly dry before using it to make aspirin. Place the salicylic acid in a 50-mL Erlenmeyer flask. Under a hood, add 2.0 mL of acetic anhydride measured with a graduated pipet. Add 5 drops of 85% phosphoric acid and mix the chemicals well by rotating the flask. Loosely stopper the flask with a cork before leaving the hood. The mixture may become warm from the exothermic reaction; allow it to stand for about 10 min. To complete the reaction, loosen the cork and heat the flask for 5 min in a 45–50°C water bath. Hot tap water may be used to make the bath.

The freshly broken surface of the scratched glass provides the nuclei for the formation of aspirin crystals.

Chill the mixture in an ice-water bath and scratch the inside wall of the flask with a stirring rod until a semicrystalline paste forms. Add 10 mL of cold water and 6 g of ice. Stir the mixture to break up the pasty solid.

When the ice has melted, collect the crystals by vacuum filtration on a small Buchner funnel [see Technique 9.5, Figure 9.4]. Wash the remaining crystals out of the flask using 2–3 mL of ice-cold water, and rinse the crystals on the funnel with several 1- to 2-mL portions of ice-cold water. Press the product with a large cork to remove as much water as possible. Set aside about 10 mg for later determination of the melting range.

The odor of acetic acid will no longer be apparent when you have washed the crystals enough.

Recrystallization of Acetylsalicylic Acid

Purify the crude aspirin by recrystallization from a mixed solvent [see Technique 9.2]. Stir your crude aspirin with 1.5 mL of ethanol in a 25-mL Erlenmeyer flask. If the crystals do not dissolve at room temperature, warm the mixture briefly on a steam bath or on a hot plate set at the lowest setting.

Probably no filtration is necessary at this stage; however, if the ethanol solution has any insoluble particles in it, you should filter it through a small fluted filter paper [see Technique 9.5]. If you carry out this filtration step, you will have to rinse the filter paper carefully with another milliliter of hot ethanol to recover all the aspirin. If some of the aspirin has crystallized during the filtration, reheat the filtrate until the crystals dissolve.

Pour 10 mL of warm water (55–60°C) into the ethanol/aspirin solution and let the solution cool at room temperature for 10–15 min. Then cool it in an ice-water bath for 5 min to complete crystallization. Collect the product by vacuum filtration. Wash the product with 1 mL of ice-cold water and remove as much liquid as possible through suction. Allow the crystals to dry thoroughly before weighing them and determining their melting points. Calculate your percent yield.

Melting Point

The melting range is a good way to assess the purity of aspirin. Take the melting ranges for both your crude aspirin and your recrystallized aspirin samples [see Technique 10.3]. What do these data tell you about the purity of your product? At the discretion of your instructor, determine the IR spectrum of your aspirin [see Technique 18.4]. Compare it to the IR spectrum shown in Figure 5.1.

CLEANUP: The filtrate from the crude aspirin should be neutralized with solid sodium carbonate before it is poured down the sink or placed in the container for aqueous inorganic waste. **(Caution: Foaming.)** The filtrate from the recrystallization can be poured down the sink.

MICROSCALE PROCEDURE

Techniques Heating Methods: Technique 6.2
Microscale Filtration: Technique 9.7a
Mixed Solvent for Recrystallization: Technique 9.2
Recrystallization Using a Craig Tube: Technique 9.7b
Melting Points: Technique 10.3

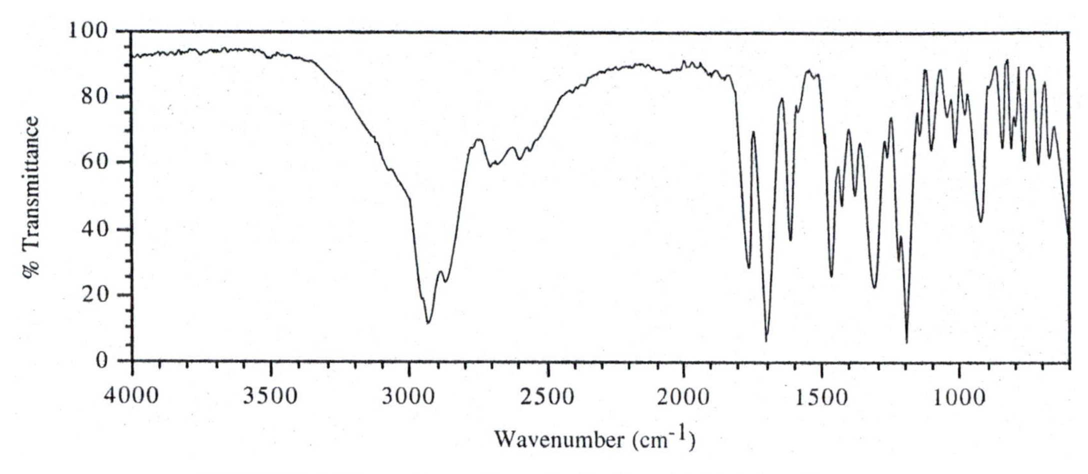

FIGURE 5.1 IR spectrum of acetylsalicylic acid (Nujol mull).

SAFETY INFORMATION

Salicylic acid is toxic and an irritant to skin, eyes, mucous membranes, and the upper respiratory tract. Avoid breathing the dust. Avoid contact with skin, eyes, and clothing.

Acetic anhydride is toxic, corrosive, and a lachrymator (it causes tears). Wear gloves, use it in a hood, and avoid contact with skin, eyes, and clothing.

Concentrated **phosphoric acid** (85%) is corrosive and causes burns. Avoid contact with skin, eyes, and clothing.

Using weighing paper, weigh a sample in the range of 220–230 mg (record the exact mass) rather than trying to weigh out exactly 225 mg.

Place 225 mg ± 5 mg of salicylic acid in a 13 × 100 mm test tube. Working under a hood, add 0.50 mL of acetic anhydride (measured with a graduated pipet) and 1 drop of 85% phosphoric acid (measured with a Pasteur pipet). Cork the test tube before leaving the hood. Loosen the cork and place the test tube in a beaker of water at 45–50°C; hot tap water may be used to make the water bath. Shake the tube gently until the salicylic acid dissolves. Continue heating for 15 min.

Remove the test tube from the water bath and add 1.4 mL of water, measured with a graduated pipet. Stir the mixture until it is a clear solution. Cool the solution at room temperature until crystallization begins, then cool it in an ice-water bath until crystallization is complete. If necessary, scratch the inside of the test tube to initiate crystallization.

The freshly broken surface of the scratched glass provides nuclei for the formation of aspirin crystals.

Collect the crude product by vacuum filtration on a Hirsch funnel [see Technique 9.7a] and wash the crystals three times with 0.5-mL

The odor of acetic acid will no longer be apparent if you have washed the crystals enough.

portions of ice-cold distilled water, using a calibrated plastic pipet to measure and deliver the water. With the vacuum turned on, draw air through the Hirsch funnel for 5 min. Then dry the crude product by pressing it between pieces of fine-grained filter paper.

Recrystallization of Acetylsalicylic Acid

Purify the crude aspirin by recrystallization from a mixed solvent [see Technique 9.2] using a Craig tube [see Technique 9.7b]. Weigh the crude aspirin on tared weighing paper and set aside about 10 mg for a melting-point determination. Transfer the crystals to a Craig tube by rolling the weighing paper into a funnel. The following solvent quantities are based on 140–160 mg of crude aspirin; if your crude product weighs more or less than this range, change the quantities of solvent proportionally. Add a boiling stick and 0.40 mL of ethanol, measured with a graduated pipet, to the Craig tube. Place the Craig tube in an aluminum block heated to 90°C on a hot plate [see Technique 6.2]. When the crystals have dissolved, add 1.0 mL (graduated pipet) of water, stir the solution, and warm it to dissolve any crystals that form.

Place the plug in the Craig tube and set the apparatus in a 25-mL Erlenmeyer flask to cool slowly to room temperature. If no crystals have formed after 15 min, add a seed crystal from the crude aspirin you have saved. To do this, pick up one or two tiny crystals on the tip of a microspatula and touch the spatula briefly to the top of the solution. When crystal growth has ceased, cool the tube in an ice-water bath for 3–4 min. Centrifuge to remove the solvent. Transfer the crystals to a small, tared watch glass to dry. After the crystals have dried, determine the yield by weighing the recrystallized aspirin, and calculate the percent yield.

Melting Point

Determine the melting range of the crude aspirin and the recrystallized aspirin [see Technique 10.3]. Has the purity been improved by the recrystallization? At the discretion of your instructor, determine the IR spectrum [see Technique 18.4]. Compare it with the spectrum shown in Figure 5.1.

Questions

1. Assuming that 1.0 g of aspirin dissolves in 450 mL of water at 10°C, how much aspirin would be lost in the 16 mL of water (miniscale preparation) or 1.4 mL of water (microscale preparation) added to the reaction mixture if the mixture were at 10°C during the filtration?
2. How much difference would the amount lost in the water (Question 1) make in your percent yield?
3. What is the purpose of adding the concentrated phosphoric acid to the reaction mixture in the synthesis of aspirin?

5.2 Thin-Layer Chromatographic Analysis of Analgesics

PURPOSE: To determine the composition of over-the-counter analgesics using thin-layer chromatography.

MICROSCALE PROCEDURE

Technique Thin-Layer Chromatography: Technique 15

SAFETY INFORMATION

Ethyl acetate and **acetone** are flammable and irritants. Do not use them near heated electrical devices and avoid contact with skin, eyes, and clothing. Keep the developing chamber closed except when inserting or removing a TLC plate.

Never look directly into an **ultraviolet lamp.** Radiation in this region of the spectrum can cause eye damage.

TLC Analysis of Reference Compounds

Standard solutions (1–2% w/v) of the following compounds will be available in the laboratory.

Read Technique 15, Thin-Layer Chromatography, before undertaking this experiment.

Aspirin (acetylsalicylic acid)
Acetaminophen (4-acetamidophenol)
Ibuprofen (4-isobutyl-α-methylphenylacetic acid)
Naproxen ((*S*)-(+)-6-methoxy-α-methyl-2-naphthaleneacetic acid)
Caffeine

Precoated silica gel plates with fluorescent indicator* cut to 2.5 × 6.7 cm dimensions work well for the chromatographic separation of the analgesics.

Prepare a developing chamber using a wide-mouthed, capped bottle. A piece of 7- or 9-cm filter paper makes a good wick to ensure solvent saturation in the developing chamber's atmosphere. The filter paper must not be so large that it will touch the TLC plate when it is placed in the jar. Add 4–5 mL of 0.5% acetic acid in ethyl acetate solution (v/v) to the developing chamber, cap it, and shake it vigorously. After the filter paper and the atmosphere have been saturated, the solvent should be 2–3 mm deep in the bottom of the chamber. If you have a greater depth, remove some of the solvent with a Pasteur pipet and save it in a corked test tube. If you do not have enough liquid, add an additional 1–2 mL of solvent to the developing chamber.

Prepare several micropipets according to the instructions in Technique 15.3 or obtain micropipets from the supply in the laboratory.

*EM Science Silica Gel 60 F-254, No. 5554-7.

Spotting technique: Just touch the micropipet against the TLC plate and immediately lift it; longer contact with the adsorbent leads to a diffuse spot, not one containing a more concentrated sample.

Make a mark about 2 mm long with a pencil at the long edge of a 2.5 × 6.7 cm TLC plate approximately 1 cm from the bottom edge, as shown in Technique 15, Figure 15.4. Dip a clean micropipet into one of the standard analgesic solutions and apply one spot of the solution at about one-third the width of the TLC plate. With another micropipet, apply one spot of another analgesic solution at about two-thirds the width of the TLC plate. Avoid holding the micropipet against the TLC plate for a prolonged time; this practice will produce only a larger diffuse spot, not one containing a more concentrated sample. Record the location on the plate of each analgesic in your notebook—sketches of the TLC plate work well for recording observations.

Do not disturb the developing chamber during the chromatography.

Develop the chromatogram by placing the TLC plate in the solvent-saturated developing chamber with a pair of tweezers. Keep the chamber capped at all times. Allow the solvent to rise up the plate until it is about 1.0 cm from the top.

Remove the plate from the chamber with a pair of tweezers and *immediately mark the solvent front* with a pencil line before the solvent begins to evaporate. After marking the solvent-front position, let the solvent evaporate in a hood for 2–3 min.

Visualize the results of your chromatogram by shining ultraviolet radiation (254 nm) on the plate in a darkened room or dark box. You will notice dark spots wherever a compound is present on the plate. Outline each spot with a pencil while the plate is under the ultraviolet lamp. Alternatively, use iodine vapor to visualize the plate (Note 1).

Calculate the R_f values for the five reference compounds under your experimental conditions [see Technique 15.1 and Figure 15.2]. You may find that aspirin and naproxen tail considerably on the thin-layer plate when ethyl acetate is used as the developing solvent, but this behavior causes no serious problem in the analysis. You can calculate the R_f value by using the center of the oval spot.

Be sure to record in your notebook all the data needed to reproduce your R_f values; draw diagrams of your chromatograms. Also estimate the size of the spots and indicate whether there was any tailing (Note 2).

TLC Analysis of Over-the-Counter Analgesics

Suggested products to test include Advil, Aleve, Anacin, Excedrin Extra Strength, Motrin, and Tylenol. Crush 1/2 tablet in a folded weighing paper. If the tablet has a colored coating, use a spatula to separate some of the crushed white powder from the coating fragments. Prepare a solution for TLC analysis by dissolving 10–15 mg in 1 mL of acetone in a 13 × 100 mm test tube. (Add one Pasteur pipet drop of acetic acid to solutions of Aleve, which is sold as the sodium salt.) Gently heat your analgesic mixture on a steam bath or in a 60–65°C hot-water bath for

2–3 min. You should heat the mixture just enough to dissolve the active ingredients of the tablet but not enough to boil away any solvent. The binder in the tablet (starch) and the inorganic buffers will not dissolve, but the analgesics and other medicinal agents will go into solution. After the solids settle, you can analyze the solution by TLC, again using 0.5% acetic acid in ethyl acetate solution as the developing solvent and UV radiation for visualization. How many components are indicated in the tablet by your TLC analysis? What are they? Calculate the R_f values and compare them with the R_f values for the known analgesic.

CLEANUP: Pour any remaining 0.5% acetic acid in ethyl acetate solution and the test solutions of analgesics into the container for flammable or organic waste. Place the solvent-saturated filter paper and the chromatograms in a waste container for solids in a hood.

Notes About TLC Procedure

1. *An iodine chamber may be used to visualize the spots on the developed chromatogram [see Technique 15.6]. The spots must be outlined with a pencil immediately after removing the plate from the iodine chamber.*
2. *If your spots turn out to be very large and show tailing, you can spot less sample on a different plate by diluting your analgesic solution with a few more drops of acetone and by using a micropipet with a smaller-diameter tip. If the spots are very small or very faint, you can spot more sample by applying repetitive spots to the same position; let the solvent evaporate before overspotting the first spot.*

Questions

1. Why can there be no breaks in the thin-layer surface of a TLC plate?
2. Two compounds have the same R_f (0.87) under identical conditions. Does this show that they have identical structures? Explain.

EXPERIMENT

6 ISOLATION OF ESSENTIAL OILS FROM PLANTS

PURPOSE: To isolate, purify, and study the essential oils from caraway seeds and orange peels.

In Experiment 6.1, a two-week experiment, you will isolate caraway seed oil by steam distillation of caraway seeds and study its major component, (*S*)-(+)-carvone, by gas-liquid chromatography and polarimetry. You will also make derivatives of (*S*)-(+)-carvone and its mirror image isomer, spearmint oil.

If you do Experiment 6.2, you will isolate (*R*)-(+)-limonene from orange peels and analyze its optical properties.

Natural products from the plant world are a fascinating group of chemicals. They serve a variety of important roles for the plants, from attracting insects for pollination to fending off predators. The specificity that nature uses in the biosynthesis of these often chiral compounds is astounding. Essential oils make up one group of these natural organic compounds. They are characterized by distinctive odors, which often depend on their exact stereochemical configuration. Many essential oils belong to the class of compounds called terpenes, whose carbon skeletons are composed of five-carbon isoprene units.

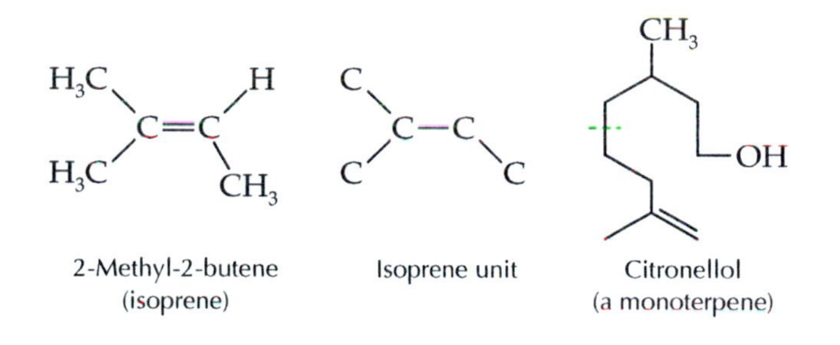

2-Methyl-2-butene (isoprene) — Isoprene unit — Citronellol (a monoterpene)

Terpenes are classified as monoterpenes (which contain 10 carbons), sesquiterpenes (C_{15}), diterpenes (C_{20}), and so on. Vitamin A is a diterpene:

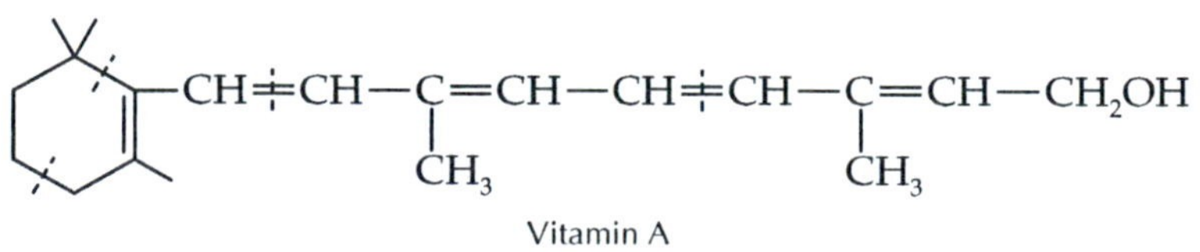

Vitamin A
(a diterpene with four isoprene units)

It is remarkable that the distinctly different smells of caraway seeds and spearmint leaves come from two isomers that differ only in their chirality. The major compound in both caraway oil and spearmint oil is carvone, an unsaturated cyclic ketone with the formula $C_{10}H_{14}O$.

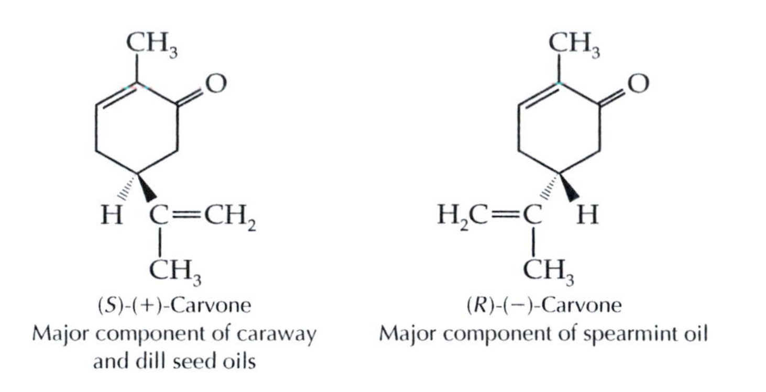

(*S*)-(+)-Carvone
Major component of caraway and dill seed oils

(*R*)-(−)-Carvone
Major component of spearmint oil

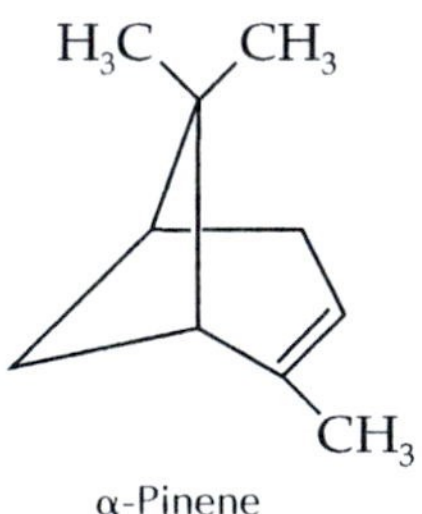

α-Pinene

Chirality plays a major role in the mechanisms of biochemical recognition. Current theories on smell hold that because odor receptors of the nose are chiral, the two enantiomers of carvone, with their mirror-image shapes, fit into different receptor sites. Curiously, it has been reported that about 1 person in 10 cannot tell the difference between the odors of caraway and spearmint. We still have much to learn about what controls our sense of smell and why different people smell the same compounds with distinctly different results.

It is also a mystery why caraway plants produce (*S*)-(+)-carvone and spearmint plants produce its mirror image, (*R*)-(−)-carvone. Other plants, such as gingergrass, produce racemic carvone. Even more curious is the fact that the α-pinenes taken from different pine trees in the same grove can have opposite optical activities. How such differences developed is still unknown.

(S)-(+)-Carvone from Caraway Seeds

In Experiment 6.1 you will investigate the differences between the carvone from spearmint and the carvone from caraway oil by using optical activity measurements, gas-liquid chromatography, and your sense of smell. Because it is far easier to work with small amounts of solids than of liquids, it is useful to make a solid derivative of the (*S*)-(+)-carvone that you isolate by steam distillation. The solid derivative of choice here is the semicarbazone. Semicarbazide is a nucleophile that can react with carvone, a ketone, at the carbonyl (C=O) double bond. The reaction mechanism for the formation of the semicarbazone is rather complex, but basically it involves a molecule of semicarbazide adding to the C=O, followed by the elimination of a molecule of water. The preparation of carvone's semicarbazone is doubly useful here because it lets you separate (+)-carvone from the other constituents of caraway seed oil. The other major component, limonene, has no carbonyl group, so it will not form a semicarbazone and will be washed away in the filtrate from the recrystallization.

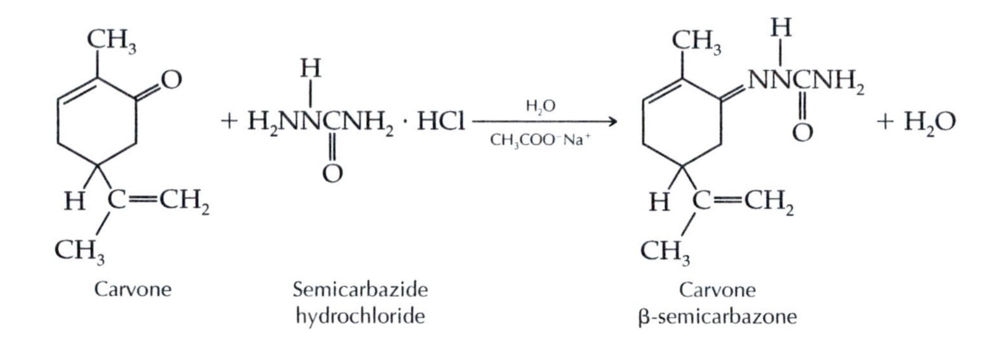

(R)-(+)-Limonene from Orange Peels

In Experiment 6.2 you will isolate limonene from orange peels using steam distillation to separate the essential oil from the solid matter of the orange. Limonene is a monoterpene whose stereochemistry depends on its source. Orange peels provide limonene that is virtually 100% (*R*)-(+)-limonene, whereas pine needles provide essentially 100% (*S*)-(−)-limonene.

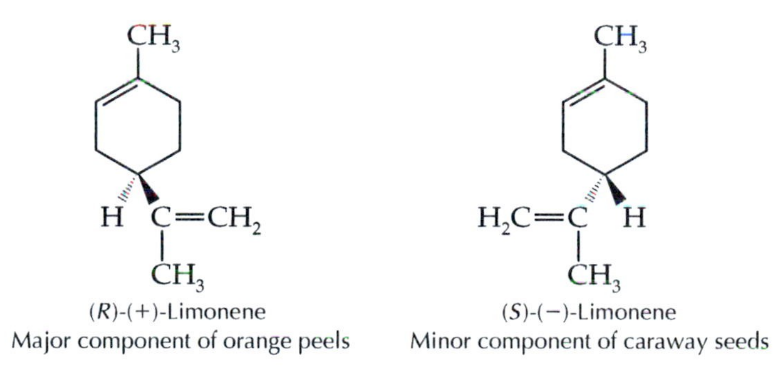

(*R*)-(+)-Limonene
Major component of orange peels

(*S*)-(−)-Limonene
Minor component of caraway seeds

The biosynthetic pathway to limonene is quite complex, but we do know in principle how different plants produce one enantiomer or the other. The stereocenter that differentiates the (+) and (−) isomers of limonene arises during the formation of the six-membered ring when the limonene precursor folds into the active site of the enzyme that catalyzes the biosynthesis.

6.1 (S)-(+)-Carvone from Caraway Seeds

PURPOSE: To extract the essential oil from caraway seeds and analyze it by gas-liquid chromatography and polarimetry.

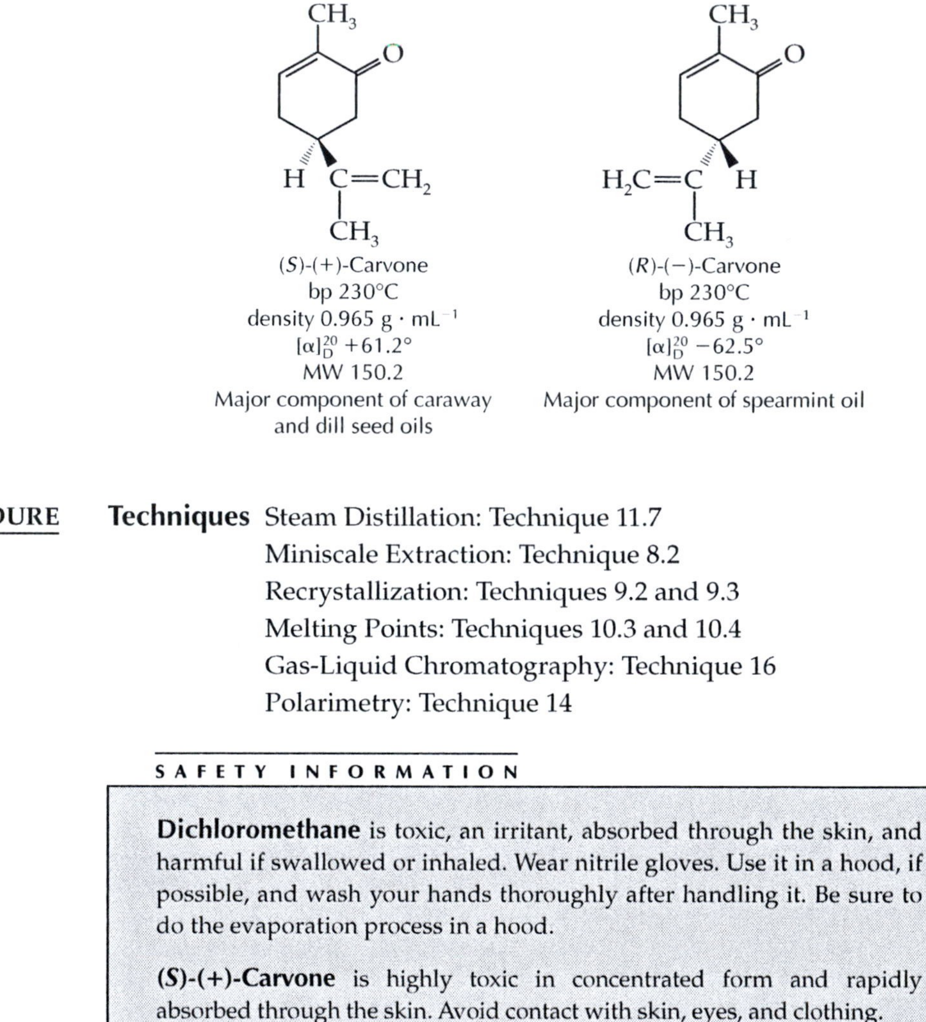

(S)-(+)-Carvone
bp 230°C
density 0.965 g · mL^{-1}
$[\alpha]_D^{20}$ +61.2°
MW 150.2
Major component of caraway and dill seed oils

(R)-(−)-Carvone
bp 230°C
density 0.965 g · mL^{-1}
$[\alpha]_D^{20}$ −62.5°
MW 150.2
Major component of spearmint oil

MINISCALE PROCEDURE

Techniques Steam Distillation: Technique 11.7
Miniscale Extraction: Technique 8.2
Recrystallization: Techniques 9.2 and 9.3
Melting Points: Techniques 10.3 and 10.4
Gas-Liquid Chromatography: Technique 16
Polarimetry: Technique 14

SAFETY INFORMATION

Dichloromethane is toxic, an irritant, absorbed through the skin, and harmful if swallowed or inhaled. Wear nitrile gloves. Use it in a hood, if possible, and wash your hands thoroughly after handling it. Be sure to do the evaporation process in a hood.

(S)-(+)-Carvone is highly toxic in concentrated form and rapidly absorbed through the skin. Avoid contact with skin, eyes, and clothing.

Methanol is toxic and flammable. Pour it only in a hood.

Steam Distillation

Weigh 25 g of fresh caraway seeds and grind them in an electric blender or a coffee grinder. Put the ground seeds into a 250-mL round-bottomed flask and add 100 mL of water. Set up a steam distillation apparatus as shown in Technique 11.7, Figure 11.24, except substitute a 100-mL graduated cylinder for the receiving flask. Be sure that the stopcock is closed before pouring 80 mL of water into the dropping funnel.

A bakery or a bakery supplier is the best source of fresh caraway seeds. Seeds purchased in a supermarket may have spent months on the shelf and still contain carvone but tend to have little or no limonene remaining in them.

Boil the mixture vigorously using a heating mantle or a sand bath as the heat source, but be careful not to let any solid material bump over into the condenser. Collect about 80 mL of distillate. During the distillation make periodic additions of approximately 10-mL portions of water from the dropping funnel to maintain the initial liquid level in the distilling flask. Faster distillation seems to make for a smoother, less bumpy process, so do not decrease the rate of heating once the distillation begins. Distill the mixture as rapidly as the cooling capacity of the condenser will permit. The distillation should take approximately 30–40 min. During this time, prepare the semicarbazone of (*R*)-(−)-carvone, spearmint oil (p. 42).

Isolation of Caraway Oil

NaCl helps to minimize emulsions during the extractions by making the organic layer less soluble in the water layer.

Pour your distillate into a 125-mL separatory funnel. Add 5 g of sodium chloride and shake the mixture to dissolve the salt. Then add a few pieces of ice to make the solution distinctly cool before adding any volatile dichloromethane.

Obtain 20 mL of dichloromethane. Place a conical funnel in the top of the separatory funnel. Hold the condenser and the vacuum adapter used to collect the distillate above the separatory funnel and rinse them with a few milliliters of the dichloromethane; let the dichloromethane drain into the separatory funnel. Use the remaining dichloromethane to rinse the 100-mL cylinder, and also add that solution to the separatory funnel. This rinse recovers all the caraway oil clinging to the glassware surfaces.

Extract the carvone from the aqueous layer by inverting the funnel and shaking it back and forth gently [see Technique 8.2]. Repeat the gentle shaking and venting for about 2 min. Allow the layers to separate before draining the bottom (organic) layer into a dry 50-mL Erlenmeyer flask. Carvone is quite soluble in dichloromethane; any organic membranes that may form at the interface should be left with the aqueous layer.

Repeat the extraction of the aqueous layer using 15 mL of fresh dichloromethane. Combine the two dichloromethane extracts and dry the solution with anhydrous magnesium sulfate for at least 10 min.

Weigh a clean, dry 50-mL Erlenmeyer flask on an analytical balance. Filter half the dichloromethane solution through a conical funnel fitted with a fluted filter paper [see Technique 9.5a, Figure 9.5] into the tared (weighed) Erlenmeyer flask. Add a boiling stick and evaporate the dichloromethane on a steam bath or hot-water bath in a hood. Cool the flask briefly and filter the remaining dichloromethane solution into it. Continue heating the flask until the solvent has completely evaporated. Alternatively, the evaporation process may be done with a stream of nitrogen or air and gentle warming of the flask in a beaker of hot tap water or with a rotary evaporator [see Technique 8.9].

Cool the flask and carefully wipe the outside dry with a tissue; then weigh it on an analytical balance. Heat the flask for several additional minutes, cool, and weigh again. The two masses should agree within 0.050 g. If they do not, repeat the heating and weighing procedures. Record the final mass of your caraway oil to the nearest milligram. The residue in the flask is caraway seed oil. Carefully compare the smell to that of (*R*)-(−)-carvone.

Gas Chromatographic Analysis

Use both nonpolar and polar columns at 170–200°C to find the retention time for a known sample of (−)-carvone on each column under your GC conditions. If you are using a capillary column chromatograph, prepare a solution containing 0.5 mL of ether and 1 drop of the compound being tested. After you have analyzed (*R*)-(−)-carvone, analyze your caraway seed oil using the same nonpolar and then polar GC columns at the same temperature and flow rate. Also chromatograph samples of known (*S*)-(+)-carvone and limonene using identical GC conditions.

Calculate the retention times of (−)- and (+)-carvone and limonene. Also calculate the percentages of (+)-carvone and limonene in caraway seed oil [see Technique 16.7].

Derivatives: Preparation of Semicarbazones

Sodium acetate provides the proper pH for rapid reaction.

Weigh 0.38–0.39 g of (*R*)-(−)-carvone (spearmint oil) into a 25-mL Erlenmeyer flask and add 4.0 mL of 95% ethanol. Dissolve 0.40 g of semicarbazide hydrochloride and 0.40 g of anhydrous sodium acetate (or 0.64 g of sodium acetate trihydrate) in 2.0 mL of water in a 13 × 100 mm test tube. Pour the resulting solution into the flask containing the carvone solution and add a boiling stone or stick. Warm the mixture on a steam bath set for a gentle steam flow or in an 80°C water bath for 15 min.

Add 2.5 mL of water to the warm solution; then set the reaction mixture aside to cool slowly to room temperature. Under these conditions, crystallization may take 30–45 min; slow crystallization gives nearly pure crystals that probably will not need to be recrystallized for the optical activity studies. After crystallization is complete, cool the solution for 5 min in an ice-water bath.

Collect the crystals by vacuum filtration using a small Buchner funnel, and wash them on the funnel with a few milliliters of cold water. Allow the solid to dry for several hours or overnight before taking the melting point [see Technique 10.3]. If it is necessary to recrystallize the semicarbazone, you may do so from an ethanol/water mixture [see Technique 9.2].

If possible, prepare this derivative on the same day that you isolate the caraway seed oil.

Prepare the semicarbazone of the carvone in your caraway seed oil in the Erlenmeyer flask containing the oil, using all your sample except the amount you need for the GC analysis. Add 4.0 mL of ethanol to your caraway seed oil. Prepare the semicarbazide solution (as directed earlier) in a 13 × 100 mm test tube and pour it into the ethanol solution of caraway seed oil in the Erlenmeyer flask. Follow the procedure given earlier for heating and crystallization. Adjust the proportions of all reagents if you have less than 0.30 g of caraway seed oil. Again, determine the melting point after the solid dries. The reported melting point of the semicarbazones that form under conditions used in this procedure is 141–142°C.

Also grind together a small amount of approximately equal portions of the carvone semicarbazones from each source and take a melting point of the mixture [see Technique 10.4].

Polarimetry

If you are using 2-dm polarimetry tubes, use two 25-mL volumetric flasks and prepare a 25-mL solution of each of your dry carvone semicarbazones in anhydrous methanol. If you are using 1-dm periscope polarimeter tubes, prepare each solution in a 10-mL volumetric flask. A concentration of 1.50% works best if you have enough material (0.15 g/10 mL of solution), but if you do not have enough for this concentration, use what you have, saving only enough for the melting points. Weigh the dry carvone semicarbazone samples to the nearest milligram. After dissolving the semicarbazone, stopper the flask and shake it a number of times to ensure a completely homogeneous solution.

If you see any undissolved particles such as paper fibers or pieces of dust in the solution, filter the solution by gravity through a small plug of glass wool or a small fluted filter paper, using a short-stemmed funnel. If you are using a periscope polarimeter tube, filter the solution directly into the tube; if you are using a straight polarimeter tube, filter the solution into a 25- or 50-mL Erlenmeyer flask; then fill the polarimeter tube. Keep the solutions tightly stoppered, except during transfer, to avoid evaporation of the solvent.

Compare the specific rotations of the carvone semicarbazones from caraway seed oil and from (*R*)-(−)-carvone. Are they of the same magnitude and opposite in sign?

CLEANUP: Filter the caraway seed residue from the aqueous liquid remaining in the distillation flask. (Do not put caraway seeds down the sink.) Wash the aqueous filtrate down the sink; dispose of the seed residue as food garbage. Allow any residual dichloromethane to evaporate from the magnesium sulfate drying agent in a hood before placing the solid in the container for inorganic waste or the container for solid hazardous waste. Pour the ether solutions used for GC analysis, the filtrate from the semicarbazone preparations, and the methanol solutions from your polarimetric measurements into the container for flammable (organic) waste.

References

1. Garin, D. L. *J. Chem. Educ.* **1976,** *53,* 105.
2. Glidewell, C. *J. Chem. Educ.* **1991,** *68,* 267–269.
3. Murov, S. L.; Pickering, M. *J. Chem. Educ.* **1973,** *50,* 74–75.

Questions

1. Critically evaluate your evidence on whether the carvone isolated from caraway seed oil is the mirror image of (−)-carvone.
2. (*S*)-(+)-carvone has a boiling point of 230°C. What made it possible to steam distill it at 100°C?
3. What caused the melting point of the mixture of (+)- and (−)-carvone semicarbazones to be higher than the melting point of either pure compound?

6.2 Isolation of (*R*)-(+)-Limonene from Orange Peels

PURPOSE: To extract the essential oil from orange peels and analyze it by polarimetry.

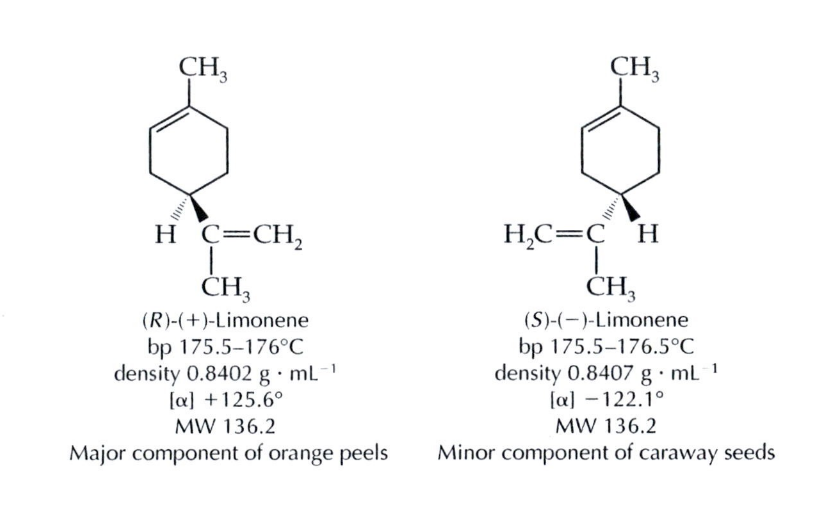

(*R*)-(+)-Limonene
bp 175.5–176°C
density 0.8402 g · mL^{-1}
[α] +125.6°
MW 136.2
Major component of orange peels

(*S*)-(−)-Limonene
bp 175.5–176.5°C
density 0.8407 g · mL^{-1}
[α] −122.1°
MW 136.2
Minor component of caraway seeds

MINISCALE PROCEDURE

Techniques Steam Distillation: Technique 11.7
Extraction: Technique 8.2
Polarimetry: Technique 14

SAFETY INFORMATION

Dichloromethane is toxic, an irritant, absorbed through the skin, and harmful if swallowed or inhaled. Wear nitrile gloves. Use it in a hood, if possible, and wash your hands thoroughly after handling. Be sure to do the evaporation process in a hood.

(*R*)-(+)-Limonene is an irritant and readily absorbed through the skin.

Steam Distillation

You will need the peels from two oranges for this experiment. Your instructor will specify whether you are to provide the oranges or whether they will be available in the laboratory. Peel the oranges just before grinding them to prevent loss of volatile limonene. Remove most of the white pulp from the orange peels with a knife or spatula before grinding the peels in a blender with 200–250 mL of water to make a slurry that can be easily poured into a 500-mL round-bottomed flask. Add 4 drops of an antifoaming agent to the flask.

Assemble the steam distillation apparatus shown in Technique 11.7, Figure 11.24, but without the separatory (dropping) funnel; close the second neck of the Claisen adapter with a glass stopper. Use a 50-mL round-bottomed flask as the receiver. Before attaching the receiving flask, pour 35 mL of water into it and mark the level on the outside of the flask; then pour out the water (the flask does not have to be dried).

Boil the mixture at a moderately rapid rate using a heating mantle as the heat source, but be careful not to let any solid material bump over into the condenser. Collect 35 mL of distillate.

Isolation of Limonene

Follow the procedure "Isolation of Caraway Oil" on pp. 41–42 to isolate your limonene from the distillate. Record the final mass of limonene to the nearest milligram. Describe its appearance and carefully note the odor (do NOT breathe the vapors of limonene).

Polarimetry

The following directions for preparation of the polarimetry solution are for 1-dm periscope polarimeter tubes. Consult your instructor if your laboratory is equipped with another type or size of polarimeter tubes.

Obtain 10 mL of 95% ethanol in a clean, dry graduated cylinder. Dissolve your limonene in 3 mL of ethanol (use a Pasteur pipet to transfer the ethanol) and quantitatively transfer the resulting solution to a 10-mL volumetric flask as follows. Set the volumetric flask in a small beaker so that it will not tip, and place a very small funnel in the neck of the volumetric flask. Carefully pour the limonene solution into the volumetric flask. Rinse the Erlenmeyer flask three times with approximately 1-mL portions of ethanol and add these rinses to the volumetric flask. Rinse the funnel with 1 mL of ethanol and remove it from the volumetric flask. Fill the volumetric flask to the calibration mark using a Pasteur pipet; stopper the flask and invert it several times until the contents are thoroughly mixed.

Calibrate the polarimeter using 95% ethanol as the reference solvent.

If the solution is cloudy, filter it directly into a polarimeter tube using a small funnel and fluted filter paper. If the solution is clear, it can be poured directly into a polarimeter tube. Determine and record the observed rotation for your limonene solution; also record the temperature [see Technique 14.4]. Calculate the specific rotation $[\alpha]_D$ of limonene and its enantiomeric excess (optical purity) [see Technique 14.5].

CLEANUP: Filter the orange peel residue from the aqueous liquid remaining in the distillation flask. (Do not put orange peels down the sink.) Wash the aqueous filtrate down the sink; dispose of the orange peel residue as food garbage. Allow any residual dichloromethane to evaporate from the calcium sulfate drying agent in a hood before placing the solid in the container for inorganic waste or the container for solid hazardous waste. Pour the ethanol solution used for your polarimetric measurements into the container for flammable (organic) waste.

Reference

1. Glidewell, C. *J. Chem. Educ.* **1991,** *68,* 267–269.

Questions

1. (*R*)-(+)-limonene has a boiling point of 176°C. What made it possible to steam distill it at 100°C?
2. Critically evaluate your evidence on whether the limonene isolated from orange peels is (*R*)-(+)-limonene.
3. Propose a method that you could use to isolate the limonene in caraway seed oil.

EXPERIMENT

7 COMPUTATIONAL CHEMISTRY

PURPOSE: To introduce you to modern methods that use the computational power of computers to calculate the structures and energies of molecules.

In Experiment 7 you will use a molecular mechanics software package to calculate the energies and conformational structures for rotamers of butane and for *axial* and *equatorial isomers* of substituted cyclohexanes. Teams of two or three students will work together on two of the computational chemistry experiments.

Computational chemistry is the calculation of physical and chemical properties of compounds using mathematical relationships derived from theory and observation. It is often referred to as molecular modeling. However, we use the term "computational chemistry" to avoid confusion with molecular model sets, which you may have already used to create three-dimensional structures of chemical compounds.

Picturing Molecules on the Computer

Computational chemistry can be used to create three-dimensional images and two-dimensional projections of chemical structures. In this way it is similar to a molecular model set, but it is also much more. In molecular model sets, the bond lengths and bond angles are fixed at certain "standard" values, such as 109.5° for the bond angle of a tetrahedral (sp^3) carbon atom. Anyone who has built a molecule containing a cyclopropane ring is well aware of the limitations of using these "standard values." The structure of a molecule created on the computer can be optimized so that it represents the lowest energy conformation of the molecule. Optimization means that the bond lengths and bond angles of the structure may deviate from the "standard" values. Thus, the molecule created on the computer is a more accurate picture of the actual molecule.

Once the exclusive domain of mainframe and supercomputers, computational chemistry has migrated to the modern microcomputer. Advances in computer hardware provide massive amounts of memory, high computational speed, and high-resolution graphics displays. Most computational chemistry programs consist of interacting modules that carry out specialized tasks such as building a molecule, displaying a graphical image of the molecule, optimizing the molecular structure, and extracting physical properties from the calculation.

The computer image of a molecule can be shown in a variety of ways—wire frame, ball and stick, and space filling, to mention a few. The rendering methods can be mixed for purposes such as emphasizing steric interactions in a specific portion of a molecule. The electron density surface of a molecule can be displayed, providing a view of its overall shape. The electrostatic potential can also be mapped onto the molecular surface, highlighting regions of potential reactivity within the molecule. Molecular orbitals can be superimposed onto a molecular structure. In many reactions, important insights can be gained by examining the

highest occupied molecular orbital (HOMO) and the ***lowest unoccupied molecular orbital (LUMO).***

Many physical and chemical properties can be extracted from an optimized molecular structure. These properties include bond lengths, bond angles, dihedral angles, interatomic distances, dipole moments, electron densities, and heats of formation.

Computational Chemistry Programs

One of the two major types of computational methods is based on quantum mechanics, which attempts to solve the Schrödinger equation of a molecule. The other type is based on a mechanical model, which treats atoms as balls and bonds as springs connecting the balls. Following are some of the packages available for modern microcomputers:

MacSpartan and PC Spartan from Wavefunction
CAChe for Macintosh and CAChe for PC from Fujitsu
HyperChem from HyperCube
PC Model from Serena Software

In this experiment we describe in brief and general terms the types of calculations and their limitations. Because the operation of the program to build a molecule and invoke the calculation modules differs from one package to another, detailed instructions are left to your instructor and to the instructions included with the computational chemistry package you use. Materials included with the packages also provide more comprehensive descriptions of the computational methods the programs use.

Ab Initio Quantum Mechanical Methods

Quantum mechanical methods are based on solving the ***Schrödinger wave equation,*** $H\Psi = e\Psi$, in which H is the Hamiltonian operator describing the kinetic energies and electrostatic interactions of the nuclei and electrons that make up a molecule, e is the energy of the system, and Ψ is the wavefunction of the system. Although simple in expression, the solution is exceedingly complex and requires extensive computational time. To date, even an organic molecule as simple as methane defies exact solution. The key to obtaining useful information from the Schrödinger relationship in a reasonable length of time lies in choosing approximations that simplify the solution. There are trade-offs, however; the more approximations used, the faster the calculation but the less accurate the result.

The quantum mechanical method with the least degree of approximation is the ***ab initio method.*** Following are some common approximations used with this method:

1. Nuclei are stationary relative to electrons, which are rapidly moving about a molecule (Born-Oppenheimer approximation).
2. The motion of any single electron is affected by the average electric field created by all the other electrons and nuclei in the molecule (Hartree-Fock approximation).
3. A molecular orbital is a linear combination of atomic orbitals (LCAO approximation).

Ab initio calculations use a collection of atomic orbitals, called a ***basis set,*** to describe the molecular orbitals of a molecule. There are numerous basis sets of varying complexity in use. The choice of a basis set affects the accuracy of the calculation and the amount of time required for a solution. The smallest basis set in common use is STO-G3. This basis set uses three Gaussian functions to describe the orbitals and works reasonably well with first- and second-row elements that incorporate s and p orbitals. An ab initio calculation using an STO-G3 basis set can often provide good equilibrium geometries. The 6-31G* basis set, using more Gaussian functions and a polarization function, provides better answers and is more flexible for elements incorporating d orbitals. However, it requires more calculation time, typically 10–20 times more than the same calculation using an STO-G3 basis set.

The Semiempirical Molecular Orbital Approach

Small organic molecules can be optimized by the ab initio method using a medium-sized 3-21G basis set with a modern microcomputer. For example, the optimization of methylcyclohexane with the methyl group in an axial position takes approximately 51 min using MacSpartan on a Macintosh G4-400. For most practical purposes, a faster method of calculation is needed. The ***semiempirical molecular orbital approach*** introduces several more approximations that dramatically speed up the calculations. A geometry optimization using a semiempirical molecular orbital method is typically 300 or more times faster than one using an ab initio method with a 3-21G basis set.

The approximations generally used with semiempirical molecular orbital methods are as follows:

1. Only valence electrons are considered. Inner shell electrons are not included in the calculation.
2. Only interactions involving one or two atoms are considered. This is called the neglect of diatomic differential overlap, or NDDO.
3. Parameter sets are used to calculate interactions between orbitals.

The parameter sets are developed by fitting calculated results with experimental data. Several popular versions of semiempirical methods follow:

MNDO or Minimum Neglect of Differential Overlap
AM1 or Austin Method 1
PM3 or Parameterized Model 3

In many cases, AM1 is the method of choice for organic chemists. The PM3 method is often used for inorganic molecules because it has been parameterized for more chemical elements. The MOPAC or Molecular Orbital PACkage combines these three semiempirical methods in a single program. As you become more familiar with computational chemistry, you will be able to experiment with the various methods to find the one that works best for the molecules you are working with.

Molecular Mechanics

The molecular mechanics method treats molecules as an assemblage of balls (atoms) and springs (bonds). The total energy of a molecule, called the ***steric energy,*** is the sum of contributions due to bond stretching, angle strain, strain due to improper torsion, steric or van der Waals interactions, and electronic charge interactions.

$$E_{\text{steric}} = E_{\text{bonds}} + E_{\text{angles}} + E_{\text{torsion}} + E_{\text{vdW}} + E_{\text{charge}}$$

The contributions are described by empirically derived equations. For example, the energy of a bond is approximated by the energy of a spring described by Hooke's Law,

$$E_{\text{bond}} = \tfrac{1}{2} k(x - x_0)^2$$

in which k is a force constant related to bond strength and $(x - x_0)$ is the displacement of an atom from its equilibrium bond length (x_0). The force constants for various types of bonds can be derived from experimental data and are incorporated in the parameter set. The energy due to the bonds in the molecule is the sum of the contributions from all of the bonds.

$$E_{\text{bonds}} = \sum_{i=1}^{i=n \text{ bonds}} \tfrac{1}{2} k_i (x - x_0)_i^2$$

Other energy contributions are developed in a similar fashion. The collection of equations describing the various energies and their associated parameter sets are called ***force fields.*** Following are some frequently used force fields:

MM2, MM3, MM4
MMFF
SYBYL

The absolute value of the steric energy of a structure has no meaning by itself. Its calculated value varies greatly from one force field to another. Steric energies are useful only for comparison purposes, and the comparisons are meaningful only if made between ***conformers***, such as chair and twist-boat cyclohexane. With caution, comparisons between diastereoisomers, such as *cis*- and *trans*-1,3-dimethylcyclohexane, can also be informative. Comparing steric energies of two structurally different molecules is fruitless, even two isomeric structures as similar as 2-methyl-1-butene and 2-methyl-2-butene.

Even though the steric energy of an individual conformer by itself is meaningless, the difference in steric energies between conformers of a molecule is significant. These energy differences can be used to estimate equilibrium values between interconverting conformers. For example, at room temperature, methylcyclohexane is a mixture of *axial*-methylcyclohexane and *equatorial*-methylcyclohexane that is rapidly interconverting by way of a ring flip.

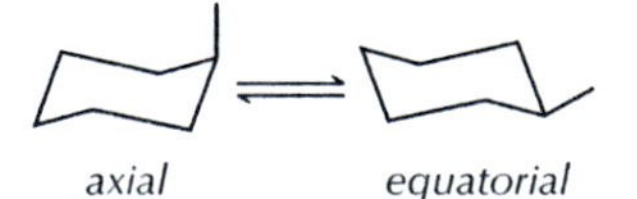

The relative amount of each conformer at equilibrium can be determined by the difference in energy between the two conformers, which is related to the equilibrium constant, K_{eq}, by the following relationships:

$$K_{eq} = \frac{\text{number of } eq\text{-methylcyclohexane molecules}}{\text{number of } ax\text{-methylcyclohexane molecules}}$$

$$\Delta G^0 = -RT \ln K_{eq} = -2.303\, RT \log K_{eq}$$

where ΔG^0 is the change in Gibbs standard free energy in going from *ax*-methylcyclohexane to *eq*-methylcyclohexane, R is the gas constant (1.986 cal deg^{-1} mol^{-1}), and T is the absolute temperature in degrees Kelvin (K).

Using the MM2 force field with CAChe, the steric energy of *axial*-methylcyclohexane is calculated to be 8.69 kcal mol^{-1} and the steric energy of *equatorial*-methylcyclohexane is calculated to be 6.91 kcal mol^{-1}. The difference in energy between conformers is -1.78 kcal mol^{-1}. The negative value for ΔG^0 signifies a release of energy in going from *ax*-methylcyclohexane to *eq*-methylcyclohexane. At room temperature (25°C, 298 K), the preceding equation becomes

$$-1.78 = -1.36 \log K_{eq}$$

At equilibrium, there would be approximately 20 molecules of *equatorial*-methylcyclohexane present for each molecule of *axial*-methylcyclohexane, quite close to the experimental value.

Which Computational Method Is Best?

The best method depends on the question you are asking and on resources at your disposal. If you are simply trying to find the optimal (lowest-energy) structures of organic molecules, molecular mechanics provides reasonable structures, and it is very fast. Good values for bond angles, bond lengths, dihedral angles, and interatomic distances can be determined from an optimized structure. In general, you are limited to typical organic compounds; for instance, there are no good parameter sets for carbon-metal bonds.

The energy differences between conformers determined by molecular mechanics are often very close to experimentally determined values. These energy differences can be used to determine equilibrium ratios of conformers. Since the calculations are fast, the energies of many conformers can be determined in a short time. This is especially useful when examining ***rotamers,*** conformations related by rotation about a single bond.

Since molecular mechanics is a mechanical model, it says nothing about electron densities and dipole moments. It also says nothing about molecular orbitals. However, the optimized structure from molecular mechanics can provide input data for other programs. In many cases, especially with large molecules, it is efficient to use molecular mechanics to get an approximation that can be further refined with a quantum mechanical method.

Semiempirical methods, which are significantly faster than ab initio calculations, provide reliable descriptions of structures, stabilities, and other properties of organic molecules. They do a reasonable job in calculating thermodynamic properties, such as heats of formation. The heats of formation can be used to compare energies of isomers such as 2-methyl-1-butene and 2-methyl-2-butene, which could not be done with steric energies from molecular mechanics calculations. The calculated heats of formation can also be used to approximate the energy changes in balanced chemical equations.

Sources of Confusion

Computational chemistry is inherently complex, but most of the commercially available packages have been "human engineered," making it relatively easy to get started. When you get to a point in the process where you have a choice, a default option is usually provided. Nevertheless, it is beneficial to acquaint yourself with the information provided with the package.

Two things can cause a good deal of confusion and should be avoided. The first occurs if you start with the wrong structure, and the second deals with the problem of local rather than global energy minima.

Starting the Computation with the Correct Structure. Starting with the correct structure is closely related to the method you use in building a molecule. In many packages, the user draws a two-dimensional projection, similar to the line formulas printed in a book, and the program translates it into a rough three-dimensional structure. However, if the projection is ambiguous, the program may create an unsuitable structure. For example, suppose you wanted to create *axial*-methylcyclohexane. The projection entered on the computer might look like this:

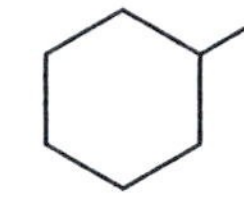

Viewing the structure created by this projection on the computer screen and then rotating it, you would probably observe a flat molecule, clearly unsuitable as the basis for optimizing the molecule's structure. To turn this projection into a three-dimensional structure usually requires invoking some sort of "cleanup" or "beautifying" routine. The routine creates a three-dimensional structure using "normal" bond lengths and bond angles. In the case of methylcyclohexane, the structure becomes a cyclohexane in the chair conformation with a methyl group in an equatorial position.

Building a cyclohexane with a methyl group in the axial position presents a different problem. The answer lies in creating the structure in stages. In this case, you need to create a chair cyclohexane and then replace one of the axial hydrogens with a methyl group. As you can see, the process involves building the framework first and then adding the necessary attachments at specific locations. Most computational chemistry packages have templates or molecular fragments to assist in creating complicated structures.

Local and Global Minima. The second potential source of confusion that is encountered in attempting optimization of a structure is the ***local minimum*** problem. During optimization, the program tries to find the structural conformation with the lowest energy. It does this by calculating the energy of the initial structure, making a small change in the structure by moving atoms, and then calculating the energy of the new structure. If the new structure is lower in energy, it continues to make small changes in the same direction. It continues this process until the energy no longer decreases. At this point, it then makes changes in a different direction until the energy no longer decreases. When it has exhausted all possible directions for structural changes, the structure should represent the lowest energy conformation of the molecule.

The local minimum problem arises because the energy profile is not a smooth curve with one minimum. The energy surface is uneven with lumps, bumps, ridges, and several low spots. The low spot that a minimization falls into depends on where you start on the energy surface. In Figure 7.1, starting from point A or B will end up at the local minimum. Starting at point C or D will end up at the desired ***global minimum.*** This is illustrated by the calculation of *axial-* and *equatorial*-methylcyclohexane. The two structures are conformers that can be interconverted by way of a ring flip. *ax*-Methylcyclohexane is a local minimum and *eq*-methylcyclohexane is the global minimum. The barrier represents the strain energy required to flip the ring. Without a barrier, minimization of *ax*-methylcyclohexane would produce *eq*-methylcyclohexane.

If you are looking for a global energy minimum for a molecule, the possibility of encountering local minima adds uncertainty and doubt to any result. How does one know if the structure represents a local minimum or a global minimum? That question is the source of many research projects. For our purposes, the answer is to create several different starting structures, carry out minimizations on each of them, and use the lowest energy as the global minimum. There are several methods for systematically creating possible starting structures. One method is conformational searching. Several conformations of a structure are created by rotating portions of the molecule connected by single bonds. Some modeling packages have routines called sequential searching that automate this process. Other packages have methods such as Monte Carlo routines for generating random structures.

Yet another method of generating candidate structures for minimization is to use a molecular dynamics simulation program. This program

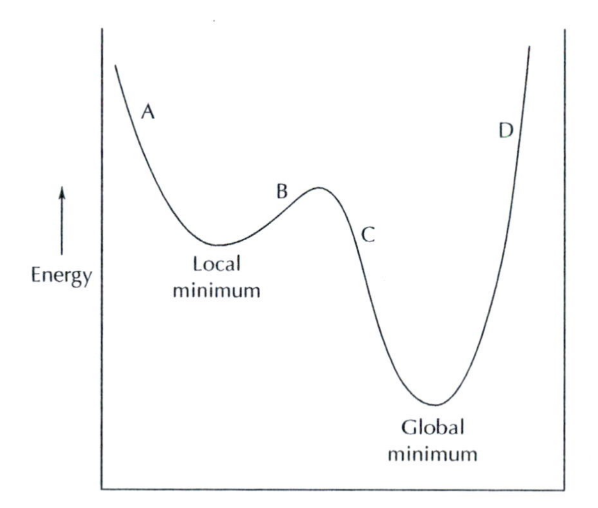

FIGURE 7.1
Local and global minima resulting from energy minimization.

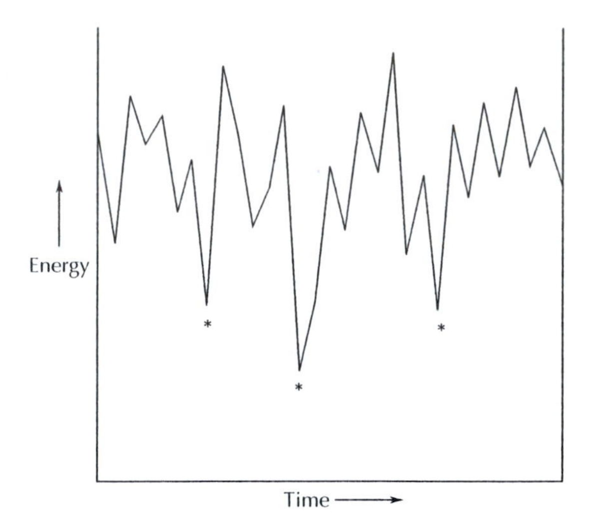

FIGURE 7.2
Output of a molecular dynamics simulation plotted as a graph of energy versus the conformation of the structure, changing with time.

simulates the motion of atoms within a structure. The molecule is given increased kinetic energy, the amount depending on the designated temperature. As the atoms move around, energy "snapshots" are taken at regular intervals. The structures with the lowest energy are used as starting structures for minimization. This method often propels molecules over energy barriers due to steric interactions, bond strain, or torsional strain. The results of a molecular dynamics simulation can be plotted as the internal energy of a molecule versus time. In Figure 7.2, structures corresponding to low-energy conformers are designated with asterisks. These conformers can be used as initial structures for energy minimizations by molecular mechanics or quantum mechanical calculations.

Relationship of Computations and Experiments

When doing computational chemistry, it is important to keep a firm grip on reality at all times. You need to evaluate the result, especially a surprising result, and determine whether it makes sense chemically and physically. All too easily, people accept the result of their calculations as physical truth. Computational chemistry is based on theoretical models using approximations and parameter sets derived from theory and experiment. One of the major concerns that researchers in computational chemistry have with commercial microcomputer computational packages is the proprietary nature of the programs and parameter sets. The computer programs may contain flaws in logic or programming, which are periodically corrected through updates. Nevertheless, computational chemistry has become a valuable tool for gaining insights into organic chemistry.

Computational Chemistry Experiments

Experiment 1: Butane. Build a molecule of butane and optimize its structure using molecular mechanics. Look at the structure using the various rendering options available with your computational system. Position the optimized butane so that you are viewing the molecule end on, along the C2—C3 bond. Print out the image from the computer screen. Notice that the bonds of C2 are staggered relative to those of C3. What is the relationship of the methyl group attached to C2 relative to the methyl group attached to C3?

Experiment 2: Conformers of Butane. There is free rotation about the C2—C3 bond in butane. The energy of the molecule changes as the groups rotate around the axis of the bond. This change can be illustrated using molecular mechanics calculations to measure the steric energy of butane as a function of the dihedral angle formed by C1, C2, C3, and C4.

Working in a team of two or three students and starting with one of the structures of butane created in Experiment 1, set the dihedral angle formed by C1, C2, C3, and C4 to 0. Constrain it to this value and minimize the energy of the structure. Record the steric energy in a table. Repeat this process using dihedral angles of 30°, 60°, 90°, 120°, 150°, and 180°. It is not necessary to calculate the values from 180° to 360° because symmetry allows you to assign the steric energies of the conformers. In some cases, the constraint may override the minimization routine. If this occurs, set the dihedral angle and determine a single point energy of the structure. Use the following format to record your calculated steric energies.

Dihedral angle (deg)	0/360	30/330	60/300	90/270	120/240	150/210	180/180
Steric energy (kcal/mol)							

Plot the steric energies versus the dihedral angles. Are there any local minima? Examine the structures and note the relationships of the methyl groups. Examine the structure of the conformer at the global minimum and note the relationship of the methyl groups.

If there are no constraints, the optimizer converges to the low-energy structure nearest the energy of the starting structure. Demonstrate this convergence by minimizing the energies of the structures of butane with starting dihedral angles formed by C1, C2, C3, and C4 at 110° and 130°.

Experiment 3: axial and equatorial Conformers of Monosubstituted Cyclohexanes. In monosubstituted cyclohexanes, we can expect the ratio of the *axial* and *equatorial* conformers to be affected by the size of the substituent. As discussed previously, at room temperature the ratio of *equatorial* to *axial* conformers in methylcyclohexane is 20 to 1.

Working in a team of two or three students, build molecules of ethylcyclohexane and *tert*-butylcyclohexane. Calculate the steric energies of the *equatorial* and *axial* conformers using molecular mechanics. Be careful to avoid local minima by using different rotational conformers about the C—C bond between the substituent carbon atom and the attached cyclohexane-ring carbon atom as starting points for energy minimizations.

References

1. Goodman, Jonathan M. *Chemical Applications of Molecular Modelling:* Royal Society of Chemistry, Cambridge, 1998.
2. Hehre, W. J.; Shusterman, A. J.; Huang, W. W. *A Laboratory Book of Computational Organic Chemistry;* Wavefunction, Irvine, 1998.
3. Cramer, Christopher J. *Essentials of Computational Chemistry: Theories and Models;* Wiley: New York, 2002.

EXPERIMENT

8

RADICAL HALOGENATION REACTIONS

QUESTION: Can the product distribution in a free-radical chain reaction be explained by statistical factors alone, or do electronic and stereochemical factors play a role?

In Experiment 8.1, you will study the free-radical chlorination of 1-chlorobutane and 2,2,4-trimethylpentane and discover what factors determine the product composition. You will be able to answer the question by using gas chromatography to determine the number and identity of the products.

In Experiment 8.2, you will study the photobromination of 1,2-diphenylethane. You will learn what effect a nearby stereocenter has on the reaction of a free radical by using the melting point of the product to discover which diastereomer of 1,2-dibromo-1,2-diphenylethane forms.

Radical Chlorination Reactions

Alkanes are relatively unreactive compounds. Other than combustion, halogenation is one of the few reactions that they undergo. The reaction of chlorine or bromine at an sp^3 carbon atom of an alkane is a free-radical chain reaction. Its energy of activation depends a great deal on the stability of the free-radical intermediate, which forms the basis of the inquiry in Experiment 8.1.

Consider first the mechanism for the chlorination of methane to produce chloromethane:

$$CH_4 + Cl_2 \xrightarrow{\Delta \text{ or } h\nu} CH_3Cl + HCl$$

The overall reaction occurs in a series of steps. In the first step, called initiation, light or heat induces the dissociation of a chlorine molecule into two free chlorine atoms. Each chlorine atom has an unpaired, or odd, electron and is called a ***free radical.***

Step 1 $$Cl_2 \xrightarrow{\Delta \text{ or } h\nu} 2Cl\cdot$$

In the second step of the chlorination reaction, the chlorine radical abstracts a hydrogen atom from methane. In the third step, the reaction of the methyl radical with a molecule of chlorine produces chloromethane and another chlorine radical.

Step 2 $$\dot{Cl} + CH_4 \longrightarrow CH_3\cdot + HCl$$

Step 3 $$CH_3\cdot + Cl_2 \longrightarrow CH_3Cl + Cl\cdot$$

Steps 2 and 3 occur over and over again. In each cycle of this propagation sequence, the highly reactive free radical produced in the step before reacts in the next step. Ultimately, often after hundreds or thousands of propagation cycles, the reaction undergoes termination steps that provide no further free radicals. Then a new initiation step must occur to start the whole process over again.

Radical halogenation reactions often lead to mixtures of products, like those you will observe in this experiment. Two major factors control the preference for isomer formation, a statistical factor and the relative stability of the carbon radical intermediates. In the reaction of 1-chlorobutane, the effect of a substrate chlorine atom on the stability of a nearby sp^2 center of a free radical must be taken into account.

If the statistical factor is the only one directing the chlorination of 2-methylpropane, then the predominant product would be the primary alkyl chloride, 2-methyl-1-chloropropane, because there are nine methyl hydrogens and only one methine (CH) hydrogen. In fact, the reaction produces a mixture of primary and tertiary chlorides. Because tertiary free radicals are more stable than primary ones, a fair amount of tertiary product (2-methyl-2-chloropropane) forms even though there is only one methine hydrogen.

$$(CH_3)_2CHCH_3 + Cl_2 \longrightarrow (CH_3)_2CHCH_2Cl + (CH_3)_3CCl$$

2-Methylpropane (isobutane)	2-Methyl-1-chloropropane (isobutyl chloride) a primary chloride	2-Methyl-2-chloropropane (*tert*-butyl chloride) a tertiary chloride

Benzylic Bromination

The reaction of 1,2-diphenylethane with bromine in Experiment 8.2 involves relatively stable secondary benzylic free radicals as intermediates. A benzylic radical is stabilized by conjugation with the π-system of the benzene ring.

Benzylic free-radical intermediates

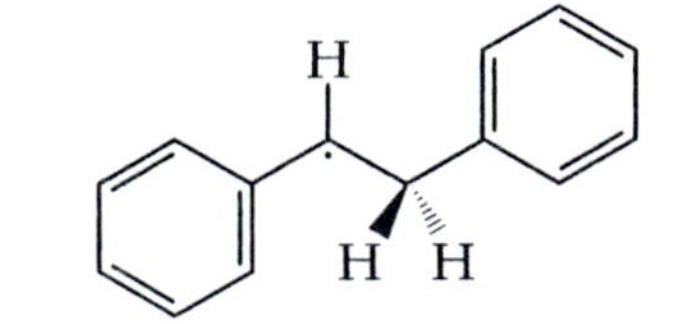

Intermediate for monobromo product

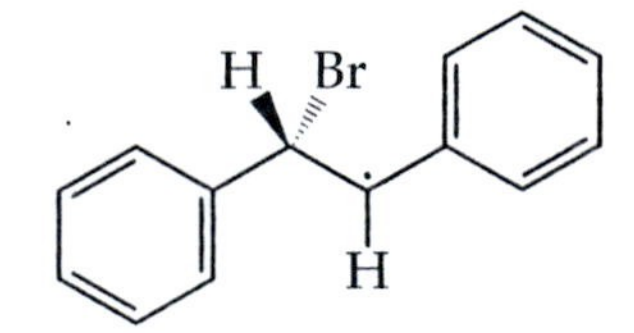

Intermediate for dibromo product

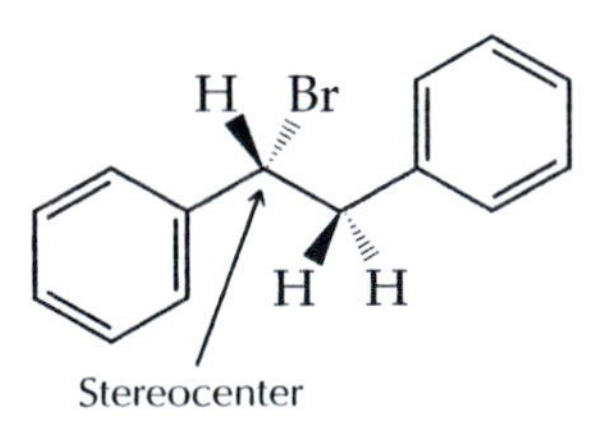

Under the conditions of Experiment 8.2 the dibromo substitution product is formed. Notice that the monobromo product, an intermediate in the reaction, has a stereocenter (an asymmetric carbon atom).

When a new propagation cycle forms the free radical that leads to the dibromo product, this new free radical is chiral. Its reaction with bromine can lead to two diastereomeric products, (2*S*,3*S*)- and (2*S*,3*R*)-1,2-dibromo-

1,2-diphenylethane. These stereoisomers have different energies and form at different rates. If the formation of the dibromo product is also a free-radical halogenation reaction, the product that forms depends on which face of the trigonal planar radical reacts with the second bromine. Reaction could occur on the same face that the neighboring bromine atom already occupies or it could occur on the opposite face. If the second bromine reacts on the same face as the bromine atom that is attached to the stereocenter, the product will be racemic chiral 1,2-dibromo-1,2-diphenylethane. If the second bromine reacts on the opposite face, *meso*-1,2-dibromo-1,2-diphenylethane will be the product.

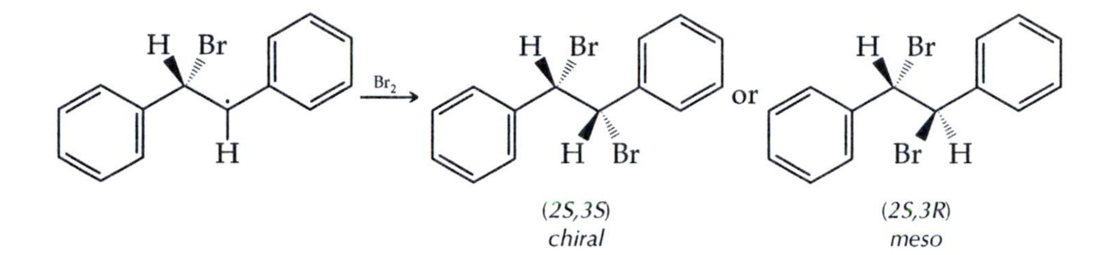

Sources of Halogen

In both Experiments 8.1 and 8.2, Cl_2 and Br_2 are not added directly to the reaction mixtures. Instead, they are generated in situ within the reaction mixtures because both halogens are highly reactive, toxic, and difficult to handle. The Cl_2 used in Experiment 8.1 comes from the decomposition of sodium hypochlorite, NaOCl (household bleach). It's interesting to note that bleach is also the source of the Cl_2 used to retard the spread of disease from swimming pools.

$$NaOCl + HCl \longrightarrow HOCl + NaCl$$

$$HOCl + HCl \longrightarrow Cl_2 + H_2O$$

The bromine used in Experiment 8.2 forms from the redox reaction between potassium bromate ($KBrO_3$) and hydrogen bromide.

$$KBrO_3 + 6HBr \longrightarrow 3Br_2 + 3H_2O + KBr$$

8.1 Radical Chlorination Reactions

QUESTION: Can the distribution of products in the radical chlorinations of 1-chlorobutane and 2,2,4-trimethylpentane be explained by a simple statistical model?

$$CH_3CH_2CH_2CH_2Cl \xrightarrow[HCl,\ H_2O]{NaOCl} \text{mixture of dichlorobutanes}$$

1-Chlorobutane
bp 78.5°C
$d = 0.886\ g \cdot mL^{-1}$

$$CH_3C(CH_3)_2CH_2CH(CH_3)CH_3 \xrightarrow[HCl,\ H_2O]{NaOCl} \text{mixture of monochlorooctanes}$$

2,2,4-Trimethylpentane
bp 99.2°C
$d = 0.692\ g \cdot mL^{-1}$

MICROSCALE PROCEDURE

PRELABORATORY ASSIGNMENT: Calculate the proportions of isomeric products that would be expected from each reaction carried out in this experiment if only statistical factors determined the course of the reaction.

Techniques Microscale Extraction: Technique 8.6c
Pasteur Filter Pipets: Technique 8.5
Gas Chromatography: Technique 16

SAFETY INFORMATION

1-Chlorobutane is harmful if inhaled, ingested, or absorbed through the skin. Wear gloves and use the reagent in a hood, if possible.

2,2,4-Trimethylpentane is flammable and a skin irritant. Wear gloves and use the reagent in a hood, if possible.

Sodium hypochlorite solution, when acidified, emits chlorine gas that is toxic as well as an eye and respiratory irritant. Use the reagent in a hood, if possible.

You will carry out the following radical chlorination procedure on 1-chlorobutane and 2,2,4-trimethylpentane. As part of your prelab preparation, draw the structures for the possible monochlorination products in each reaction and give the IUPAC name of each.

Freshly purchased household bleach such as Clorox or a supermarket brand works well.

Label the bottoms of two conical vials fitted with screw caps and septa. Place 1.0 mL of the first compound to be tested in one of the vials.

Working in a hood, add 1.0 mL of 5% sodium hypochlorite solution followed by 0.5 mL of 3 M hydrochloric acid and *immediately* cap the vial. Shake the vial until the yellow color of the chlorine has moved from the aqueous layer to the organic layer (about 30 s). Repeat this procedure with the other compound to be tested.

Place the vials at a distance of 5–8 cm from an unfrosted light bulb, and irradiate them until the yellow color of the chlorine disappears from the organic layer (usually within 1–2 min). Shake each vial occasionally during the irradiation period.

Working with one vial at a time, carry out the following procedure on each. Add 100 mg of anhydrous sodium carbonate in several portions; wait for the foaming to cease before making the next addition. When the addition is complete and foaming ceases, draw the mixture into a Pasteur pipet fitted with either a syringe or a rubber bulb and expel it back into the vial at least six times. Remove the lower aqueous layer with the Pasteur pipet [see Technique 8.6c]. Add 100 mg of anhydrous calcium chloride and allow the solution to dry for at least 10 min.

Prepare two Pasteur filter pipets while you are waiting for the solutions to dry [see Technique 8.5]. Using one of these filter pipets, transfer one of the dried solutions to a labeled sample vial or small test tube fitted with a cork. Repeat this procedure for the other solution, using the other filter pipet.

Product Analysis by Gas Chromatography

Analyze the solutions on a gas chromatograph that has a nonpolar column, such as SE-30 or OV-101 [see Technique 16]. Consult your instructor about specific operating instructions and sample size for the gas chromatographs in your laboratory. Record the operating parameters, including column temperature, injector temperature, detector temperature, helium flow rate, chart speed, and the voltage or attenuation for the recorder. See Question 1 for the elution order of the dichlorobutanes from a nonpolar column. The following order of elution from a nonpolar column has been reported by Russell and Haffley (Ref. 3) for the monochlorination of 2,2,4-trimethylpentane:

1. 2-chloro-2,4,4-trimethylpentane
2. 3-chloro-2,2,4-trimethylpentane
3. 1-chloro-2,4,4-trimethylpentane
4. 1-chloro-2,2,4-trimethylpentane.

CLEANUP: The aqueous phase separated from the reaction mixture may be washed down the sink or placed in the container for aqueous inorganic waste. The calcium chloride drying agent should be placed in the container for hazardous solid waste because it is coated with halogenated hydrocarbons. After performing the GC analysis, pour the remaining product mixture into the halogenated waste container.

Interpretation of Experimental Data

For each reaction, analyze the chromatogram to determine the percentage of each product [see Technique 16.7]. Compare your experimental results with the product distributions calculated in the prelaboratory assignment. From your experimental results, do statistical factors determine the composition of the product mixture in the chlorination of each substrate, or do electronic factors play a role? Explain your reasoning.

OPTIONAL EXPERIMENT Additional Substrates

Carry out the preceding procedure using 3-methylpentane and 2-chlorobutane as the substrates. Both compounds give diastereomers for one of the products that have boiling points different enough to be partially separated on a capillary column chromatograph.

OPTIONAL COMPUTATIONAL CHEMISTRY EXPERIMENT

Build a molecule of 2,2,4-trimethylpentane. Find the lowest-energy conformer using the AM1 semiempirical quantum mechanical method. To reduce the time required for the calculation, optimize the structure by using a molecular mechanics package; then use that structure as the starting point of the AM1 calculation.

Measure the lengths of the C—H bonds of the methyl, methylene, and methine groups. If we assume that shorter bonds are stronger bonds, which C—H bond will break most easily? Which will dissociate least easily?

Save the structure of the optimized molecule of 2,2,4-trimethylpentane so that you can use it as the starting point for building radicals. Create a tertiary radical by deleting the hydrogen atom at C4. The method for creating the radical may differ slightly depending on the specific software. For example, with CAChe you should not "beautify" the structure, or the program will replace the deleted hydrogen atom. With Spartan packages, you will also need to delete the valence at C4. Using the AM1 method, optimize the structure of the radical and record its heat of formation. In setting up the calculation, specify doublet multiplicity because there is an unpaired electron in the radical.

In a similar fashion, optimize the structures of the radicals formed by removing hydrogen atoms at C1, C3, and C5, and record their heats of formation. Are there energy differences between primary, secondary, and tertiary radicals? Do the stabilities of the radicals correlate with the C—H bond lengths determined previously?

Do your experimental results show a preference for formation of any one product, or is the product distribution purely statistical? Are your computed results consistent with your experimental results? Explain.

References

1. Gilow, H. M. *J. Chem. Educ.* **1991,** *68,* A122–124.
2. Reeves, P. C. *J. Chem. Educ.* **1971,** *48,* 636–637.
3. Russell, G. A.; Haffley, P. G. *J. Org. Chem.* **1966,** *31,* 1869–1871.

Questions

1. A handbook will reveal the boiling points of the four dichlorobutanes formed by the chlorination of 1-chlorobutane. The dichlorobutanes elute in the following order: (1) 1,1-dichlorobutane, (2) 1,2-dichlorobutane, (3) 1,3-dichlorobutane, and (4) 1,4-dichlorobutane. Use the boiling points to explain the GC elution order.
2. Using your experimental results for each compound tested, calculate the relative reactivity (rate) for abstraction of each type of hydrogen by a chlorine radical.
3. 1-Chlorobutane has 2° hydrogens on two different carbon atoms. Did your results show an equal amount of chlorination at each 2° site? Give a possible explanation for any difference observed in the amount of chlorination at each site.
4. From the study of radical chlorination reactions using many hydrocarbons, chemists have found that the relative reactivity for 1°, 2°, and 3° hydrogen atoms is 1:3.8:5.0. Compare your experimental values for relative reactivity to these values and suggest a reason for any possible differences.

8.2 Photobromination of 1,2-Diphenylethane

QUESTION: Which diastereomer is formed from the photobromination of 1,2-diphenylethane?

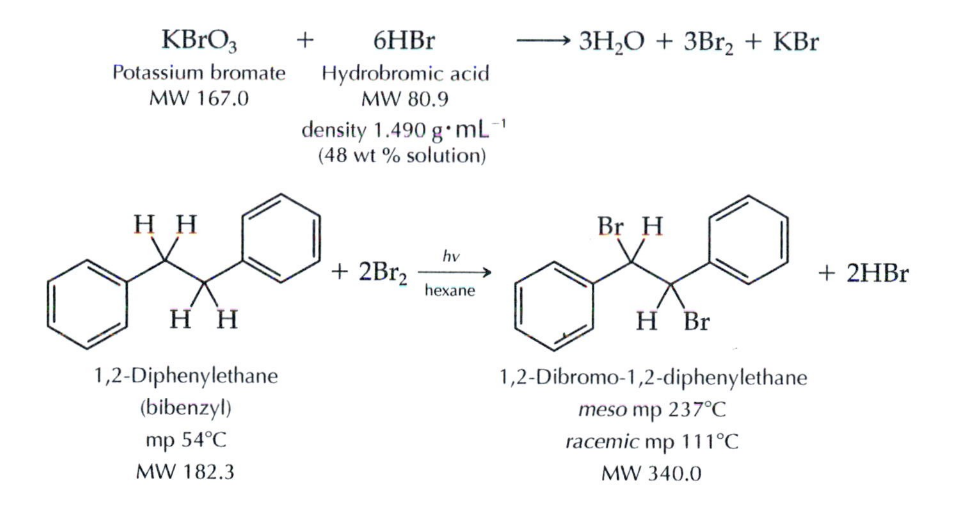

MICROSCALE PROCEDURE

PRELABORATORY ASSIGNMENT: An intermediate in the formation of 1,2-dibromo-1,2-diphenylethane is 1-bromo-1,2-diphenylethane. If this intermediate were isolated, would it be (a) nonchiral, (b) optically active, or (c) a racemic mixture? Explain your reasoning.

Techniques Removal of Noxious Vapors: Technique 7.4
Microscale Filtration and Recrystallization: Technique 9.7a

SAFETY INFORMATION

Wear gloves and, if possible, conduct the entire experiment in a hood.

Potassium bromate is a strong oxidant and a cancer-suspect reagent. No potassium bromate should be discarded in contact with paper because of a potential fire hazard. Wet the weighing paper after use and place it in a designated container. Do not put the paper in a wastebasket.

Hydrogen bromide solution (48%) and its vapor are corrosive and toxic, and they cause severe burns. Wear gloves and measure the solution in a hood.

The reaction mixture contains **bromine,** the vapors of which are toxic and cause severe burns. Be sure that the gas trap is in place during the reaction period.

Hexane and **diethyl ether** are extremely volatile and flammable.

Xylene is flammable and an irritant.

If your lab is equipped with water aspirators, use the apparatus shown in Technique 7.4, Figure 7.7b, to remove noxious bromine vapor. If your lab is not equipped with water aspirators, prepare a gas trap using a 25-cm piece of 2-mm Teflon tubing and two rubber fold-over septa. Thread one end of the tubing through a rubber fold-over septum that fits the top of the air condenser and the other end of the tubing through a rubber fold-over septum that fits the top of a 25-mL filter flask, as shown in Technique 7.4, Figure 7.7a. Put about 10 mL of water in the filter flask and close the top with the larger septum. Adjust the level of the tubing so that its end is *just above* the surface of the water. Fit the smaller septum over the top of the air condenser.

Place 0.200 g of 1,2-diphenylethane (bibenzyl) and 2.7 mL of hexane in a 10-mL round-bottomed flask. Add a magnetic stirring bar and stir the solution until the solid dissolves. Add 0.135 g of potassium bromate to the stirred solution. Obtain 0.50 mL of 48% (wt) hydrobromic acid solution in a small test tube; cork the test tube before carrying it to your workstation. Using a Pasteur pipet, transfer the HBr solution to the stirred reaction mixture. Immediately fasten the air condenser to the flask. Irradiate the reaction mixture with a 100-watt incandescent light bulb located 5–6 inches from the reaction flask until the orange color of bromine is gone.

When the reaction is complete, remove the air condenser. Using a Pasteur pipet fitted with a rubber bulb or a syringe, insert the tip of the pipet to the bottom of the flask and draw the lower aqueous phase into the pipet. Set the aqueous phase aside in a labeled flask. Add 1.5 mL of water to the reaction flask and thoroughly stir the mixture using the magnetic stirrer. Again remove the aqueous phase from the reaction flask with the Pasteur pipet and add it to the flask containing the first aqueous phase. Collect the solid remaining in the reaction flask by vacuum filtration on a Hirsch funnel [see Technique 9.7a]. Rinse the flask with approximately 1 mL of cold ethanol to transfer any residual solid from the flask to the Hirsch funnel. Disconnect the vacuum and rinse the crystals on the Hirsch funnel with approximately 1 mL of cold diethyl ether.

The product is sufficiently pure for characterization without recrystallization; however, if necessary, the product can be recrystallized from xylene. Allow the crystals to dry completely before determining the melting point.

CLEANUP: Combine the solution in the filter flask used as the gas trap with the aqueous phase removed from the reaction mixture; neutralize the combined solution with solid sodium carbonate before washing it down the sink or pouring it into the container for aqueous waste. Pour the filtrate from the reaction mixture into the container for halogenated waste. If your product was recrystallized, pour the recrystallization filtrate into the container for halogenated waste.

Interpretation of Experimental Data

From your observed melting range, which diastereomer formed in the reaction? Why?

Questions

1. Why is the *meso* diastereomer of 1,2-dibromo-1,2-diphenylethane nonchiral?
2. If by mistake a student used HCl rather than HBr in the photobromination reaction, what would be the consequence?
3. The bond-dissociation energy for the C—H bond to the α-carbon of methylbenzene (toluene) is 88 kcal/mol, whereas the bond-dissociation energy for the C—H bond of benzene is 111 kcal/mol. How does this information fit with the experimental results in this experiment, namely that the bromination of 1,2-diphenylethane gives 1,2-dibromo-1,2-diphenylethane?
4. If (2*S*,3*S*)-1,2-dibromo-1,2-diphenylethane were formed in the photobromination reaction, why would it always be accompanied by an equal amount of the (2*R*,3*R*)-enantiomer, giving a racemic mixture?

EXPERIMENT

9 NUCLEOPHILIC SUBSTITUTION REACTIONS

QUESTION: Which structural factors favor S_N1 reactions and which ones favor S_N2 reactions?

In Experiment 9.1 you will investigate the relationship of structure and reactivity in nucleophilic substitution reactions using simple qualitative tests for your evidence.

In Experiment 9.2 you will carry out the synthesis of an alkyl halide from a primary alcohol by an S_N2 reaction and identify the structure of your product by boiling point and NMR spectroscopy.

Perhaps the most-studied and well-established mechanisms of organic chemistry are those for nucleophilic substitution reactions. In these reactions a ***nucleophile*** (literally, a nucleus-loving reagent) is used to displace a ***leaving group*** from a carbon atom of an organic substrate:

$$\underset{\text{Substrate}}{R{-}G} + \underset{\text{Nucleophile}}{Nu^-} \longrightarrow \underset{\text{Product}}{R{-}Nu} + \underset{\text{Leaving group}}{G^-}$$

The nucleophile provides both electrons of the new bond to carbon, and when it leaves, the leaving group takes away both electrons of its bond to the substrate. The nucleophile need not be negatively charged, but it must have at least one or more nonbonded electron pairs at the nucleophilic atom. A typical example of a nucleophilic substitution reaction is the displacement of an iodide anion from iodomethane by a hydroxide ion:

$$HO^- + CH_3{-}I \longrightarrow HO{-}CH_3 + I^-$$

There are two limiting mechanisms for nucleophilic substitution: a direct displacement (S_N2) process and a carbocation (S_N1) process. (In the earlier literature, carbocations are called carbonium ions.) The notation ***S_N2*** means that a reaction is a substitution (S) induced by a nucleophile (N) in a bimolecular rate-determining step. Similarly, ***S_N1*** denotes a nucleophilic substitution reaction with only one molecule involved in the rate-determining step, typically the step in which the carbocation intermediate forms.

S_N2 Pathway

The direct displacement (S_N2) process, favored by primary substrates, involves backside attack by a nucleophile, which displaces the leaving group in one concerted step. Effective nucleophiles include the hydroxide and alkoxide anions, as well as other strong bases. Good leaving groups include bromide and iodide ions that have readily polarizable bonds to carbon, as well as other weak bases. The ease with which the leaving group is displaced strongly influences the rate of an S_N2 reaction.

$$CH_3O^- + CH_3CH_2{-}Br \longrightarrow CH_3CH_2{-}OCH_3 + Br^-$$

Any change in the substrate that increases the steric hindrance about the reactive center makes the S_N2 transition state more difficult to form and slows the reaction. A greater number of alkyl groups or an increase in the branching of one or more of the R groups decreases the rate of reaction:

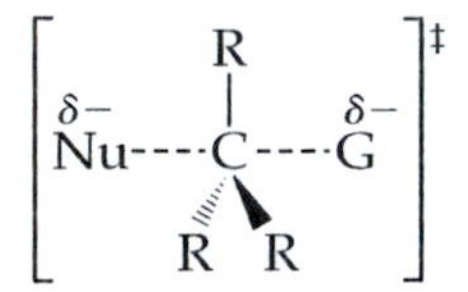

Thus we find that S_N2 reactivity has the following order of dependence on substrate type:

$$\text{methyl} > \text{primary} > \text{secondary} > \text{tertiary}$$

S_N1 Pathway

An interesting general aspect of nucleophilic substitution is the alternative pathways provided by the S_N2 and S_N1 mechanisms. For example, tertiary halides, which are very poor substrates for an S_N2 reaction, are able to undergo substitution by another pathway, the S_N1 mechanism. This S_N1 pathway takes advantage of the fact that tertiary carbocations are reasonably stable. Thus the reaction of *tert*-butyl bromide (2-bromo-2-methylpropane) with water takes place by a process in which the transition state in the rate-determining step is unimolecular:

Rate-determining step: $$(CH_3)_3CBr \rightleftharpoons (CH_3)_3C^+ + Br^-$$

$$(CH_3)_3C^+ + H_2O \rightleftharpoons (CH_3)_3COH_2^+ \xrightleftharpoons{-H^+} (CH_3)_3COH$$

As in the S_N2 pathway, the ability of a leaving group to stabilize the electron pair that it carries away strongly influences the rate of an S_N1 reaction. Also, it is only when the carbocation is reasonably stable that a leaving group can be lost quickly enough for an S_N1 reaction to take place. Otherwise, competing reaction pathways win out, and other products form. The relative rates of S_N1 reactions are consistent with carbocation formation:

$$\text{tertiary} > \text{secondary} > \text{primary} > \text{methyl}$$

Relative Rates of S_N1 and S_N2 Reactions

In Experiment 9.1 you will use qualitative chemical tests to study structure-reactivity relationships in S_N2 and S_N1 reactions. You will be able to use the relative rates of reaction to arrange a group of eight alkyl halides according to their relative reactivities in the following reactions.

$$S_N1:\quad RX + AgNO_3 \xrightarrow{CH_3CH_2OH} ROCH_2CH_3 + AgX\,(ppt) + HNO_3$$
X = Cl, Br

$$S_N2:\quad RX + NaI \xrightarrow{\text{acetone}} RI + NaX\,(ppt)$$
X = Cl, Br

You will use these two reagents to test the ability of substrates to undergo S_N1 and S_N2 reactions. The reaction conditions that favor an S_N2 pathway are a solution of sodium iodide in acetone. The iodide anion is an excellent nucleophile, and acetone, the solvent, has a very limited ability to stabilize a carbocation intermediate. These two factors strongly favor S_N2 reactions and inhibit the S_N1 pathway.

Sodium iodide is soluble in acetone, but when a bromide or chloride anion is the leaving group in an S_N2 reaction, sodium bromide or sodium chloride precipitates as an insoluble salt. Precipitation of the sodium halide drives the reaction to the right.

$$RX + NaI \longrightarrow [\overset{\delta-}{I}\text{----}R\text{----}\overset{\delta-}{X}]^{\ddagger} \longrightarrow I\text{—}R + NaX$$

The reaction conditions that favor an S_N1 pathway are a silver nitrate solution in ethanol. First of all, ethanol is a polar solvent that can effectively stabilize charged species such as carbocations and the transition states leading to them, but it is a poor nucleophile, so S_N2 reactions have little chance to occur. The silver ion coordinates with the halide ion in the organic substrate and enhances the ability of the carbon-halogen bond to break, producing a halide ion and a carbocation. Compounds that undergo rapid reaction with ethanolic silver nitrate solutions are substrates from which carbocations are stable enough to form.

$$AgNO_3 \longrightarrow Ag^+ + NO_3^-$$

$$RX + Ag^+ \longrightarrow \left[\overset{\delta+}{R}\text{----}\overset{\delta-}{X}\text{----}\overset{\delta+}{Ag}\right]^{\ddagger} \longrightarrow R^+ + XAg\ (ppt)$$

$$R^+ + CH_3CH_2OH \longrightarrow ROCH_2CH_3 + H^+$$

Synthesis of Alkyl Bromides Through S_N2 Chemistry

Experiment 9.2 is the acid-catalyzed synthesis of a primary alkyl bromide from a primary alcohol. The reaction uses a mixture of sodium bromide and a strong mineral acid, sulfuric acid. The acid is essential because it turns R—OH into $R\text{—}OH_2^+$, which converts a very poor leaving group, hydroxide, into a reasonably good one, water. Because alcohols are weak bases, a strong acid is required to protonate them. Sodium bromide provides the bromide ion, an excellent nucleophile, which directly displaces the protonated hydroxyl group in an S_N2 reaction.

$$R\text{—}OH + H^+ \rightleftharpoons R\text{—}OH_2^+$$

$$\text{Rate-determining step: } R\text{—}OH_2^+ + Br^- \longrightarrow R\text{—}Br + H_2O$$

Examination of the S_N2 transition state in the rate-determining step shows that for an instant there are effectively five bonds to the central carbon. This cluttered state is achievable with primary alcohols because two of the three groups originally bonded to the primary carbon are hydrogen atoms. The concerted fashion in which the bonds are made and broken provides no opportunity for any rearrangement to occur; the alkyl halide has the same carbon structure as the alcohol used as the starting reagent.

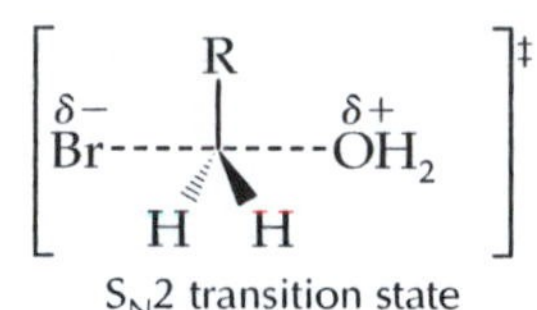

S_N2 transition state

You will characterize the alkyl group in the bromoalkane product by determining its boiling point and interpreting its ^{1}H NMR spectrum.

9.1 S_N1/S_N2 Reactivity of Alkyl Halides

QUESTION: Which structural factors favor S_N1 reactions and which ones favor S_N2 reactions?

$$S_N1:\quad RX + AgNO_3 \xrightarrow{CH_3CH_2OH} ROCH_2CH_3 + AgX\,(ppt) + HNO_3$$
X = Cl, Br

$$S_N2:\quad RX + NaI \xrightarrow{\text{acetone}} RI + NaX\,(ppt)$$
X = Cl, Br

MICROSCALE PROCEDURE

PRELABORATORY ASSIGNMENT: List the experimental conditions that favor S_N1 reactions and those that favor S_N2 reactions.

SAFETY INFORMATION

All **alkyl halides** are harmful if inhaled, ingested, or absorbed through the skin. Wear gloves, dispense the compounds only in a hood, and conduct the experiment in a hood, if possible.

Silver nitrate solutions are toxic and will discolor the skin.

Acetone is flammable and an irritant. Do not use it near open flames or hot electrical devices. Avoid contact with skin, eyes, and clothing.

S_N2 Reaction Conditions (15% NaI in Acetone)

Do the tests with 15% NaI in acetone first because these tests require dry test tubes.

The following compounds will be tested with each reagent:

1-Chlorobutane
1-Bromobutane
Bromocyclopentane
Bromocyclohexane
2-Chlorobutane (*sec*-butyl chloride)
2-Bromobutane (*sec*-butyl bromide)
2-Chloro-2-methylpropane (*tert*-butyl chloride)
2-Bromo-2-methylpropane (*tert*-butyl bromide)

Label eight dry 10 × 75 mm test tubes before obtaining any reagents and fit each with a cork. Place 3 drops, measured with a Pasteur pipet, of each halide in its own test tube. Immediately close each tube with a cork and keep the tubes stoppered except while adding the test reagent. Obtain 10 mL of 15% NaI/acetone solution in a small flask fitted with a cork.

Using a graduated pipet, add 1.0 mL of NaI solution to the first test tube; recork the tube, record the time, and shake the tube to ensure complete mixing. Continue to monitor the reaction and record the time required for a precipitate to form. In the meantime, continue adding 1.0 mL of the test reagent to each of the other tubes. If no precipitate forms after 5 min at room temperature, arrange the cork loosely in the tube and heat it in a warm-water (50°C) bath. Check the reaction periodically and record the time that any turbidity (cloudiness) or precipitate appears. Consider the alkyl halide unreactive under the test conditions if no change occurs after 15 min of heating.

Periodically inspect any tube in which a precipitate has not formed for signs of changes.

CLEANUP: Pour the contents of all test tubes into the container for halogenated organic waste. Wash and rinse the test tubes thoroughly before using them for the S_N1 reactions.

S_N1 Reaction Conditions (1% $AgNO_3$ in Ethanol)

Label eight clean (but not necessarily dry) test tubes. Use 3 drops of each halide, one halide per test tube. Obtain 10 mL of 1% silver nitrate/ethanol solution in a clean small flask. Using a graduated pipet, add 1.0-mL portions of silver nitrate to each test tube in the same way you did in the S_N2 tests. Record the time of addition, cork the tubes, and shake to mix thoroughly. Note the time at which the first signs of turbidity or distinct precipitate formation occurs. Any test tube that has not shown signs of reactivity after 5 min at room temperature should be heated in a warm-water (50°C) bath for 15 min. Loosen the cork before heating the tubes and periodically check each one for precipitation.

CLEANUP: Pour the contents of all test tubes into the container for halogenated organic waste. Pour any remaining silver nitrate solution into the container for inorganic waste.

Interpretation of Experimental Data

1. List the alkyl halides in order of decreasing reactivity toward each of the reaction conditions. Briefly discuss why this order of reactivity was observed.
2. Order the reactivity of the primary halides to each reagent. Briefly explain.
3. Order the reactivity of the secondary halides to each reagent. Briefly explain.
4. Order the reactivity of the tertiary halides to each reagent. Briefly explain.
5. Did you observe any difference in the reactivity of bromocyclopentane and bromocyclohexane to each reagent? Briefly explain.

9.2 Synthesis and Identification of Alkyl Bromides from Unknown Alcohols

PURPOSE: Synthesize an alkyl bromide and identify it on the basis of its boiling point and NMR spectrum.

$$R{-}CH_2OH + NaBr + H_2SO_4 \longrightarrow R{-}CH_2Br + H_2O + NaHSO_4$$

MINISCALE PROCEDURE

Techniques Reflux: Technique 7.1
Simple Distillation: Technique 11.3
Microscale Extractions: Technique 8.6b
Short-Path Distillation: Technique 11.3a
NMR Spectroscopy: Technique 19

SAFETY INFORMATION

Wear gloves while doing the experiment.

The **primary alcohols** used in this experiment are flammable and irritants to skin and eyes. Avoid contact with skin, eyes, and clothing.

The **alkyl halides** synthesized are harmful if inhaled, ingested, or absorbed through the skin.

Sodium bromide is a skin irritant.

Sulfuric acid solution (8.7 M) is corrosive and causes severe burns.

Concentrated hydrochloric acid is corrosive, causes burns, and emits HCl vapors. Use it in a hood.

Pour 25 mL of 8.7 M sulfuric acid into a 50-mL Erlenmeyer flask and place the flask in an ice-water bath to cool.

Weigh your vial of unknown alcohol, then pour all the liquid except a few drops into a 100-mL round-bottomed flask and weigh the vial again. Save the vial and the remaining alcohol for NMR analysis.

Obtain 10.4 g of sodium bromide and transfer it to the round-bottomed flask, using a powder funnel so that salt granules do not stick to the neck of the flask. Attach a water-cooled condenser positioned for refluxing [see Technique 7.1]. Pour the chilled sulfuric acid solution slowly (over a period of 15–20 s) down the condenser while the flask is gently swirled.

Loosen the clamp holding the flask while you are swirling it.

After retightening the clamp holding the flask, place an electric heating mantle or a sand bath under the flask and bring the mixture to a boil (the two phases should be mixing thoroughly); reflux for 45 min. Adjust the rate of heating so that the vapors condense in the lower third of the condenser; otherwise, some of the organic materials may be lost. Consult your instructor about performing the NMR analysis of your unknown alcohol during this time.

At the end of the reflux period, lower the heating mantle and cool the reaction mixture for 10 min. Add 20 mL of water and assemble the apparatus for a simple distillation [see Technique 11.3, Figure 11.6]. Use a 50-mL Erlenmeyer flask for the receiver. Distill until the condensate no longer contains water-insoluble droplets (about 15–25 mL of condensate). To test for completeness of the distillation, collect 1 or 2 drops of distillate in a clean test tube, add 0.5 mL of water, and look for any insoluble droplets. The distillation is complete when no droplets are visible. The final temperature of the distillation will range from 100–115°C.

The boiling point gradually rises during the azeotropic distillation of the alkyl bromide and water. The ternary azeotrope of HBr, H_2SO_4, and H_2O boils at 115°C.

Extractions

SAFETY PRECAUTION

Wear gloves while doing the extractions and, if possible, work in a hood while using concentrated hydrochloric acid.

In all the following extractions, the organic phase is the lower layer.

Transfer the distillate to a 15-mL centrifuge tube fitted with a tight cap. Add 3 mL of water, cap the tube and shake it, then allow the layers to separate. Record the volume of the lower phase. Remove the upper layer with a Pasteur pipet fitted with a syringe or a Pasteur filter pipet and transfer it to a 100-mL beaker [see Technique 8.6b].

The distillate contains water, your alkyl halide, and any alcohol that failed to react. To separate the alkyl bromide, we take advantage of three facts: (1) the alkyl bromide is insoluble in water; (2) it is insoluble in concentrated hydrochloric acid, whereas any remaining alcohol dissolves in the acid; (3) its density is greater than that of either water or concentrated HCl.

Chill 8 mL of concentrated hydrochloric acid in an ice-water bath. Add about half of the cold HCl to the crude product, swirl the centrifuge tube briefly in the ice-water bath, then cap the tube tightly and shake to thoroughly mix the layers. If the phases do not separate cleanly, spin the tube in a centrifuge for 1 min. Record the volume of the lower phase. Carefully remove the upper HCl layer (all the aqueous layers can be combined in the same beaker). Wash the organic phase with the rest of the cold HCl. Record the volume of the lower (organic) layer after the phases separate (see Question 7). Remove the upper (acid) layer.

Wash the alkyl halide with 5 mL of ice-cold water, allow the phases to separate, and remove the upper aqueous phase. Repeat the washing procedure with 3 mL of 5% sodium bicarbonate. **(Caution: Foaming may occur.)** Allow the layers to separate and transfer the lower organic phase to a clean, dry centrifuge tube or 10-mL Erlenmeyer flask.

Drying the product twice is more effective than using twice as much drying agent once.

Add anhydrous calcium chloride pellets to the product in several small portions until a few pellets do not immediately clump together or form a milky liquid. Cap the tube and shake it. Allow the mixture to stand for 5 min. Using a clean Pasteur pipet, transfer the product to a dry 10-mL Erlenmeyer flask and add small portions of anhydrous calcium chloride pellets until several move freely in the liquid. Cork the flask and allow it to stand at least 15 min.

Final Distillation

Place a dry 25-mL round-bottomed flask in a 100-mL labeled beaker and determine the combined mass. You will use this flask as the receiving flask. Set the beaker aside until you are ready to weigh the product.

Using a Pasteur pipet, carefully transfer the crude product to a clean, dry 25- or 50-mL round-bottomed flask. Add a boiling stone. Assemble the apparatus for short-path distillation as shown in Technique 11.3a, Figure 11.7, using the tared 25-mL round-bottomed flask as the receiver. Hold the apparatus together by placing Keck clips on both ends of the vacuum adapter. Submerge the receiving flask in a beaker of cold tap water. Record the boiling range from the temperature at which the first drop falls into the receiver to the highest temperature noted. Weigh the receiving flask and contents to determine your yield.

NMR Spectrum

Consult your instructor about the correct sample preparation for the NMR spectrometer available in your laboratory. A general procedure used with a midfield (300 MHz) spectrometer is to place one Pasteur-pipet drop of the compound being analyzed in an NMR tube and add 0.6 mL of $CDCl_3$ [see Technique 19.2]. Cap the tube and tap it with your finger several times to mix the solution thoroughly. Determine the ^1H NMR spectrum for both your unknown alcohol and the alkyl bromide synthesized from it.

CLEANUP: Carefully pour the sulfuric acid solution remaining in the reaction flask into a large beaker that contains about 200 mL of water. Rinse the flask with water and add the rinse water to the beaker. Also pour all the aqueous phases from the extractions into this beaker. Add sodium carbonate in small portions with stirring **(Caution: Foaming!)** until the acid is neutralized. The solution can then be washed down the sink or placed in the container for aqueous inorganic waste. Place the calcium chloride drying agent in the container for hazardous solid waste or the container for inorganic solid waste. Pour the residue remaining in the distillation flask after the final distillation into the container for halogenated waste.

Identification of the Alkyl Halide

Table 9.1 lists the molecular weights and boiling points of the alkyl bromides that could be synthesized in this experiment. Use your observed boiling point and the ^1H NMR spectra to identify your alkyl bromide and the unknown alcohol from which you synthesized it. Analyze your spectra using the chemical shifts of the proton signals, the splitting pattern, and the integration ratio to support your assignment of all peaks [see Techniques 19.4–19.9]

TABLE 9.1 Boiling points of alkyl bromides

Alkyl bromide	Molecular weight	Boiling point, °C
1-Bromopropane	123	71
1-Bromobutane	137	101
1-Bromo-2-methylpropane	137	90–92
1-Bromopentane	151	130
1-Bromo-3-methylbutane	151	120–121
1-Bromohexane	165	154–158
1-Bromo-4-methylpentane	165	146–147

After you have identified the alcohol and the alkyl bromide, calculate the theoretical yield of alkyl bromide from the mass of alcohol that you used in the synthesis. Calculate your percent yield.

Questions

1. What is the upper layer that forms in the reaction flask during the reflux period? Why does it separate from the aqueous layer?
2. Why was the product washed with aqueous sodium bicarbonate?
3. Treatment of 1-butanol with phosphoric acid (H_3PO_4) mixed with sodium chloride should result in the formation of 1-chlorobutane. Write a mechanism for this reaction.
4. Treatment of *tert*-butyl alcohol (2-methyl-2-propanol) with HCl leads to a reasonable yield of *tert*-butyl chloride (2-chloro-2-methylpropane). Write a mechanism for this reaction.
5. With an eye on the process used here for preparing a 1-bromoalkane, comment on whether each of the following reactions might lead to 1-bromobutane:
   a. 1-butanol + NaBr
   b. 1-butanol + $NaBr/H_3PO_4$
   c. 1-butanol + KBr/H_2SO_4
   d. 1-butanol + NaBr/HCl
6. In the extraction steps, why was the crude product washed with concentrated hydrochloric acid?
7. You may have observed a decrease in the volume of the crude organic product during the extractions with HCl. What causes this decrease?

EXPERIMENT

10

E2 ELIMINATION OF 2-BROMOHEPTANE: INFLUENCE OF THE BASE

QUESTION: What effect does the bulkiness of the base have on the product composition in the E2 dehydrobromination of 2-bromoheptane?

> You will investigate the elimination of HBr from 2-bromoheptane using sodium methoxide and potassium *tert*-butoxide and determine by gas chromatography how the ratio of isomeric heptenes changes.

The chemical reaction of a strong base and a secondary or tertiary alkyl bromide produces alkenes by an E2 elimination reaction. Secondary and tertiary alkyl halides have too much steric hindrance for an S_N2 substitution reaction to compete very effectively, especially with a bulky base. When the substrate is an alkyl bromide, the loss of HBr is called a dehydrobromination reaction. It is a 1,2-elimination reaction, in that when the base abstracts a proton, the elimination of bromide occurs from the adjacent carbon atom. The reaction is called E2 because the rate-determining step is bimolecular, first-order in base and first-order in alkyl bromide. The E2 reaction occurs in a single step. As the proton is removed by the base, the carbon-carbon double bond (C=C) forms and Br^- is eliminated simultaneously.

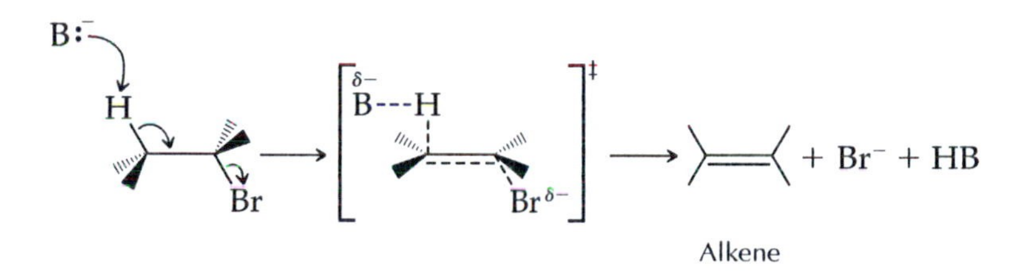

If there is considerable C=C character in the transition state, double-bond stability could be an important factor in determining which products form. In many elimination reactions the more highly substituted alkene is the favored product. The following order of alkene stability has been observed:

$$R_2C{=}CR_2 > R_2C{=}CHR > trans\text{-}RCH{=}CHR > R_2C{=}CH_2 > cis\text{-}RCH{=}CHR > RCH{=}CH_2 > H_2C{=}CH_2$$

In the reaction that you will carry out in this experiment, the base abstracts a proton (called a β-hydrogen) from either carbon-1 or carbon-3 of 2-bromoheptane. The bromide ion leaves carbon-2 (the α-carbon) at the same time, resulting in the formation of 1-heptene and the cis and trans isomers of 2-heptene.

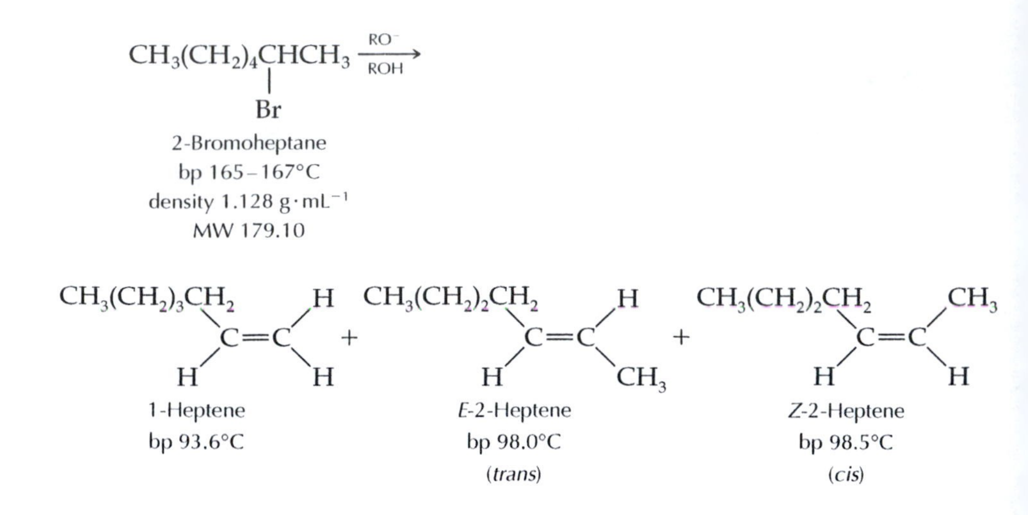

The question being addressed deals with the ratio of the three possible 1,2-elimination products. Disubstituted alkenes, such as the isomers of 2-heptene, are more stable than monosubstituted C═C. A large amount of double-bond character in the E2 transition state predicts that *cis*- and *trans*-2-heptene will be major products. Given the greater stability of the trans isomer, one would expect it to dominate.

On the other hand, the base must also be in the process of abstracting a β-proton in the E2 transition state. Steric factors may be in play. The methyl protons are more out in the open than the methylene protons. Perhaps, even though disubstituted-alkene products are more stable, they have a difficult time forming for steric reasons. The steric hindrance would be greater if a bulky base, such as *tert*-butoxide, is used.

$CH_3O^-Na^+$

Sodium methoxide

$(CH_3)_3C{-}O^-K^+$

Potassium *tert*-butoxide
(potassium 2-methyl-2-propoxide)

In this experiment you need to consider both the size of the base and the stability of the alkene as factors that may influence the product composition. You will use gas chromatography to discover whether the ratio of alkene isomers obtained in the dehydrobromination reaction of 2-bromoheptene with sodium methoxide in methanol and with potassium *tert*-butoxide in *tert*-butyl alcohol reflect product stability or steric effects. Your experimental data will tell.

MICROSCALE PROCEDURE

TEAMWORK: Your instructor may decide that you will work with a laboratory partner. If so, each of you will do one of the two reactions and share your results with the other.

PRELABORATORY ASSIGNMENT: Explain what makes potassium *tert*-butoxide more sterically hindered than sodium methoxide and why sodium methoxide could also exhibit steric hinderance when it acts as a base.

Techniques Pasteur Filter Pipet: Technique 8.5
Microscale Reflux/Anhydrous Conditions: Technique 7.2
Microscale Extractions: Technique 8.6c
Gas Chromatography: Technique 16

SAFETY INFORMATION

2-Bromoheptane is flammable and an irritant. Wear gloves. Avoid contact with skin, eyes, and clothing.

Sodium methoxide in methanol solution is flammable, corrosive, and moisture sensitive. **Potassium *tert*-butoxide** is corrosive and moisture sensitive. Avoid contact with skin, eyes, and clothing. Keep the containers tightly closed and store both reagents in desiccators.

Methanol is flammable and toxic. **2-Methyl-2-propanol** (*tert*-butyl alcohol) is flammable and an irritant.

Pentane is extremely flammable.

Set the conical vial in a small beaker while adding the reagents.

Sodium Methoxide Reaction

Prepare a microscale drying tube containing anhydrous calcium chloride [see Technique 7.2, Figure 7.2b]. Pipet 1.65 mL of a 25% (weight) solution of sodium methoxide in methanol into a 5-mL conical vial. Add 0.35 mL of dry methanol and 0.22 mL of 2-bromoheptane. Put a magnetic spin vane into the vial. Fit the vial with a water-cooled condenser and place the drying tube containing calcium chloride in the top of the condenser. Begin stirring and heat the reaction for 30 min under reflux in a hot-water (75°C) bath.

Potassium tert-Butoxide Reaction

Prepare a microscale drying tube containing anhydrous calcium chloride [see Technique 7.2, Figure 7.2b]. Begin heating an aluminum block on a hot plate/stirrer to 95°C. Put a magnetic spin vane into a 5-mL conical vial. Pipet 2.0 mL of dry 2-methyl-2-propanol (*tert*-butyl alcohol) into the vial. Weigh 0.825 g of potassium *tert*-butoxide and immediately add it to the vial containing the alcohol. Add 0.22 mL of 2-bromoheptane. Fit the vial with a water-cooled condenser and place the drying tube containing calcium chloride in the top of the condenser. Begin stirring and heat the reaction under reflux in the aluminum block for 30 min.

Isolation of Product Mixture for Both Reactions

At the end of the reflux period, cool the reaction to room temperature. Remove the spin vane with tweezers. Add 1.3 mL of pentane and 1.0 mL of water to the reaction vial. Cap the vial tightly and shake the mixture gently but thoroughly until the white solid dissolves. A few additional drops of water may be necessary to completely dissolve the solid. Remove the lower aqueous phase with a Pasteur filter pipet or a Pasteur pipet fitted with a syringe [see Technique 8.5] and transfer it to a 50-mL Erlenmeyer flask, leaving the organic phase in the conical vial [see Technique 8.6c]. Repeat the extraction procedure twice with 1.0-mL portions of water, removing the lower aqueous phase before adding the next portion of water; all the aqueous phases should be combined in the Erlenmeyer flask. After removing the last portion of water, add 200 mg of anhydrous sodium sulfate to the pentane solution remaining in the vial. Cap the vial and allow the mixture to dry for at least 10 min. Transfer the dried solution to a clean dry test tube, using a new Pasteur filter pipet.

Gas Chromatography

Determine the composition of the product mixture using a gas chromatograph equipped with a nonpolar column, such as SE-30 or OV-101 [see Technique 16]. If the product solution is too concentrated for capillary gas chromatography, prepare a GC solution of 0.5 mL of pentane and 5 drops of the product mixture in a small test tube. The products elute in order of increasing boiling point. A small peak due to the methyl ether (S_N2 product) with a retention time greater than the alkenes may occur in the sodium methoxide reaction product. Calculate the percent of each heptene isomer in your product mixture [see Technique 16.7].

CLEANUP: Determine the pH of the combined aqueous solution from the extractions using pH test paper. Add 5% hydrochloric acid dropwise until the pH is about 7; the neutralized solution may be washed down the sink or poured into the container for aqueous inorganic waste. The pentane solution from the GC analysis should be poured into the container for halogenated waste because it contains some unreacted 2-bromoheptane.

Interpretation of Experimental Data

Compare your team's gas chromatographic data for the product mixtures of the two reactions. What effects do the two different bases have on the course of E2 dehydrobromination of 2-bromoheptane? Consider relative alkene stabilities and the bulkiness of the two bases in analyzing your data.

OPTIONAL COMPUTATIONAL CHEMISTRY EXPERIMENT

Build a molecule of 1-heptene. Using the AM1 semiempirical quantum mechanical method, optimize the structure that you have built. The energy surface for the structure of 1-heptene has many local minima because the hydrocarbon chain is flexible. Adjust the dihedral angles along the carbon backbone so that the carbon atoms are all in an *anti* staggered relationship to one another. Use this structure as a starting point for the optimization. Record its heat of formation.

In a similar fashion, build and optimize molecules of *cis*-2-heptene and *trans*-2-heptene. Record their heats of formation. Are your experimental results consistent with the product ratios calculated from the thermodynamic heats of formation of the alkenes?

References

1. Sayed, Y.; Ahlmark, C. A.; Martin, N. H. *J. Chem. Educ.* **1989**, *66*, 174–175.
2. Leone, S. A.; Davis, J. D. *J. Chem. Educ.* **1992**, *69*, A175–A176.

Questions

1. Compare the results of the two E2 reactions you ran and discuss how using different bases affects the product composition.
2. Which factor predominates in each reaction, the size of the base or the stability of the alkenes formed? Explain.
3. Because alcohol protons are reasonably acidic and easily undergo exchange, it is important that the base be matched to the appropriate solvent. Explain why it would be virtually impossible to sort out the factors in Question 2 if a reaction were carried out using methanol as the solvent and potassium *tert*-butoxide as the base.
4. A methyl ether (S_N2 product) can form as a byproduct in the sodium methoxide reaction. What is its structure?
5. Internal alkenes are usually found to be somewhat more stable in the trans (*E*) configuration than the cis (*Z*) configuration. Explain why.
6. In contrast to the internal alkenes in Question 5, cyclohexene occurs as the cis cycloalkene rather than the trans isomer. Explain.

EXPERIMENT

11 DEHYDRATION OF ALCOHOLS

QUESTION: What factors control the products that form in acid-catalyzed dehydration reactions?

In Experiment 11.1, you will study the acid-catalyzed dehydration of a tertiary alcohol, 2-methyl-2-butanol, and use gas chromatography to discover whether the relative stability of the alkenes is the factor that determines the ratio of the 2-methylbutene products.

In Experiment 11.2, you will investigate the acid-catalyzed dehydration of a cyclic secondary alcohol, 3-methylcyclohexanol. By determining the number and identity of the alkene products using gas chromatography, you will be able to determine the importance of carbocation rearrangements in the reaction.

Acid-Catalyzed Dehydration

When an alcohol is heated in the presence of a strong acid, the major product is an alkene or a mixture of alkenes:

$$-\underset{\text{H}}{\overset{|}{\underset{|}{\text{C}}}}-\underset{\text{OH}}{\overset{|}{\underset{|}{\text{C}}}}- \xrightarrow[\text{heat}]{H^+} \text{>C=C<} + H_2O$$

The elimination of a water molecule from an alcohol is called a dehydration reaction. Elimination reactions are one of the fundamental classes of chemical reactions. They require a leaving group that departs with its bonding electrons. The second group lost, from the adjacent carbon atom, is often a proton. The acid is essential because it converts R—OH into $R{-}OH_2^+$, which converts a very poor leaving group into a reasonably good one: water. Because alcohols are weak bases, strong acids such as sulfuric acid or phosphoric acid are required to protonate them.

When secondary or tertiary alcohols are used, protonation and heating provide the driving forces for the loss of water to form a positively charged carbon species called a carbocation:

$$\text{H}-\overset{|}{\underset{|}{\text{C}}}-\underset{\text{OH}}{\overset{|}{\underset{|}{\text{C}}}}- + H_2SO_4 \rightleftharpoons HSO_4^- + \text{H}-\overset{|}{\underset{|}{\text{C}}}-\underset{{}^{+}\text{O}-\text{H},\ \text{H}}{\overset{|}{\underset{|}{\text{C}}}}- \rightleftharpoons \text{H}-\overset{|}{\underset{|}{\text{C}}}-\text{C}^{+}\!\!< + H_2O$$

Carbocation

The rate of formation of a carbocation by loss of a water molecule depends heavily on the stability of the carbocation being formed. Secondary and especially tertiary carbocations are normally stable enough to be important reaction intermediates. Carbocation stability increases markedly with an increase in the number of alkyl or aryl substituents attached to the carbon atom bearing the positive charge. These substituents allow the charge to be dispersed throughout a larger space and thereby stabilize it.

In all but the most highly acidic environments, even tertiary carbocations are too unstable to persist. They react with nucleophiles to give substitution products (S_N1) or lose a proton to give elimination products (E1). The term E1 simply means unimolecular elimination. The reaction is said to be unimolecular because only one molecule is involved in the slowest or rate-determining step, which is the loss of a water molecule from the protonated alcohol. Loss of a proton from the carbocation is fast because carbocations are very strong acids.

Sulfuric acid and phosphoric acid are chosen over hydrochloric acid and hydrobromic acid as catalysts for the dehydration of alcohols, in part because the conjugate bases of sulfuric and phosphoric acids are poor nucleophiles and do not lead to the formation of large amounts of substitution products. Naturally, when a catalyst is selected for the synthesis of an alkene, we want to minimize yield-reducing substitution reactions.

The entire pathway for the E1 elimination of water from an alcohol in the presence of strong acid is

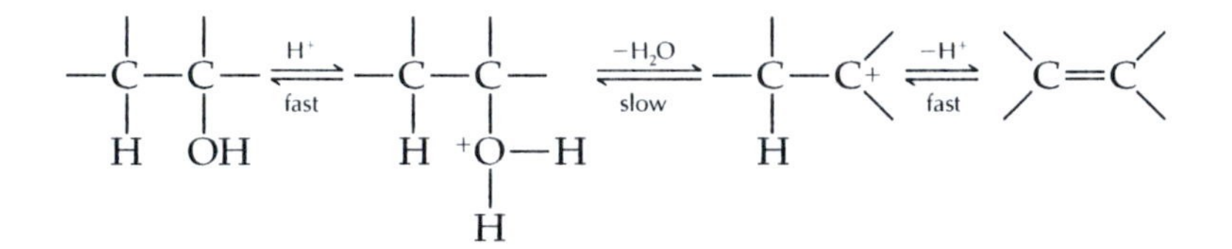

All the reaction steps in the E1 mechanism are reversible. Thus, the mechanism for dehydration of alcohols is just the reverse of the mechanism for hydration of alkenes. This reversibility also means that alkenes formed under these conditions can revert to alcohols unless proper experimental conditions are used. To drive the elimination to completion, the alkene is distilled from the reaction mixture as it is formed. This strategy allows the first two equilibria to be shifted continually in favor of the alkene.

Dehydration of 2-Methyl-2-Butanol

Two possible products from the dehydration of 2-methyl-2-butanol in Experiment 11.1 are 2-methyl-2-butene and 2-methyl-1-butene:

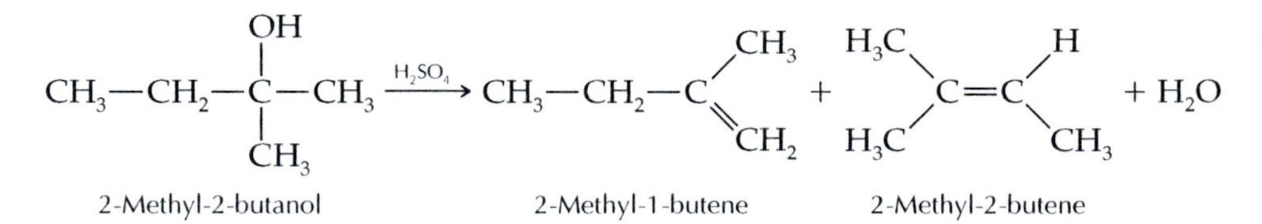

Because 2-methyl-2-butene has three alkyl groups attached to the carbon-carbon double bond (C=C), it is more stable than 2-methyl-1-butene, which has only two alkyl groups attached to the C=C. If the product mixture reflects these different stabilities, it should be composed mainly of 2-methyl-2-butene. This will certainly be the case if the alkene products are in equilibrium with each other under the reaction conditions. Even if the products do not rearrange to each other, the energies of the transition states for their formation may reflect product stability. In other words, the more stable product may form faster.

On the other hand, it is possible that the factor controlling the ratio of the two alkenes is not product stability. It could be a statistical factor, and 2-methyl-1-butene might form faster because there are six methyl protons, any of which could be lost to form 2-methyl-1-butene. By comparison, there are only two methylene protons that could be lost in the formation of 2-methyl-2-butene. In addition, there may be a steric factor that inhibits the removal of a methylene proton, which has a more congested environment. You will be able to discover the most important factor by determining the ratio of the two products using gas chromatography.

Dehydration of 3-Methylcyclohexanol

In Experiment 11.2 the question concerns the importance of carbocation rearrangements in the dehydration of 3-methylcyclohexanol. Protonation of the alcohol followed by loss of H_2O gives rise to a secondary carbocation, which can undergo three separate reactions. The carbocation can lose a proton from either carbon adjacent to the positively charged carbon atom, forming either 3-methylcyclohexene or 4-methylcyclohexene. The carbocation can also rearrange by a 1,2-hydride shift to an isomeric secondary carbocation that would rapidly rearrange further to a more stable tertiary carbocation, which would only then lose a proton to form 1-methylcyclohexene. If 1-methylcyclohexene is a component of the product mixture, carbocation rearrangements must have occurred in the reaction pathway.

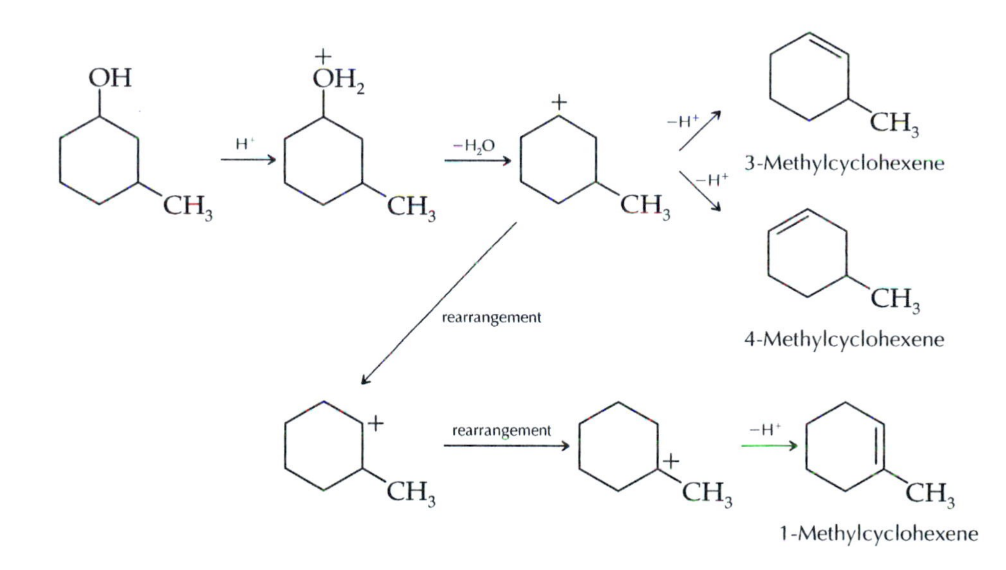

Such rearrangements are common for carbocations, but in this case we do not have a strong driving force because the initial rearrangement is from a secondary carbocation to another secondary carbocation. Only in the second rearrangement could the substantially greater stability of the tertiary carbocation come into play. Does rearrangement from one secondary carbocation to another compete with direct loss of a proton to form alkene products? You will be able to tell by analyzing the mixture of alkenes from the acid-catalyzed dehydration of 3-methylcyclohexanol.

Gas Chromatographic Analysis of Products

Quantitative separation and characterization of the alkene products can be demanding because their structures are similar. Even separation by gas chromatography may be a challenge. GC analysis of the mixture using packed (1/4–3/4 in.) columns that separate on a boiling-point basis will probably only separate 1-methylcyclohexene from the other two possible alkenes. GC analysis of the mixture using packed columns intended for separations based on polarity differences will probably be no more successful. Capillary GC works best, and we recommend conditions in the experiment.

Qualitative Tests for Alkenes

Two simple identification tests can quickly distinguish alkenes from other classes of compounds. Almost all alkenes react rapidly and smoothly with a dilute solution of bromine in dichloromethane to form dibromoalkanes. The molecular bromine is consumed, usually instantaneously, and its characteristic red-brown color disappears.

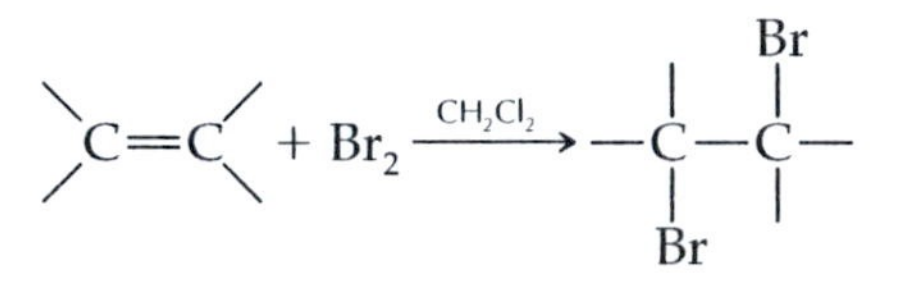

The second simple qualitative test for alkenes involves their rapid oxidation by a solution of potassium permanganate. The purple color of the permanganate solution disappears within 2 or 3 min. In its place, a brown precipitate of manganese dioxide appears. Because alcohols may also be oxidized by potassium permanganate, both a positive bromine addition test and a positive permanganate oxidation test are necessary for a compound to be characterized as an alkene.

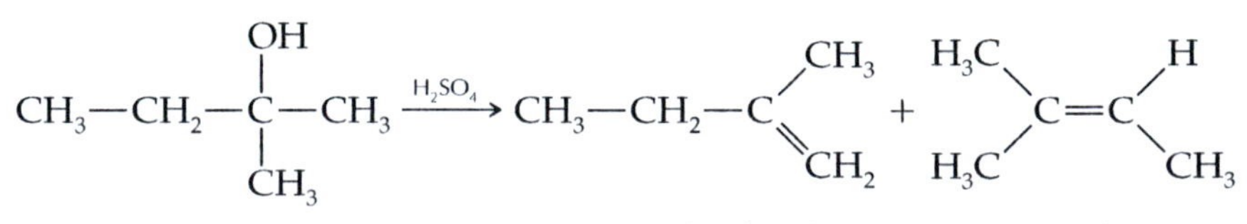

11.1 Acid-Catalyzed Dehydration of 2-Methyl-2-butanol

QUESTION: Does the relative stability of the products determine their ratio in the acid-catalyzed dehydration of 2-methyl-2-butanol?

$$CH_3{-}CH_2{-}C(OH)(CH_3){-}CH_3 \xrightarrow{H_2SO_4} CH_3{-}CH_2{-}C(CH_3){=}CH_2 + (H_3C)_2C{=}C(H)CH_3$$

Ratio of products to be determined

2-Methyl-2-butanol	2-Methyl-1-butene	2-Methyl-2-butene
	bp 31.1°C	bp 38.5°C
	MW 70.1	MW 70.1
	density 0.650 g · mL^{-1}	density 0.662 g · mL^{-1}

MINISCALE PROCEDURE

PRELABORATORY ASSIGNMENT: Explain how the product distribution might change if the alkene products were left in the reaction medium for a period of time rather than removed by distillation as they form.

Techniques Fractional Distillation: Technique 11.4
Gas Chromatography: Technique 16

SAFETY INFORMATION

Wear gloves while dispensing the reagents and transferring them to the reaction flask.

2-Methyl-2-butanol is flammable and toxic. Avoid contact with skin, eyes, and clothing.

Sulfuric acid solution (6 M) is corrosive and causes burns. Notify the instructor if any acid is spilled.

Heptane is volatile and flammable.

In this experiment, the product is removed from the reaction mixture by fractional distillation as the reaction proceeds.

The low boiling points of the products make them extremely volatile. Be certain that all ground glass joints are greased and fastened with Keck clips.

Before you set up the apparatus, obtain the tare mass of a labeled 150-mL beaker holding a 25- or 50-mL round-bottomed flask fitted with a cork.

Place 3.0 mL of 2-methyl-2-butanol in another 25-mL round-bottomed flask containing a magnetic stirring bar. Carefully add 15 mL of 6 M sulfuric acid to the material in the flask. Fit an air condenser to the flask and complete the apparatus for fractional distillation as shown in Technique 11.4, Figure 11.15, using the tared 25- or 50-mL round-bottomed flask as the receiving flask. Set aside the tared beaker for use later. Immerse the receiving flask in an ice-water bath.

Begin stirring the reaction mixture and gently heat the reaction mixture in a 85–90°C water bath. When distillation begins, regulate the heating so that the temperature of the distilling vapor does not exceed 45°C. The distillate should be collected at a rate of 1 to 2 drops per second. Collect all distillate that boils below 45°C. Turn off the hot plate and remove the hot-water bath from underneath the distillation flask. Remove the receiving flask, immediately cork it, and carefully dry the outside with a paper towel. Set the flask in the beaker used to tare it and allow the contents to warm to room temperature before determining the mass of product collected. Calculate your percent yield.

Product Analysis

Determine the composition of your product by gas chromatography. For a capillary column chromatograph with a nonpolar polydimethylsiloxane column, such as SE-30, OV-1, or DB-1, use a column temperature no higher than 50°C. Prepare the sample by dissolving 2 drops of the product in 0.5 mL of heptane; inject 1 μL of this solution into the GC. The alkenes elute in order of increasing boiling point, with the solvent peak (heptane) occurring after the product peaks rather than before, as is usually the case.

CLEANUP: Cool the residue in the distilling flask to room temperature before pouring it into a large beaker containing approximately 100 mL of water. Add solid sodium carbonate in small portions until the acid is neutralized. **(Caution: Foaming.)** Pour the neutralized solution into the container for aqueous inorganic waste or wash it down the sink, as directed by your instructor. Pour the heptane solution remaining from the GC analysis into the container for flammable (organic) waste.

Interpretation of Experimental Data

What are the relative amounts of each product present in the distillate? What factors controlled the course of the reaction? Explain the formation of the products and discuss their observed ratio.

OPTIONAL EXPERIMENT Qualitative Tests for Alkenes

Perform the bromine addition and the potassium permanganate oxidation tests on your product. Do these tests indicate the presence of unsaturated carbon-carbon bonds? Also perform these tests on 2-methyl-2-butanol and ethanol. Compare the results with those that you obtained for your product.

SAFETY INFORMATION

Bromine solutions cause burns and emit toxic bromine vapors. Wear gloves and use the solution only in a hood.

Bromine Addition

Put 3 drops of the product in a small test tube and add 1.0 mL of dichloromethane. Make dropwise additions of 5 drops of a solution of 5% bromine in dichloromethane, shaking the test tube after each addition. Record your observations. Repeat the test on 2-methyl-2-butanol. Alkenes react with bromine, and the characteristic red-brown color of bromine disappears. For most alkenes, this reaction occurs so rapidly that the reaction solution never acquires the red color of the bromine until the alkene is completely brominated.

CLEANUP: The entire reaction mixture should be poured into the halogenated-waste container.

Oxidation with Potassium Permanganate

Place 3 drops of the product and 2.0 mL of water in a small test tube. Then add 1 drop of a 2% aqueous solution of potassium permanganate. Cork the test tube and shake it vigorously. Record your observations. Repeat the test on 2-methyl-2-butanol. Alkenes are oxidized, thereby causing the purple color of the permanganate solution to be replaced within 2–3 min by a brown precipitate of manganese dioxide.

CLEANUP: Rinse the reaction mixtures down the sink or place them in the inorganic-waste container.

Questions

1. Why does the solvent peak (heptane) show a longer GC retention time than do the alkenes? Why is heptane a better choice for the solvent than diethyl ether in this GC analysis?
2. Acid-catalyzed dehydration of an alcohol is an equilibrium situation. How was the reaction forced to completion in this experiment?
3. Why might a poor yield of alkenes be realized if the dehydration were carried out with HCl instead of H_2SO_4?
4. What alkene products would you predict from the acid-catalyzed dehydration of (a) 3-methyl-3-pentanol? (b) 3-methyl-2-butanol? (c) 4-methyl-2-pentanol?

11.2 Acid-Catalyzed Dehydration of 3-Methylcyclohexanol

QUESTION: Does rearrangement of the carbocation intermediate occur in acid-catalyzed dehydration of 3-methylcyclohexanol?

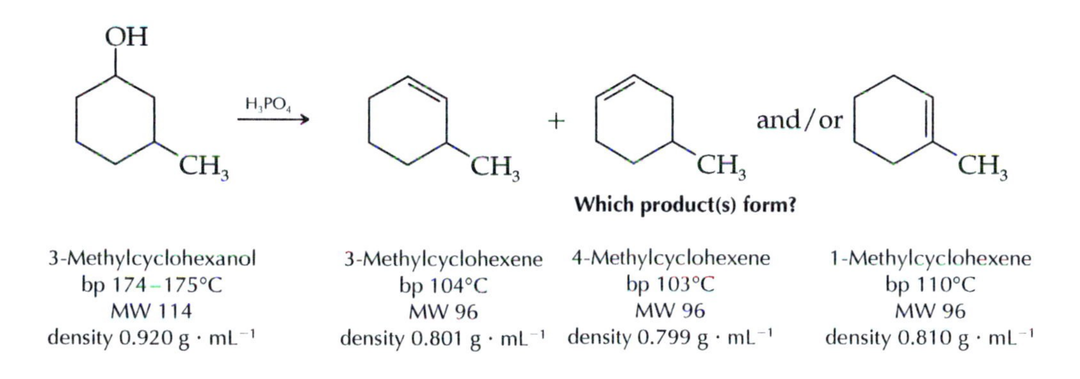

MICROSCALE PROCEDURE

PRELABORATORY ASSIGNMENT: Explain why the rearrangement of a secondary carbocation to a tertiary carbocation is usually faster than the rearrangement of a secondary carbocation to another secondary carbocation.

Techniques Microscale Distillation: Technique 11.3b
Drying Agents: Technique 8.7
Gas Chromatography: Technique 16

SAFETY INFORMATION

3-Methylcyclohexanol is harmful if inhaled and may cause skin or eye irritation. Avoid contact with skin, eyes, and clothing.

Concentrated (85%) phosphoric acid is very irritating to the skin and mucous membranes. Wear gloves. If you spill any phosphoric acid on your skin, wash it off immediately with copious amounts of water.

Place 2.0 mL of 3-methylcyclohexanol in a 5-mL conical vial. Add 0.34 mL of 85% phosphoric acid and a magnetic spin vane to the vial. Assemble the apparatus for short-path distillation as shown in Technique 11, Figure 11.9, using a 5-mL conical vial as the receiving vessel. Set the assembled distillation apparatus in an aluminum block or sand bath and begin heating the block or bath to 170°C. Immerse the receiving vial in a 100-mL beaker containing ice and water. Wrap the portion of the distillation head above the flask with aluminum foil. As the distillation proceeds, increase the temperature of the aluminum block or sand bath to 220°C. When the reaction is complete and the products have distilled, the phosphoric acid remaining in the vial usually turns yellow. Carefully raise the apparatus away from the heat source. **(Caution: The apparatus is very hot.)** Allow it to cool to room temperature before removing the distillation vial.

Add 1.0 mL of water to the vial containing the distillate, cap the vial tightly, and shake it to mix the layers. Allow the layers to separate and remove the aqueous layer with a Pasteur filter pipet or a Pasteur pipet fitted with a syringe [see Technique 8.5 for the Pasteur pipet and Technique 8.6c for microscale extraction]. Then repeat the washing process with 1.0 mL of 5% sodium carbonate solution. After removing the aqueous sodium carbonate layer, add 6–10 pellets of anhydrous calcium chloride to the product layer remaining in the conical vial. Cap the vial and allow the product to dry for 10–15 min.

Prepare another Pasteur filter pipet. Weigh a small sample vial and transfer the dried product to this vial with the Pasteur filter pipet. Weigh the product and calculate the percent yield. Carry out the product analysis; then submit any remaining product to your instructor.

Gas Chromatographic Analysis

Determine the ratio of products by gas chromatographic analysis on a nonpolar column, such as SE-30, OV-1, or DB-1, heated to 50–55°C [see Technique 16]. For a capillary GC, prepare a solution containing 1 drop of your product in 1.0 mL of diethyl ether. Inject a 1-μL sample of this solution into the GC. Consult your instructor about the appropriate sample dilution and sample volume if you are using a packed column gas chromatograph. The methylcyclohexenes elute in order of increasing boiling point. The three methylcyclohexenes are all commercially available; confirm the identification of each product by the peak enhancement method [see Technique 16.6].

Your instructor may have you work with other students to run the peak-enhancement chromatograms.

CLEANUP: Pour the phosphoric acid remaining in the reaction vial and the aqueous solutions saved from the extractions into a beaker containing about 50 mL of water; neutralize this solution with solid sodium carbonate before washing it down the sink or pouring it into the container for aqueous inorganic waste. Place the spent drying agent in the container for solid inorganic waste. Pour the ether solution from the GC analysis into the container for flammable (organic) waste.

Interpretation of Experimental Data

When you have identified the product(s) of your acid-catalyzed dehydration reaction, determine the product distribution. You are now ready to use your results to answer the question the experiment addresses: Does rearrangement of the carbocation intermediate occur in the acid-catalyzed dehydration of 3-methylcyclohexanol?

OPTIONAL EXPERIMENT Qualitative Tests for Alkenes

Perform the bromine addition and the potassium permanganate oxidation tests described in Experiment 11.1, p. 88, on your product and compare the results to the same tests on 3-methylcyclohexanol.

Questions

1. Why was the product washed with 5% sodium carbonate solution?
2. Acid-catalyzed dehydration of an alcohol is an equilibrium situation. How was the reaction forced to completion in this experiment?
3. Why might a poor yield of alkenes be realized if the dehydration were carried out with HCl instead of H_3PO_4?
4. What alkene products would you predict from the acid-catalyzed dehydration of (a) 3-methylcyclopentanol? (b) 1-*tert*-butylcyclohexanol?

EXPERIMENT

12 SYNTHESIS OF ESTERS FROM ALCOHOLS

PURPOSE: To study the conversion of a carboxylic acid and an alcohol to an ester using acidic catalysis.

In Experiment 12.1 you will use an esterification procedure to prepare isopentyl acetate (3-methyl-1-butyl acetate), a compound that is a major component of banana oil and used commercially to mimic the flavor of bananas. In an optional experiment, the same procedure is used to prepare the acetate of an "unknown" alcohol; the structures of the unknown compounds will be identified by boiling point and NMR spectroscopy.

In Experiment 12.2 you will synthesize benzocaine (ethyl *p*-aminobenzoate) in an acid-catalyzed esterification of *p*-aminobenzoic acid.

In Experiment 12.3 you will use gas chromatography to determine the equilibrium constant for an acid-catalyzed esterification reaction and compare your data with that of your lab mates.

Acid-Catalyzed Esterification

A strong acid, such as sulfuric acid, is an effective catalyst for the conversion of a carboxylic acid and an alcohol into the corresponding ester.

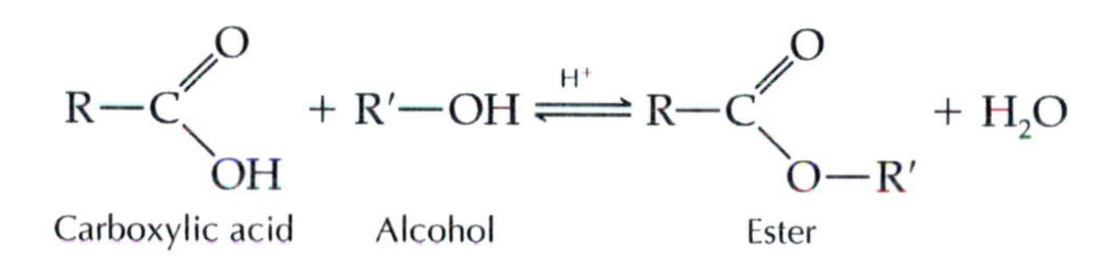

Because the reaction is readily reversible, we must also consider the magnitude of the equilibrium constant. When it has a value close to 1, often the case for ***esterification*** (ester-forming) reactions, the use of a 1:1 stoichiometric mixture of the two starting materials can give only a modest yield of product. It is, however, possible to make use of the Le Chatelier principle to manipulate the equilibrium and increase the yield of ester. For example, synthesis of an ethyl or methyl ester can be carried out in ethanol or methanol as the solvent. When the alcohol is in high concentration, the equilibrium is driven to the product side, toward formation of the ester and water. This strategy is used in Experiment 12.2 where an excess of ethanol drives the equilibrium toward the product. If the yield is then measured in terms of moles of carboxylic acid converted to ester, an excellent yield can be realized. Moreover, ethanol is a convenient solvent, because excess ethanol is easily removed after the reaction is done.

Acetic acid is also a good solvent, and it has a convenient boiling point (118°C). In Experiment 12.1 acetic acid is both solvent and reactant. Using an excess of acetic acid helps to drive the equilibrium toward the desired product.

You will be able to determine the equilibrium constant of an acid-catalyzed esterification reaction in Experiment 12.3, using an equimolar mixture of acetic acid and an alcohol you select.

$$\text{R—OH} + \underbrace{\text{CH}_3\text{—C(=O)—OH}}_{\text{Acetic acid}} \overset{\text{H}^+}{\rightleftharpoons} \underbrace{\text{CH}_3\text{—C(=O)—O—R}}_{\text{Acetate ester}} + \text{H}_2\text{O}$$

Equilibrium and Mechanism

At equilibrium, all four compounds are present. Knowing their molar amounts allows calculation of the equilibrium constant (K_{eq}), which can be expressed as

$$K_{eq} = \frac{[\text{ester}]\,[\text{water}]}{[\text{acid}]\,[\text{alcohol}]}$$

The mechanism of acid-catalyzed ester formation is well understood. The first step is protonation of the carboxylic acid by the sulfuric acid catalyst to produce the conjugate acid of the carboxylic acid:

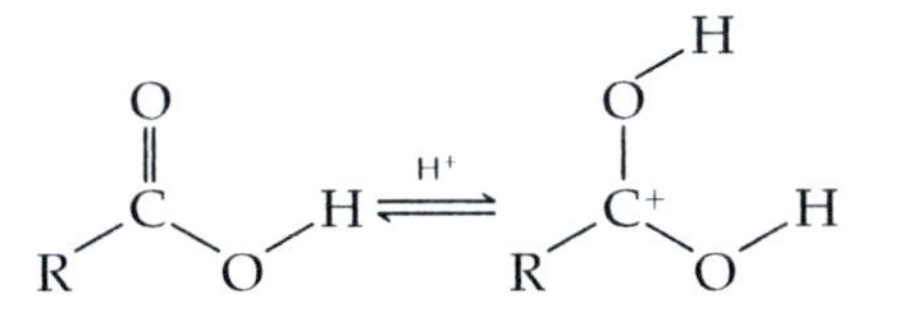

This protonated intermediate is very reactive and highly vulnerable to attack by an alcohol molecule:

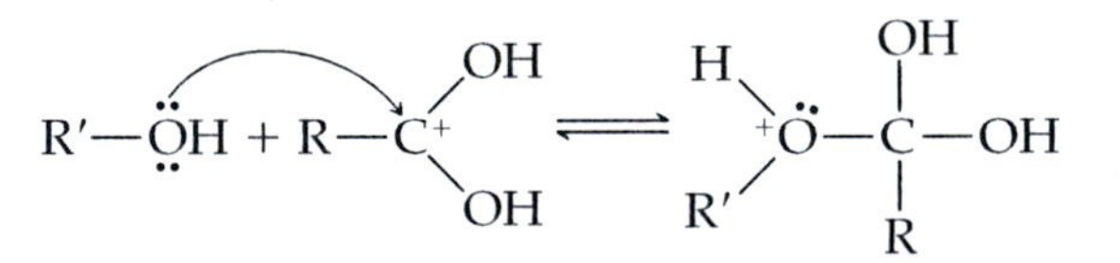

A proton shift then produces a new intermediate from which water can be lost:

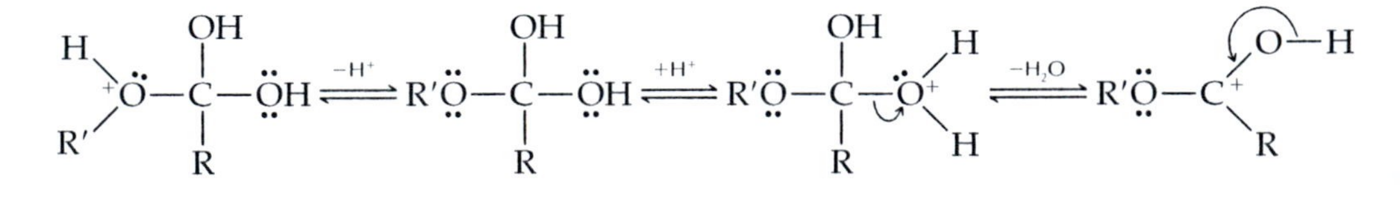

Finally, the remaining protonated species loses a proton to form the ester:

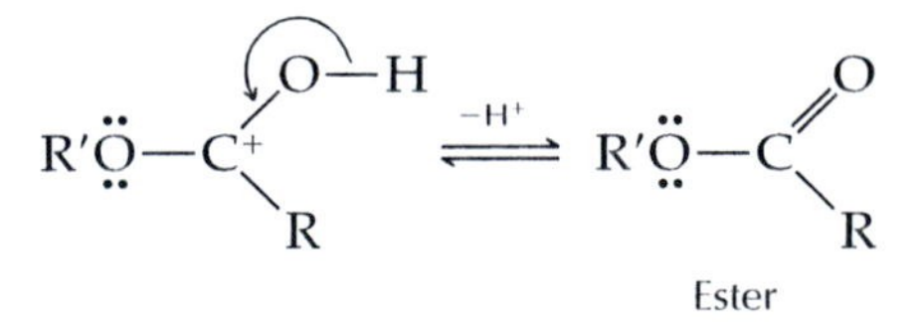

12.1 Synthesis of Isopentyl Acetate

PURPOSE: To make the ester isopentyl acetate, which is the major component of banana oil, using a variety of standard lab techniques.

$$CH_3C(=O)-OH + HO-CH_2CH_2CH(CH_3)_2 \overset{H^+}{\rightleftharpoons} CH_3C(=O)-OCH_2CH_2CH(CH_3)_2 + H_2O$$

Acetic acid (ethanoic acid)	Isopentyl alcohol (3-methyl-1-butanol)	Isopentyl acetate (3-methyl-1-butyl ethanoate)
bp 118°C	bp 129°C	bp 142°C
MW 60.0	MW 88.2	MW 130.2
density 1.05 g · mL^{-1}	density 0.81 g · mL^{-1}	density 0.867 g · mL^{-1}

MINISCALE PROCEDURE

PRELABORATORY ASSIGNMENT: Explain how the reaction conditions favor formation of a high yield of isopentyl acetate.

Techniques
Reflux: Technique 7.1
Extraction: Technique 8.2
Simple Distillation: Technique 11.3
Gas Chromatography: Technique 16
IR Spectroscopy: Techniques 18.4 and 18.5

SAFETY INFORMATION

Isopentyl alcohol (3-methyl-1-butanol) and the **alcohols used in the optional experiments** are flammable and are irritants to the skin and eyes. Avoid contact with skin, eyes, and clothing.

Both **glacial acetic acid** and **sulfuric acid** are corrosive and cause severe burns, especially when they are hot. Acetic acid fumes are also irritating, so the acid should be dispensed in a hood.

Glacial acetic acid gets its name from the fact that when it is frozen (mp 16°C) it looks like an icy glacier.

A brownish color may develop during the reflux period because of acid-catalyzed polymerization reactions.

Pour 16 mL of isopentyl alcohol (3-methyl-1-butanol) and 22 mL of glacial acetic acid into a 100-mL round-bottomed flask. Carefully, with swirling, add 1.0 mL of concentrated sulfuric acid to the contents of the flask. Add one or two boiling stones.

Assemble a reflux apparatus [see Technique 7.1], and heat the reaction mixture at reflux for 1 h. Remove the heat source and let the mixture cool to room temperature. You can speed the cooling process by using a water bath of room-temperature water after the flask has cooled slightly.

Pour the cooled mixture into a separatory funnel and carefully add 50 mL of cold distilled water. Rinse the reaction flask with 5–10 mL of cold water and also add this rinse to the separatory funnel. Stopper the funnel and invert it several times before separating the lower aqueous layer from the upper organic layer [see Technique 8.2]. Set the aqueous phase aside for treatment later.

Be sure to vent the separatory funnel frequently because CO_2 evolution will cause a pressure buildup.

When a drop of the lower aqueous extract turns red litmus paper blue, you have neutralized all the acid in the product layer.

Pour 25 mL of a 0.5 M sodium bicarbonate solution into the separatory funnel. Stir the two phases with a stirring rod until the evolution of CO_2 gas nearly stops. Then stopper the funnel, carefully turn it upside down, and immediately vent the CO_2 gas that forms. Shake and frequently vent the funnel until no more gas is evolved when the funnel is vented. Remove the lower aqueous layer and repeat the extraction with another 25 mL of 0.5 M sodium bicarbonate solution. Continue doing $NaHCO_3$ extractions until the lower layer remains basic (blue) to red litmus paper after the extraction. Wash the organic layer a last time with 20 mL of 4 M sodium chloride solution. Remove the aqueous layer again and pour the ester into a dry 50-mL Erlenmeyer flask. Dry the product with anhydrous calcium chloride and allow it to stand over the drying agent for 10–15 min.

Remove the heat source to stop the distillation before the distilling flask reaches dryness.

Decant the product from the $CaCl_2$ or filter it through a small plug of glass wool into a 50 mL round-bottomed flask. Rinse the condenser with acetone to remove residual water from the reflux process before assembling a simple distillation apparatus [see Technique 11.3]. Collect the fraction boiling between 135°C and 145°C in a tared 50-mL round-bottomed flask. Weigh the receiving flask and product, determine your yield of isopentyl acetate, and calculate the percent yield.

GC and IR Analysis of the Product

Determine the purity of your product by gas chromatographic analysis, using a nonpolar methylsilicone column such as SE-30, OV-1, or DB-1. For a packed column GC, a column temperature of 120–140°C works well. Inject a 1- to 2-mL sample of your product. For a capillary column GC, the column temperature should be 100–110°C; inject 1 mL of a solution containing 1 drop of your product in 0.5 mL of diethyl ether. If your product is impure, the impurity is most probably unreacted isopentyl alcohol, which should have a shorter retention time than the ester on a nonpolar column. You may check this by the peak enhancement method, using a known sample of isopentyl alcohol [see Technique 16.7].

Determine the IR spectrum of your product [see Technique 18.4]. Is there evidence of starting material in your spectrum? If this were the case, what band would be present? What band indicates the formation of product [see Technique 18.5]?

CLEANUP: Combine the aqueous phases left from the extractions. If the pH as determined by pH test paper is below 7, neutralize the solution with solid sodium carbonate before washing it down the sink or pouring it into the container for aqueous inorganic waste. Place the spent drying agent in the container for nonhazardous solid waste. Pour the residue remaining in the distilling flask into the container for flammable (organic) waste.

MICROSCALE PROCEDURE

PRELABORATORY ASSIGNMENT: Explain how the reaction conditions favor formation of a high yield of isopentyl acetate.

Techniques Microscale Reflux: Technique 7.1
Microscale Extraction: Technique 8.6c
Microscale Distillation: Technique 11.3b
Gas Chromatography: Technique 16
IR Spectroscopy: Techniques 18.4 and 18.5

SAFETY INFORMATION

Isopentyl alcohol (3-methyl-1-butanol) and the **alcohols used in the optional experiments** are flammable and are irritants to the skin and eyes. Avoid contact with skin, eyes, and clothing.

Both **glacial acetic acid** and **sulfuric acid** are corrosive and cause severe burns, especially when they are hot. Acetic acid fumes are also irritating, so the acid should be dispensed in a hood.

Glacial acetic acid gets its name from the fact that when it is frozen (mp 16°C) it looks like an icy glacier.

Use graduated pipets to measure 1.6 mL of isopentyl alcohol (3-methyl-1-butanol) and 2.2 mL of glacial acetic acid into a 10-mL round-bottomed flask. Carefully, with swirling, add 1 drop of concentrated sulfuric acid to the flask. Also add a boiling stone.

Assemble a reflux apparatus [see Technique 7.1], using a water-cooled condenser, and heat the reaction mixture at reflux for 50–60 min on an aluminum block or a sand bath heated to 150°C. At the end of the reflux period, remove the apparatus from the heating block, let it cool briefly, then put the flask in a beaker of water until the reaction mixture reaches room temperature.

Add 5.0 mL of water to the reaction mixture and carefully stir the contents of the flask. Transfer approximately half of the reaction mixture to a 5-mL conical vial, using a Pasteur pipet fitted with a rubber bulb or a syringe [see Technique 8.5]. Allow the phases to separate and transfer the lower aqueous layer to a 100-mL beaker [see Technique 8.6c]. Transfer the rest of the reaction mixture to the conical vial and again remove the lower aqueous layer, adding it to the beaker.

Wash the organic phase remaining in the conical vial with 2.5-mL portions of 0.5 M sodium bicarbonate solution until the lower aqueous layer remains basic to litmus paper after the extraction. Vent the vial frequently by loosening the cap. After each washing, remove the lower aqueous phase. Wash the organic layer a last time with 2.0 mL of 4 M sodium chloride solution. Remove the aqueous layer and add a small amount of anhydrous calcium chloride. Allow the product to dry for 10–15 min.

Read Technique 11.3b before beginning the distillation.

Use a Pasteur filter pipet [see Technique 8.5] to transfer the dried product to a dry 5-mL conical vial containing a spin vane. Assemble the apparatus for short-path distillation [see Technique 11.3b, Figure 11.9] or for distillation using an air condenser above the Hickman distilling head [see Technique 11.3b, Figure 11.11]. For the short-path distillation use a tared conical vial submerged in an ice-water bath as the receiving vessel; for a Hickman still distillation, use a tared vial to hold the distillate removed from the stillhead port. Place the auxiliary blocks around the vial as shown in Technique 6, Figure 6.3a. Turn on the stirrer and begin heating the aluminum block to 180–185°C. Collect the distillate in a tared vial. Record the boiling point. Stop the distillation by lifting the apparatus out of the aluminum block before the vial reaches dryness. Weigh your product and determine the percent yield. Carry out the GC and IR analysis described in the miniscale experimental procedure.

CLEANUP: Combine the aqueous phases left from the extractions. If the pH as determined by pH test paper is below 7, neutralize the solution with solid sodium carbonate before washing it down the sink or pouring it into the container for aqueous inorganic waste. Place the spent drying agent in the container for nonhazardous solid waste. Pour the residue remaining in the distilling flask into the container for flammable (organic) waste.

OPTIONAL EXPERIMENT Synthesis and Identification of an Ester from an Unknown Alcohol and Acetic Acid

QUESTION: What are the identities of an unknown alcohol and its esterification product?

In this experiment you will be provided with a sample of an unknown alcohol having a molecular formula of either $C_5H_{11}OH$ or $C_6H_{13}OH$. You will prepare the acetate ester of your alcohol unknown by treating it with acetic acid in the presence of sulfuric acid using either the miniscale or microscale procedure in Experiment 12.1. You will use the boiling point and the ^1H NMR spectrum of the alcohol and the boiling point of the ester you synthesize to identify the structure of the alcohol and the ester product.

MINISCALE PROCEDURE

PRELABORATORY ASSIGNMENT: Draw the structures for all the 1° and 2° isomeric alcohols defined by the formulas $C_5H_{12}O$ and $C_6H_{14}O$ (7 and 14, respectively). List the isomers in groups of primary and secondary alcohols.

Techniques NMR Spectroscopy: Technique 19
Microscale Boiling Point Determination: Technique 11.1

You will be given a vial containing a volume of an unknown alcohol (either a $C_5H_{12}O$ or a $C_6H_{14}O$ alcohol) equivalent to 0.15 mol plus 1.0 mL. Transfer 1.0 mL of the alcohol to a small test tube and use the rest for the synthesis; cork the test tube to prevent the alcohol from evaporating.

Follow the miniscale experimental procedure given in Experiment 12.1, substituting your unknown alcohol for isopentyl alcohol.

During the reflux period, use the remaining amount of your unknown alcohol to prepare a sample for NMR analysis and to determine the boiling point of the alcohol. Prepare the sample for NMR analysis as directed by your instructor, using $CDCl_3$ as the solvent. Ascertain the boiling point of the unknown alcohol by a microscale boiling-point determination [see Technique 11.1, Figure 11.2].

The boiling point of the acetate ester will be 12–16°C above the boiling point that you determine for the alcohol. Use this information to decide whether a water-cooled or an air-cooled condenser should be used for the final distillation of the ester.

Identification of the Unknown Alcohol

Use all your evidence, including the boiling points of both the alcohol and the ester plus the NMR interpretation, to decide the identity of your alcohol. You should consult a handbook such as the *CRC Handbook of Chemistry and Physics* for the reported boiling point of a possible alcohol and its corresponding acetate ester. For some unknowns, it may not be possible to narrow the choice to one alcohol. In that case, list all the possible structures that fit your experimental data and NMR spectrum.

Consider the chemical shift, relative integration value, and multiplicity of each signal in the NMR spectrum to determine a possible structure for your unknown alcohol [see Techniques 19.4–19.9]. Use the CH or CH_2 multiplet farthest downfield as the base integral and the relative integration values for the other signals to determine whether your unknown is a 5-carbon or a 6-carbon alcohol. The integration of this multiplet relative to the other proton signals will also enable you to classify the alcohol as primary or secondary. For many of the possible alcohols, several protons may contribute to a complex multiplet of overlapping signals rather than show distinct multiplets for each type of proton.

If you have a C_5 alcohol, use a density of 0.81 g · mL^{-1} in the yield calculations. If you have a C_6 alcohol, use a density of 0.82 g · mL^{-1}.

Once you have identified your unknown alcohol, calculate the theoretical yield for your synthesis and determine your percent yield.

MICROSCALE PROCEDURE

PRELABORATORY ASSIGNMENT: Draw the structures for all the 1° and 2° isomeric alcohols defined by the formulas $C_5H_{12}O$ and $C_6H_{14}O$ (7 and 14, respectively). List the isomers in groups of primary and secondary alcohols.

Techniques NMR Spectroscopy: Technique 19
Microscale Boiling Point Determination: Technique 11.1

You will be given a vial containing 3 mL of an unknown alcohol (either a $C_5H_{12}O$ or a $C_6H_{14}O$ alcohol). Using a graduated pipet, measure 1.6 mL of the alcohol for your synthesis. Follow the microscale experimental procedure given in Experiment 12.1, p. 97, substituting your unknown alcohol for isopentyl alcohol.

During the reflux period, use the remaining amount of your unknown alcohol to prepare a sample for NMR analysis and to determine the boiling point of the alcohol. Prepare the sample for NMR analysis as directed by your instructor. Ascertain the boiling point of the unknown alcohol by a microscale boiling-point determination [see Technique 11.1, Figure 11.2b]. The boiling point of the acetate ester will be 12–16°C above the boiling point that you determine for the alcohol.

If you have a C_5 alcohol, use a density of $0.81\ g \cdot mL^{-1}$ in the yield calculations. If you have a C_6 alcohol, use a density of $0.82\ g \cdot mL^{-1}$.

Follow the procedure given in the miniscale directions for identifying the unknown alcohol. Once you have identified the alcohol, calculate the theoretical yield for the product.

Reference

1. Branz, S. E. *J. Chem. Educ.* **1985,** *62,* 899–900.

Questions

1. The ester is washed with 0.5 M sodium bicarbonate. Why? Explain with chemical reactions.
2. Ethyl acetate is conveniently prepared by using ethanol as both solvent and reactant. Explain why this procedure would be reasonable.
3. What physical properties of methanol suggest that it would be a good solvent to use for an esterification reaction?
4. Assume that you have a sample of 3-methyl-1-butanol that has its hydroxyl group enriched with ^{18}O. Use the mechanism for the esterification reaction to find out where the ^{18}O enrichment will appear in the final products.
5. Odor and volatility are related. Conversion of a long-chain carboxylic acid to an ethyl ester often enhances its volatility. For example, acetic acid has a bp of 118°C, whereas the bp of ethyl acetate is 77°C. Why?
6. All acetate esters show a ^{1}H NMR singlet at roughly 2.0 ppm. What causes this signal?
7. Samples of 2-methyl-1-propyl acetate and 1-butyl acetate are subjected to ^{1}H NMR analysis. One sample shows an upfield (approximately 1.0 ppm) doublet and the other does not. Which is which? Explain.

12.2 Synthesis of Benzocaine

PURPOSE: To synthesize the ester ethyl *p*-aminobenzoate—the local anesthetic benzocaine—by reaction of *p*-aminobenzoic acid with ethanol in the presence of a catalytic amount of sulfuric acid.

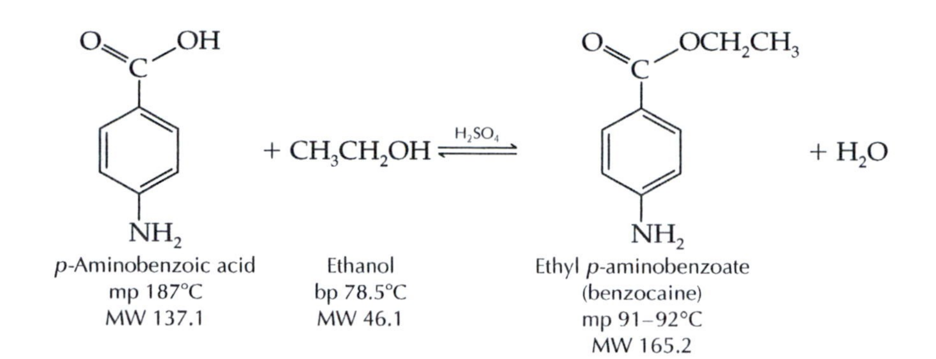

MINISCALE PROCEDURE

PRELABORATORY ASSIGNMENT: Explain how the reaction conditions favor formation of a high yield of benzocaine.

Techniques Recrystallization from Mixed Solvent Pairs: Technique 9.2
Recrystallization: Technique 9.5
Extraction: Technique 8.2
IR Spectroscopy: Techniques 18.4 and 18.5
NMR Spectroscopy: Technique 19

SAFETY INFORMATION

Ethanol and **diethyl ether** are very flammable.

***p*-Aminobenzoic acid** is an irritant. Wear gloves while handling it.

Concentrated sulfuric acid is corrosive and causes severe burns.

Place 0.50 g of *p*-aminobenzoic acid, a boiling stone, and 7.0 mL of absolute ethanol in a dry 25-mL round-bottomed flask. Carefully add 0.50 mL of concentrated sulfuric acid; precipitate will form but will dissolve as the mixture is heated. Fit the flask with a water-cooled condenser and heat the mixture under reflux for 75 min in a boiling-water bath.

At the end of the reflux period, carefully pour the hot solution into a 100-mL beaker and cool the solution to room temperature. Slowly add small portions of 2 M sodium carbonate solution until the acid is neutralized and the pH reaches 9. **(Caution: Foaming will occur.)** At this point a precipitate will begin to form. Add 10 mL of water and cool the mixture in an ice-water bath for 15 min to complete the crystallization process. Collect the crude product by vacuum filtration on a Buchner funnel.

Recrystallize the crude product from an ethanol/water solution [see Technique 9.2] in a 25-mL Erlenmeyer flask. While the flask is being heated on a steam bath or in a boiling-water bath, add 95% ethanol dropwise until the crude product, which forms an oil as it heats, dissolves [see Technique 9.5]. Then add hot water dropwise until the solution becomes cloudy; the volume of water required may nearly equal the volume of ethanol. Stir the mixture while cooling it in an ice-water bath. Collect the crystals of benzocaine by vacuum filtration on a Buchner funnel. Allow the product to dry thoroughly before determining the percent yield and the melting point.

IR or NMR Analysis of the Product

Obtain an IR or NMR spectrum of the dried product, as directed by your instructor [see Technique 18.4 for IR or Technique 19.2 for NMR sample preparation]. Analyze the peaks in your spectrum. Does your spectrum show evidence of starting material or was the conversion to product complete?

CLEANUP: The aqueous filtrate from the crude product may be washed down the sink or poured into the container for aqueous inorganic waste. Place the filtrate from the recrystallization in the container for flammable (organic) waste.

MICROSCALE PROCEDURE

PRELABORATORY ASSIGNMENT: Explain how the reaction conditions favor formation of a high yield of benzocaine.

Techniques
Recrystallization from Mixed Solvent Pairs: Technique 9.2
Microscale Recrystallization: Technique 9.7a
Microscale Extraction: Techniques 8.6a and 8.6c
IR Spectroscopy: Technique 18
NMR Spectroscopy: Technique 19

SAFETY INFORMATION

Ethanol and **diethyl ether** are very flammable.

***p*-Aminobenzoic acid** is an irritant. Wear gloves while handling it.

Concentrated sulfuric acid is corrosive and causes severe burns.

Place 0.200 g of *p*-aminobenzoic acid, a boiling stone, and 5.0 mL of absolute ethanol in a dry 10-mL round-bottomed flask. Carefully add 0.20 mL of concentrated sulfuric acid; precipitate will form but will dissolve as the mixture is heated. Fit the flask with a water-cooled condenser and heat the mixture under reflux for 75 min in a boiling-water bath.

At the end of the reflux period, carefully pour the hot solution into a 50-mL beaker and cool the solution to room temperature. Slowly add small portions of 2 M sodium carbonate solution until the acid is neutralized and the pH reaches 9. **(Caution: Foaming will occur.)** At this point a precipitate will begin to form. Add 3 mL of water and cool the mixture in an ice-water bath for 15 min to complete the crystallization process. Collect the crude product by vacuum filtration on a Hirsch funnel.

Recrystallize the crude product from an ethanol/water solution [see Technique 9.2] in a 10-mL Erlenmeyer flask. While the flask is being heated on a steam bath or in a boiling-water bath, add 95% ethanol dropwise until the crude product, which forms an oil as it heats, dissolves [see Technique 9.7a]. Then add hot water dropwise until the solution becomes cloudy; the volume of water required may nearly equal the volume of ethanol. Stir the mixture while cooling it in an ice-water bath. Collect the crystals of benzocaine by vacuum filtration on a Hirsch funnel. Allow the product to dry thoroughly before determining the percent yield and the melting point. Obtain and analyze an IR or NMR spectrum of the dried product, as directed in the miniscale procedure on p. 102.

CLEANUP: The aqueous filtrate from the crude product may be washed down the sink or poured into the container for aqueous inorganic waste. Place the filtrate from the recrystallization in the container for flammable (organic) waste.

Questions

1. The ester is washed with 2 M sodium carbonate. Why? Explain with chemical reactions.
2. What is the reason for using anhydrous ethanol and concentrated (98%) sulfuric acid rather than 95% ethanol (95/5 ethanol:water) or more dilute sulfuric acid (6 M) in the esterification of *p*-aminobenzoic acid?
3. Ethyl acetate is conveniently prepared by using ethanol as both solvent and reactant. Explain why this procedure would be reasonable.
4. Should you be concerned if denatured alcohol (prepared by adding the poisonous compound methanol to 100% ethanol) is inadvertently used in the esterification reaction?
5. The analysis of a ^1H NMR spectrum of benzocaine in $CDCl_3$ showed the following signals: 7.9 ppm, 2H, d; 6.25 ppm, 2H, d; 4.25 ppm, 2H, q; 4.0 ppm, 2H, s; 1.3 ppm, 3H, t (s = singlet, d = doublet, t = triplet, q = quartet). Assign all the signals. Which one of the signals would be expected to show a significant concentration dependence? Why?

12.3 Determination of the Equilibrium Constant for an Esterification Reaction

QUESTION: What effect does the structure of an alcohol have on the equilibrium constant of an esterification reaction?

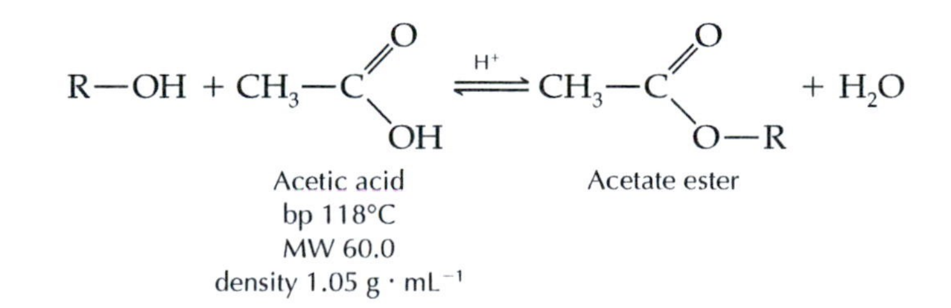

The reaction mixture needs to stand at room temperature for one week for equilibration to occur. Suggested alcohols for this experiment include

1-pentanol, 2-pentanol, 3-pentanol, 3-methyl-1-butanol, 2-methyl-1-butanol, and 3-methyl-2-butanol.

PRELABORATORY ASSIGNMENT: Explain why knowing the equilibrium constant for an esterification reaction is an important factor in designing the synthesis of an ester.

MICROSCALE PROCEDURE

Techniques Microscale Extraction: Technique 8.6c
Gas Chromatography: Technique 16

SAFETY INFORMATION

The **alcohols** used in this experiment are flammable and are irritants to the skin and eyes. Wear gloves, and avoid contact with skin, eyes, and clothing.

Both **glacial acetic acid** and **sulfuric acid** are corrosive and cause severe burns. Acetic acid also emits irritating fumes, so it should be dispensed in a hood.

Pipet 0.44 mL of the your assigned alcohol and 0.24 mL of acetic acid that contains 0.5% sulfuric acid into a clean, dry 3-mL conical vial. Close the vial with a screw cap and septum, placed with the Teflon (dull) side toward the inside of the vial. Shake the vial briefly to ensure complete mixing. Set the vial in a 30-mL beaker and store it for one week.

One Week Later

Add 1 mL of saturated sodium bicarbonate solution to the conical vial. **(Caution: CO_2 evolution.)** Draw the mixture into a Pasteur pipet fitted with a rubber bulb or syringe [see Technique 8.5] and expel it back into the vial at least three times. Remove the lower aqueous layer with the Pasteur pipet [see Technique 8.6c]. Wash the organic layer with 1 mL of saturated sodium chloride solution. Dry the organic layer with a few pellets of anhydrous calcium chloride for 10–15 min. Transfer the dried product mixture to a clean test tube or vial with a Pasteur filter pipet [see Technique 8.5].

Gas Chromatographic Analysis

Analyze the dried product solution by gas chromatography on a methyl silicone-5% phenylmethyl silicone column, such as AT-5, OV-5, or SE-52, with a column temperature of 70–80°C. For a capillary column chromatograph, prepare a solution containing 1 drop of the product mixture in 0.5 mL of diethyl ether; use 1 μL for injection.

The response of the detector may be different for the alcohol and the ester. It is necessary to analyze a standard mixture of your alcohol and its acetate ester to determine their relative response factors [see Technique 16.7].

CLEANUP: The aqueous solutions remaining from the extractions can be washed down the sink or poured into the container for inorganic waste.

Place the spent pellets of calcium chloride in the container for solid inorganic waste. Pour the remaining solutions used for GC analysis into the container for flammable organic waste.

Interpretation of Experimental Data

Analyze the chromatogram to find the relative ratio of alcohol to ester using the response factors that you determined. Calculate the equilibrium constant.

In this experiment an equimolar mixture of acetic acid and the alcohol you selected are allowed to react for one week under acid catalyzed conditions to form an acetate ester and water.

At equilibrium all four compounds will be present. From the molar amounts of the compounds present at equilibrium, an equilibrium constant can be calculated. The equilibrium constant for the reaction can be expressed as

$$K_{eq} = \frac{[\text{ester}]\,[\text{water}]}{[\text{acid}]\,[\text{alcohol}]}$$

Starting with equal molar amounts, Y, of acid and alcohol at equilibrium, X moles of ester and water will be present. Then the amounts of alcohol and acid remaining at equilibrium will be (Y–X) moles.

	Acetic acid	Alcohol	Ester	Water
At start	Y	Y	0	0
At equilibrium	$Y - X$	$Y - X$	X	X

The equilibrium constant can be expressed as

$$K_{eq} = \frac{(X)(X)}{(Y - X)(Y - X)} = \left[\frac{(X)}{(Y - X)}\right]^2$$

The workup of the equilibrium mixture uses sodium bicarbonate to neutralize the mineral acid catalyst and the remaining organic acid (acetic acid in this experiment). Drying agent removes the water present in the reaction mixture. Thus the organic material remaining after the workup is a mixture of ester and alcohol. By using gas chromatography to analyze the ester-alcohol mixture, the ratio of ester to alcohol, which corresponds to $[X/(Y–X)]$, can be determined. This quantity is substituted into the equilibrium constant expression.

Questions

1. Why is the reaction mixture allowed to react for a week before it is analyzed?
2. The product mixture is washed with a saturated sodium bicarbonate solution. Why? Explain with chemical reactions.
3. Why is the product mixture washed with saturated NaCl solution rather than with pure water?
4. How would you change the reaction conditions used in this equilibrium study to maximize the yield of the ester, assuming that the alcohol is the limiting reagent?
5. Suppose that the alcohol you selected has a small but significant solubility in water, while the ester has absolutely no solubility in water. Would this cause the calculated equilibrium constant to be too large or too small? Refer explicitly to the equilibrium constant expression in your answer.

EXPERIMENT

13

GREEN CHEMISTRY

Oxidation of Cyclohexanol Using Sodium Hypochlorite

PURPOSE: To investigate the oxidation of a secondary alcohol to a ketone using an environmentally friendly reaction.

In Experiment 13 you will oxidize cyclohexanol to cyclohexanone with household bleach in a mildly acidic environment and assess the purity of your product with IR spectroscopy.

"Green chemistry" is chemistry that causes minimum pollution, and it is a field that is increasingly attracting the attention of chemists in both industry and academia as concern for the environment grows. The goal of green chemistry is to be as environmentally friendly as possible in the synthesis and utilization of chemicals. Green chemistry involves the development of synthetic methods that use nontoxic reagents and nonharmful solvents and produce virtually no wasteful by-products. In green organic chemistry, the goal is to synthesize materials, such as plastics, that easily biodegrade after they have served their purpose, rather than occupy space for years in landfills.

Experiment 13 is an example of green chemistry in which water is used as the reaction medium and household bleach as the oxidizing agent in the synthesis of a ketone. In the reaction, bleach is converted into sodium chloride, which can be safely washed down the sink. There is very little material that must go into specialized waste disposal. Let's consider the chemistry to see how it works.

The reaction is the oxidation of cyclohexanol with NaOCl, forming cyclohexanone.

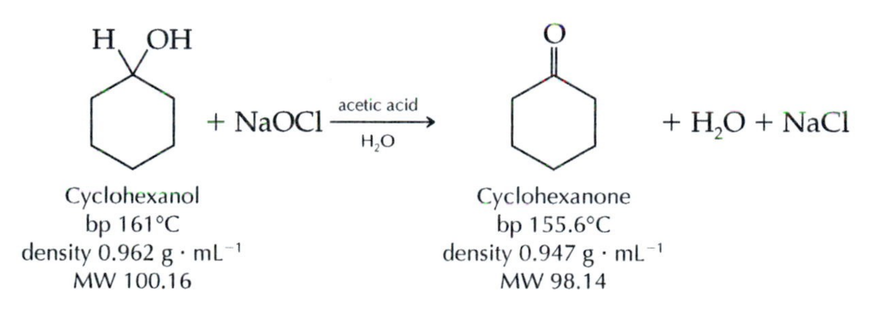

Oxidation-Reduction Reactions

Oxidation-reduction reactions form an extremely diverse group, and they proceed by a variety of mechanisms. The unifying factor is that one substrate is oxidized and another is reduced. In the inorganic chemistry of metal cations, the reaction can most easily be considered as an electron transfer phenomenon; that is, ***oxidation*** is a loss of electrons and ***reduction*** is a gain of electrons. While this approach also applies to covalent organic systems, it is awkward to use. It is often easier to think of oxidation as the loss of hydrogen (the equivalent of H_2) or gain of oxygen atoms.

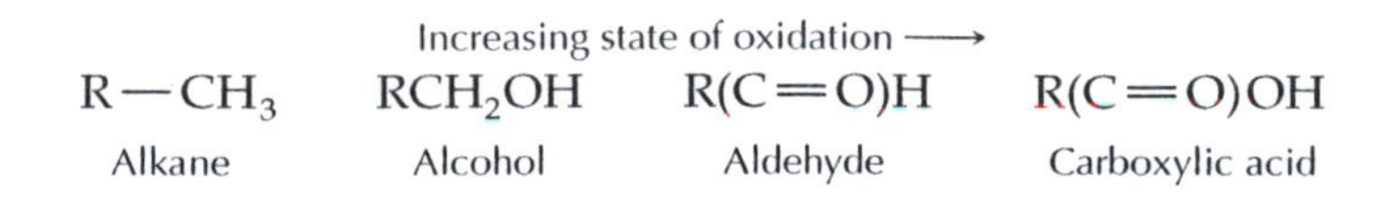

In this oxidation of a secondary alcohol to a ketone, the α-hydrogen atom is lost from carbon and the hydrogen on the oxygen atom is also lost.

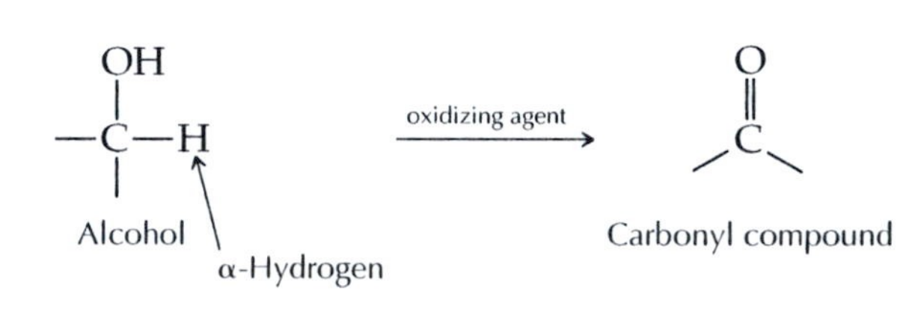

Many alcohols can be oxidized in this manner; primary alcohols can be oxidized to aldehydes and secondary alcohols to ketones. Tertiary alcohols, because they bear no α-hydrogens, do not easily undergo oxidation.

When oxidation of an organic substrate is needed, the oxidizing agent is often an inorganic compound. In academic chemistry laboratories, chromium (VI) compounds have probably been the most popular oxidizing agents over the last 60 years. When CrO_3 is the oxidizing agent, Cr(VI) gains three electrons and becomes reduced to Cr(III). Although chromium trioxide (CrO_3) is still used as an oxidizing agent, it has major drawbacks.

Chromium compounds are highly toxic to humans and to freshwater aquatic life. Some Cr(VI) compounds have been indicted as carcinogens. They are dangerous to handle, and they pose a hazard to the environment. They must be disposed of carefully, and there is no way that the chromium can biodegrade. Except in special situations, alternatives to chromium (VI) are now used in the undergraduate organic chemistry laboratory.

Oxidation with Household Bleach

An excellent alternative oxidizing agent is sodium hypochlorite (NaOCl), which is the active component of household bleach. This reagent has low toxicity if it is handled with care, and it reduces to products that are nontoxic as well.

In this experiment you will use household bleach (for example, Clorox) in the presence of acetic acid as the oxidizing agent. Most bleach is an aqueous solution of sodium hypochlorite, NaOCl, prepared by adding Cl_2 to aqueous sodium hydroxide solution:

$$Cl_2 + 2NaOH \rightleftharpoons NaOCl + NaCl + H_2O$$

When sodium hypochlorite is added to acetic acid, the following acid-base reaction produces hypochlorous acid (HOCl):

$$NaOCl + CH_3{-}C(=O)OH \rightleftharpoons HOCl + CH_3{-}C(=O)O^{-}Na^{+}$$

In acidic solution, HOCl is in equilibrium with Cl_2:

$$HOCl + HCl \rightleftharpoons Cl_2 + H_2O$$

Chlorine gas is toxic, but when household bleach is used to oxidize alcohols, Cl_2 is present in very low and safe concentrations in the atmosphere if one takes simple precautions.

Both Cl_2 and hypochlorous acid are sources of positive chlorine, which has two fewer electrons than does chloride anion. A key step in reactions of positive chlorine reagents is the transfer of Cl^+ to the substrate. It is reasonable to expect that the first step in the oxidation of cyclohexanol is exchange of positive chlorine with the hydroxyl proton. Subsequent E2 elimination of HCl from the resulting alkyl hypochlorite forms the ketone, cyclohexanone. If HOCl is not present, the first step cannot take place and the overall oxidation is very slow.

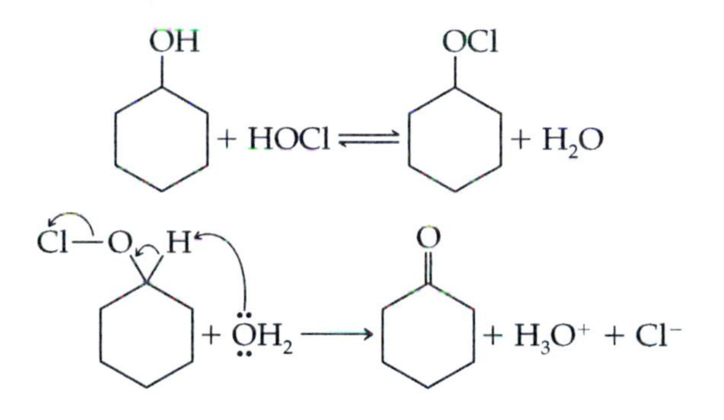

In the first reaction, Cl^+ is transferred to the substrate, and in the second reaction, Cl^- is lost. The change is a reduction by two electrons. Cyclohexanol provides the two electrons as it is oxidized to cyclohexanone.

There are a few additional experimental strategies in making the experiment as "green" as possible. Because this is a two-phase reaction, it is necessary to mix the organic and aqueous layers during the reaction to bring the reactants together. The rate of the reaction depends in part on effective mixing of the reactants. It is also important to have enough acid present in the reaction mixture to bring the pH of the aqueous solution to less than pH 7 so that the alkyl hypochlorite intermediate can form.

Steam distillation of the reaction mixture to separate cyclohexanone from the inorganic salts avoids the need for extractions with organic solvents. Even though the yield of cyclohexanone is somewhat greater if extractions with dichloromethane are used, CH_2Cl_2 is difficult to recycle and becomes an organic waste product. We have chosen to accept a yield of 50–60% without extractions rather than 70–80% using extractions. In the microscale version of the experiment, extractions with a small amount of dichloromethane are necessary for recovering an acceptable amount of product. The compensating factor that allows it to be green chemistry is that it is done on a very small scale, using only 10% as much substrate and oxidizing agent as the miniscale experiment.

Infrared Spectroscopic Analysis of Cyclohexanone

Infrared (IR) spectroscopy is an especially good method for assessing the purity of your product. The boiling points of the substrate, cyclohexanol, and the product, cyclohexanone, are only 5° apart; therefore, any unreacted cyclohexanol will not be separated by distillation. Figure 13.1 shows the IR spectrum of cyclohexanol. The strong, broad O—H stretching band in the 3550–3250 cm^{-1} region is diagnostic of all alcohols. If the oxidation has gone to completion, this peak will not be present in the spectrum of your product. The IR spectrum of cyclohexanone will show an intense C=O stretching band near 1715 cm^{-1}. The rest of the IR spectrum will yield a fingerprint that should allow straightforward identification of the ketone.

One caveat is important to note, however. If your product is not dried properly, the O—H stretching vibration of water can lead you to think that unreacted cyclohexanol is present in your product. The only way by which you could prove the point is to do a gas chromatographic analysis on your product and use the peak-enhancement method to identify cyclohexanol.

MINISCALE PROCEDURE

Techniques Steam Distillation: Technique 11.7
Gas Chromatography: Technique 16
IR Spectroscopy: Technique 18

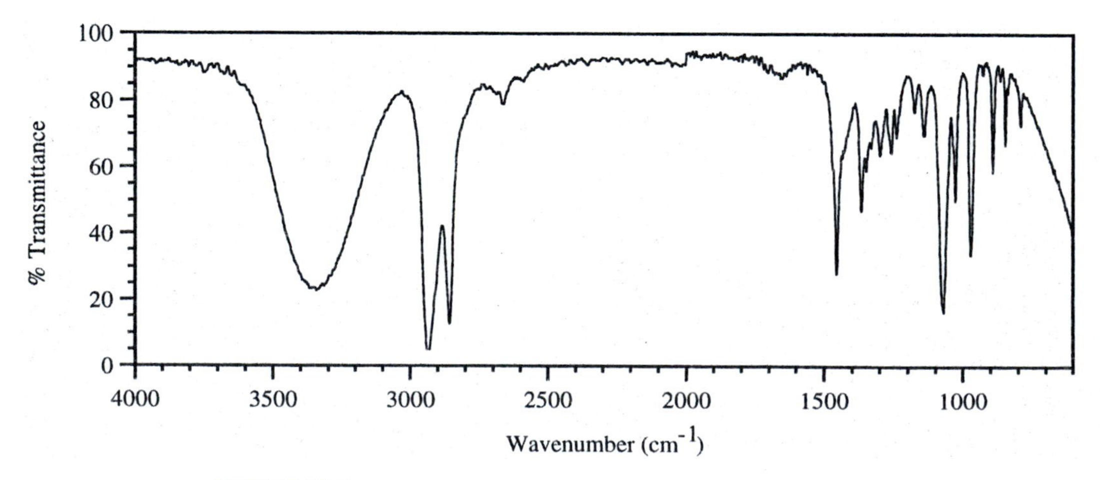

FIGURE 13.1
IR spectrum of cyclohexanol (thin film).

SAFETY INFORMATION

Cyclohexanol is an irritant. Avoid contact with skin, eyes, and clothing.

Glacial acetic acid is a dehydrating agent and an irritant, and it causes burns. Dispense it in a hood and avoid contact with skin, eyes, and clothing.

Sodium hypochlorite solution emits chlorine gas, which is a respiratory and eye irritant. Dispense it in a hood.

Sodium hydroxide solutions are corrosive and cause burns. Solutions as dilute as 2.5 M can cause severe eye damage. Avoid contact with skin, eyes, and clothing.

Clorox or a supermarket brand of household bleach works well; use only newly opened bleach.

Pour 16 mL of cyclohexanol and 8.0 mL of glacial acetic acid into a three-necked 500-mL round-bottomed flask, and add a magnetic stirring bar. In a hood, pour 115 mL of 5.25% (0.74 M) sodium hypochlorite (bleach) solution into a 125-mL dropping funnel. (Your separatory funnel may be used as a dropping funnel.) Stopper the dropping funnel before carrying it to your work station. Insert the dropping funnel in a neck of the flask, then put a small piece of paper between the stopper and neck of the funnel. Place a condenser in the middle neck. Insert a thermometer through a thermometer adapter in the third neck of the flask, taking care that the thermometer bulb is positioned above where it could be hit by the magnetic stirring bar. Prepare an ice-water bath for cooling the flask if the reaction becomes too warm.

The small piece of paper prevents a buildup of vacuum and the solution can drip smoothly.

Add approximately one-fourth of the bleach solution to the reaction flask and begin to stir the reaction mixture. Add the rest of the sodium hypochlorite solution over a period of 15 min. Adjust the rate of addition so that the temperature remains between 40°C and 45°C during the addition. Cool the reaction flask briefly with the ice-water bath if the temperature exceeds 45°C; however, allowing the temperature of the reaction mixture to fall below 40°C for any period of time slows down the reaction and may result in a lowered yield.

When the addition of sodium hypochlorite is complete, remove the dropping funnel, and temporarily stopper the reaction flask. Pour another 115 mL of sodium hypochlorite solution into the funnel and add this portion of bleach to the reaction flask over a period of 15 min, taking care that the temperature again remains between 40°C and 45°C during the addition. After the sodium hypochlorite addition is complete, remove the dropping funnel and stopper the flask opening. Continue stirring the reaction mixture for 20 min.

The yellow color of chlorine should no longer disappear near the end of the addition period.

If the bleach is too concentrated, colorless iodate may form on the indicator paper.

Test the reaction mixture for excess hypochlorite by placing a drop of the reaction solution on a piece of wet starch-iodide indicator paper. The appearance of a blue-black color from the formation of the triiodide-starch complex on the indicator paper signifies the presence of excess hypochlorite. Add 2.0 mL of saturated sodium bisulfite solution to the reaction flask and swirl the flask to mix the contents thoroughly. Again test the reaction solution with starch-iodide paper. If necessary, continue adding bisulfite solution and testing with starch-iodide paper until excess oxidant is removed.

The blue color of thymol blue appears at about pH 9.

Add 2.0 mL of thymol blue indicator solution to the mixture in the reaction flask. Place a conical funnel in one neck of the flask and add 6 M sodium hydroxide solution over 3 min until the indicator changes to blue (30–40 mL will be needed). Swirl the flask during the addition of NaOH. Use a magnet retriever to remove the magnetic stirring bar from the reaction flask.

You will separate the organic product from the reaction mixture by steam distillation. Add 2 or 3 boiling stones to the reaction flask. Rinse the thermometer, condenser, and dropping funnel with water at a sink, and rinse again with sodium bisulfite solution before using them for the steam distillation. Set up the distillation apparatus as shown in Technique 11.7, Figure 11.24, omitting the dropping funnel and closing the second neck of the Claisen adapter plus the other two necks of the round-bottomed flask with glass stoppers. Alternatively, instead of the Claisen adapter, a short column can be used between the reaction flask and the distilling head.

The dropping funnel can be omitted, because the reaction mixture already contains a substantial amount of water and it is unnecessary to add more.

Collect 70–80 mL of distillate consisting mainly of product and water. Add 10 g of solid sodium chloride to the distillate to decrease the solubility of cyclohexanone in the aqueous phase. Stir the mixture until most of the NaCl dissolves. Decant the liquid from undissolved NaCl into a separatory funnel. Allow the phases to separate, and remove the lower aqueous layer. Pour the upper product layer into a clean 50-mL Erlenmeyer flask and dry it with anhydrous magnesium sulfate or potassium carbonate for 20 min [see Technique 8.7].

The dried product is fairly pure at this point. If time permits, you can distill the final product using a simple distillation apparatus [see Technique 11.3, Figure 11.6]; collect the fraction boiling from 150–156°C. To separate the product from the drying agent, decant the product through a conical funnel containing a small piece of cotton into a tared (weighed) vial or small Erlenmeyer flask [see Technique 8.7]. If you are doing the final simple distillation, use a tared 50-mL round-bottomed flask as the receiving flask. Weigh your product and calculate the percent yield.

Analysis of the Product

Obtain and analyze the IR spectrum of your product [see Technique 18.4]. Does your spectrum show evidence of cyclohexanol? What band indicates that you have the desired product? Your instructor may also ask you to determine the purity of your product by gas chromatographic analysis using a 5% diphenyl/95% dimethylpolysiloxane column, such as AT-5, DB-5, or OV-5 [see Technique 16]. For a capillary column chromatograph, prepare a solution containing 1 drop of your product in

0.5 mL of diethyl ether; inject 0.5–1.0 μL of the ether solution. Use the peak-enhancement method to identify cyclohexanone and to see if any cyclohexanol is present [see Technique 16.6]. Submit the remaining product as directed by your instructor.

CLEANUP: Adjust the pH of the solution remaining in the reaction flask to 7 with dilute hydrochloric acid before washing the solution down the sink or pouring it into the container for aqueous inorganic waste. Place the drying agent in the container for nonhazardous solid waste or the container for inorganic waste. Pour the residue in the boiling flask used for the simple distillation of the product and any solution remaining from the GC analysis into the container for flammable organic waste.

MICROSCALE PROCEDURE

Techniques Microscale Extraction: Technique 8.6b
Drying Organic Liquids: Techniques 8.7–8.9
Gas Chromatography: Technique 15
IR Spectroscopy: Technique 18

SAFETY INFORMATION

Cyclohexanol is an irritant. Avoid contact with skin, eyes, and clothing.

Glacial acetic acid is a dehydrating agent and an irritant, and it causes burns. Dispense it in a hood and avoid contact with skin, eyes, and clothing.

Sodium hypochlorite solution emits chlorine gas, which is a respiratory and eye irritant. Dispense it in a hood.

Sodium hydroxide solutions are corrosive and cause burns. Solutions as dilute as 2.5 M can cause severe eye damage. Avoid contact with skin, eyes, and clothing.

Dichloromethane is toxic, an irritant, absorbed through the skin, and harmful if inhaled. Use it in a hood and wear neoprene gloves while doing the extractions and evaporation.

Clorox or a supermarket brand of household bleach works well; use only newly opened bleach.

Place 1.51 g of cyclohexanol, 0.75 mL of acetic acid, and a magnetic stirring bar in a 50-mL Erlenmeyer flask. While stirring the reaction mixture rapidly, add 24 mL of 5.25% (0.74 M) sodium hypochlorite (bleach) solution over a period of 5 min. Continue stirring the reaction mixture for 20 min after the addition of NaOCl is complete.

If the bleach is too concentrated, colorless iodate may form on the indicator paper.

Test the reaction mixture for excess hypochlorite by placing a drop of the reaction solution on a piece of wet starch-iodide indicator paper. The appearance of a blue-black color from the formation of the triiodide-starch complex on the indicator paper signifies the presence of excess hypochlorite. Add 0.5 mL of saturated sodium bisulfite solution, swirl the flask, and again test a drop of the reaction mixture with starch-iodide paper. If necessary, continue adding bisulfite solution in 0.5 mL increments and testing with starch-iodide paper until excess oxidant is removed.

Add 1.5 mL of 6 M sodium hydoxide solution to the reaction mixture. Continue adding 6 M NaOH dropwise until the reaction solution shows a pH of 6–8.

Add 5 mL of dichloromethane and stir the mixture rapidly for 3–4 min. Remove the magnetic stirring bar with tweezers or a magnetic retrieving rod. Holding the flask slightly tipped, remove the lower dichloromethane layer with a Pasteur filter pipet or a Pasteur pipet fitted with a syringe [see Technique 8.5] pressed against the "corner" of the flask and transfer it to a 15-mL centrifuge tube.

Add 5 mL of dichloromethane to the aqueous phase remaining in the Erlenmeyer flask, return the magnetic stirring bar to the flask, and stir the mixture for 3–4 min. Again remove the stirring bar and transfer the lower organic layer to the centrifuge tube containing the first dichloromethane layer. Add anhydrous potassium carbonate, cap the tube, and shake the contents briefly. Dry the product solution for at least 10 min.

Transfer the dried product to a tared 25-mL Erlenmeyer flask using a new Pasteur filter pipet. Working in a hood, blow off the dichloromethane with a stream of nitrogen or air. Alternatively, remove the dichloromethane on a rotary evaporator. Weigh the flask and calculate your percent yield. Analyze the product as directed in the miniscale procedure.

Placing the flask in a beaker of 40–45°C tap water speeds the evaporation process.

CLEANUP: Wash the solution remaining in the reaction flask and the aqueous sodium chloride solution from the extraction down the sink or pour them into the container for aqueous inorganic waste. Place the drying agent in the container for nonhazardous solid waste or the container for inorganic waste.

References

1. Mohrig, J. R.; Neinhuis, D. M.; Linck, C. F.; Van Zoeren, C.; Fox, B. G.; Mahaffy, P. G. *J. Chem. Educ.* **1985,** *62,* 519–521.
2. Perkins, R. A.; Chau, F. *J. Chem. Educ.* **1982,** *59,* 981.

Questions

1. Describe what your IR spectrum would show if any cyclohexanol remained in the product.
2. Balance the equation for the oxidation-reduction reaction that occurs between bisulfite (HSO_3^-) and hypochlorite (OCl^-) to give sulfate and chloride ions.
3. What is the purpose of adding sodium hydroxide to the reaction mixture before the steam distillation in the miniscale version of the experiment or before the extractions in the microscale version? Write an equation for the reaction that occurs when the base is added.
4. Mohrig and coworkers (Ref. 1) suggest that 2-propanol, rather than sodium bisulfite ($NaHSO_3$), can be used to destroy any excess sodium hypochlorite. What organic compound would be formed from the reaction of 2-propanol and NaOCl and why should it not be a contamination problem in this synthesis?
5. Predict the product that will be obtained if *trans*-2-methylcyclohexanol is oxidized with NaOCl. What will be the product if the cis isomer is oxidized?

EXPERIMENT

14

OXIDATION OF CINNAMYL ALCOHOL USING PYRIDINIUM CHLOROCHROMATE

PURPOSE: To oxidize a primary alcohol to an aldehyde under anhydrous conditions.

> In Experiment 14 you will synthesize cinnamaldehyde by oxidizing cinnamyl alcohol with a pyridinium chlorochromate/sodium acetate/molecular sieve mixture. The reaction progress will be monitored by thin-layer chromatography, and purification of the product will be done by column chromatography.

Aldehydes are important compounds, but special methods often have to be used in their synthesis because of their great reactivity. Many reactions that might be used to oxidize a primary alcohol to an aldehyde produce over-oxidation and the carboxylic acid is formed.

Increasing state of oxidation ⟶

$R-CH_3$	RCH_2OH	$R(C=O)H$	$R(C=O)OH$
Alkane	Alcohol	Aldehyde	Carboxylic acid

To understand the need for special methods in the synthesis of aldehydes, we need to review the basic ideas of oxidation-reduction chemistry. Oxidation-reduction reactions are extremely diverse, and they proceed by a variety of mechanisms. The unifying factor is that one substrate is oxidized and another is reduced. In the inorganic chemistry of metal cations this process can be thought of most easily as an electron transfer phenomenon; that is, ***oxidation*** is a loss of electrons and ***reduction*** is a gain of electrons. While this approach also applies to covalent organic systems, it is awkward to use. It is often easier to think of oxidation as the loss of hydrogen (the equivalent of H_2) or gain of oxygen atoms. In the oxidation of an alcohol to an aldehyde, the α-hydrogen atom is lost from carbon and the hydrogen is also lost from the oxygen atom.

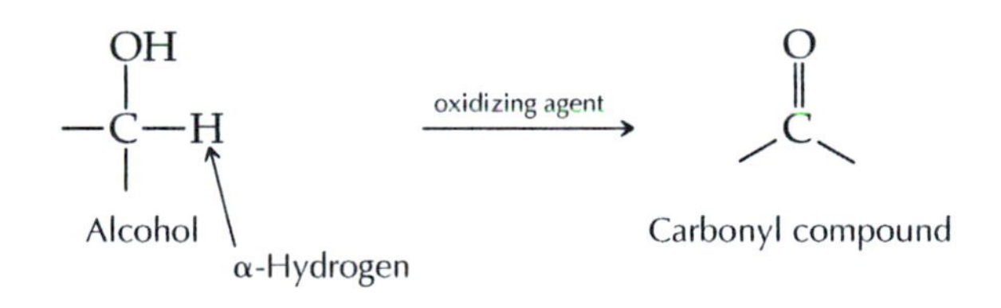

When oxidation of an organic substrate is needed, the oxidizing agent is often an inorganic compound. In academic chemistry laboratories, chromium (VI) compounds have probably been the most popular oxidizing agents over the last 60 years. When chromium trioxide (CrO_3) is the oxidizing agent, Cr(VI) gains three electrons and becomes reduced to Cr(III).

Although chromium compounds are toxic, they are used in special situations at the microscale level. The controlled oxidation of primary alcohols to aldehydes is such a special situation. This oxidation requires anhydrous conditions so that the aldehyde is not oxidized to a carboxylic acid. Under aqueous conditions aldehydes become hydrated to form 1,1-diols (also called geminal diols). The 1,1-diol reacts with chromium trioxide to form the chromate ester. Subsequent E2 elimination of the Cr(IV) compound H_2CrO_3 forms the carboxylic acid. Eventually, Cr(IV) is transformed into Cr(III).

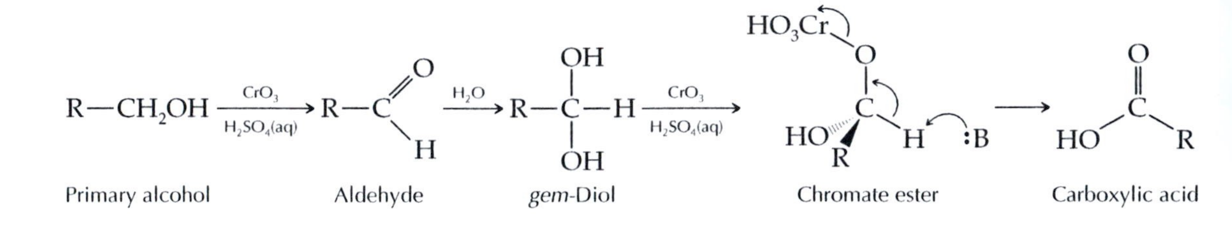

Primary alcohol Aldehyde *gem*-Diol Chromate ester Carboxylic acid

In 1975 Corey and Suggs discovered that pyridinium chlorochromate (PCC) was soluble in dichloromethane and other organic solvents. This reagent has proved effective for the chromium (VI) oxidation of a primary alcohol to an aldehyde under anhydrous conditions. Pyridinium chlorochromate is prepared by treating pyridine with chromium trioxide in the presence of HCl:

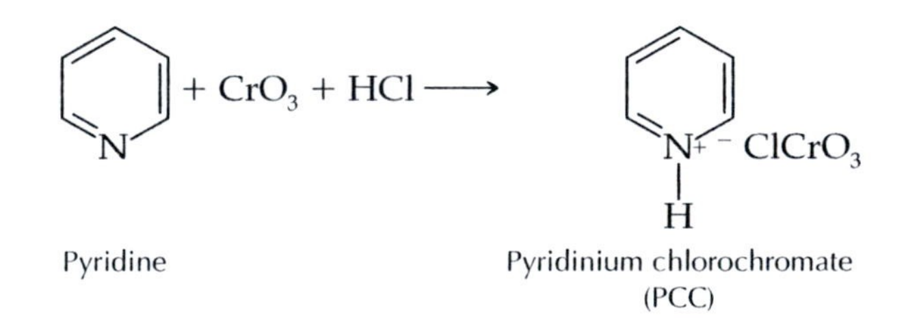

Pyridine Pyridinium chlorochromate (PCC)

In this experiment you will oxidize cinnamyl alcohol to cinnamaldehyde, using a PCC/sodium acetate/molecular sieve mixture in dichloromethane as the oxidant. Cinnamaldehyde is a major component of cinnamon oil, and both cinnamyl alcohol and cinnamaldehyde are used in the flavor and fragrance industry.

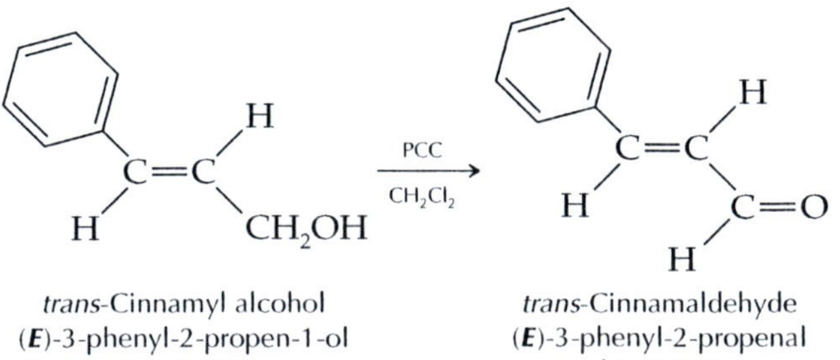

trans-Cinnamyl alcohol
(*E*)-3-phenyl-2-propen-1-ol
mp 33°C
MW 134.18

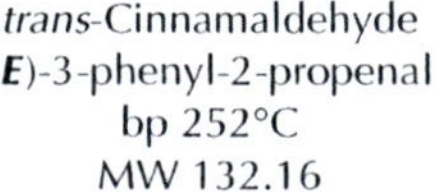
trans-Cinnamaldehyde
(***E***)-3-phenyl-2-propenal
bp 252°C
MW 132.16

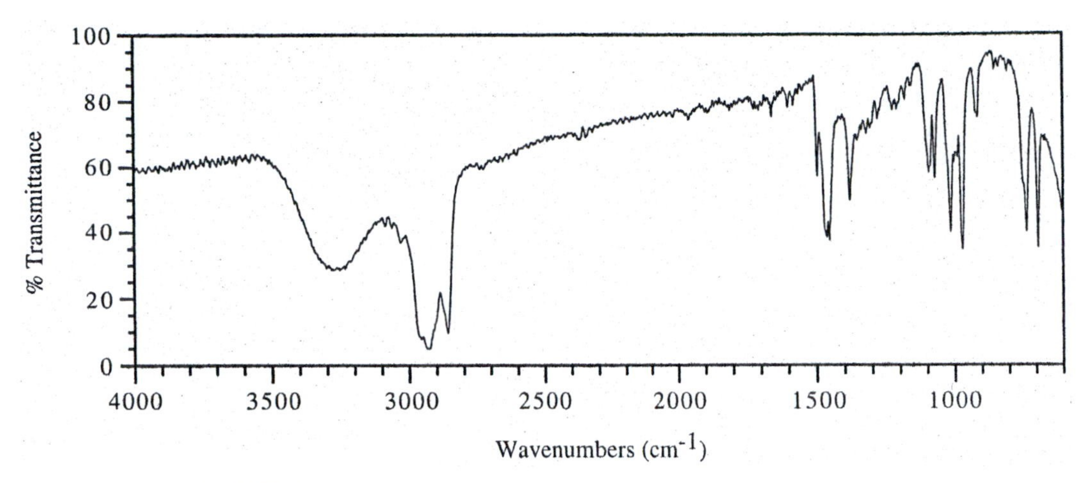

FIGURE 14.1
IR spectrum of cinnamyl alcohol (thin film).

The course of the reaction will be monitored by thin-layer chromatography, and purification of the cinnamaldehyde will be done by column chromatography.

Infrared (IR) spectroscopy is an especially good method to assess the purity of your product. Figure 14.1 shows the IR spectrum of cinnamyl alcohol. The strong, broad O—H stretching band in the 3550–3250 cm^{-1} region is diagnostic of all alcohols. If the oxidation has gone to completion, this peak will not be present in the spectrum of your product. The IR spectrum of cinnamaldehyde will show an intense C=O stretching band. The rest of the IR spectrum will yield a fingerprint that should allow straightforward identification of the aldehyde.

MICROSCALE PROCEDURE

Techniques Thin-Layer Chromatography: Technique 15
Column Chromatography: Technique 17
IR Spectroscopy: Technique 18

SAFETY INFORMATION

Cinnamyl alcohol is an irritant. Avoid contact with skin, eyes, and clothing.

Pyridinium chlorochromate is a suspected carcinogen. Wear gloves while handling it and avoid breathing the dust.

Dichloromethane is toxic, an irritant, absorbed through the skin, and harmful if inhaled. Use it only in a hood, wear neoprene gloves, and wash your hands thoroughly after handling it.

Diethyl ether and **pentane** are extremely volatile and flammable.

Silica gel is a lung irritant. Avoid breathing the dust.

Following the Reaction by Thin-Layer Chromatography

Alternatively, plastic-backed TLC plates can be used with iodine visualization [see Technique 15.6].

A standard solution of 2% cinnamyl alcohol in dichloromethane will be available as a reference for monitoring the reaction by thin-layer chromatography [see Technique 15]. Use silica gel on glass or aluminum TLC plates and 30:70 (v/v) ethyl acetate/pentane (or petroleum ether) as the developing solvent. Visualize the chromatograms by dipping the TLC plate in *p*-anisaldehyde reagent solution and then heating the TLC plate on a hot plate (in a hood) adjusted to a medium-heat setting until the color appears [see Technique 15.6].

Oxidation and Chromatography

Place 1.8 g of the pyridinium chlorochromate/sodium acetate/4-Å molecular sieve mixture (1:1:1 by wt) and 10 mL of dichloromethane in a dry 50-mL Erlenmeyer flask. Suspend the oxidant mixture with magnetic stirring and add 250 mg of cinnamyl alcohol. Continue stirring the reaction mixture, and monitor the course of the reaction by TLC at approximately 20-min intervals. When TLC indicates that the reaction is complete, add 2.0 g of Florisil (a filter aid), stir vigorously, add 10 mL of ether, and stir 5 min longer. Vacuum filter the mixture, using a few milliliters of ether to rinse the Erlenmeyer flask; pour the rinse over the solid in the Buchner funnel. Transfer the filtrate to a 50-mL round-bottomed flask, add 1.0 g of silica gel (70–230 mesh column chromatography grade), and evaporate the solvent in a hood, using a gentle stream of nitrogen or air. Warm the flask in a beaker of water at 40–45°C during the evaporation. Alternatively, the solvent can be removed on a rotary evaporator [see Technique 8.9].

Silica gel may spatter out of the flask during the evaporation if the flow of nitrogen or air is too vigorous.

Use a dry-packed chromatographic column. Push a small amount of glass wool to the bottom of a 19 × 200 (or 300) mm chromatography column using a long stirring rod. Clamp the column in a vertical position and add enough sand to cover the glass wool. Obtain 7.0 g of silica gel (70–230 mesh for column chromatography) and pour it slowly into the column, tapping the column gently to settle the particles. The top of the silica gel must be level. Add the silica gel/product mixture to the top of the column and tap the column to again level the silica gel. Cover the silica gel with a thin layer of sand.

It is critical that the silica gel surface not be disturbed while solvent is added to the column.

Open the stopcock at the bottom of the chromatography column and slowly pour 60 mL of 20:80 (v/v) ethyl acetate/pentane (or petroleum ether) solution (the eluting solvent) onto the column, using a small funnel to direct the flow against the inner wall of the column so that the silica gel surface is not disturbed. Collect 10-mL fractions of eluent. Check each fraction by TLC. Combine the fractions containing pure cinnamaldehyde in a tared round-bottomed flask and evaporate most of the solvent on a steam bath in a hood, using a boiling stick or stone to prevent bumping. Then, using a stream of nitrogen or air, and working in a hood, continue the evaporation of solvent until a constant weight (within 0.03 g) is reached. Alternatively, the solvent can be removed on a rotary evaporator [see Technique 8.9]. Weigh the product and determine the percent yield.

IR Analysis of the Product

Obtain an IR spectrum of your product as directed by your instructor [see Technique 18.4]. Was your column chromatography successful in separating any unreacted cinnamyl alcohol from the product? What band would be present if you have unreacted cinnamyl alcohol? What bands confirm the formation of cinnamaldehyde?

CLEANUP: Place the mixture of Florisil and pyridinium chlorochromate in the container for chromium (or hazardous inorganic) waste. Pour the developing solvent from the TLC chamber and any fractions from the column chromatography that did not contain cinnamaldehyde into the container for flammable (organic) waste.

References

1. Corey, E. J.; Suggs, J. W. *Tetrahedron Lett.* **1975,** 2647–2650.
2. Taber, D. F.; Wang, Y.; Liehr, S. *J. Chem. Educ.* **1996,** *73,* 1042–1043.
3. Herscovici, J.; Antonakis, K. *J. Chem. Soc. Chem. Comm.* **1980,** 561.

Questions

1. Compare this oxidation procedure with that using aqueous Cr(VI) oxidation reagents. What are the advantages of the reagent used in this experiment?
2. Why does cinnamaldehyde have a higher R_f than cinnamyl alcohol in TLC?
3. Analyze the IR spectrum of cinnamyl alcohol shown in Figure 14.1 by identifying the vibrations that cause the strong bands.
4. Predict the product that will be obtained if *cis*-4-*tert*-butylcyclohexanol is oxidized (a) with $CrO_3/H_2SO_4/H_2O$; (b) with pyridinium chlorochromate.

EXPERIMENT

15 GRIGNARD SYNTHESES

PURPOSE: To investigate how organometallic Grignard reagents can be used to make carbon-carbon bonds.

In Experiment 15.1 you will synthesize a Grignard reagent, phenylmagnesium bromide.

In Experiment 15.2, you will react the Grignard reagent that you made in Experiment 15.1 with a ketone, acetophenone, to produce a tertiary alcohol, 1,1-diphenylethanol.

In Experiment 15.3 you will synthesize benzoic acid from carbon dioxide and the phenylmagnesium bromide that you made in Experiment 15.1.

Organometallic compounds are versatile intermediates in the synthesis of many organic compounds, and their reactions form the basis of some of the most useful methods in synthetic organic chemistry. The use of organometallic reagents permits the synthesis of highly specific carbon-carbon bonds in excellent yields.

Among the most important organometallic reagents are the alkyl- and arylmagnesium halides, which are almost universally called ***Grignard reagents*** after the French chemist Victor Grignard, who first realized their tremendous potential in organic synthesis. Grignard received the 1912 Nobel prize in chemistry.

Formation of Grignard Reagents

Grignard reagents are substances containing carbon-metal bonds. Their synthesis requires the reaction of an alkyl or aryl halide with magnesium metal in the presence of an ether solvent:

$$\mathrm{R{-}X + Mg \xrightarrow{\text{ether}} \overset{\delta-}{R}{-}\overset{\delta+}{Mg}{-}\overset{\delta-}{X}}$$

Formation of a Grignard reagent takes place in a heterogeneous reaction at the surface of the solid magnesium metal, and the surface area and reactivity of the magnesium are crucial factors in the rate of the reaction. It is thought that the alkyl or aryl halide reacts with the surface of the metal to produce a carbon free radical and a magnesium-halogen radical, which combine to form the Grignard reagent. This reaction gives the magnesium atom two covalent bonds, but it must somehow acquire two more since magnesium has a coordination number of 4. In an ether solvent, ether molecules occupy the other two coordination sites.

The oxygen atom of the ether molecule provides both electrons in the magnesium-ether bond:

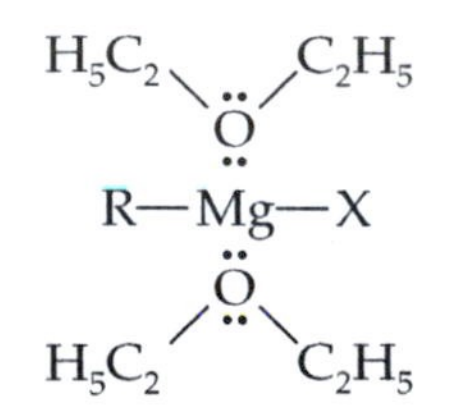

These complexes are quite soluble in ether. In the absence of the ether solvent, the reaction of magnesium and an alkyl or aryl halide takes place rapidly but soon stops because the surface of the metal becomes coated with polymeric Grignard reagent. In the presence of ether, the surface of the metal is kept clean and the reaction proceeds until the limiting reagent is entirely consumed.

Treatment of an ether solution of bromobenzene with magnesium metal in Experiment 15.1 results in the formation of the Grignard reagent phenylmagnesium bromide. Insertion of the magnesium between carbon and bromine greatly enhances the reactivity of the aromatic structure and reverses the polarity of the carbon formerly bonded to bromine. In the Grignard reagent, the carbon-metal bonds have a high degree of ionic character, with a good deal of negative charge on the carbon atom, making it a strong nucleophile:

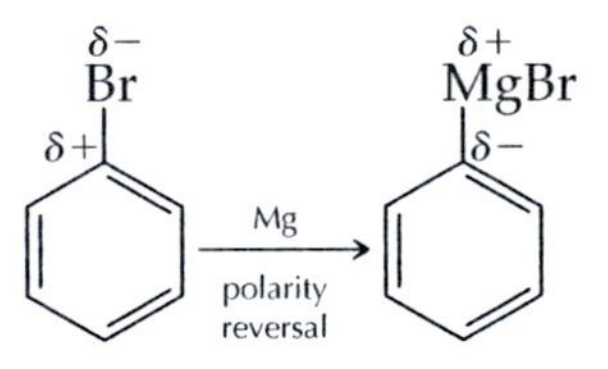

Reaction with Carbonyl Compounds

The mechanism of the Grignard reaction with carbonyl compounds, such as aldehydes, ketones, esters, and carbon dioxide, is actually quite complex, but it can easily be rationalized as a simple nucleophilic addition reaction:

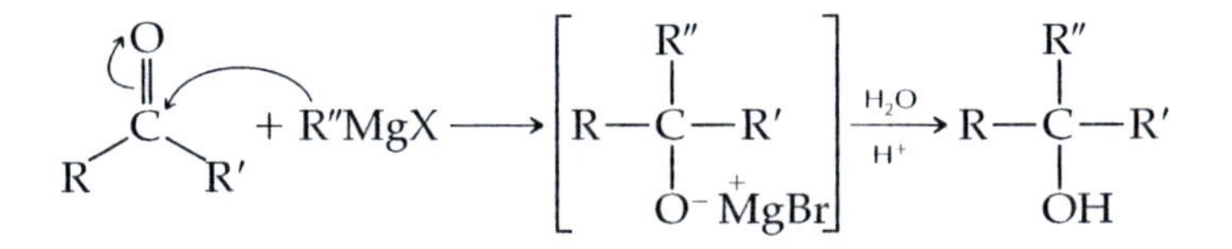

Usually a Grignard reagent is used immediately after synthesis. In Experiment 15.2 you will react the phenylmagnesium bromide that you have made with acetophenone to produce the tertiary alcohol

1,1-diphenylethanol. In Experiment 15.3 the Grignard reagent will undergo reaction with carbon dioxide (dry ice) to form benzoic acid:

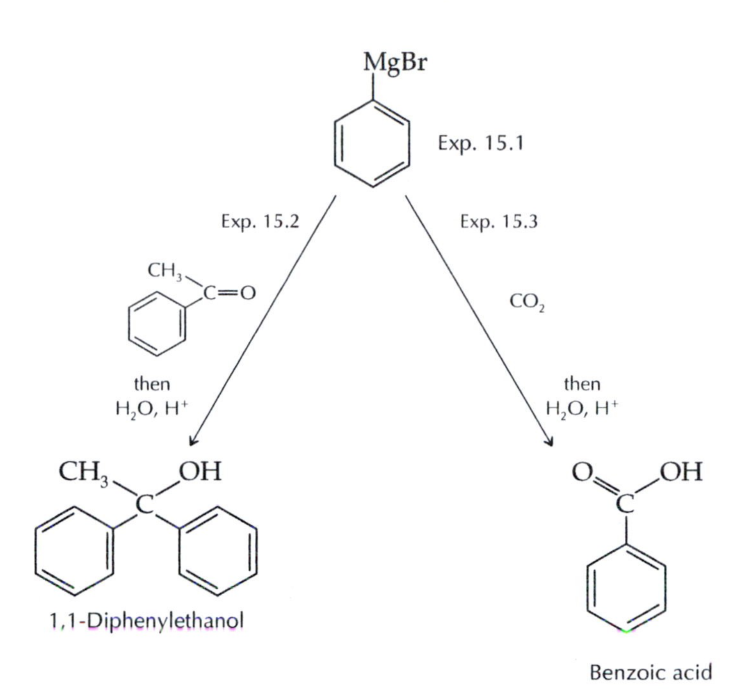

In both experiments, the reaction produces a magnesium salt of the organic product that you want. These magnesium salts are readily hydrolyzed to the organic product by reaction with an acidic water solution. The use of an aqueous mineral acid, such as sulfuric or hydrochloric acid, expedites hydrolysis. Not only does the acid cause the protonation to occur more readily, but Mg(II) is converted from insoluble salts to water-soluble sulfates or chlorides. For products, such as tertiary alcohols, that might dehydrate with strong acids, the weaker acid ammonium chloride is an excellent alternative.

Anhydrous Reaction Conditions

Because Grignard reagents have such tremendous reactivity as nucleophiles, it is not surprising that they are also strong bases. They react quickly and easily with acids, even relatively weak acids such as water and alcohols. Although this reaction can be useful for the reduction of a carbon-halogen bond, it is more often a nuisance because it destroys the Grignard reagent before the carbonyl compound is added and therefore reduces the yield of the desired product.

$$RX \xrightarrow{Mg} RMgX \xrightarrow{H_2O} RH + HOMgX$$

The presence of water or other acids also inhibits the formation of the Grignard reagent. **Thus, it is important that the reaction conditions be completely anhydrous.** All glassware and reagents must be thoroughly dry before beginning a Grignard experiment.

Side Reactions

Other reactions that may interfere with Grignard syntheses include Grignard coupling and reaction with molecular oxygen:

$$2\,RX + Mg \longrightarrow R{-}R + MgX_2$$

$$RMgX + O_2 \longrightarrow ROOMgX \xrightarrow{RMgX} 2\,ROMgX$$

The coupling reaction takes place at the surface of the solid magnesium when two free radicals react to form a stable hydrocarbon dimer. This side reaction is favored by a high concentration of the bromobenzene, where the concentration of free radicals is greater. Combination of the Grignard reagent with molecular oxygen is a side reaction that occurs in the presence of air. Low-boiling diethyl ether (bp 35°C) helps to exclude air near the surface of the reaction solution because of its great volatility. Alternatively, Grignard reagents can be used under nitrogen or argon to avoid reactions with molecular oxygen.

15.1 Synthesis of Phenylmagnesium Bromide for Use as a Synthetic Intermediate

PURPOSE: To synthesize a Grignard reagent from bromobenzene and magnesium metal in diethyl ether.

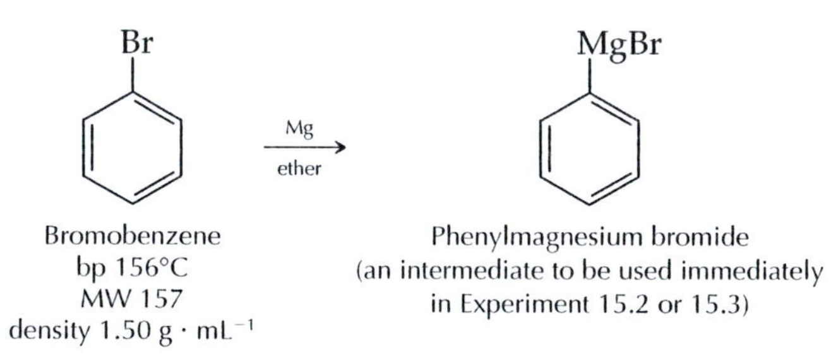

MINISCALE PROCEDURE

PRELABORATORY ASSIGNMENT: Anhydrous conditions are vital for a successful Grignard synthesis. To demonstrate this point, calculate the mass of water it would take to destroy the phenylmagnesium bromide that you will synthesize. What is the volume of this amount of water?

Technique Anhydrous Reaction Conditions: Technique 7.2

SAFETY INFORMATION

Diethyl ether, commonly called ether, is extremely flammable. Be certain that no flames are used in the laboratory and that hot electrical devices are not in the vicinity where ether is being used. Work in a hood, if possible.

Bromobenzene is an irritant. Wear gloves. Avoid contact with skin, eyes, and clothing.

Make sure that all your glassware and reagents are thoroughly dry before beginning the experiment. Refer to p. 123 concerning the anhydrous conditions required for a Grignard reaction.

Assemble the apparatus shown in Technique 7.2, Figure 7.3a or 7.3b, using a 100-mL round-bottomed flask, a water-jacketed condenser, and a 125-mL separatory funnel. For a one-necked flask, use a Claisen adapter to make two openings; for a three-necked flask, close the third opening with a glass stopper. Prepare two drying tubes containing anhydrous calcium chloride. Fit the tubes to the top of the condenser and the separatory funnel with thermometer adapters or one-hole rubber stoppers.

Reaction is indicated by a cloudy appearance in the solution near the fresh surface of magnesium.

Obtain 0.300 g of magnesium turnings and 2.5 mL of anhydrous ether. Open the round-bottomed flask and add the magnesium and the ether to the flask, then add the iodine crystal. With a dry stirring rod, gently break one or two magnesium pieces to expose a fresh surface. Close the flask.

SAFETY PRECAUTION

> Hold the round-bottomed flask firmly by the neck, and be careful not to press so hard on a magnesium turning that you crack the flask.

Prepare a solution of bromobenzene in ether by dissolving 1.20 mL of bromobenzene in 3.0 mL of anhydrous ether. Pour the solution into the separatory (dropping) funnel with the stopcock closed. Add approximately half of this solution to the reaction flask and swirl the flask to mix the reagents thoroughly.

The precipitate is probably magnesium hydroxide, which results from any moisture in the flask or the reagents.

Place the flask in a warm-water bath (45–50°C) and swirl the flask from time to time. The ether will begin to boil. A small amount of cloudy precipitate may appear. Formation of the organometallic reagent is exothermic. Periodically remove the water bath and note whether the reaction has started. This process is indicated by refluxing of the ether, which results from the heat produced by the reaction.

When the reaction is well under way, remove the warm-water bath, and add 5.0 mL of anhydrous ether through the condenser to dilute the solution. Then add the rest of the bromobenzene/ether solution from the dropping funnel at a speed that maintains a moderate reflux rate. Have a cold-water bath available to briefly cool the reaction in case the condensation ring in the condenser is more than halfway up the condenser. Do not cool the reaction so much that it stops. When all the bromobenzene has been added, pour 3.0 mL of ether into the dropping funnel and add the ether to the reaction mixture. When the reaction stops refluxing spontaneously, heat the reaction in a warm-water bath (45–50°C) for 20 min. Cool the reaction flask in a cold-water bath before proceeding immediately to Experiment 15.2 for the miniscale preparation of 1,1-diphenylethanol.

MICROSCALE PROCEDURE

PRELABORATORY ASSIGNMENT: Anhydrous conditions are vital for a successful Grignard synthesis. To demonstrate this point, calculate the mass of water it would take to destroy the phenylmagnesium bromide that you will synthesize. What is the volume of this amount of water?

Technique Anhydrous Reaction Conditions: Techniques 7.2 and 7.3

SAFETY INFORMATION

Diethyl ether, commonly called ether, is extremely flammable. Be certain that no flames are used in the laboratory and that hot electrical devices are not in the vicinity where ether is being used. Work in a hood, if possible.

Bromobenzene is an irritant. Wear gloves. Avoid contact with skin, eyes, and clothing.

Prepare a microscale drying tube using two small pieces of cotton and anhydrous calcium chloride, as shown in Technique 7.3, Figure 7.4a. Place the drying tube, a 10-mL round-bottomed flask, a water-jacketed condenser, a Claisen adapter, and two small screw-capped vials (without the caps) in a 250-mL beaker and dry the glassware in a 110°C oven for at least 30 min. You also need a rubber fold-over septum or a screw cap and septum for the Claisen adapter and a 1-mL syringe; be sure that these items are dry.

While the glassware is cooling, sand or scrape the oxide coating from a 9- to 10-cm piece of magnesium ribbon and cut the cleaned ribbon into 1-mm pieces using scissors. Weigh 75 mg of magnesium, using tweezers or forceps to pick up the metal pieces. Place the magnesium pieces in the dried round-bottomed flask; add a magnetic stirring bar to the flask. Assemble the apparatus shown in Technique 7.3, Figure 7.4a, as quickly as possible.

Ether tends to squirt out of a syringe when it is printed downward. Perform the transfer from the vial to the septum as quickly as possible.

Draw 0.40 mL of anhydrous diethyl ether into the 1-mL syringe. Inject the ether through the septum on the Claisen adapter.

Prepare a solution consisting of 0.32 mL of bromobenzene and 0.65 mL of anhydrous ether in one of the oven-dried vials. Draw this solution into the same syringe used previously and insert the needle through the septum. Add another 0.20 mL of ether to the vial as a rinse, cap the vial, and set it aside.

The precipitate is probably magnesium hydroxide, which results from any moisture in the flask or the reagents.

Begin stirring the reaction mixture and inject about 0.1 mL of the bromobenzene/ether solution into the reaction flask. It may be necessary to warm the flask in a beaker of warm water (45–50°C) for a few minutes to initiate the reaction. A cloudiness in the reaction solution indicates that the reaction has begun; remove the warm-water bath. The heat of reaction will cause the ether to reflux. Once the reaction has begun, add the remaining bromobenzene solution dropwise from the syringe at a rate of 2–3 drops per min. After the addition is complete, draw the ether rinse solution remaining in the vial into the syringe and inject it into the reaction mixture in one portion. Heat the reaction flask in a warm-water bath (45–50°C) for 15 min, then cool the mixture to room temperature.

Proceed immediately to Experiment 15.2 for the microscale preparation of 1,1-diphenylethanol or to Experiment 15.3 for the microscale preparation of benzoic acid.

Questions

1. Benzene is often produced as a by-product during the synthesis of phenylmagnesium bromide. How can its formation be explained? Write a balanced chemical equation for the formation of benzene.
2. If, by mistake, a chemist used 100% ethanol rather than diethyl ether as the reaction solvent, would the Grignard synthesis still proceed as expected?
3. A small crystal of I_2 is often used to clean the surface of the magnesium metal in the syntheses of Grignard reagents. Write a balanced chemical equation for the reaction of magnesium and molecular iodine.
4. What might be the reason why the bromobenzene substrate is added to the reaction flask in two portions rather than all at once?

15.2 Synthesis of 1,1-Diphenylethanol

PURPOSE: Use a Grignard reagent to prepare a tertiary alcohol and then purify and characterize the product.

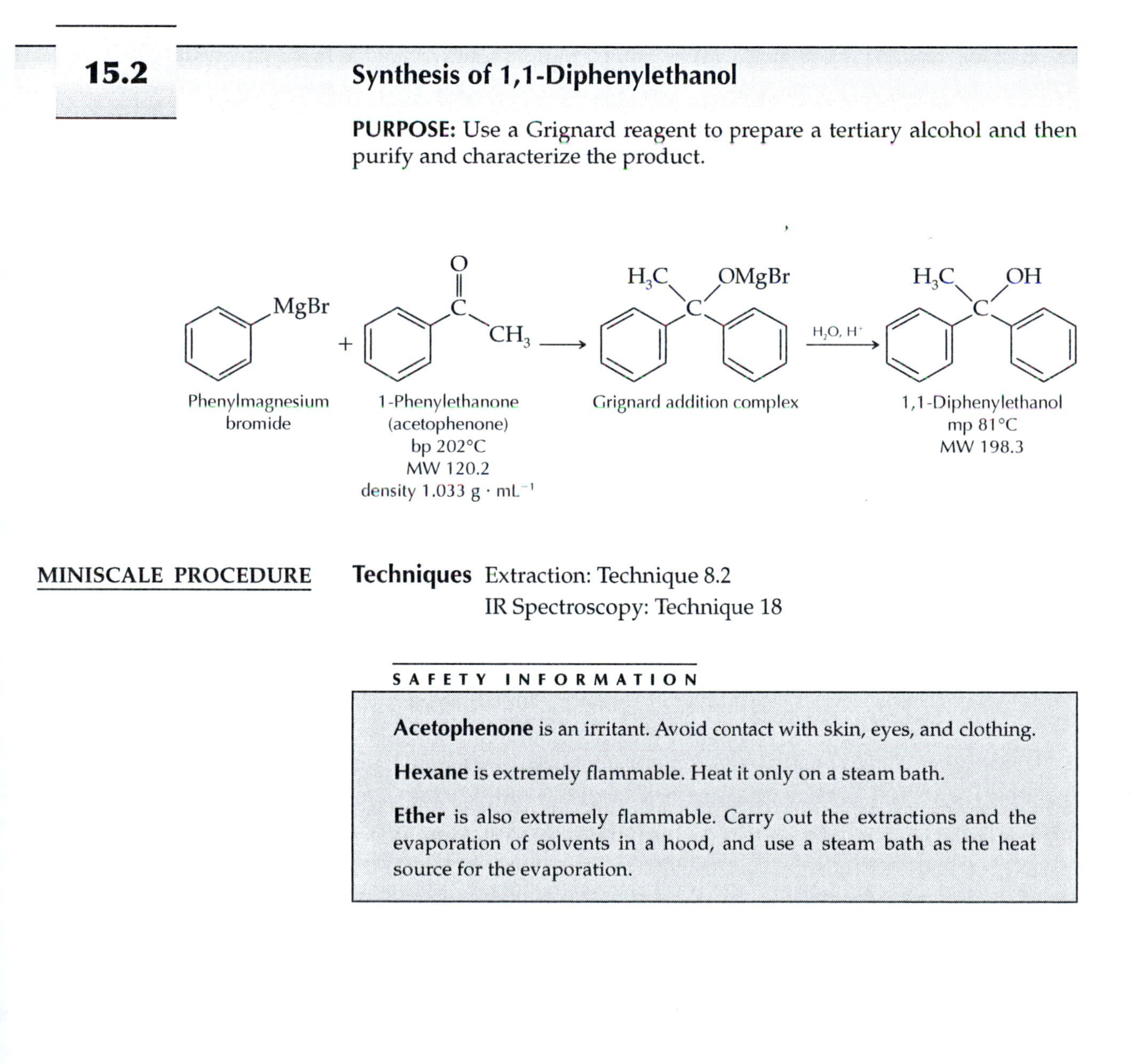

MINISCALE PROCEDURE

Techniques Extraction: Technique 8.2
IR Spectroscopy: Technique 18

SAFETY INFORMATION

Acetophenone is an irritant. Avoid contact with skin, eyes, and clothing.

Hexane is extremely flammable. Heat it only on a steam bath.

Ether is also extremely flammable. Carry out the extractions and the evaporation of solvents in a hood, and use a steam bath as the heat source for the evaporation.

Prepare a solution of 1.2 mL of acetophenone and 6.0 mL of ether; pour the solution into the dropping funnel. With the magnetic stirrer running, add the acetophenone solution dropwise to the Grignard reagent at such a rate that the reaction mixture refluxes gently.

When spontaneous reflux ceases, place a 45–50°C water bath under the reaction flask and heat the reaction mixture at a gentle reflux for 15 min. Place a cold-water bath under the reaction mixture to cool it. Pour 12 mL of 10% (wt) ammonium chloride solution into the dropping funnel; add the ammonium chloride solution dropwise to the stirred solution. After the addition of ammonium chloride solution is complete, determine the pH of the lower aqueous phase by removing a drop of the solution and placing it on pH test paper. If the pH is above 5 and there are still undissolved magnesium salts in the lower aqueous layer, continue stirring the mixture and add 6 M hydrochloric acid dropwise until the pH is below 4. Continue to stir the mixture until the white solids in the aqueous phase have dissolved.

A few magnesium fragments are likely to be present in the reaction mixture that will undergo reaction with the acid to produce hydrogen gas. Place a funnel lined with filter paper in a separatory funnel and carefully pour the reaction mixture into the filter paper. Rinse the round-bottomed flask with a few milliliters of ether and add the rinse to the separatory funnel. Remove the lower aqueous phase from the separatory funnel, then pour the upper organic phase into a labeled flask. Return the aqueous phase to the separatory funnel and extract it with 15 mL of ether [see Technique 8.2]. Drain the aqueous phase from the separatory funnel and combine the original ether solution with the ether solution remaining in the funnel. Wash the combined ether solution with 10 mL of saturated sodium bicarbonate solution. **(Caution: Foaming.)** Then wash the ether solution with 10 mL of saturated sodium chloride solution. Transfer the organic phase to a dry 50-mL Erlenmeyer flask and dry the solution with anhydrous magnesium sulfate for at least 10 min.

Filter the dried product solution into a dry 50-mL Erlenmeyer flask and add 10 mL of hexane. Place a boiling stick in the flask and evaporate the solution on a steam bath in a hood until the volume is approximately 10 mL. Alternatively, the solvent volume can be reduced on a rotary evaporator. Cool the solution and stir it with a glass rod if crystallization does not occur spontaneously. Collect the solid product on a Buchner funnel by vacuum filtration; use small portions of hexane to assist the transfer and to wash the crystals.

Analysis of the Product

Allow the product to dry before ascertaining its melting point, mass, and percent yield. Determine the IR spectrum [see Technique 18.4] or NMR spectrum [see Technique 19.2] of the dry product, as directed by your instructor. Does the spectrum show evidence of peaks for acetophenone or bromobenzene, or was the conversion to product complete?

CLEANUP: Combine the aqueous solutions remaining from the extractions and determine the pH. Add solid sodium carbonate or 6 M hydrochloric acid to adjust the pH to 6–8 before washing the solution down the sink or placing it in the container for aqueous inorganic waste. Place the filtrate from the crystallization in the container for flammable waste.

MICROSCALE PROCEDURE

Techniques Microscale Extraction: Technique 8.6c
IR Spectroscopy: Technique 18

SAFETY INFORMATION

Acetophenone is an irritant. Avoid contact with skin, eyes, and clothing.

Hexane is extremely flammable. Heat it only on a steam bath.

Ether is also extremely flammable. Carry out the extractions and the evaporation of solvents in a hood, and use a steam bath as the heat source for the evaporation.

Prepare a solution of 0.23 mL of acetophenone and 0.60 mL of anhydrous diethyl ether in the second oven-dried vial; cap the vial. Draw the solution into the same syringe you used previously. With the magnetic stirrer running, add the acetophenone solution dropwise to the Grignard reagent at such a rate that the reaction mixture refluxes gently. Add another 0.20 mL of ether to the vial as a rinse; draw the rinse into the syringe and add it slowly to the reaction mixture.

When spontaneous reflux ceases, heat the reaction mixture at a gentle reflux for 15 min in a 45–50°C water bath. Place a cold-water bath under the reaction mixture to cool it.

Obtain 3.0 mL of 10% (wt) ammonium chloride solution. Remove the syringe, screw cap, and septum from the Claisen adapter. Add the ammonium chloride solution dropwise to the stirred solution. After the addition of ammonium chloride solution is complete, determine the pH of the lower aqueous phase by removing a drop of the solution and placing it on pH test paper. If the pH is above 5 and there are still undissolved magnesium salts in the lower aqueous layer, continue stirring the mixture and add 6 M hydrochloric acid dropwise until the pH is below 4. Continue to stir the mixture until the white solids in the aqueous phase and all magnesium fragments have dissolved.

Using a small conical funnel, carefully pour the contents of the reaction flask into a 15-mL centrifuge tube. Rinse the flask with about 2 mL of ether and add the rinse to the centrifuge tube. Separate the lower aqueous phase with a Pasteur pipet fitted with a rubber bulb or a syringe [see Technique 8.5] and transfer it to a second centrifuge tube [see Technique 8.6c]. Save the ether solution remaining in the first centrifuge tube. Add 3.0 mL of ether to the aqueous solution in the second tube. Mix the two phases by drawing the mixture into the Pasteur pipet and expelling it back into the tube 4–5 times. Remove the aqueous phase with a Pasteur pipet and combine the original ether solution in the first tube with the ether solution remaining in the second tube. Wash the combined ether solution with 2 mL of saturated sodium bicarbonate solution. **(Caution: Foaming.)** Remove the aqueous phase, and then wash the ether solution with 2 mL of saturated sodium chloride solution. Remove the NaCl solution and dry the remaining ether solution with anhydrous magnesium sulfate for at least 10 min.

Filter the dried product solution into a dry 25-mL Erlenmeyer flask through a Pasteur pipet with cotton packed at the top of the tip, as shown in Technique 8, Figure 18.17b. Rinse the drying agent with about 1 mL of ether and filter the rinse into the Erlenmeyer flask. Add 2.5 mL of hexane. Place a boiling stick in the flask and evaporate the solution on a steam bath in a hood until the volume is approximately 2.5 mL. Alternatively, the solvent volume can be reduced on a rotary evaporator. Cool the solution and stir it with a glass rod if crystallization does not occur spontaneously. Collect the solid product on a Hirsch funnel by vacuum filtration; use a few drops of hexane to complete the transfer and to wash the crystals. Analyze your product as directed in the miniscale procedure.

CLEANUP: Combine the aqueous solutions remaining from the extractions and determine the pH. Add solid sodium carbonate or 6 M hydrochloric acid to adjust the pH to 6–8 before washing the solution down the sink or placing it in the container for aqueous inorganic waste. Place the filtrate from the crystallization in the container for flammable waste.

Questions

1. What is the structure of the white precipitate that forms when acetophenone is added to a solution of phenylmagnesium bromide?
2. An aqueous ammonium chloride solution, rather than sulfuric acid, is used in the hydrolysis of the magnesium complex of 1,1-diphenylethanol to ensure that no acid-catalyzed dehydration occurs. Write the chemical structure of the dehydration product from 1,1-diphenylethanol.
3. If acetone, rather than acetophenone, were reacted with phenylmagnesium bromide, followed by hydrolysis of the intermediate magnesium complex, what would the organic product be?

15.3 Synthesis of Benzoic Acid

PURPOSE: Investigate the addition of phenylmagnesium bromide to carbon dioxide to produce benzoic acid.

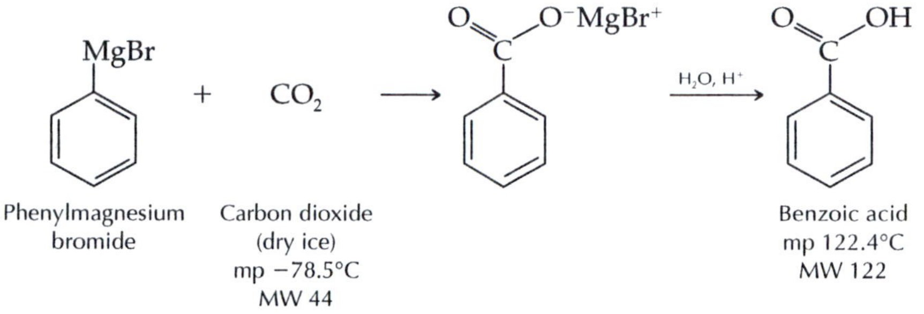

MICROSCALE PROCEDURE

Techniques: Microscale Extraction: Technique 8.6c
Microscale Recrystallization: Technique 9.7a

SAFETY INFORMATION

Handle **solid CO_2** (dry ice) with a towel or wear heavy cloth gloves. Contact with the skin can cause frostbite.

Diethyl ether is extremely flammable.

Aqueous **hydrochloric acid** solutions are a skin irritant. Wash your hands thoroughly if any acid spills on your hands. Aqueous **sodium hydroxide** solutions are corrosive and cause burns. Solutions as dilute as 2.5 M can cause severe eye injury.

The **crystallization solution** is acidic. Avoid contact with skin, eyes and clothing.

Any moisture on the dry ice will lower the yield of benzoic acid.

Obtain 2.0 mL of anhydrous ether in a test tube fitted with a cork. Obtain a small piece of dry ice (solid CO_2), wipe the condensed frost off the surface with a dry towel, and place the piece in a dry 30-mL beaker. Open the reaction flask containing the phenylmagnesium bromide prepared in Experiment 15.1, quickly draw the solution into a dry Pasteur pipet, and drip the solution over the dry ice. Using the same Pasteur pipet, rinse the reaction flask with the ether in the test tube and transfer the rinse to the beaker containing the dry ice. Stir the contents of the beaker occasionally until the excess dry ice sublimes.

Add 2.0 mL of 3 M hydrochloric acid and stir the reaction mixture to hydrolyze the magnesium salt. Transfer the mixture to a centrifuge tube, using a Pasteur pipet. Rinse the beaker with 0.5 mL of ether and add this rinse to the centrifuge tube. Cap the tube and shake it thoroughly. This procedure should produce two clear layers. It may be necessary to add a few more drops of ether or hydrochloric acid. Transfer the lower aqueous layer with a Pasteur pipet fitted with a rubber bulb or a syringe to a small flask [see Technique 8.6c]. Wash the ether phase with 1.0 mL of water; transfer the water layer to the container that already holds the previous aqueous phase, and set the container aside, leaving the ether layer in the centrifuge tube.

Extract the ether layer with 1.5 mL of 3 M sodium hydroxide solution by capping the tube and shaking the mixture thoroughly. Remove the lower aqueous phase to a 30-mL beaker. Repeat the extraction procedure with 1.0 mL of 3 M NaOH, then with 1.0 mL of water, each time transferring the lower aqueous phase to the 30-mL beaker containing the first sodium hydroxide extract. Set aside the ether layer remaining in the centrifuge tube; it contains some biphenyl, a reaction by-product.

Warm the beaker containing the sodium hydroxide extracts for several minutes on a steam bath in a hood to expel any dissolved ether. Cool the solution to room temperature and add 6 M hydrochloric acid dropwise until the solution is acidic according to pH paper (pH 2–3) and a thick white precipitate of benzoic acid forms. Chill the mixture in an ice-water bath before collecting the benzoic acid by vacuum filtration on a Hirsch funnel [see Technique 9.7a]. Wash the crystals twice with 1-mL portions of ice-cold water.

Analysis of the Product

Allow the benzoic acid to dry overnight or longer before determining the melting point and percent yield. Determine the IR spectrum [see Technique 18.4] or NMR spectrum [see Technique 19.2] of the dry product, as directed by your instructor. What peaks in the spectrum indicate formation of benzoic acid?

CLEANUP: Combine the aqueous extracts and the aqueous filtrate; neutralize the solution with sodium carbonate before washing it down the sink or pouring it into the container for aqueous inorganic waste. Pour the ether remaining in the centrifuge tube into the container for flammable (organic) waste.

Questions

1. The reactions of carbonyl compounds with Grignard reagents are highly exothermic, yet in the reaction of phenylmagnesium bromide with CO_2 the Grignard reagent can be added quickly to the dry ice without much concern for the heat evolved. Why?
2. Write a balanced equation for the reaction of benzoic acid with 3 M sodium hydroxide solution.
3. Some biphenyl ($C_6H_5—C_6H_5$) is produced as a by-product in the synthesis of benzoic acid. Where does it come from?
4. How did the extraction with sodium hydroxide purify your product from any unreacted bromobenzene and biphenyl?
5. Write a balanced equation for the reaction that precipitated benzoic acid when HCl was added to the solution from the sodium hydroxide extraction.
6. Write a chemical equation showing how benzoic acid would react with phenylmagnesium bromide.

EXPERIMENT

16 ADDITION REACTIONS OF ALKENES

QUESTIONS: What effect do the substrate and the reaction environment have on the regiochemistry of HBr addition to an alkene, and what is the stereochemistry of the addition of bromine to an alkene?

> In Experiment 16.1 you will use gas chromatography to compare the relative rates of Markovnikov and anti-Markovnikov addition of HBr to 1-hexene and 2-methyl-2-butene. Your data will determine the relative importance of ionic and free-radical pathways in the two reactions.
>
> In Experiment 16.2 you will use the melting point of the reaction product to discover whether the ionic addition of Br_2 to *trans*-cinnamic acid is a *syn* or an *anti* addition.

The major reaction of alkenes is ***1,2-addition,*** where one bond of the carbon-carbon double bond (C=C) is broken and an atom or group of atoms adds to each carbon. A wide range of reagents can add by different mechanisms to the C=C:

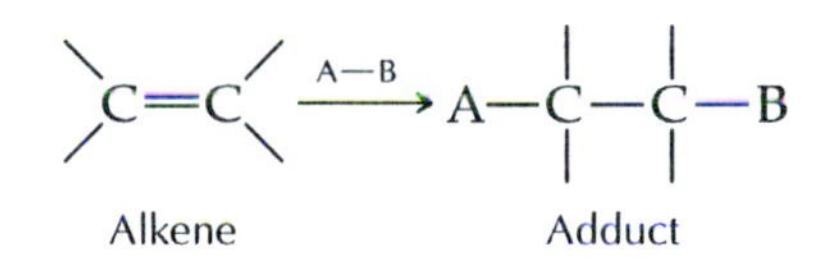

Electrophilic addition is the most common reaction of alkenes. It is an ionic reaction that involves attack of the electron-rich π-cloud of the double bond by an ***electrophile,*** an electron-loving agent. Electrophiles that can initiate these addition reactions include protons (H^+) and halonium ions (X^+) contributed by Br_2 and Cl_2.

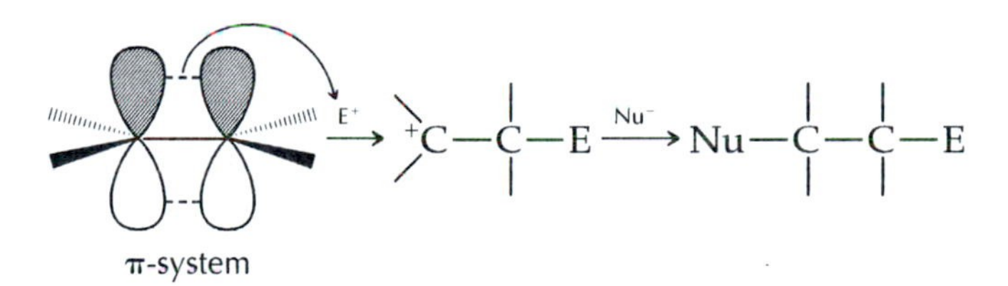

Many ionic additions to double bonds proceed through carbocation intermediates. In general, the more stable the carbocation intermediate, the faster it forms. For example, the fact that HCl adds to propene to produce only 2-chloropropane is readily explained by the formation of the more stable secondary carbocation. The alternative primary carbocation, $CH_3CH_2CH_2^+$, is far too unstable to be formed. Therefore, when the chloride ion traps the positively charged intermediate, only 2-chloropropane is formed. This process is called ***Markovnikov addition*** after the Russian chemist who first described

Vladimir V. Markovnikov (1838–1904) was a chemist at Odessa and later at Moscow University who carried out early studies of additions to alkenes.

the regiochemistry of ionic additions to alkenes. Markovnikov addition occurs when the hydrogen from HCl adds to the less substituted carbon atom of the C═C.

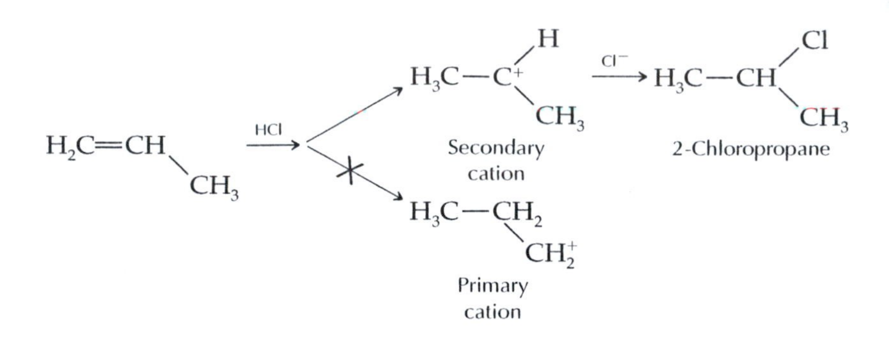

Understanding the addition of HBr to alkenes is more complex. It requires an appreciation and comparison of two types of mechanisms, ionic electrophilic addition and free-radical addition:

Ionic addition

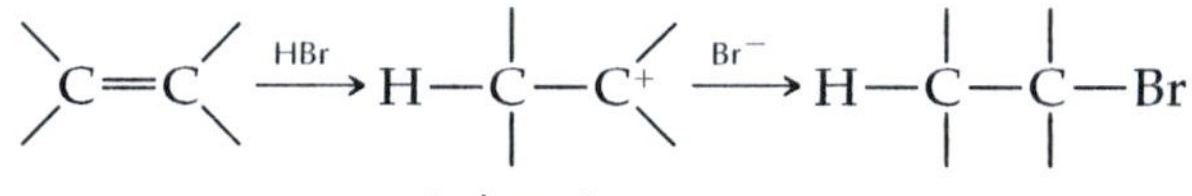

Radical addition

$$\mathrm{>C{=}C<} \xrightarrow[\mathrm{ROO\cdot}]{\mathrm{HBr}} \mathrm{Br{-}\overset{|}{\underset{|}{C}}{-}\overset{\cdot}{C}<} \xrightarrow{\mathrm{HBr}} \mathrm{Br{-}\overset{|}{\underset{|}{C}}{-}\overset{|}{\underset{|}{C}}{-}H + Br\cdot}$$

Carbon radical

Free-Radical Addition of HBr

In the addition of HBr to an alkene, a free-radical chain reaction can compete with the ionic reaction. The chain reaction is only efficient enough to compete with HBr; addition of HCl and HI are always ionic additions. Adding HBr by a free-radical process proceeds through the most stable radical intermediate. If propene is the alkene, the secondary free radical is more stable than the primary. Therefore, 1-bromopropane rather than 2-bromopropane is formed. In this process, called anti-Markovnikov addition, the hydrogen from HBr adds to the more substituted carbon atom of the C═C.

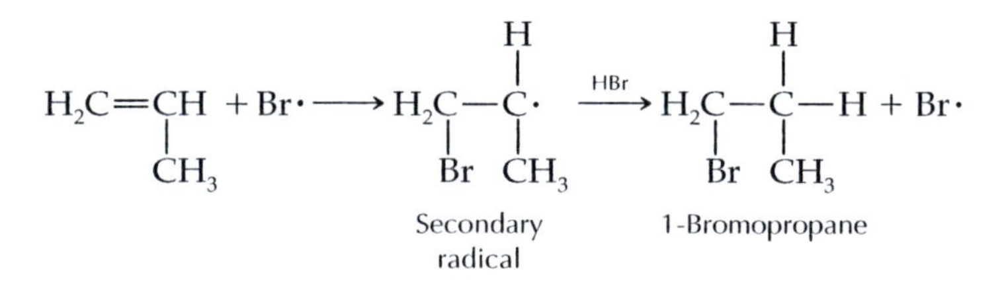

Before 1933, chemists did not understand why addition of HBr to alkenes occurred with inconsistent results. Sometimes Markovnikov addition was the major process, and sometimes anti-Markovnikov addition predominated. The pioneering work of Morris Kharasch and Frank Mayo led to the discovery that unless the reagents for this reaction are scrupulously purified, HBr addition could yield mixed results. For example, O_2 in the atmosphere has two unpaired electrons, and small amounts of dissolved oxygen in the reaction mixture, especially when light or heat are present, can form alkyl peroxy radicals from hydrocarbons:

$$O_2 + RH \longrightarrow R\cdot + H{-}O{-}O\cdot$$
$$O_2 + R\cdot \longrightarrow R{-}O{-}O\cdot$$

The alkyl peroxy radicals can initiate anti-Markovnikov addition by a free-radical pathway:

Initiation $$R{-}O{-}O\cdot + HBr \longrightarrow Br\cdot + ROOH$$

Propagation $$H_2C{=}CH(CH_3) + Br\cdot \longrightarrow H_2C(Br){-}\dot{C}H(CH_3)$$

$$H_2C(Br){-}\dot{C}H(CH_3) + HBr \longrightarrow H_2C(Br){-}CH(CH_3){-}H + Br\cdot$$

Termination $$R\cdot + R\cdot \longrightarrow R{-}R$$

Experiment 16.1 is the addition of HBr to 1-hexene and 2-methyl-2-butene in acetic acid. Gas chromatography is used to find the ratio of the addition products 2-bromohexane and 1-bromohexane, which in turn determines the ratio of ionic to radical addition of HBr to 1-hexene.

Experiment A $$CH_2{=}CH(CH_2)_3CH_3 + HBr \longrightarrow CH_3CH(Br)(CH_2)_3CH_3 \text{ and/or } BrCH_2CH_2(CH_2)_3CH_3$$

1-Hexene; 2-Bromohexane; 1-Bromohexane

In the same way, the product mixture of 2-bromo-2-methylbutane and 2-bromo-3-methylbutane from the addition of HBr to 2-methyl-2-butene serves as a measure of the relative importance of ionic and free-radical processes:

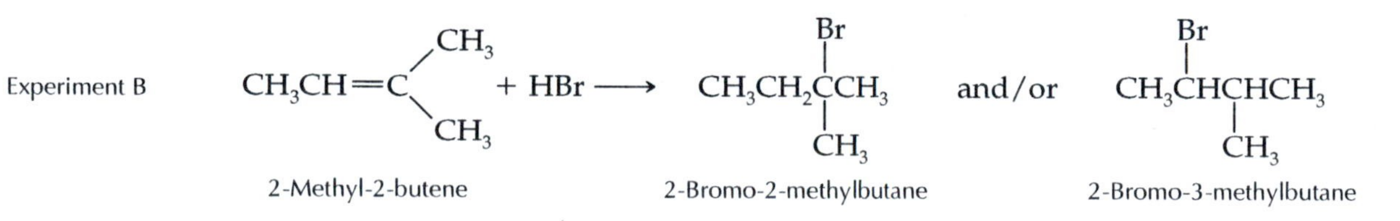

Your experimental data will allow you to discover if the structure of the alkene affects the relative importance of the ionic and free-radical addition pathways of HBr.

Addition of Br_2 to trans-Cinnamic Acid

In Experiment 16.2 you will study the addition of molecular bromine to the double bond of an alkene. Molecular bromine acts as a source of electrophilic Br^+, which adds to the π-system of the double bond to produce a bridged bromonium ion. This bridged ion is in turn captured by a bromide-ion nucleophile to produce the dibromide product:

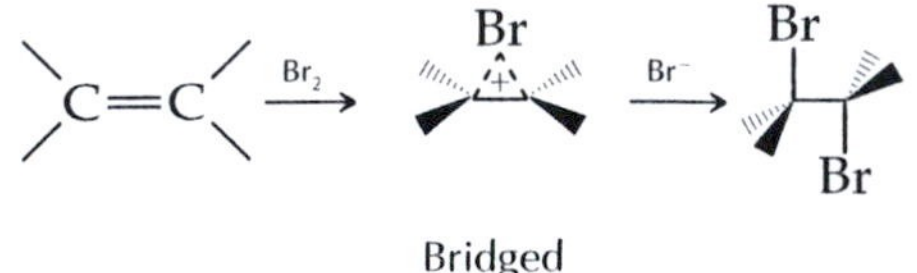

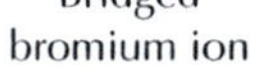

The addition of bromine to *trans*-cinnamic acid gives a dibromide with two stereocenters (chiral centers). Thus, there are four stereoisomers possible; these are shown here as their Fischer projections:

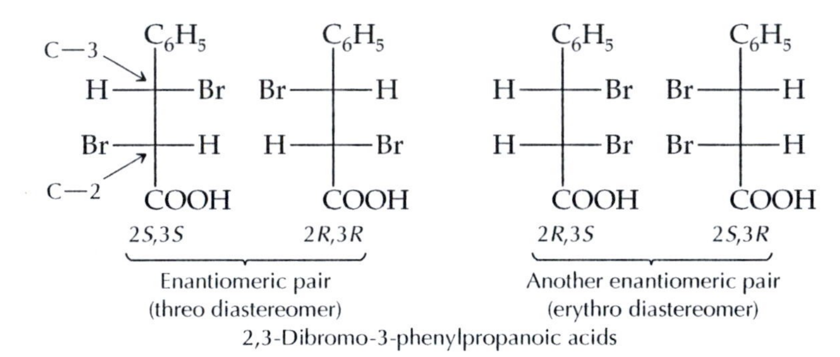

In this experiment, you will determine the melting point of the dibromide product and use this result to draw a conclusion as to which enantiomeric pair has formed. You can then use the configuration of the product to deduce the stereochemistry of the bromination reaction, that is, whether it is a *syn* or an *anti* addition.

The threo-erythro nomenclature is based on the structures of two simple sugars, threose and erythrose, which have two stereocenters.

The threo diastereomer, the 2*S*,3*S* and 2*R*,3*R* racemic mixture, has a melting point more than 100° lower than the erythro diastereomer, the 2*R*,3*S* and 2*S*,3*R* racemic mixture. A reasonably narrow melting range will tell you which diastereomer forms in the bromine addition to cinnamic acid. A mixture of both diastereomeric products would melt over a broad range that should not coincide with the melting point of either pure dibromide.

16.1 Free-Radical Versus Ionic Addition of Hydrobromic Acid to Alkenes

QUESTION: What effect do the substrate and the reaction environment have on the regiochemistry of HBr addition to 1-hexene and 2-methyl-2-butene?

PRELABORATORY ASSIGNMENT: What is the order of elution for the reaction products from a nonpolar GC column?

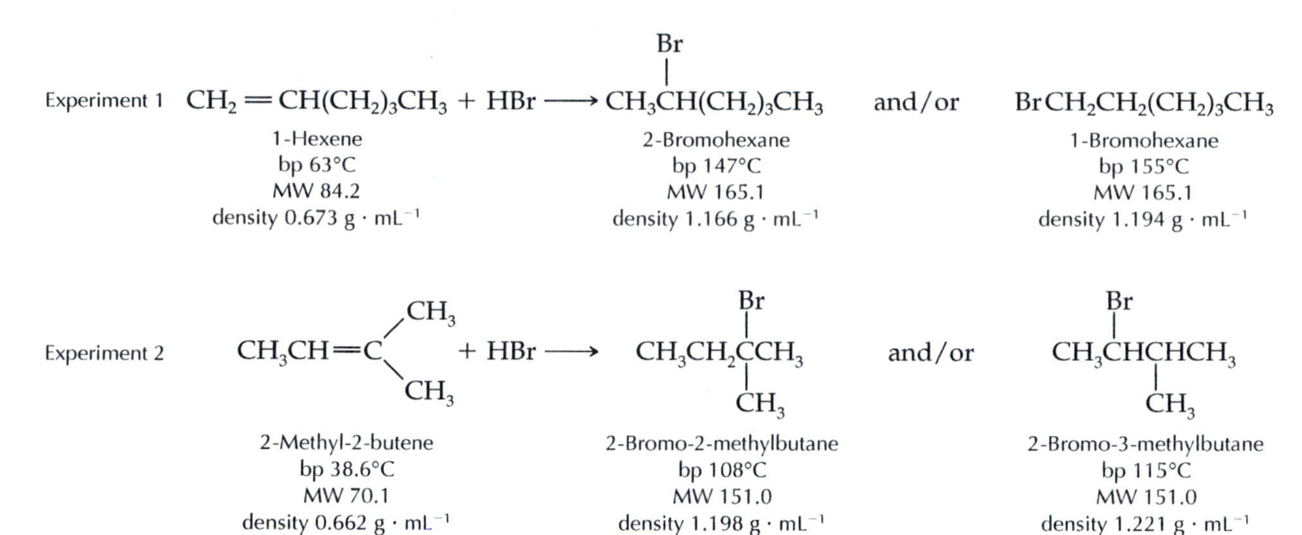

MICROSCALE PROCEDURE

Techniques Microscale Extractions: Technique 8.6c
Gas Chromatography: Technique 16

SAFETY INFORMATION

Wear gloves while conducting all these experiments.

1-Hexene and **2-methyl-2-butene** are skin irritants and very flammable. Avoid contact with skin, eyes, and clothing. Use them in a well-ventilated area.

30% hydrogen bromide in acetic acid is toxic and corrosive. Avoid breathing the vapors. Avoid contact with skin, eyes, and clothing. Use the reagent in a hood and wear gloves.

Diethyl ether is very volatile and flammable. Use it in a hood and keep it away from flames and electrical heating devices.

Experiment A

Measure 0.35 mL of 1-hexene into a 5-mL conical vial. Working in a hood, add 1.0 mL of 30% HBr in acetic acid (5 M) to the vial. Cap the vial with the Teflon (dull) side of the septum down and shake it frequently for 10 min. Occasionally loosen the cap to release any buildup of pressure.

Set the conical vial in a small beaker so that it does not tip over.

After the reaction period, allow the phases to separate and remove the lower acetic acid layer with a Pasteur pipet fitted with a rubber bulb or a syringe attached with a piece of Tygon tubing [see Technique 8.6c]. Place the acid layer in a 150-mL beaker containing about 50 mL of water. Add 2.0 mL of ether and 2.0 mL of water to the organic phase remaining in the conical vial. Cap the vial and shake it to mix the phases.

Remove the lower aqueous layer, adding it to the beaker containing the previously removed acid layer. Wash the ether layer with 2.0 mL of 5% sodium bicarbonate solution. **(Caution: Foaming.)** Remove the lower aqueous phase. Add anhydrous calcium chloride to the remaining ether solution, cap the vial, and allow the solution to dry for 10–15 min.

Experiment B

Repeat the procedure for Experiment A using a dry 5-mL conical vial and substituting 0.35 mL of 2-methyl-2-butene for 1-hexene.

Gas Chromatography

Prepare two Pasteur filter pipets and label two small test tubes, one for each reaction. Transfer each dried ether/product solution to its labeled test tube. Use a nonpolar column such as OV-1 or SE-30 with a column temperature of 60–70°C for the analyses of your products.

If you are analyzing the product mixture on a capillary column gas chromatograph, simply inject 1 μL of the ether solution into the chromatograph. If your laboratory is equipped with packed column chromatographs, you will probably need to evaporate some of the ether with a stream of nitrogen or air or in a hot-water bath or on a steam bath before injecting 1 μL into the chromatograph. Consult your instructor about specific sample preparation techniques and sample size for the chromatographs in your lab.

Identification of Products

Because you are using a nonpolar column on the gas chromatograph, the products usually elute in order of increasing boiling point. Having known samples of the products available will allow you to verify the identity of the products by using the peak enhancement method [see Technique 16.6]. After you have taken a gas chromatogram of the product of Experiment A, add 1 drop of 1-bromohexane to the ether (product) solution. Inject a 1-μL sample of this "enhanced" solution into the same chromatograph (at the same parameters) already used for Experiment A. The peak whose height has increased relative to the others corresponds to the known compound that you added. In the same way, add 1 drop of 2-bromo-2-methylbutane to the ether/product solution from Experiment B after you have analyzed it by GC. Obtain the chromatogram of the "enhanced" solution.

CLEANUP: Combine all the aqueous solutions from the extractions in the beaker containing the acetic acid/HBr extracts. Neutralize the acid by adding solid sodium carbonate in small portions. **(Caution: Foaming.)** Wash the neutralized solution down the sink or pour it into the container for aqueous inorganic waste. Pour the product solutions into the container for halogenated organic waste. Place the calcium chloride drying agent in the container for hazardous solid waste or in the container for inorganic waste.

Interpretation of Experimental Data

Analyze the chromatograms to determine the percentage of each product formed in the two experiments. Compare the results of Experiments A and B. Which reaction pathway predominated in each experiment?

Reference

1. Brown, T. M.; Dronsfield, A. T.; Ellis, R. *J. Chem. Educ.* **1990**, *67*, 518.

Questions

1. What function does the 5% sodium bicarbonate wash of the ether layer serve?
2. Hydrogen chloride adds ionically to alkenes. In each case, predict the course of the reaction of HCl with the following hydrocarbons by writing the structure of the product: (a) 1-butene, (b) 2-methylpropene, (c) 1-methyl-1-phenylethene.
3. Predict the course of the reaction of HBr with the substrates listed in Question 2 by providing the structure of the organic product. Assume that a free-radical process applies in all cases.

16.2 Stereochemistry of Bromine Addition to *trans*-Cinnamic Acid

QUESTION: Is the stereochemistry of the ionic addition of bromine to *trans*-cinnamic acid a *syn* or *anti* addition?

PRELABORATORY ASSIGNMENT: Construct molecular models of threo- and erythro-2,3-dibromo-3-phenylpropanoic acid and show which diastereomer would be formed by *anti* addition of bromine to *trans*-cinnamic acid.

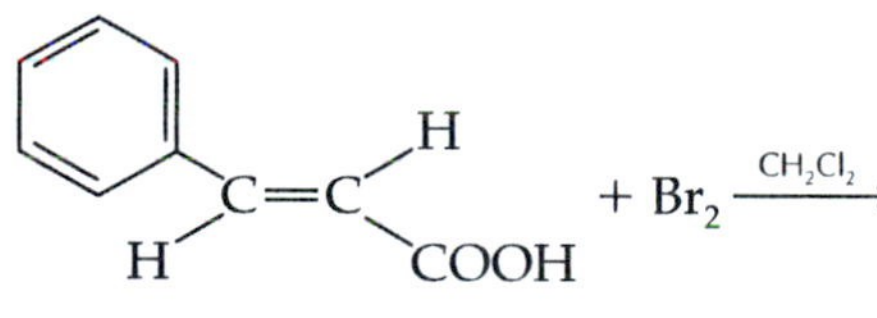

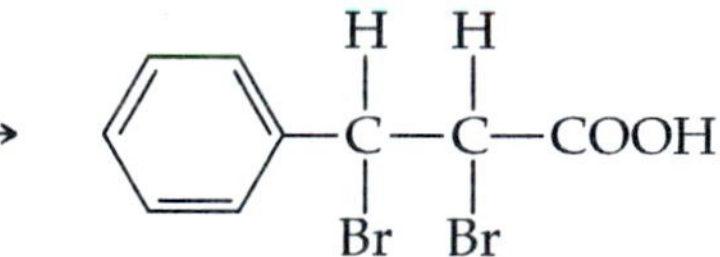

trans-Cinnamic acid
(***E***-3-phenyl-2-propanoic acid)
MW 148
mp 133°C

2,3-Dibromo-3-phenylpropanoic acid
MW 308
Which racemic mixture is formed?
enantiomers (2*R*,3*S*) and (2*S*,3*R*)
mp 202–204°C
enantiomers (2*R*,3*R*) and (2*S*,3*S*)
mp 93.5–95°C

MINISCALE PROCEDURE

Techniques Reflux: Technique 7.1
Mixed Solvent Recrystallization: Technique 9.2

SAFETY INFORMATION

Dichloromethane is toxic, an irritant, absorbed through the skin, and harmful if swallowed or inhaled. Wear neoprene gloves, use it in a hood, and wash your hands thoroughly after handling it.

Bromine is very corrosive and causes serious burns. Its vapors are toxic and irritating to the eyes, mucous membranes, and respiratory tract. A solution of bromine also emits bromine vapor and should be used only in a well-ventilated hood. Wear gloves impermeable to Br_2 while measuring and transferring the Br_2 solution.

***trans*-Cinnamic acid** is a mild irritant. Avoid skin contact and wash your hands after handling it.

Ethanol is flammable.

Place 0.60 g of *trans*-cinnamic acid in a 25-mL round-bottomed flask. Add 3.5 mL of dichloromethane and 2.0 mL of 10% bromine in dichloromethane solution. Put a boiling stone in the flask and attach the water-cooled condenser [see Technique 7.1]. Clamp the flask in a beaker of water at a temperature of 45–50°C. Reflux the reaction mixture gently for 30 min. If the bromine color disappears during the reflux period, add 10% bromine in dichloromethane solution dropwise through the top of the condenser until a light orange color persists.

The product will begin to precipitate as the reaction proceeds.

Cool the reaction flask to room temperature, then cool it further in an ice-water bath for 10 min to ensure complete crystallization of the product. Collect the crude product by vacuum filtration on a Buchner funnel. Wash the crystals three times with 2.0-mL portions of cold dichloromethane by disconnecting the vacuum, pouring the solvent over the crystals, and then restarting the vacuum.

Transfer the crystals to a 50-mL Erlenmeyer flask and add 2.0 mL of ethanol. Heat until boiling on a steam bath or hot plate. If the crystals do not all dissolve, continue adding ethanol in 0.5-mL increments, boiling briefly after each addition, until the crystals dissolve. Add a volume of water equal to the total amount of ethanol used and warm the mixture until any crystals that formed when the water was added dissolve. Cool the flask to room temperature and place it in an ice-water bath. Recover the crystals by vacuum filtration. Allow the crystals to dry overnight before determining the melting point and mass. Calculate the percent yield.

CLEANUP: Pour the filtrates from the reaction mixture and from the recrystallization into the container for halogenated waste.

Interpretation of Experimental Data

From the melting point of your product, decide whether the addition of bromine followed a *syn* or *anti* mechanism (or both). Propose a mechanism that explains your results.

MICROSCALE PROCEDURE

Techniques Microscale Reflux: Technique 7.1
Mixed Solvent Recrystallization: Technique 9.2
Microscale Filtration: Technique 9.7a

SAFETY INFORMATION

Dichloromethane is toxic, an irritant, absorbed through the skin, and harmful if swallowed or inhaled. Wear neoprene gloves, use it in a hood, and wash your hands thoroughly after handling it.

Bromine is very corrosive and causes serious burns. Its vapors are toxic and irritating to the eyes, mucous membranes, and respiratory tract. A solution of bromine also emits bromine vapor and should be used only in a well-ventilated hood. Wear gloves impermeable to Br_2 while measuring and transferring the Br_2 solution.

***trans*-Cinnamic acid** is a mild irritant. Avoid skin contact and wash your hands after handling it.

Ethanol is flammable.

Place 100 mg of *trans*-cinnamic acid in a 3-mL reaction vial. Add 0.7 mL of dichloromethane and 0.35 mL of 10% bromine in dichloromethane solution. Put a boiling stone in the vial and fit it with a water-cooled condenser [see Technique 7, Figure 7.1b]. Clamp the apparatus so that the lower portion of the vial is submerged in a beaker of water at a temperature of 45–50°C. Reflux the reaction mixture gently for 20 min. The product will begin to precipitate as the reaction proceeds. If the bromine color disappears during the reflux period, add 10% bromine in dichloromethane solution dropwise through the top of the condenser until a light orange color persists.

The product will begin to precipitate as the reaction proceeds.

Cool the reaction vial in an ice-water bath for 5 min to ensure complete crystallization of the product. Collect the crude product by vacuum filtration on a Hirsch funnel [see Technique 9.7a, Figure 9.7]. Wash the crystals three times with 0.5-mL portions of cold dichloromethane.

Transfer the crystals to a 10-mL Erlenmeyer flask and add 0.4 mL of ethanol. Heat until boiling on a steam bath or hot plate. If the crystals do not all dissolve, continue adding ethanol in 0.1-mL increments, boiling briefly after each addition, until the crystals do dissolve. Add a volume of water equal to the total amount of ethanol used and warm the mixture until any crystals that formed when the water was added dissolve. Cool the flask to room temperature and place it in an ice-water bath. Recover the crystals by vacuum filtration. Allow the crystals to dry overnight before determining the melting point and mass. Calculate the percent yield.

CLEANUP: Pour the filtrates from the reaction mixture and the recrystallization into the container for halogenated waste.

Interpretation of Experimental Data

From the melting point of your product, decide whether the addition of bromine followed a *syn* or *anti* mechanism (or both). Propose a mechanism that explains your results.

Questions

1. Why is it necessary to maintain excess bromine in the reaction mixture? How can you tell that an excess of bromine is present?
2. When ionic bromination of cyclopentene is carried out, the product is *trans*-1,2-dibromocyclopentane. Gas chromatographic analysis with a heated injection port can result in the elimination of Br_2 from the dibromo product to regenerate the starting alkene. Thus, the GC properties of both the starting material and the adduct are of interest. Predict the relative retention times of cyclopentene and *trans*-1,2-dibromocyclopentane on a nonpolar column. (**Hint:** Look up the physical properties of these two compounds in a handbook.)
3. Based on the stereochemistry of the addition of bromine to cinnamic acid, write the structure of the dibromide product(s) that you think will be formed as the result of the reaction of *Z*-2-pentene with bromine. Also draw the structure of the dibromide product(s) predicted when *E*-2-pentene reacts with bromine.

EXPERIMENT

17

SYNTHESIS OF POLYSTYRENE

PURPOSE: To compare the properties of polystyrene that you have synthesized.

In Experiment 17 you will synthesize polystyrene samples by free-radical and cationic polymerization reactions and compare their physical properties and IR spectra.

Polymers (from Greek *poly* = many and *meros* = part) are compounds containing long chains of atoms. A single polymer molecule can have thousands or even millions of atoms, all covalently bonded together. Polymers are conveniently divided into two major groups, biopolymers and synthetic polymers. Biopolymers occur naturally in living organisms. They include nucleic acids (DNA and RNA), proteins, and polysaccharides. Synthetic polymers are an important part of materials science, a significant area of technology in which chemistry plays an important role. Synthetic polymers are made from relatively small organic compounds called ***monomers,*** which are chemically linked together to produce the polymer molecules.

Under the right conditions, a large number of ethylene molecules can combine to form polyethylene:

$$n\mathrm{CH_2{=}CH_2} \xrightarrow{\text{catalyst}} \mathrm{+\!\!\!\!\!\!\!\!\ \ CH_2{-}CH_2 +}_n$$

Polyethylene is an addition polymer, formed by the linkage of ethylene monomer units. The ***subscript n*** refers to the number of identical repeating units in the polymer, that is, the degree of polymerization. It is not uncommon to find synthetic polymers that have molecular weights in the tens to hundreds of thousands or higher.

Polymers are significant commercial materials. Polystyrene is one that we have all encountered in daily living as Styrofoam, used in insulated coffee cups, in the windows of mailing envelopes, and as packing "peanuts." Polystyrene is an excellent electrical insulator and is used in insulating panels and appliance components. Polystyrene can be extruded, molded, and made into a foam for its various applications.

Treatment of the monomer styrene (vinylbenzene) with either a free-radical or an ionic catalyst can produce polystyrene. The properties of the polymer that is formed vary significantly with the nature of the catalyst and the conditions used to form it.

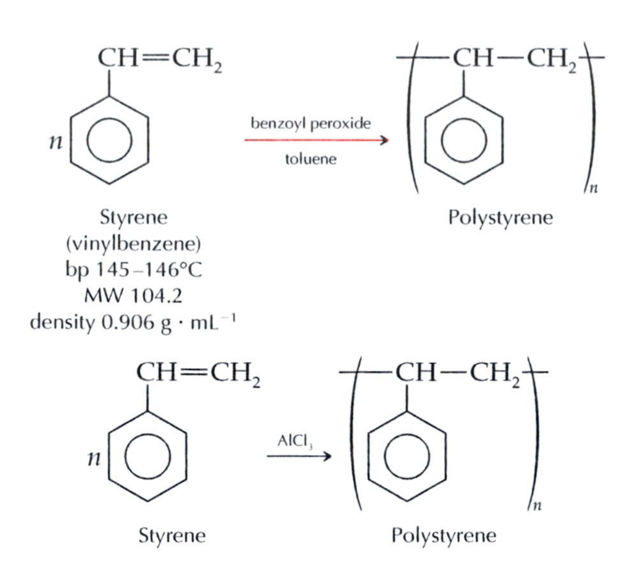

Free-Radical Polymerization

The reaction of styrene with benzoyl peroxide produces free-radical polymerization. The process is initiated by homolytic cleavage of the weak oxygen-oxygen bond of benzoyl peroxide to form two equivalent benzoyloxy radicals:

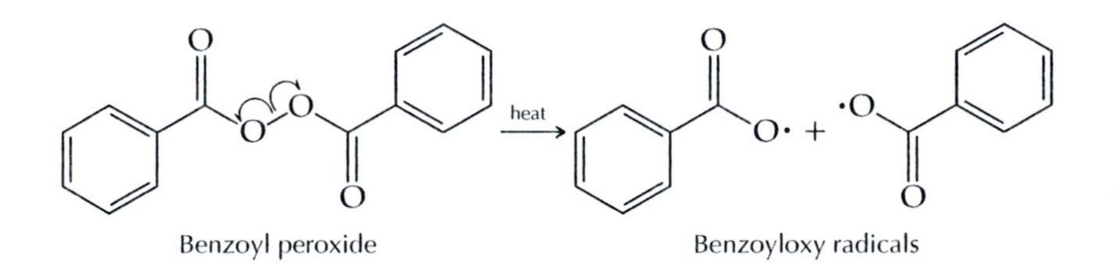

A benzoyloxy radical initiates further radical formation by adding to the terminal carbon of a styrene molecule, forming the relatively stable secondary benzylic free radical. This radical is stabilized by resonance with the delocalized aromatic π-system. The benzylic radical continues the free-radical propagation sequence by reacting with another styrene molecule to form a dimer radical:

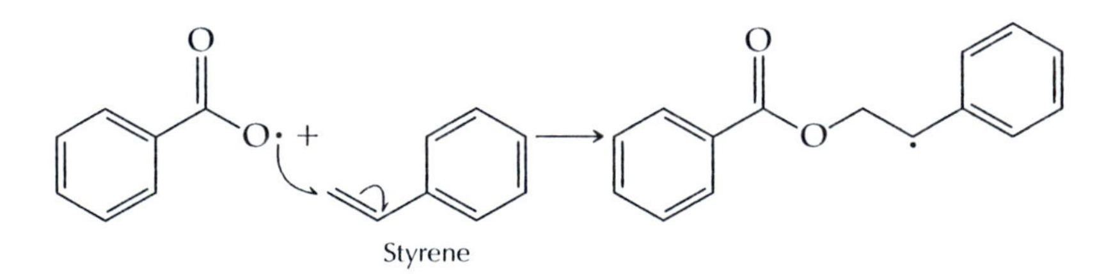

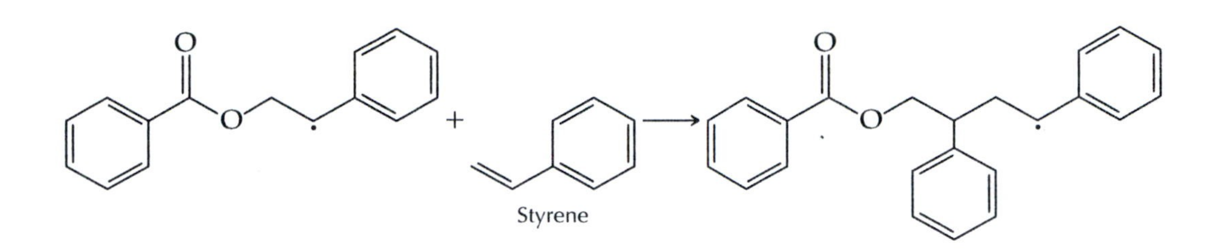

The dimer radical can now react with another molecule of styrene, forming a trimer radical, which reacts with another molecule of styrene to form a tetramer radical, and so on and so on. The result is a polymeric radical. The polymerization reaction is terminated when the polymeric radical reacts with any other radical in the reaction mixture:

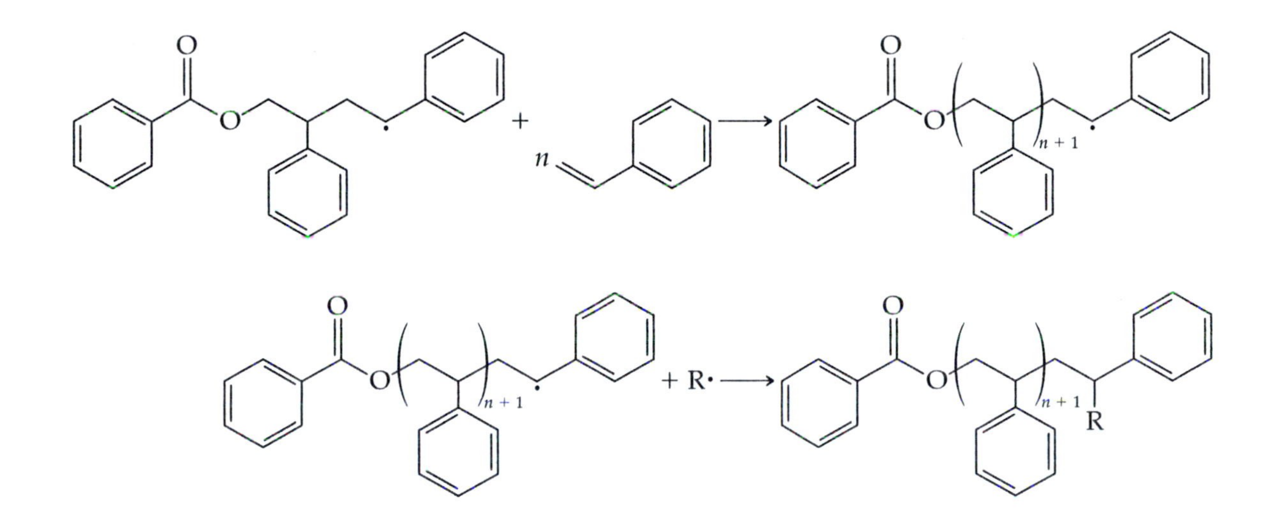

Cationic Polymerization

Polystyrene can also be formed by ionic methods. When styrene is treated with aluminum trichloride, a vigorous reaction occurs. The mechanism of the reaction is probably initiated by autoionization of the aluminum trichloride:

$$2AlCl_3 \rightleftharpoons AlCl_2^+ + AlCl_4^-$$

The $AlCl_2^+$ electrophile is highly reactive and will attack styrene to produce a secondary carbocation, again stabilized by resonance with the benzene ring:

$$C_6H_5-CH=CH_2 + AlCl_2^+ \longrightarrow C_6H_5-\overset{+}{C}H-CH_2-AlCl_2$$

The newly formed benzylic carbocation can react with a series of styrene molecules to form a long polymer chain. This structure can then react with any Brønsted acid present to cleave the carbon-aluminum bond and form a C—H bond in its place:

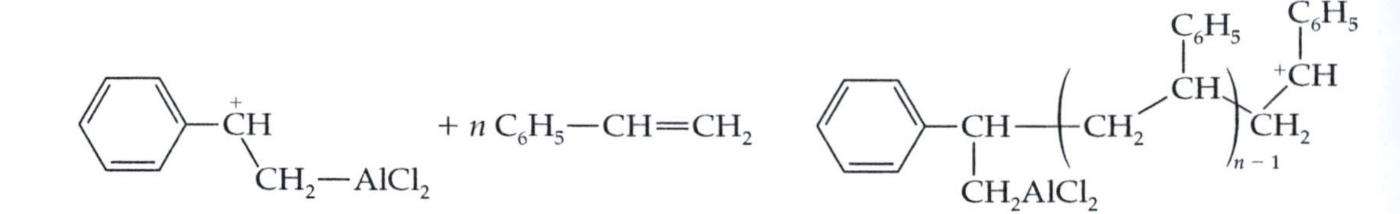

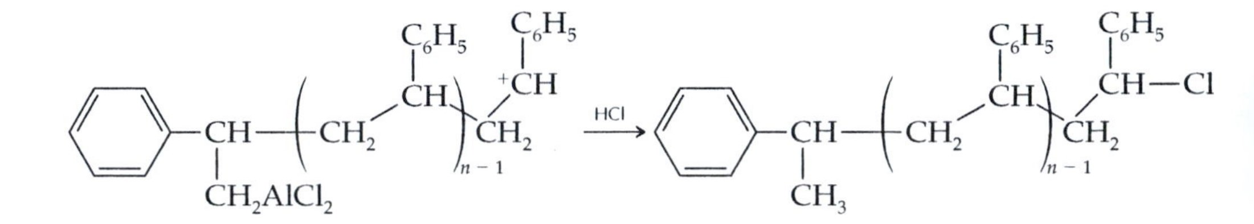

Unlike biopolymers, which usually are homogeneous with every molecule having the same molecular weight, synthetic polymer samples normally have a range of molecular weights because there is a probability range within which different polymer chains undergo termination of the chain propagation sequence. The molecular weights of individual polystyrene molecules differ as the number of their monomer units differs. The average molecular weight of polystyrene is often from 100,000 to 400,000, but the MW range can be from 10,000 to 1,000,000. The average MW and its range depend on the conditions used in the polymerization reaction.

Properties of polymers also differ with their degree of polymerization and molecular weight. In addition, the relative orientation of the long covalent polymer chains to each other is important. Physical properties for polystyrene range from highly crystalline, where its melting point is 240°C, to amorphous gooey solids with no definite melting point. The IR spectrum of polystyrene is less susceptible to its molecular weight distribution since the fundamental vibrations of its chemical bonds do not change with molecular weight.

MINISCALE PROCEDURE

Techniques Liquid Chromatographic Column: Technique 17.6
IR Spectroscopy: Technique 18.4

SAFETY INFORMATION

Wear gloves and work in a hood, if possible, while doing these procedures.

Styrene is a flammable liquid and an irritant to eyes and mucous membranes. Use it only in a hood, and avoid contact with skin, eyes, and clothing.

Benzoyl peroxide is a strong oxidizer and may explode if it is heated above its melting point or if the dry compound is subjected to friction or shock. Avoid contact with skin, eyes, and clothing. Do not transfer it with a metal spatula.

Aluminum chloride is corrosive and moisture sensitive, and it forms HCl on contact with atmospheric moisture. Measure the required amount quickly and store the reagent container tightly closed in a desiccator.

Methanol is volatile, toxic, and flammable. Use it only in a hood.

2-Butanone (methyl ethyl ketone) is flammable and an irritant.

Polymerization of Styrene with Benzoyl Peroxide

Reactive alkenes are doped with inhibitors to prevent their polymerization.

The styrene available in your laboratory probably contains a free-radical inhibitor such as an alkylated phenol. The inhibitor must be removed before the styrene is used in this procedure. To do this, prepare a microscale chromatography column in a Pasteur pipet [see Technique 17.6]. Pack a small plug of cotton in the upper stem of the pipet and add about 1.8 g of dry alumina on top of the cotton. Tap the pipet gently on the bench top to settle the alumina. Clamp the pipet in an upright position and place a tared 18 × 150 mm test tube under the tip. Transfer about 2 mL of styrene to the top of the column with a Pasteur pipet and allow the styrene to flow through the column into the test tube. You may use a rubber bulb on the top of the column to exert a moderate pressure. Weigh the test tube and the eluted styrene that it contains; you should have 1.0–1.2 g of styrene. Record the mass of styrene for your yield calculation.

Add 6 mL of toluene and 50 ± 5 mg of benzoyl peroxide to the test tube containing the styrene. Clamp the test tube in a hot-water (90–95°C) bath and heat it for 1 h (carry out the polymerization of styrene with aluminum chloride during this time). Cool the test tube for 5 min, note the viscosity of the solution, and pour the solution into a 100-mL beaker containing 40 mL of methanol. Collect the precipitated white polymer by vacuum filtration on a Buchner funnel and wash the precipitate with 10 mL of methanol. Spread the polymer on a piece of filter paper to dry.

CLEANUP: Pour the methanol filtrate into the container for flammable (organic) waste. Place the alumina from the chromatography column in the container for solid inorganic waste.

Polymerization of Styrene with Aluminum Chloride

Pour 2.0 mL of styrene (the inhibitor does not need to be removed) into a 50-mL beaker. Obtain 80 ± 5 mg of anhydrous aluminum chloride in a small tared test tube. Cork the test tube immediately after adding the aluminum chloride. Set the beaker toward the back of the hood and add the aluminum chloride in four or five portions, stirring the reaction mixture after each addition. The reaction is extremely vigorous, and the mixture becomes dark brown. Allow the mixture to stand for 15 min, then add 15 mL of methanol and stir to mix. Heat the mixture to boiling on a steam bath or in an 80°C water bath while continuing to stir, and then set the beaker aside until it cools to room temperature. Decant the methanol and allow the polymer to dry a few minutes in the hood. Describe the properties of the product and compare them to the properties of polystyrene prepared by benzoyl peroxide catalysis.

CLEANUP: Pour the methanol decanted from the reaction mixture into the container for flammable (organic) waste.

IR Analysis of Polystyrene

Weigh a sample of polymer from each procedure that is approximately 50 mg. Place each sample in a separate labeled 13 × 100 mm test tube and add 1 mL of 2-butanone (methyl ethyl ketone). Stir until the polymer dissolves (this may take several minutes).

A thin film (0.01–0.03 mm) of polystyrene can be cast directly onto a sodium chloride IR plate [see Technique 18.4]. Work in a hood to prepare the thin film. Mount the salt plate (do not use a second plate on top of the film) in the spectrometer sample holder and obtain the infrared spectrum. Analyze and compare the IR spectra of your two polymerization products. Assign the stretching vibrations that produce the major IR bands [see Techniques 18.5 and 18.6 and Tables 18.2 and 18.3].

CLEANUP: The polystyrene film can be removed by rinsing the salt plate with dichloromethane. Pour the residual dichloromethane into the container for halogenated waste. Pour any remaining polystyrene solutions into the container for flammable (organic) waste.

OPTIONAL EXPERIMENT **IR Analysis of Commercial Polystyrene Products**

TEAMWORK: To minimize the number of samples to be run on the IR spectrometer, work with another student to obtain the spectra of your commercial products and share your data with each other.

Bring to the laboratory commercial products that you think might be made of polystyrene, such as transparent food containers, hot-drink cups, packing "peanuts," or any other product that is labeled polystyrene (PS). For each product, prepare a thin film casting and obtain the IR spectrum [see Technique 18.4]. The "window" in mailing envelopes is frequently made from polystyrene. Mount a piece of the window over a hole in a piece of cardboard that is sized to fit in the spectrometer sample holder. Obtain the IR spectrum. Compare the spectra of the various products and describe any similarities or differences.

Reference

1. Selinger, B. *Chemistry in the Market Place;* 4th ed.; Harcourt Brace: Sydney, 1994.

Questions

1. What would happen if the benzoyl peroxide were omitted in the first experiment?
2. What might happen if the inhibitor were not removed from the styrene in the first experiment?
3. Why must the free-radical polymerization reaction be heated when it is not necessary to heat the ionic polymerization?
4. If a small amount of *p*-divinylbenzene is added to styrene, a cross-linked polymer can be formed. Show why this would be the case with structures and a brief mechanism. Predict how cross-linking of the polymer chains might change the polymer's physical properties.

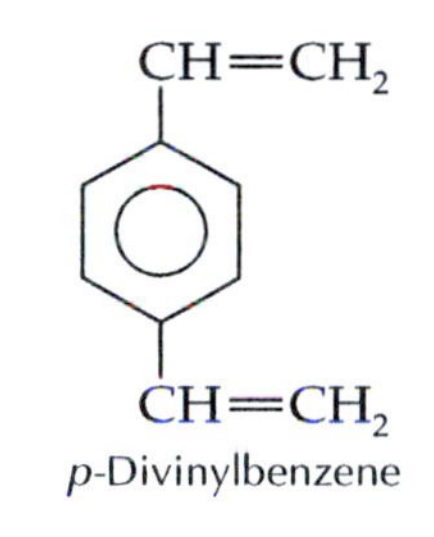

p-Divinylbenzene

EXPERIMENT

18

PHASE TRANSFER CATALYSIS

Synthesis of Butyl Benzoate

PURPOSE: To discover how a phase transfer catalyst can be used to facilitate the synthesis of an ester.

In Experiment 18 you will use an S_N2 reaction, catalyzed by a quaternary ammonium salt in a two-phase system, to synthesize butyl benzoate; then you will characterize it by IR spectroscopy.

Chemists usually run reactions in a homogeneous reaction medium, but doing so when a reaction involves a nonpolar organic substrate and a polar inorganic reagent can be difficult. The inorganic reagent is soluble in water but not in many organic solvents, whereas the organic compound is soluble in nonpolar organic solvents but not in water.

One approach to achieving a homogeneous reaction medium is to use a mixture of two miscible solvents of different polarities, such as the organic solvent tetrahydrofuran (THF) and water. This method requires that water and the nonpolar organic solvent readily dissolve in each other, a situation that does not occur for many common organic solvents. Another problem with the use of a mixed solvent is its environmental cost. Often it is very expensive to recover organic solvents that are miscible with water, so they become a waste disposal problem. It would be much simpler and safer if we could just use water and no organic solvent at all.

The development of phase transfer catalysis has provided this opportunity. Phase transfer catalysts are reagents that react with inorganic salts, forming compounds that are soluble to an appreciable degree in both polar and nonpolar solvents. They can move from one liquid phase to the other, allowing reagents to be moved between aqueous solutions and nonpolar organic liquids.

The phase transfer catalyst in Experiment 18 is trioctylmethylammonium chloride, $([CH_3(CH_2)_7]_3NCH_3)^+Cl^-$, sold under the trade name Aliquat 336. Trioctylmethylammonium chloride is a quaternary ammonium salt, in which the nitrogen atom bears a formal positive charge and has bonds to four nonpolar alkyl groups. Three of the hydrophobic alkyl groups are quite large. Consequently, it has both polar and nonpolar character and can act as a detergent.

Aliquat 336 catalyzes the reaction by increasing the concentration of the benzoate anion in the organic layer where the S_N2 reaction occurs.

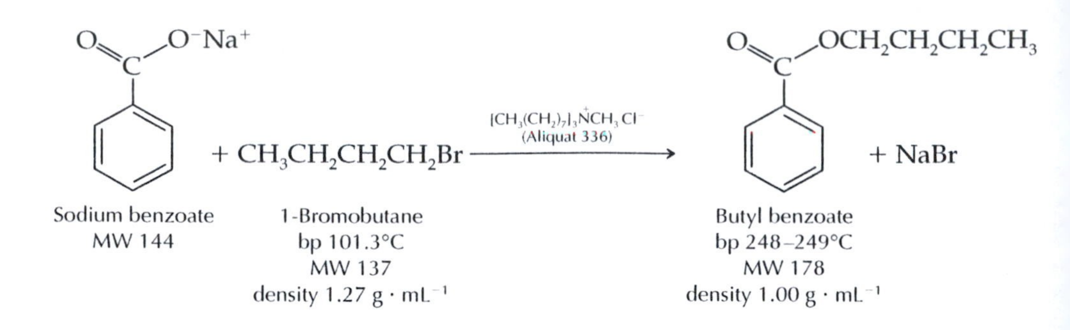

In this S_N2 reaction the benzoate anion acts as a nucleophile that displaces the bromide ion, a good leaving group, from 1-bromobutane. The benzoate anion is not a particularly strong nucleophile in aqueous solution, where it is stabilized by hydrogen bonding with water. The hydrogen bonding is not present in 1-bromobutane, so the phase transfer catalyst not only moves the benzoate ion into the organic layer but also accelerates the S_N2 reaction by making benzoate a stronger nucleophile.

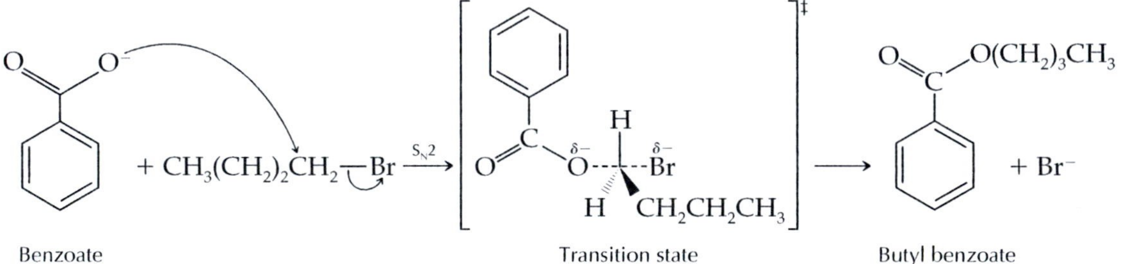

The catalytic phase transfer process begins when a positively charged Aliquat 336 cation forms an ion pair with the benzoate anion. Because the phase transfer catalyst has a large proportion of nonpolar character, it can then carry the benzoate anion into the organic layer where it reacts with 1-bromobutane and then recycles back into the water layer:

The solvent appears in parentheses in these equations.

1. Ion exchange in the water layer ($R = CH_3(CH_2)_7$):

$$R_3\overset{+}{N}CH_3\ Cl^-_{(H_2O)} + C_6H_5COO^-Na^+_{(H_2O)} \rightleftharpoons R_3\overset{+}{N}CH_3\ C_6H_5COO^-_{(H_2O)} + Na^+Cl^-_{(H_2O)}$$

2. Phase transfer from water to the organic layer:

$$R_3\overset{+}{N}CH_3\ C_6H_5COO^-_{(H_2O)} \rightleftharpoons R_3\overset{+}{N}CH_3\ C_6H_5COO^-_{(organic)}$$

3. S_N2 reaction in the organic layer:

$$R_3\overset{+}{N}CH_3\ C_6H_5COO^-_{(organic)} + CH_3(CH_2)_2CH_2Br \longrightarrow$$

$$C_6H_5COOCH_2(CH_2)_2CH_3 + R_3\overset{+}{N}CH_3\ Br^-_{(organic)}$$

Butyl benzoate

4. Phase transfer (catalyst recycling) to the water layer:

$$R_3\overset{+}{N}CH_3\ Br^-_{(organic)} \rightleftharpoons R_3\overset{+}{N}CH_3\ Br^-_{(H_2O)}$$

After step 4, the cycle of steps 1–4 begins again in the water layer when the quaternary ammonium ion ($R_3NCH_3^+$) exchanges the bromide ion for another benzoate ion. The vigorous mixing of reflux facilitates the exchanges between layers.

You will use IR spectroscopy to differentiate the product from the starting alkyl bromide.

MINISCALE PROCEDURE

Techniques Reflux: Technique 7.1
Extraction: Technique 8.2
IR Spectroscopy: Techniques 18.4–18.6

SAFETY INFORMATION

1-Bromobutane is a skin irritant and harmful if inhaled, ingested, or absorbed through the skin. Wear neoprene gloves, and use it in a hood, if possible.

Aliquat 336 (trioctylmethylammonium chloride) is toxic and an irritant. Avoid contact with skin, eyes, and clothing.

Dichloromethane is toxic, an irritant, absorbed through the skin, and harmful if inhaled. Use it only in a hood and wear neoprene gloves. Wash your hands thoroughly after handling it.

Place 2.0 mL of 1-bromobutane, 3.0 g of sodium benzoate, 5.0 mL of water, 4 drops of Aliquat 336, and a boiling stone in a 50-mL round-bottomed flask. Fit a reflux condenser above the flask and heat the reaction mixture under reflux for 1 h [see Technique 7.1].

Cool the flask in a beaker of room-temperature water. If solid forms in the cooled reaction mixture, add 2–3 mL of water, stopper the flask, and shake the mixture until the solid dissolves. Transfer the cooled contents of the flask to a separatory funnel. Rinse the flask with 20 mL of dichloromethane and add the dichloromethane rinse to the separatory funnel. Add 10 mL of water to the funnel and gently shake or swirl it to mix the layers [see Technique 8.2]. Drain the lower organic phase into a

Extractions with CH_2Cl_2 can lead to emulsions. Vigorous shaking of the two phases in the separatory funnel should be avoided.

clean, labeled Erlenmeyer flask; pour the aqueous phase out the top of the funnel into another labeled Erlenmeyer flask. Return the organic phase to the separatory funnel and wash it with 5 mL of 15% NaCl solution. Drain the organic phase into a clean, dry Erlenmeyer flask and add a small amount of anhydrous sodium sulfate as the drying agent [see Technique 8.7]. Allow the solution to dry for at least 15 min.

Alternatively, evaporate the CH_2Cl_2 on a rotary evaporator or in a hood using a 70–75°C water bath.

Weigh a dry 25-mL Erlenmeyer flask to the nearest 0.01 g. Prepare a Pasteur filter pipet [see Technique 8.5] and use this pipet to transfer the solution from the drying agent to the 25-mL flask. Working in a hood, remove the dichloromethane with a stream of nitrogen or air while warming the flask in a beaker of water at 50°C. When the volume is reduced to about 2 mL, dry the outside of the flask and weigh it. Blow nitrogen or air over the solution for another minute and then weigh the flask again. Continue this process until two weighings agree within 0.03g.

IR Spectroscopic Analysis of the Product

Liquids need to be completely dry before you take an IR spectrum. If your product appears to have any water droplets in it after the evaporation of the dichloromethane, add a few grains of anhydrous sodium sulfate and wait 10 min before determining the IR spectrum. Use a clean Pasteur filter pipet to remove the product from the drying agent.

Obtain the IR spectrum of your product [see Technique 18.4]. Compare your product spectrum to the IR spectrum of 1-bromobutane and determine whether there is evidence of starting material in the product. Analyze the IR spectrum of your product and assign its major absorption bands to specific bond vibrations in butyl benzoate [see Techniques 18.5 and 18.6, including Table 18.2].

CLEANUP: The aqueous solutions from the extractions may be washed down the sink or poured into the container for aqueous inorganic waste. Place the used sodium sulfate in the container for solid inorganic waste.

MICROSCALE PROCEDURE

Techniques Microscale Reflux: Technique 7.1
Microscale Extraction: Technique 8.6b
IR Spectroscopy: Techniques 18.4–18.6

SAFETY INFORMATION

1-Bromobutane is a skin irritant and harmful if inhaled, ingested, or absorbed through the skin. Wear neoprene gloves, and use it in a hood, if possible.

Aliquat 336 is toxic and an irritant. Avoid contact with skin, eyes, and clothing.

Dichloromethane is toxic, an irritant, absorbed through the skin, and harmful if inhaled. Use it only in a hood and wear neoprene gloves. Wash your hands thoroughly after handling it.

Place 0.75 mL of 1-bromobutane, 2.0 mL of water, 1.15 g of sodium benzoate, and two drops of Aliquat 336 in a 5-mL conical vial. Add a magnetic spin vane and fit a water-jacketed condenser to the top of the vial [see Technique 7.2]. Clamp the apparatus in a heating block or sand bath, turn on the stirrer, and heat the reaction mixture at reflux for 1 h at a block or bath temperature of 160–170°C.

At the end of the reflux period, cool the vial to room temperature and add 2 mL of water to it. Stir the contents of the vial until the solid that precipitated on cooling dissolves. Transfer the solution to a centrifuge tube that has a tight-fitting cap, using a Pasteur pipet fitted with a rubber bulb or a syringe [see Technique 8.5]. Rinse the vial with 3 mL of dichloromethane and transfer this rinse to the centrifuge tube with the same Pasteur pipet.

Cap the centrifuge tube and shake it to thoroughly mix the two phases. Alternatively, the phases can be mixed by drawing the mixture into a Pasteur pipet (with no cotton in the tip) and expelling it back into the tube at least six times. Allow the layers to separate. Transfer the lower organic phase to another centrifuge tube with the Pasteur filter pipet or the Pasteur pipet and syringe used previously [see Technique 8.6b]. Add 2 mL of dichloromethane to the aqueous phase remaining in the first centrifuge tube and mix the phases. Remove the lower organic phase and combine it with the previous organic phase in the second centrifuge tube. Wash the combined organic phase with 2 mL of 15% sodium chloride solution. Transfer the organic layer to a clean centrifuge tube or test tube. Add 200 mg of anhydrous sodium sulfate and dry the solution for at least 10 min.

Weigh a dry 25-mL Erlenmeyer flask to the nearest 0.001 g. Using a clean Pasteur filter pipet, transfer the product solution to the Erlenmeyer flask. Working in a hood, remove the dichloromethane by blowing a stream of nitrogen or air over the solution while the flask is warmed in a 40–45°C water bath. Weigh the flask when the volume of liquid no longer appears to be decreasing. Blow nitrogen or air over the product for another minute, then weigh the flask again. Continue this process until two weighings agree within 0.025 g. Calculate the percent yield.

Obtain the IR spectrum of your product as directed in the miniscale procedure.

CLEANUP: The aqueous solutions from the extractions may be washed down the sink or poured into the container for aqueous inorganic waste. Place the used sodium sulfate in the container for solid inorganic waste.

Questions

1. What is the solid that precipitates when the reaction mixture is cooled to room temperature?
2. How is the phase transfer catalyst removed from the product?
3. What purpose does washing with 15% (half-saturated) NaCl solution serve?
4. There is a slight excess of sodium benzoate in the reaction. How is it removed from the product?
5. Why is the benzoate anion a stronger nucleophile in 1-bromobutane than in water solution?
6. What peak in the IR spectrum suggests the presence of water in the product? What could be done to remove this water?

EXPERIMENT

19

DIELS-ALDER REACTION

PURPOSE: To study one of the most important chemical reactions used to synthesize organic compounds containing six-membered rings.

In Experiment 19 you will carry out a Diels-Alder synthesis of cyclohex-4-ene-1,2-*cis*-dicarboxylic anhydride and hydrolyze the product to cyclohex-4-ene-1,2- *cis*-dicarboxylic acid.

Diels-Alder Reaction of Butadiene and Maleic Anhydride

The conjugated diene 1,3-butadiene, used as reactant in this experiment, is a gas at room temperature (bp 24°C) and inconvenient to handle. Therefore, butadiene is formed in situ, within the reaction mixture, by the extrusion of sulfur dioxide from the solid reactant, butadiene sulfone. The butadiene then reacts quickly with maleic anhydride before it can escape from the reaction mixture.

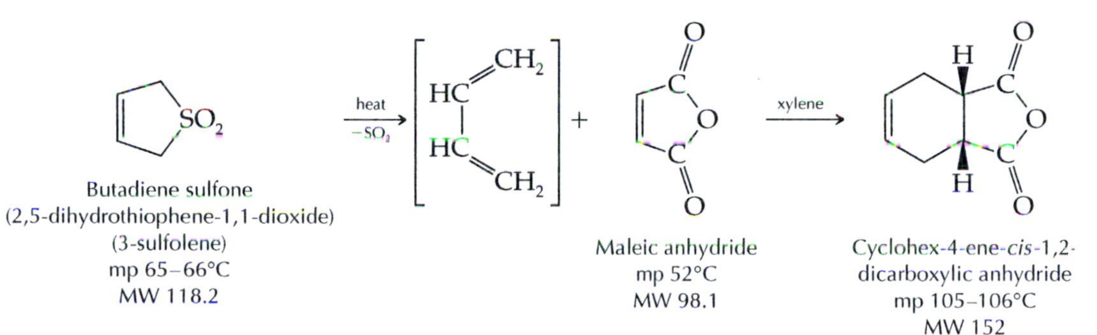

The cis configuration of the alkene, maleic anhydride, and the stereospecificity of the Diels-Alder reaction determine the cis geometry of the carbonyl groups in the Diels-Alder product. The trans-like conformation of butadiene must rearrange to the cis-like conformation before reaction with maleic anhydride can proceed because a C=C double bond in a cyclohexene ring must have a cis configuration (*trans*-cyclohexene is unknown).

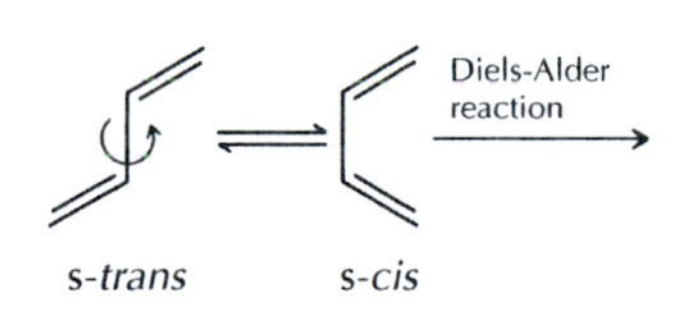

Formation of Diels-Alder Adduct

The Diels-Alder reaction has great utility in the chemical synthesis of six-membered ring compounds from simple cyclohexanes to complex steroids. It is a ***cycloaddition reaction*** in which two molecules add together to form a cyclic compound. The reaction involves the rearrangement of six π-electrons in a cyclic array of six atoms. The Diels-Alder reaction is a [4 + 2] cycloaddition that takes place in one step. It is a ***pericyclic reaction*** in which the *p*-orbitals of all six carbon atoms overlap simultaneously in the transition state.

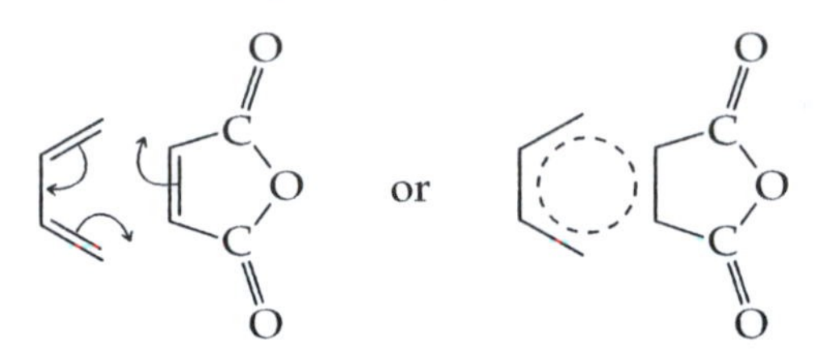

The conjugated diene provides four π-electrons, and the alkene provides the other two π-electrons. This arrangement allows a concerted reaction with a stabilized aromatic-like transition state having six π-electrons. The alkene is often called a ***dienophile*** because it loves to react with the diene.

The Diels-Alder reaction has a much faster rate when the carbon-carbon double bond of the dienophile is electron deficient. Electron-withdrawing groups on the dienophile speed up the reaction significantly. The two electron-withdrawing carbonyl groups attached to its C=C make maleic anhydride an excellent dienophile. Often only gentle heating of a mixture of the diene and dienophile is required for almost quantitative yields of cyclic products. The reaction of butadiene and maleic anhydride will be carried out at 140°C, the boiling point of dimethylbenzene (xylene). Interestingly, electron-donating groups on the diene also speed up the reaction.

Diels-Alder reactions can be modified to allow preparation of many different compounds by using various substituents on the diene and dienophile. The dienophile can also be a carbon-oxygen double bond or a carbon-carbon triple bond. Diels-Alder reactions have been of great synthetic utility in organic synthesis. Otto Diels and Kurt Alder received the Nobel prize in 1950 for discovering the reaction that bears their names.

Hydrolysis of Diels-Alder Product

The second step of Experiment 19 is the hydrolysis of the anhydride produced in the Diels-Alder reaction of 1,3-butadiene and maleic anhydride.

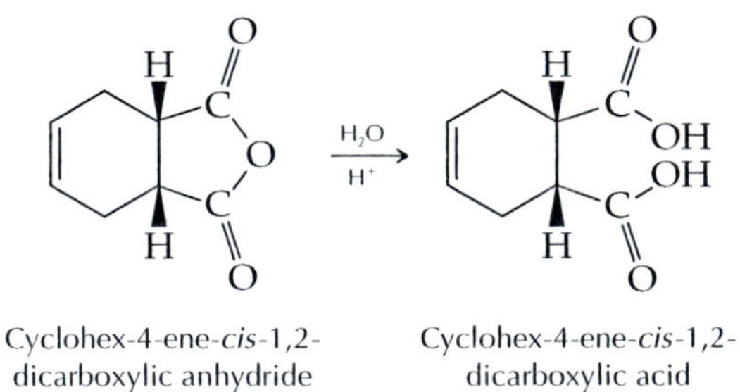

Cyclohex-4-ene-*cis*-1,2-dicarboxylic anhydride

Cyclohex-4-ene-*cis*-1,2-dicarboxylic acid
mp 165°C
MW 170.2

The initial Diels-Alder product is a bicyclic anhydride, which can be isolated. However, you will hydrolyze the anhydride directly to the dicarboxylic acid, which can easily be recovered in a pure state without recrystallization. The hydrolysis involves the addition of water to the carbonyl group and breaking of the anhydride linkage. In water solution the equilibrium lies on the side of the dicarboxylic acid product.

MINISCALE PROCEDURE

Techniques Removal of Noxious Vapors: Technique 7.4
IR Spectroscopy: Techniques 18.4–18.6
NMR Spectroscopy: Techniques 19.2, 19.4–19.9

SAFETY INFORMATION

Conduct this experiment in a hood, if possible.

Butadiene sulfone (3-sulfolene) is an irritant. Wear gloves and avoid contact with skin, eyes, and clothing. This compound emits toxic corrosive sulfur dioxide when it is heated. Be sure that the gas trap is positioned before you begin heating the reaction mixture.

Maleic anhydride is toxic and corrosive. Avoid breathing the dust and avoid contact with skin, eyes, and clothing.

Xylene is flammable.

Synthesis of Diels-Alder Adduct

Combine 2.0 g of butadiene sulfone, 1.2 g of finely ground maleic anhydride, 0.80 mL of xylene (a mixture of isomers is all right), and a boiling stone in a 25-mL round-bottomed flask. Attach a Claisen adapter and a water-cooled condenser to the flask; close the second neck of the Claisen adapter with a glass stopper. If your lab is not equipped with water aspirators, assemble the gas trap as shown in Technique 7.4, Figure 7.5a, using a thermometer adapter, bent glass tubing, and a 125-mL filter flask. If your lab is equipped with water aspirators, place a vacuum adapter at the top of the condenser and connect the side arm to a water aspirator with heavy-walled rubber tubing as shown in Technique 7.4, Figure 7.5b; turn on the water.

Begin heating the mixture gently with a heating mantle or sand bath. After the solids dissolve, continue heating the mixture at a gentle reflux for 30 min. Remove the heat source and cool the reaction mixture for about 5 min before proceeding immediately with the hydrolysis of the anhydride product.

Hydrolysis of Diels-Alder Product to the Dicarboxylic Acid

Remove the gas trap from the top of the condenser. Pour 4 mL of water down the condenser, add another boiling stone, and heat the mixture under reflux for 30 min. Cool the solution to room temperature. If crystallization of the product does not occur, add 3 or 4 drops of concentrated sulfuric acid, stir the contents of the flask, and cool the resulting mixture in an ice-water bath for 5 min. Collect the product by vacuum filtration. Wash the crystals twice with 1-mL portions of ice-cold water. The dicarboxylic acid is usually quite pure without recrystallization. If time permits, the product can be recrystallized from water [see Technique 9.5]. Allow the product to dry overnight before determining the melting point and percent yield.

Spectroscopic Analysis of the Product

Obtain and analyze an IR or NMR spectrum of the dried product, as directed by your instructor [see Technique 18.4 for IR sample preparation, Technique 19.2 for NMR sample preparation]. What IR bands [see Techniques 18.5 and 18.6 and Tables 18.2 and 18.3] indicate that you have the desired product? Does your NMR spectrum fit your prediction of the NMR spectrum for the desired product [see Techniques 19.4 and 19.9 and Tables 19.2 and 19.3]?

CLEANUP: Pour the filtrate from the reaction mixture into the container for flammable organic waste. Pour the water in the gas trap into a large beaker. Neutralize the solution with solid sodium carbonate before washing it down the sink or pouring it into the container for aqueous inorganic waste.

MICROSCALE PROCEDURE

Techniques Removal of Noxious Vapors: Technique 7.4
IR Spectroscopy: Techniques 18.4–18.6
NMR Spectroscopy: Techniques 19.2, 19.4–19.9

SAFETY INFORMATION

Conduct this experiment in a hood, if possible.

Butadiene sulfone (3-sulfolene) is an irritant. Wear gloves and avoid contact with skin, eyes, and clothing. This compound emits toxic corrosive sulfur dioxide when it is heated. Be sure that the gas trap is positioned before you begin heating the reaction mixture.

Maleic anhydride is toxic and corrosive. Avoid breathing the dust and avoid contact with skin, eyes, and clothing.

Xylene is flammable.

Synthesis of Diels-Alder Adduct

Prepare a gas trap as shown in Technique 7, Figure 7.7, using a piece of Teflon tubing with one end threaded through a rubber fold-over septum [see Technique 7.4, Figure 7.6] that fits the top of a water-jacketed condenser. Thread the other end through a rubber septum that fits a 25-mL filter flask containing approximately 10 mL of water. Be sure that the tip of the tubing is *just above* the surface of the water in the filter flask.

Combine 250 mg of butadiene sulfone, 150 mg of finely ground maleic anhydride, and 0.15 mL of xylene (a mixture of isomers is all right) in a 5-mL conical vial containing a magnetic spin vane. Fit the water-jacketed condenser with its accompanying gas trap to the vial. Clamp the apparatus on an aluminum heating block or in a sand bath set on a hot plate. Begin heating the mixture gently; allow the temperature of the heat source to rise slowly to 150°C. After the solids dissolve, continue heating the mixture at a gentle reflux for 30 min. Carefully remove the flask from the heat source and cool the reaction mixture for about 5 min before proceeding immediately with the hydrolysis of the anhydride product.

Hydrolysis of Diels-Alder Adduct

Remove the gas trap. Pour 0.5 mL of water down the condenser, add a new boiling stone, and heat the mixture under reflux for 30 min. The aluminum block or sand bath should be heated to 125–140°C. Cool the solution to room temperature. If crystallization of the product does not occur, add 1 drop of concentrated sulfuric acid, stir the contents of the vial, and cool the resulting mixture in an ice-water bath for 5 min.

Collect the product by vacuum filtration on a Hirsch funnel. Wash the crystals twice with a few drops of ice-cold water. The dicarboxylic acid is usually quite pure without recrystallization. If time permits, the product can be recrystallized from water [see Technique 9.7a]. Allow the product to dry overnight before determining the melting point and percent yield.

Carry out the spectroscopic analysis of the product as described in the miniscale procedure.

CLEANUP: Pour the filtrate from the reaction mixture into the container for flammable organic waste. Pour the water in the gas trap into a large beaker. Neutralize the solution with solid sodium carbonate before washing it down the sink or pouring it into the container for aqueous inorganic waste.

Reference

1. Sample, T. E.; Hatch, L. F. *J. Chem. Educ.* **1968,** *45,* 55–56.

Questions

1. In what major ways will the IR spectrum of cyclohex-4-ene-*cis*-1,2-dicarboxylic anhydride differ from the IR spectrum of cyclohex-4-ene-*cis*-1,2-dicarboxylic acid?
2. Predict the relative rates for the reaction of butadiene and (a) ethene, (b) methyl propenoate, (c) maleic anhydride.
3. What product is expected from the reaction of fumaric acid (*trans*-2-butenedioic acid) with 1,3-butadiene?
4. Which of the following two compounds much more readily takes the part of the diene in the Diels-Alder reaction? Explain.

EXPERIMENT

20

SELECTIVITY IN THE BROMINATION OF ACETANILIDE AND 4-METHYLACETANILIDE

QUESTION: At which carbon of the aromatic ring does electrophilic substitution occur in the bromination of acetanilide (*N*-phenylacetamide) and 4-methylacetanilide?

> In Experiment 20 you will study the regiochemistry of electrophilic aromatic substitution. Is the acetamido group primarily an ortho, meta, or para directing group? Is a methyl group or the acetamido group the dominant directing group? Will bromination occur *ortho* to the acetamido group or ortho to the methyl group? You may do the experiment as part of a two-person team, using melting points and NMR analysis to identify the major product of each reaction.

Acetanilide, on which this experiment is based, is a mild analgesic and antipyretic. *para*-Hydroxyacetanilide, also called acetaminophen, is one of the most popular pain killers currently used in the United States; Tylenol is one of the major brands of acetaminophen.

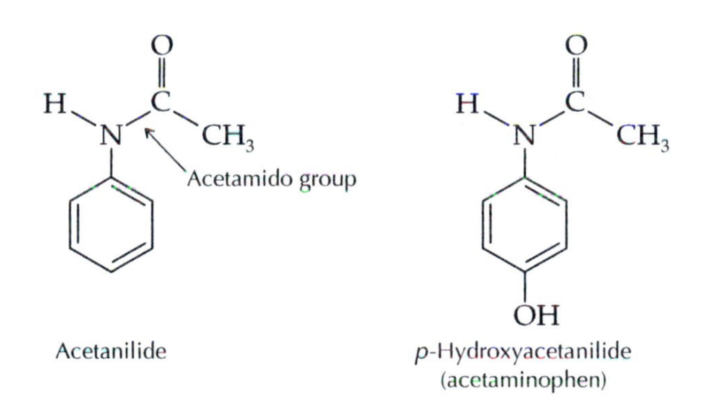

Acetanilide

p-Hydroxyacetanilide (acetaminophen)

Electrophilic Aromatic Substitution

Even though the delocalized π-systems of aromatic compounds are much less vulnerable to attack by electrophiles (E^+) than the localized π-double bonds of alkenes, aromatic compounds undergo electrophilic substitution reactions. Because the conjugated 6 π-electron system of the aromatic ring is so stable, the carbocation intermediate loses a proton and regenerates the aromatic ring rather than undergoing reaction with a nucleophile:

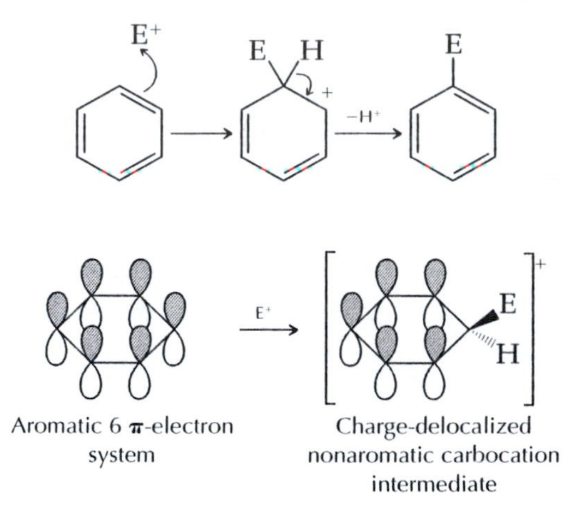

The electrophile in Experiment 20 is molecular bromine. It forms from the oxidation-reduction reaction between hydrogen bromide and potassium bromate ($KBrO_3$):

$$KBrO_3 + 6HBr \longrightarrow 3H_2O + 3Br_2 + KBr$$

$KBrO_3$	$6HBr$
Potassium bromate	Hydrobromic acid
MW 167.0	MW 80.9
	density 1.490 $g \cdot mL^{-1}$
	(48 wt % solution)

The reaction conditions are designed to limit the amount of bromine present to ensure that only monobromination occurs.

The mechanism of electrophilic aromatic substitution has been thoroughly studied. Ring substituents strongly influence the rate and position of electrophilic attack. Electron-donating groups on the benzene ring speed up the substitution process by stabilizing the carbocation intermediate. If, however, the first substituent is an electron-withdrawing group, the aromatic substitution becomes slower because formation of the carbocation intermediate is more difficult.

Because of the greater electronegativity of nitrogen compared to that of carbon and also the electron-withdrawing properties of the carbonyl group, one would expect that inductive effects would make the acetamido group electron withdrawing, thereby making aromatic substitution slower. However, the acetamido group could activate the benzene ring toward electrophilic substitution by donating electron density to the aromatic ring by a resonance effect:

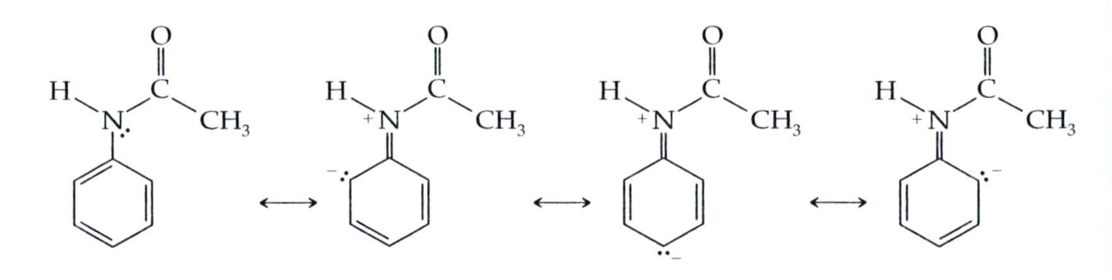

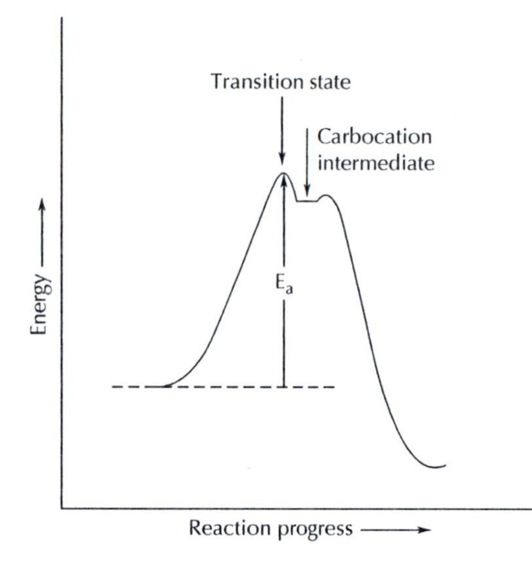

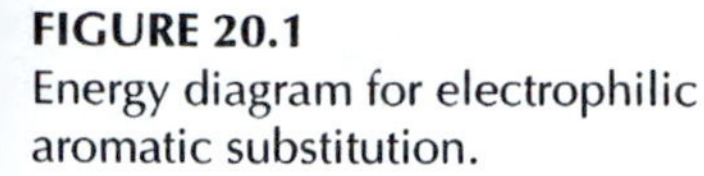
FIGURE 20.1
Energy diagram for electrophilic aromatic substitution.

An electron-donating resonance effect would provide additional stabilization of the carbocation intermediate in electrophilic aromatic substitution by delocalizing the charge into the substituent group. Because the carbocation intermediate closely resembles the transition state in the rate-determining endothermic step, stabilizing the intermediate would lower the activation energy for the reaction (Figure 20.1).

The inductive effect of the acetamido group predicts that it will be a deactivating group, and its resonance effect predicts that it will be an activating group. Moreover, if the acetamido group is activating, our understanding of the reaction mechanism predicts that it will be an ortho, para directing group, which can be demonstrated by examining the structure of the cationic intermediate in the bromination of acetanilide:

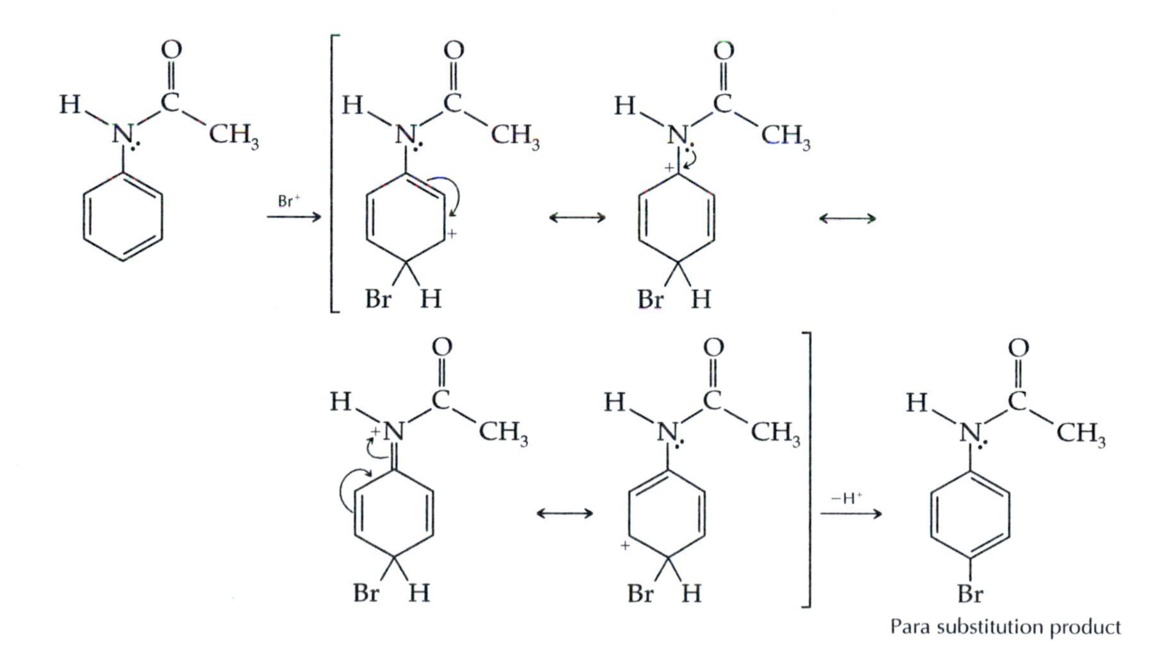

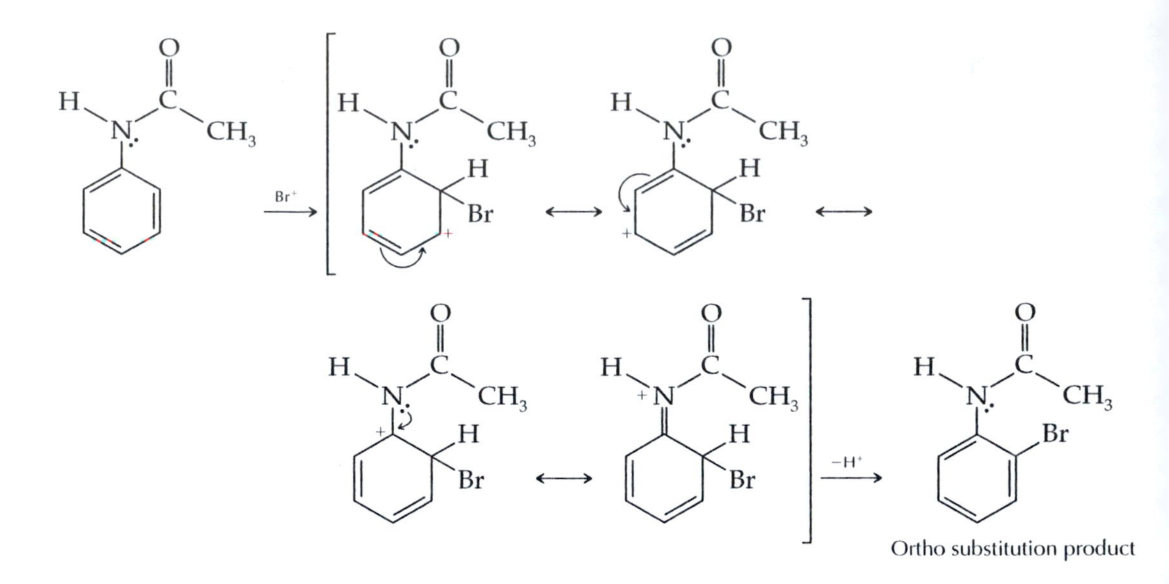

Bromination of Acetanilide

Now we can return to the original question: What is the major product obtained from the bromination of acetanilide in aqueous acetic acid, a polar solvent?

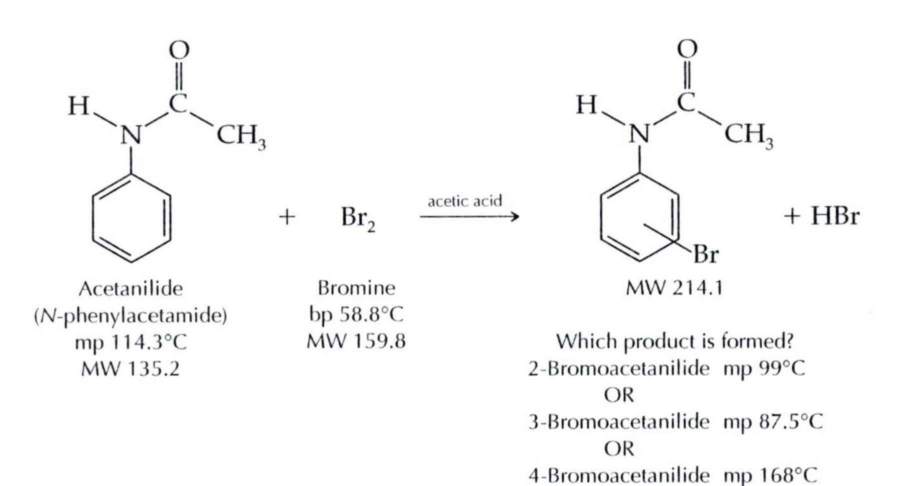

You can deduce the structure of the bromoacetanilide that you isolate by using its melting point or by NMR spectroscopy. If ^{1}H NMR is used to deduce the structure of the product, its symmetry is probably the most useful consideration. The symmetry of a para compound is such that the aromatic proton region at 7–8 ppm would have two signals, each integrating for two protons. Moreover, the two signals should appear to be a doublet or a distorted doublet. If the ortho or meta isomer is obtained, the aromatic proton region of the ^{1}H NMR spectrum is less symmetrical; both of these disubstituted benzene isomers would have four peaks in the aromatic region, which would range from singlets to triplets.

In the bromination of 4-methylacetanilide, there are two possible products:

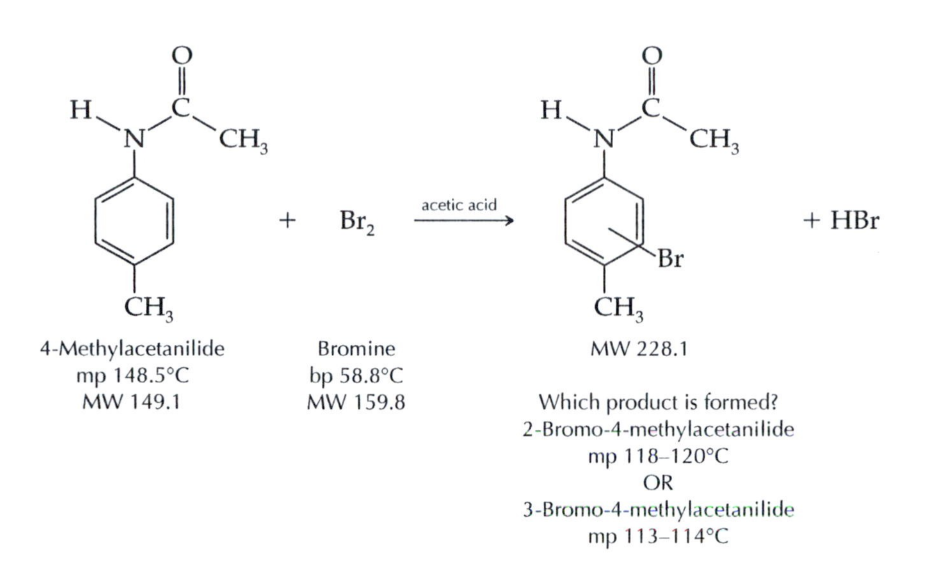

The determining factor in product formation is whether the acetamido group or the methyl group is the dominant directing group. In other words, the strongest activator wins out. It is necessary to do a careful recrystallization and melting-point determination to distinguish between 2-bromo-4-methylacetanilide and 3-bromo-4-methylacetanilide, which melt only 5–6° apart. The chemical shift of the C-3 proton in 2-bromo-4-methylacetanilide and that of the C-2 proton in 3-bromo-4-methylacetanilide are useful in distinguishing which isomer forms. Use Table 19.6 in Technique 19 to calculate the expected chemical shift values for these two protons.

MICROSCALE PROCEDURE

TEAMWORK: Carry out the experiment in teams of two students with one student doing the bromination of acetanilide and the other student using 4-methylacetanilide as the substrate.

PRELABORATORY ASSIGNMENT: Predict the likely products from the bromination of acetanilide and 4-methylacetanilide by using your understanding of the electronic principles governing electrophilic aromatic substitution.

Techniques Removal of Noxious Gases: Technique 7.4
Mixed Solvent Recrystallization: Technique 9.2
Microscale Recrystallization: Technique 9.7a
NMR Spectroscopy: Technique 19

SAFETY INFORMATION

Wear gloves and conduct the experiment in a hood.

Acetanilide and **4-methylacetanilide** are toxic and irritants. Avoid contact with skin, eyes, and clothing.

Acetic acid is a dehydrating agent and an irritant, and it causes burns. Dispense it only in a hood. Avoid contact with skin, eyes and clothing.

Potassium bromate is a strong oxidant and a cancer suspect reagent. No potassium bromate should be discarded in contact with paper because a fire could result. Wet the weighing paper after use and place it in a designated container, not in a wastebasket.

Hydrogen bromide solution (48%) and its vapor are corrosive and toxic, and they cause severe burns. Wear gloves and measure the solution in a hood.

The reaction mixture contains **bromine,** the vapors of which are toxic. Be sure that the gas trap is in place during the reaction period.

Preparation of the Gas Trap

Prepare a gas trap using a 20-inch piece of 2-mm Teflon tubing threaded through a rubber fold-over septum that fits the top of a 10-mL Erlenmeyer flask on one end and through a septum that fits the top of a 25-mL filter flask on the other end [see Technique 7.4, Figures 7.6 and 7.7a]. Put approximately 10 mL of water in the filter flask and close the top with the larger septum. Adjust the level of the tubing so that it is just above the surface of the water in the filter flask; the end of the tubing inside the 10-mL Erlenmeyer (reaction) flask should not extend below the neck of the flask.

Running the Reaction

Place 1.8 mmol of acetanilide or 4-methylacetanilide in a 10-mL Erlenmeyer flask containing a magnetic stirring bar. Add 0.10 g of potassium bromate and 2.4 mL of acetic acid to the flask. Stir the mixture rapidly while adding 0.45 mL of 48% (8.8 M) hydrobromic acid solution. Stir the reaction mixture at room temperature for 30 min.

At the end of the reaction period, pour the mixture into a beaker containing 25 mL of cold water. Rinse any residual solid from the Erlenmeyer flask with a few milliliters of water and add the rinse to the beaker. Stir the aqueous mixture for at least 10 min before collecting the crude product by vacuum filtration on a Hirsch funnel. Disconnect the vacuum source and pour 5 mL of saturated sodium bisulfite solution over the solid to remove any orange color due to residual bromine. Reconnect the vacuum and draw off the liquid. Repeat the washing process with 5 mL of water. Recrystallize the crude product from a 1:1 (v/v) aqueous ethanol solution in a 10-mL Erlenmeyer flask [see Technique 9.7a]. Cool the recrystallization mixture in an ice-water bath before collecting the product by vacuum filtration.

Allow the product to dry overnight before determining the yield, melting point, and NMR spectrum; alternatively, the product can be dried by heating it in a 60°C oven for 20 min. Prepare the NMR sample as directed by your instructor using $CDCl_3$ as the solvent [see Technique 19.2].

Interpretation of Experimental Data

Analyze the NMR spectrum of each bromoacetanilide and assign all the peaks [see Techniques 19.4 and 19.9, Table 19.2, and especially Table 19.6]. Where did the substitution of bromine occur in each acetanilide? Do your melting point and NMR data support substitution at this position?

CLEANUP: If the filtrate from the crude product still shows the orange color of bromine, add saturated sodium bisulfite solution dropwise until the color disappears. Neutralize both the filtrate solution and the solution remaining in the gas trap with solid sodium carbonate before washing them down the sink or pouring them into the container for aqueous inorganic waste. Pour the filtrate from the recrystallization into the container for halogenated waste.

OPTIONAL COMPUTATIONAL CHEMISTRY EXPERIMENT (Group Project)

Because this experiment involves a number of calculations, it makes a good group project. You will use the AM1 semiempirical method to predict the regiochemistry in the monobromination of acetanilide. The energy of activation is the heat of formation of the transition state, leading to the carbocation intermediate minus the heats of formation of the reactants.

$$E_a = \Delta H_f(\text{transition state}) - \Delta H_f(\text{acetanilide}) - \Delta H_f(\text{Br}^+)$$

Because the reactants are the same in the three reactions leading to ortho, meta, and para products, the one with the lowest-energy transition state has the lowest energy of activation and the fastest rate. Modeling a transition state is actually a complex procedure. The computation is considerably simplified if the heats of formation of the carbocation intermediates are calculated instead. This is a safe approximation because the structure of a transition state in an endothermic process is very similar to the structure of the product, which is the carbocation intermediate in this case.

Build the carbocation intermediate that would form from the reaction of Br^+ and benzene. Start with 1,3-cyclohexadiene. Replace one of the methylene hydrogen atoms with a bromine atom. Delete a hydrogen atom on the second methylene group. With MacSpartan you will also need to delete the valence. If you are using CAChe, place a positive charge on the carbon atom from which the hydrogen atom was removed.

Optimize the molecule using the AM1 semiempirical method. With MacSpartan, place a positive charge on the molecule when you set up the calculation. Record the heat of formation of the carbocation. Save the molecule so that you can use it as a template for building the intermediates for bromination of acetanilide.

Build the carbocations for the three possible acetamido carbocation intermediates. Optimize the molecules and record their heats of formation. Which carbocation intermediate has the lowest heat of formation? Is your experimental result consistent with the product predicted by your computations?

Build and optimize molecules of benzene and acetanilide. Record their heats of formation. Calculate the difference in energy between benzene and its carbocation intermediate and also the difference in energy between acetanilide and its lowest-energy carbocation intermediate. Is acetanilide predicted to be more or less reactive than benzene in electrophilic substitution? Is the acetamido group an electron-withdrawing group or electron-donating group?

Reference

1. Schatz, P. *J. Chem. Educ.* **1996,** *73,* 267.

Questions

1. Calculate the number of moles of HBr and $KBrO_3$ you used in the bromination reaction. Which compound is the limiting reagent?
2. Sodium bisulfite is a reducing agent. Write a balanced oxidation-reduction equation for the reaction that occurs between excess bromine and $NaHSO_3$ after the synthesis of the bromoacetanilide product.
3. What is the advantage of recrystallizing your product from a 1:1 aqueous ethanol solution rather than using ethanol alone?

EXPERIMENT

21 ACYLATION OF FERROCENE

PURPOSE: To carry out the acylation of a colorful organometallic compound and purify the product by chromatography.

In this two-week experiment you will study the acid-catalyzed reaction of acetic anhydride with the unusual aromatic compound, ferrocene. You will assay the composition of your crude product with thin-layer chromatography, and after purification of the product by column chromatography, you can characterize it by IR and NMR spectroscopy.

Electrophilic Aromatic Substitution

The delocalized π-systems of aromatic compounds are much less vulnerable to attack by electrophiles (E^+) than the localized π-double bonds of alkenes. However, as the following equation shows, aromatic compounds do undergo electrophilic substitution reactions. Because the conjugated 6 π-electron system of the aromatic ring is so stable, the carbocation intermediate loses a proton and regenerates the aromatic ring rather than undergoing reaction with a nucleophile:

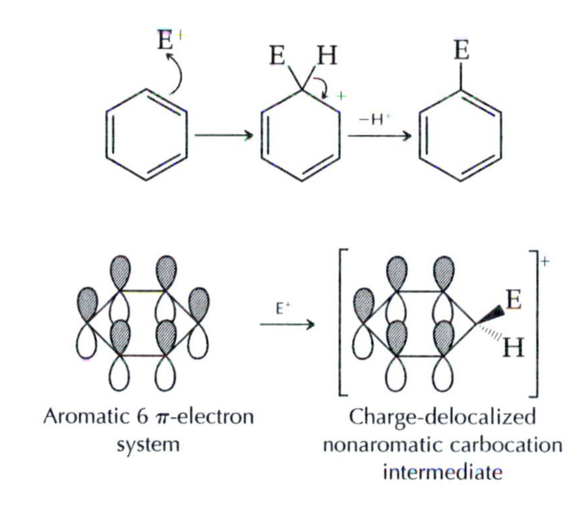

Friedel-Crafts acylation of aromatic compounds is an important example of an electrophilic aromatic substitution reaction. Named after Charles Friedel and James Mason Crafts, who discovered and developed it over 100 years ago, the Friedel-Crafts reaction has great utility in the synthesis of carbon-carbon bonds. In a Friedel-Crafts acylation reaction, an acyl anhydride or acyl chloride reacts with the aromatic compound in the presence of an acid catalyst, such as H_3PO_4 or $AlCl_3$. The product is an aromatic ketone:

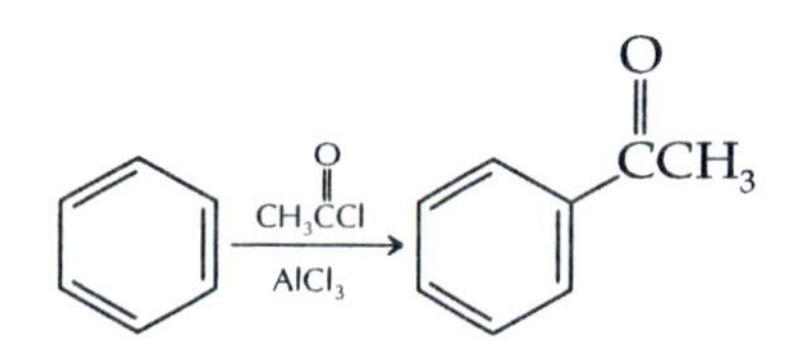

Ferrocene

Experiment 21 uses the unusual aromatic compound ferrocene, an organometallic compound that is composed of two planar five-membered rings that "sandwich" an iron ion:

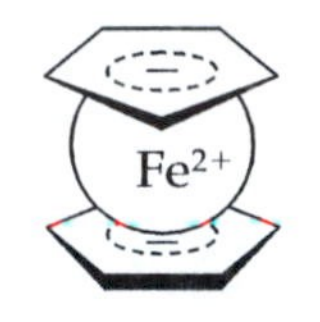

Ferrocene can be thought of as a compound formed by the bonding of an Fe^{2+} cation to two cyclopentadienide ligands, each bearing a negative charge. Each cyclopentadienyl anion is aromatic because it has 6 π-electrons in a completely conjugated ring:

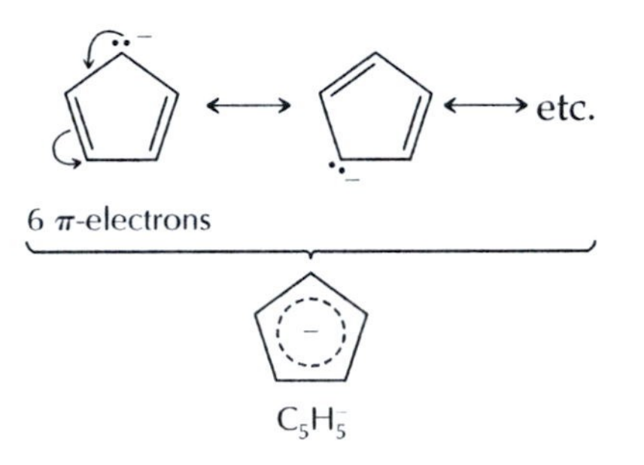

This Friedel-Crafts acylation of ferrocene produces acetylferrocene by using acetic anhydride in the presence of a catalytic amount of phosphoric acid:

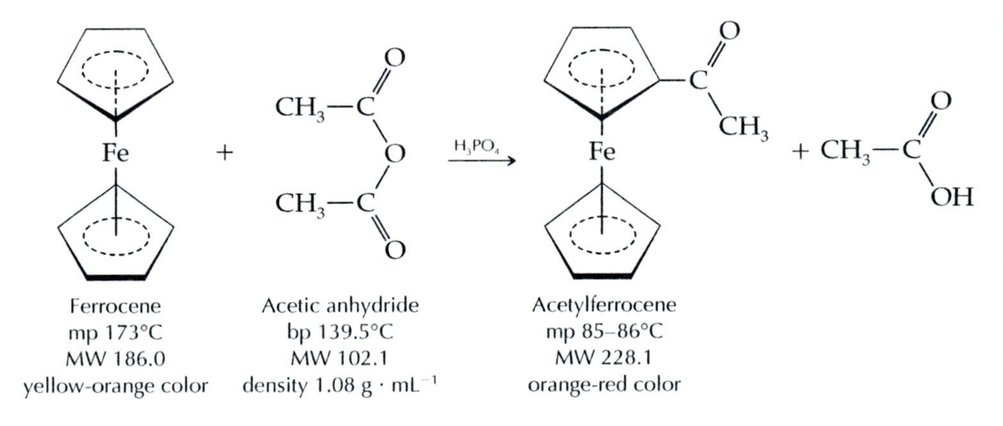

The reaction conditions generate the potent acylium-ion electrophile by protonation of the acetic anhydride, followed by loss of acetic acid:

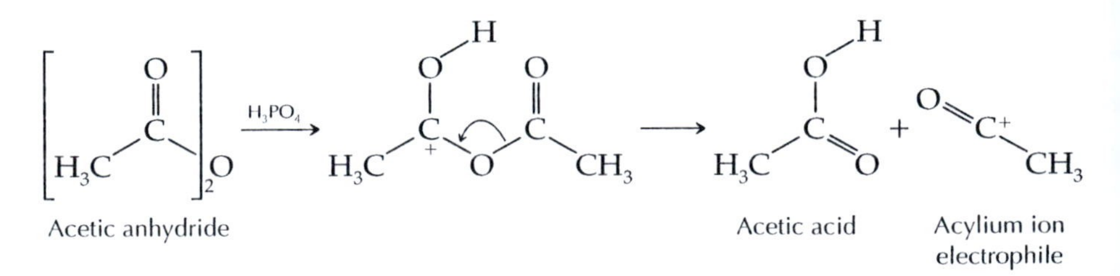

The electrophile then attacks the ring, resulting in substitution of the acetyl group for a ring proton:

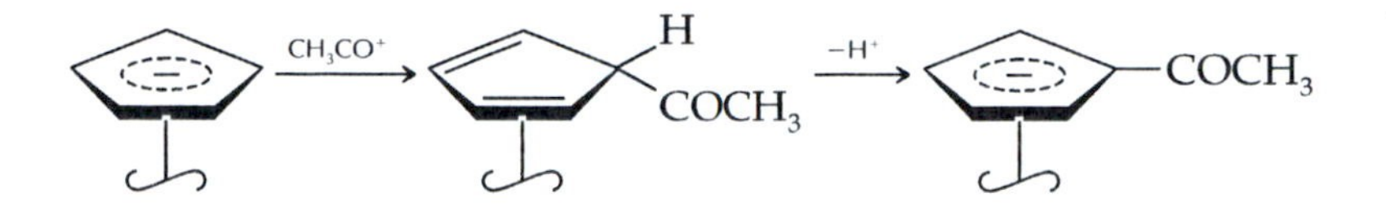

In your Friedel-Crafts reaction you will probably isolate a mixture of ferrocene, acetylferrocene, and diacetylferrocene. You will use column chromatography to separate the mixture. It is quite easy to see the colored bands that develop on the silica gel column. Acetylferrocene can be characterized by examining its IR and ^{1}H NMR spectra, and it also gives well-defined melting points after recrystallization from methanol.

You will see an intense band in the IR spectrum at about 1660 cm^{-1}, which corresponds to the conjugated carbonyl-stretching absorption. The ^{1}H NMR spectrum of ferrocene shows its 10 equivalent aromatic protons as a singlet at approximately 4.15 ppm, so you can expect the aromatic-ring protons of acetylferrocene to appear in the 4–5 ppm region of its ^{1}H NMR spectrum.

MICROSCALE PROCEDURE*

Techniques Thin-Layer Chromatography: Technique 15
Column Chromatography: Technique 17
IR Spectroscopy: Technique 18
NMR Spectroscopy: Technique 19

SAFETY INFORMATION

Ferrocene is relatively nontoxic, but avoid contact with the skin. The product, **acetylferrocene,** is highly toxic. Wear gloves and avoid contact with skin, eyes, and clothing.

Acetic anhydride is corrosive and a lachrymator (causes tears). Wear gloves and avoid contact with skin, eyes, and clothing. Dispense it in a hood.

Concentrated (85%) phosphoric acid is irritating to the skin and mucous membranes. Wear gloves. If you spill any phosphoric acid on your skin, wash it off immediately with copious amounts of water.

Aqueous **sodium hydroxide** solutions are corrosive and cause burns. Solutions as dilute as 9% (2.5 M) can cause severe eye injury. Avoid contact with skin, eyes, and clothing.

Hexane and **diethyl ether** are extremely volatile and flammable.

*This procedure was developed by David Alberg, Department of Chemistry, Carleton College, Northfield, MN.

Preparation of Acetylferrocene

Fit a dry 10-mL round-bottomed flask with a screw cap and drying tube containing anhydrous calcium chloride [see Technique 7.2, Figure 7.2b (omit the air condenser)]. Keep the drying tube on the flask except while you are adding reagents.

Place 200 mg (1.07 mmol) of ferrocene and 2.0 mL (21 mmol) of acetic anhydride in the flask. Swirl the flask to mix the reagents. Slowly add 0.4 mL of 85% phosphoric acid (about 10 drops with a Pasteur pipet; the exact amount is not critical). Put the drying tube on the flask and swirl the reaction mixture to thoroughly mix the reagents. Heat the flask on a steam bath or in a beaker of boiling water for 10 min with occasional swirling.

Remove the flask from the heat source and check the progress of the reaction by thin-layer chromatography on silica gel plates [see Technique 15]. Also spot the plate with a 2% solution of ferrocene in ether. Use 25:75 (v/v) anhydrous diethyl ether/hexane as the TLC elution solvent. A UV lamp allows you to visualize traces of ferrocene. A trace amount of ferrocene is likely, but if you can see a substantial yellow spot of ferrocene without the aid of the UV lamp, heat your reaction mixture for an additional 2–5 min. If the amount of ferrocene is minimal, cool the reaction flask for a total of 10 min.

Pour the reaction mixture over about 10 g of ice in a 50-mL beaker. Use an additional 1–2 mL of water to complete the transfer of your mixture to the ice. Partially neutralize the mixture by adding 5 mL of 6 M sodium hydroxide in at least three portions. Determine the pH with pHydrion paper or other pH paper. Continue adding 6 M NaOH dropwise until the pH is 7–8. Swirl the beaker after each addition to mix the contents. Cool the mixture to room temperature and collect the product by vacuum filtration on a Hirsch funnel. Use a few milliliters of water to complete the transfer of the tarry solid. With the vacuum on, pull air over the crude product on the Hirsch funnel for 15 min to dry the product while you prepare for the column chromatography.

Purification by Column Chromatography

Assemble all the equipment and reagents that you will need for the entire chromatography procedure before you begin to prepare the column.

Large-volume Pasteur pipets, available from Fisher Scientific, catalog no. 13678-8, have a capacity of 4 mL.

Read this procedure completely and review Technique 17 before you undertake this part of the experiment.

Obtain about 25 mL of hexane in a 50-mL Erlenmeyer flask fitted with a cork. Transfer your air-dried crude product to a 13 × 100 mm test tube and add about 1 mL of hexane. Much of the material will not dissolve in the hexane. Spot a TLC plate with this hexane mixture, then set the test tube and the TLC plate aside while you prepare the column.

Obtain a large-volume Pasteur pipet to use as the chromatography column and pack a small plug of glass wool down into the stem, using a wood applicator stick or a thin stirring rod. Clamp this pipet in a vertical position and place a 25-mL Erlenmeyer flask underneath it to collect the hexane that you will be adding to the column. Weigh approximately

1.7–1.8 g of silica gel (70–230 mesh) in a tared 50-mL beaker; add approximately 15 mL of hexane to make a thin slurry. Transfer the silica gel slurry to the column, using a regular Pasteur pipet. Continue adding slurry until the column is one-half to two-thirds full of silica gel. Fill and drain the column four or five times with hexane to pack it well. **(Note: Do not let the hexane level fall below the top of the silica gel.)** The eluted hexane can be reused for this purpose. After the silica gel is packed, add a 2–3 mm layer of sand above the silica gel by letting it settle through the hexane.

Be sure that the silica gel is covered with solvent at all times during the chromatographic procedure.

Allow the hexane level to almost reach the top of the sand, and place a flask labeled "Fraction 1" under the column. Transfer your crude product mixture to the top of the column, using a Pasteur pipet and as many small portions of hexane as necessary to transfer all your material (do not use the eluted hexane). When all the crude product is on the column and the hexane level is just above the top of the silica gel, elute the column with 15 mL of hexane. You may see a faint yellow color in the hexane solution collecting in fraction 1; the color is due to ferrocene that did not react.

Next elute with 10 mL of 50:50 (v/v) hexane/anhydrous diethyl ether solution. You will see the orange-red acetylferrocene move rapidly down the column. Collect the eluent in fraction 1 until you see the orange-red solution in the column tip, then quickly change the collection flask to a clean, tared 50-mL Erlenmeyer flask labeled "Fraction 2." Continue adding 50:50 hexane/ether until the orange-red product has eluted from the column. (This elution requires about 10–15 mL of 50:50 hexane/ether.) Spot fraction 1 and fraction 2 on the same TLC plate that you have already spotted with your crude acetylferrocene. Develop the TLC plate as you did previously. Record the results in your notebook.

Recover your purified acetylferrocene by evaporating the solvent from fraction 2 on a steam bath or with a stream of nitrogen in a hood. Alternatively, if a rotary evaporator is available, transfer fraction 2 to a tared 25- or 50-mL round-bottomed flask and remove the solvent under reduced pressure. If you will be analyzing your product by NMR, do not heat the product solutions extensively in the presence of air, as O_2 can oxidize the Fe^{2+} of ferrocene to Fe^{3+}, which is paramagnetic. Even a small amount of this oxidation can cause broadening of the NMR peaks.

Weigh your purified product, calculate the percent yield, and determine the melting point of your acetylferrocene. Prepare a sample for NMR analysis [see Technique 19.2] and IR analysis [see Technique 18.4] as directed by your instructor.

Interpretation of Experimental Data

Analyze both spectra and assign all major peaks [see Technique 18.5 and Table 18.2; Techniques 19.4 and 19.9 and Table 19.2]. How would the IR spectrum of ferrocene differ from that of acetylferrocene? How would the NMR spectrum of ferrocene differ from that of acetylferrocene?

CLEANUP: The aqueous filtrate from the crude product may be washed down the sink or placed in the container for aqueous inorganic waste. Pour any remaining TLC solvent and fraction 1 into the container for flammable (organic) waste. Place the TLC plates and the silica gel from the column in the container for inorganic solid waste.

Questions

1. The diacetylferrocene produced in this reaction remains near the top of the column under the chromatographic conditions used in the experiment. Explain the order of elution of ferrocene and acetylferrocene and why the diacetylferrocene is retained by the column.
2. In the ^1H NMR spectrum of most aromatic compounds, the aromatic protons exhibit a chemical shift of 7–8 ppm. However, in ferrocene, the chemical shift of the aromatic protons is 4.15 ppm. Explain what factors cause the upfield shift.
3. Explain how the ^1H NMR spectrum of ferrocene supports the assigned sandwich structure rather than a structure in which the iron atom is bound to only one carbon atom of each ring.

ALKYLATION OF 1,4-DIMETHOXYBENZENE

QUESTION: What is the regiochemistry of alkylation with a para disubstituted benzene compound?

In Experiment 22 you will synthesize a tetrasubstituted benzene derivative, using a tertiary alcohol and H_2SO_4 to produce the electrophile in the aromatic substitution reaction. The substitution pattern of the product will be proved by its melting point and NMR spectrum.

Electrophilic Aromatic Substitution

Even though the delocalized π-systems of aromatic compounds are much less vulnerable to attack by electrophiles (E^+) than are the localized π-double bonds of alkenes, aromatic compounds undergo electrophilic substitution reactions. Their mechanism has been thoroughly studied. Because the conjugated 6π-electron system of the aromatic ring is so stable, the carbocation intermediate loses a proton and generates the aromatic ring rather than undergoing reaction with a nucleophile:

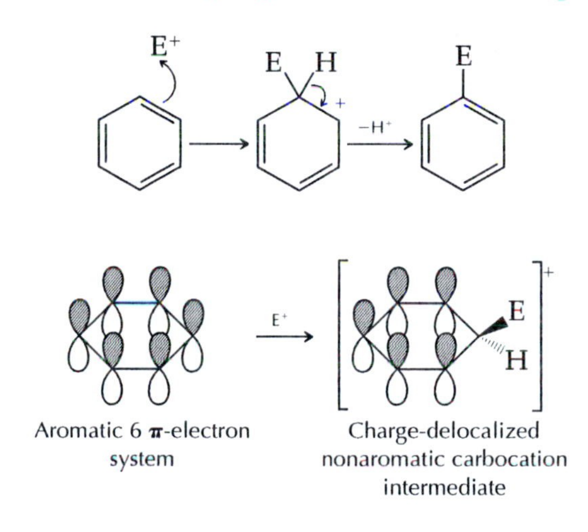

Aromatic 6 π-electron system

Charge-delocalized nonaromatic carbocation intermediate

The alkylation of aromatic compounds is an example of the Friedel-Crafts reaction, which is an important industrial process. Named after Charles Friedel and James Mason Crafts, who discovered its synthetic importance over 100 years ago, the Friedel-Crafts reaction has great utility in the synthesis of carbon-carbon bonds. Nearly 10 billion pounds of ethylbenzene are produced in the United States each year by the reaction of ethylene and benzene in the presence of an acidic catalyst. Most of the ethylbenzene is dehydrogenated to form styrene, from which polystyrene is made:

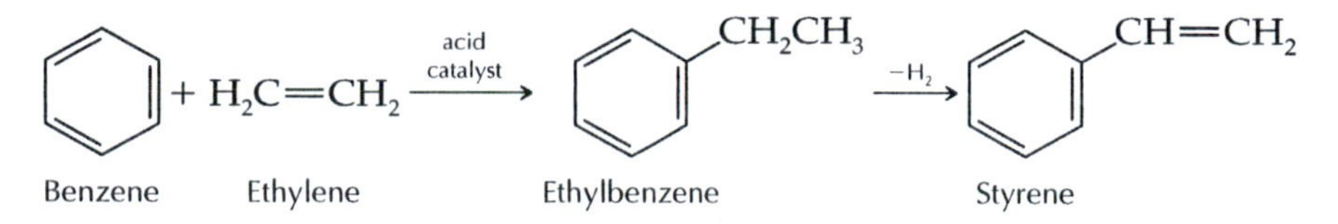

Alkylation with tert-Butyl Carbocations

In the alkylation of benzene the electrophile is a carbocation. In this experiment it is a *tert*-butyl carbocation, produced by the protonation of *tert*-butyl alcohol with sulfuric acid, followed by loss of a water molecule:

$$(CH_3)_3C{-}OH + H_2SO_4 \longrightarrow (CH_3)_3C{-}\overset{+}{O}H_2 \xrightarrow{-H_2O} (CH_3)_3C^+$$

tert-Butyl cation

The carbocation is especially easy to produce because it is tertiary. The mechanism for the electrophilic aromatic substitution of benzene itself would be

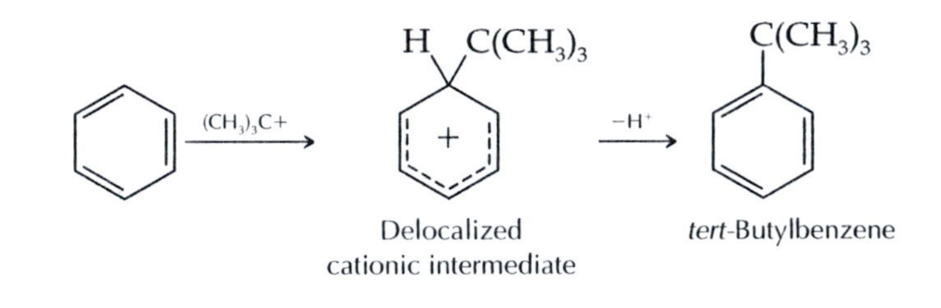

Delocalized cationic intermediate

tert-Butylbenzene

The delocalized cationic intermediate corresponds to three localized resonance forms:

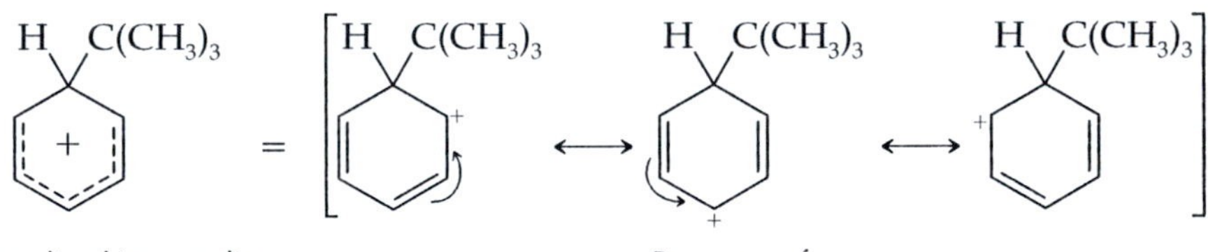

Delocalized intermediate

Resonance forms

The delocalization of the positive charge over a large portion of the ring system stabilizes the cationic intermediate. In addition, electron-donating groups on the benzene ring speed up the substitution process by stabilizing the carbocation intermediate, which is produced in the slow step of the reaction. Because the two methoxy groups of 1,4-dimethoxybenzene are both activating, the rate of the reaction is relatively fast. The positive charge in the rate-determining transition state can be delocalized not only over much of the ring but also onto the methoxy groups, which donate electron density by resonance effects:

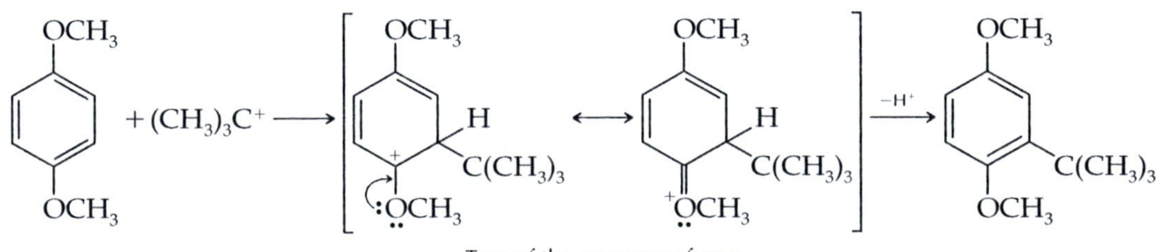

Two of the resonance forms

Under the reaction conditions that you will use, there are 2.4 moles of *tert*-butyl alcohol for every mole of 1,4-dimethoxybenzene. Since the *tert*-butyl group is also mildly activating and plenty of *tert*-butyl alcohol is present, the product you will isolate is the result of dialkylation. When two electrophilic substitutions occur, a tetrasubstituted product forms.

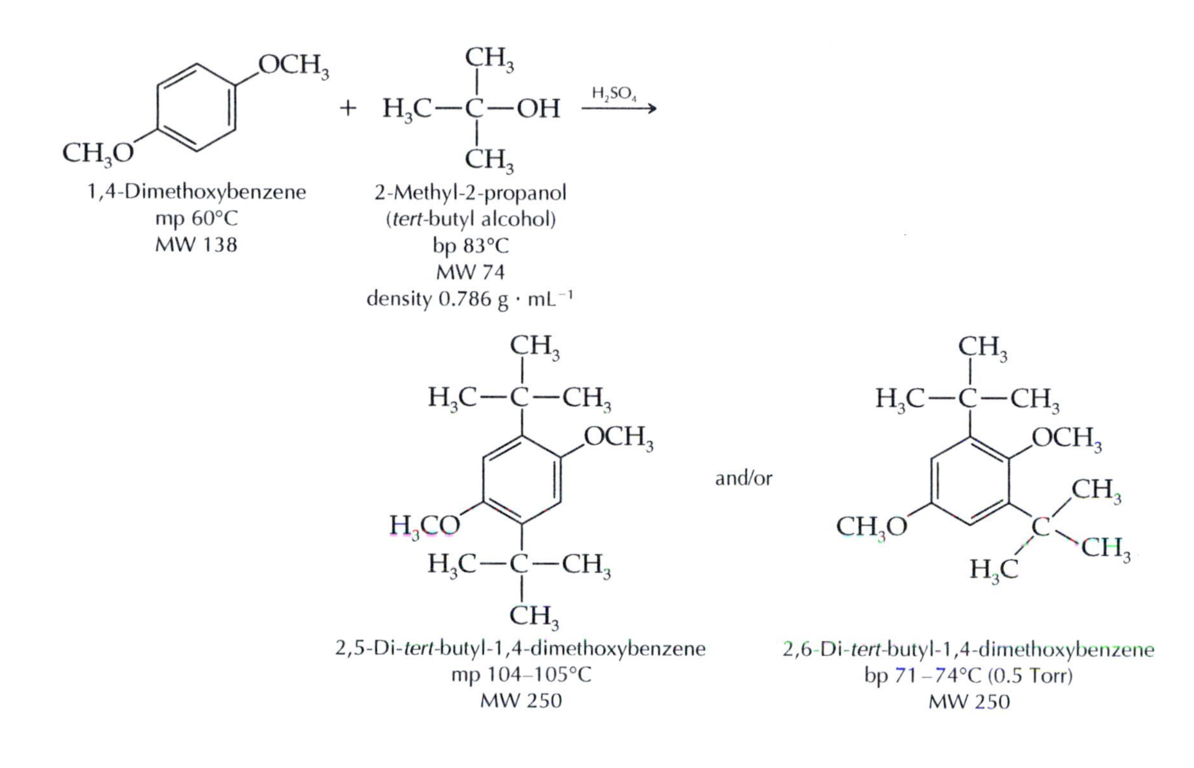

The question in Experiment 22 is which of the two possible products is the major one. The *tert*-butyl group is quite bulky, and this may have an effect on which product forms faster. It might inhibit the formation of the 2,6-di-*tert*-butyl isomer because this would have three large groups right next to one another. Steric hindrance in the transition state might retard its formation.

It should not be difficult to determine which di-*tert*-butyl isomer forms in the reaction. One of them is a solid at room temperature whereas the other is a high-boiling liquid. The ^{1}H NMR spectrum of the product should also be revealing. The 2,5-di-*tert*-butyl isomer has an element of symmetry, which makes the methyl protons on each methoxy group identical. Thus, there will be only one signal for the methoxy protons in its NMR spectrum. It is a different story for the 2,6-di-*tert*-butyl isomer, where there is no symmetry that would make the two methoxy groups magnetically equivalent; this isomer will have two separate singlets for the two different methoxy groups in its NMR spectrum.

MICROSCALE PROCEDURE

Techniques Recrystallization: Technique 9.7b
Microscale Extraction: Technique 8.6c
NMR Spectroscopy: Technique 19

PRELABORATORY ASSIGNMENT: Using your model kit, make models of the two possible dialkylation products of 1,4-dimethoxybenzene. How much more steric hindrance does the 2,6-di-*tert*-butyl isomer manifest than the 2,5-di-*tert*-butyl isomer?

OPTIONAL PRELABORATORY ASSIGNMENT: If the reaction is not at equilibrium and the product is determined by the rate at which it forms, it would be more accurate to look at the carbocation intermediate, which is very like the transition state in the slow step of the reaction. Using your model kit, make models of the two isomeric carbocation intermediates for the formation of 2,5- and 2,6-di-*tert*-butyl-1,4-dimethoxybenzene and evaluate the relative steric hindrance in their formation.

SAFETY INFORMATION

Sulfuric acid is corrosive and causes severe burns. Measure the required amount carefully. Notify the instructor if any acid is spilled.

2-Methyl-2-propanol (*tert*-butyl alcohol) causes skin and eye irritation. Avoid contact with skin, eyes, and clothing. Wash thoroughly after handling it.

1,4-Dimethoxybenzene is a skin irritant. Wash thoroughly after handling it.

Acetic acid is a dehydrating agent and an irritant and causes burns. Dispense it in a hood and avoid contact with skin, eyes, and clothing.

Diethyl ether is extremely volatile and flammable.

Methanol is toxic and flammable. Pour it only in a hood.

Place 150 mg of 1,4-dimethoxybenzene in a 10-mL Erlenmeyer flask, and add 0.25 mL of 2-methyl-2-propanol (*tert*-butyl alcohol) and 0.5 mL of acetic acid. Warm the mixture briefly on a hot plate until the solid dissolves. Clamp the flask in a small ice-water bath and stir the contents with a short stirring rod. Obtain 1.0 mL of concentrated sulfuric acid in a 13×100 mm test tube and chill the acid in the ice-water bath. While stirring the mixture in the flask, add the sulfuric acid dropwise with a Pasteur pipet over a period of 1 min. Leave the reaction mixture in the ice-water bath for an additional 5 min, then allow the flask to stand at room temperature for 10 min.

Again cool the flask in the ice-water bath and add 5 mL of ice-cold water while stirring the contents. If the product is a solid, proceed to the work-up for a solid product. If the product separates from the reaction mixture as a liquid, use the procedure for a liquid product.

For a Liquid Product

For a liquid product, add 2 mL of diethyl ether to the mixture in the reaction flask, then transfer the contents of the flask to a 15-mL centrifuge tube using a Pasteur pipet fitted with a syringe. Rinse the flask with another 2 mL of ether and transfer the rinse to the centrifuge tube. Remove the lower aqueous phase from the centrifuge tube with the Pasteur pipet and place it in a 250-mL beaker containing approximately 50 mL of water. The aqueous layers from the following washes can be combined in the same 250-mL beaker. Wash the remaining ether layer first with 2 mL of water and then with 2 mL of saturated sodium bicarbonate solution. **(Caution: Foaming.)** Finally, wash the ether layer with 2 mL of water. After removing the last aqueous wash, dry the ether solution with anhydrous sodium sulfate for at least 10 min.

Transfer the dried ether solution to a tared 13 × 100 mm test tube using a Pasteur filter pipet [see Technique 8.5]. Working in a hood, evaporate the ether using a stream of nitrogen or air until a constant mass (within 0.030 g) is reached. Determine the percent yield. Prepare a sample of your product for NMR analysis as directed by your instructor [see Technique 19.2].

For a Solid Product

If a solid product has precipitated from the reaction mixture, collect the crude product by vacuum filtration on a Hirsch funnel. Chill 1 mL of methanol in a small test tube in the ice-water bath. Wash the crude product three times with 1-mL portions of water. Then wash the product twice with 0.5-mL portions of ice-cold methanol; to do this, disconnect the vacuum, cover the crystals with the cold methanol, then reconnect the vacuum to remove the methanol.

Transfer the crude product to a 10-mL Erlenmeyer flask. Recrystallize the product from a minimum amount of methanol, using a boiling stick or stone to prevent bumping and a steam bath or a 70–75°C water bath as the heat source [see Technique 9.7a]. Collect the product on a Hirsch funnel by vacuum filtration. After the crystals are dry, determine the melting point and percent yield. Prepare a sample of your dried product for NMR analysis as directed by your instructor [see Technique 19.2].

Interpretation of Experimental Data

Analyze your NMR spectrum and assign all the signals [see Techniques 19.4 and 19.9 and Tables 19.2 and 19.6]. What is the regiochemistry of the alkylation?

CLEANUP: For a solid product, pour the filtrate from the reaction mixture into a 250-mL beaker containing approximately 50 mL of water and neutralize the solution in the 250-mL beaker with solid sodium carbonate **(Caution: Foaming.)** before washing it down the sink or pouring it into the container for aqueous inorganic waste. Dispose of the methanol filtrate from the recrystallization in the container for flammable (organic) waste.

For a liquid product, neutralize the combined aqueous extracts in the 250-mL beaker with solid sodium carbonate **(Caution: Foaming.)** before washing it down the sink or pouring it into the container for aqueous inorganic waste. Place the spent drying agent in the container for solid inorganic waste.

References

1. Hammond, C. N.; Tremelling, M. J. *J. Chem. Educ.* **1987**, *64*, 440–441.
2. Eaton, D. *Laboratory Investigations in Organic Chemistry*; McGraw-Hill: New York, 1989, pp. 454–456.

Questions

1. Draw all the resonance forms that account for the ring substitution of the second *tert*-butyl group on the 1,4-dimethoxybenzene ring system.
2. When 2-methylpropene is treated with sulfuric acid, the resulting reaction mixture can be used to convert benzene to *tert*-butylbenzene. Write a mechanism that accounts for this conversion.
3. What product(s) would be formed if 1-butanol were used instead of *tert*-butyl alcohol in this experiment? Would this be the best way to substitute a butyl group in a benzene ring? Why?

EXPERIMENT

KINETIC VERSUS EQUILIBRIUM CONTROL IN THE ALKYLATION OF CHLOROBENZENE

QUESTION: What is the ratio of ortho, meta, and para products in the electrophilic substitution of chlorobenzene under conditions of kinetic and thermodynamic or equilibrium control?

In this two-week experiment you will study the effect of a solvent and the reaction time on the regioselectivity of the aluminum chloride-catalyzed reaction of 2-chloropropane with chlorobenzene, using gas chromatography to assay the product mixture. Including the optional experiment provides an opportunity for teamwork.

Kinetic and Thermodynamic Control

When two different products are possible in a chemical reaction and the isolated product is the one that forms faster, the reaction is said to be under ***kinetic control.*** Often the more stable product forms faster, but sometimes a less stable product forms quickly and then gets trapped. In both situations, the product that forms faster is the one that is isolated, so both reactions are under kinetic control. However, if the two products are in equilibrium and the most stable product mixture results, especially if it is not the mixture formed initially, the reaction is said to be under ***thermodynamic*** or ***equilibrium control.***

The energy diagram shown in Figure 23.1 may help you visualize the concepts of thermodynamic and kinetic control where the products of the two processes are different. The energy barrier to the product on the left is lower, and it is formed faster (kinetic control). The product on the right is formed more slowly because it has a higher-energy transition state, but it has lower energy than the product on the left. If the product on the right is the one isolated, the reaction is under thermodynamic control.

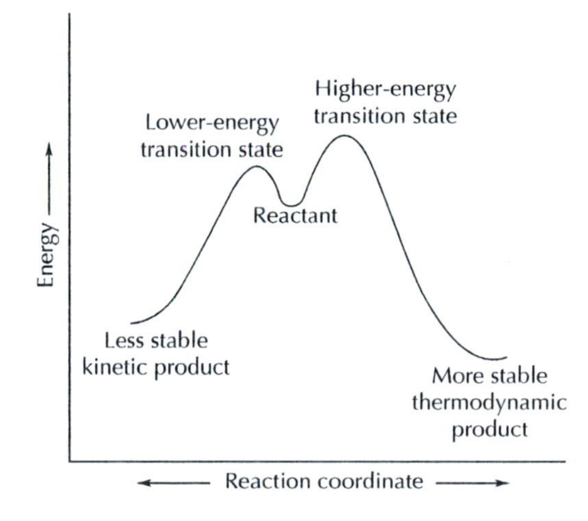

FIGURE 23.1
Energy diagram for a reaction illustrating both kinetic and thermodynamic control.

In a process that behaves as shown in Figure 23.1, the lower-energy product will not be formed first. It takes reaction conditions that allow an equilibrium to develop so that the product of kinetic control can return to the reactants many times. Every so often the thermodynamic product forms, and once it does, it is much harder for it to get over the high-energy barrier of the reverse reaction and return to the starting materials. Reactions run at low temperature and for short times favor kinetically controlled products. At equilibrium, when there is sufficient energy to overcome all transition-state barriers, the outcome of the reaction is determined by the relative energies of the products.

Electrophilic Aromatic Substitution

Even though the delocalized π-systems of aromatic compounds are much less vulnerable to attack by electrophiles (E^+) than the localized π-double bonds of alkenes, aromatic compounds undergo electrophilic substitution reactions. Their mechanism has been thoroughly studied. Because the conjugated 6 π-electron system of the aromatic ring is so stable, the carbocation intermediate loses a proton and regenerates the aromatic ring rather than undergoing reaction with a nucleophile:

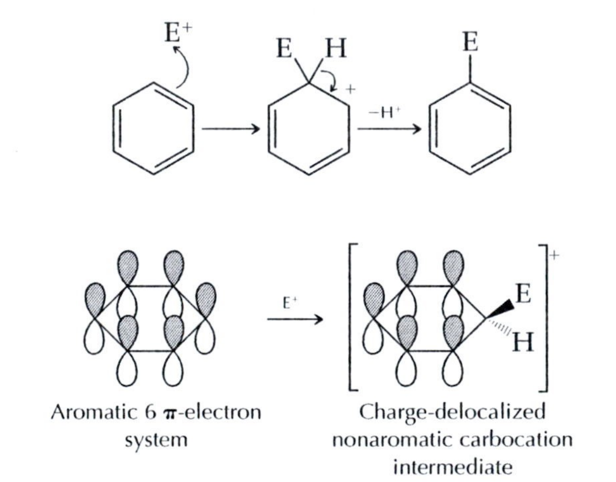

The alkylation of aromatic compounds is an example of the Friedel-Crafts reaction that is an important industrial process. Named after the French and American chemists who discovered its synthetic importance over 100 years ago, the Friedel-Crafts reaction has great utility in the synthesis of carbon-carbon bonds. Nearly 10 billion pounds of ethylbenzene are produced in the United States each year by the reaction of ethylene and benzene under Friedel-Crafts conditions. Most of the ethylbenzene is dehydrogenated to form styrene, from which polystyrene is made:

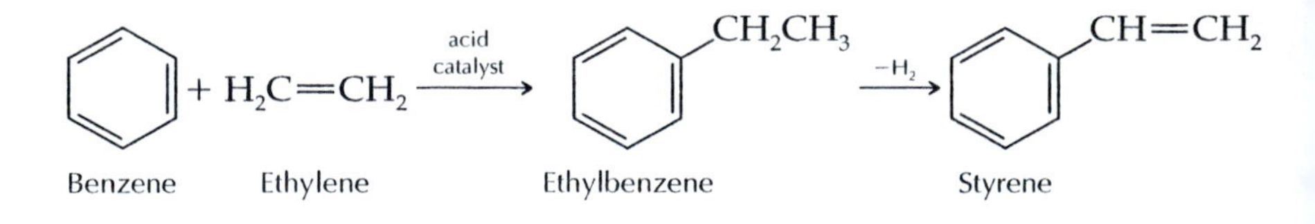

In the alkylation of benzene with 2-chloropropane and $AlCl_3$, the electrophile is most probably the isopropyl carbocation. $AlCl_3$ is a strong, electron-deficient Lewis acid, which can bond to the chlorine atom of 2-chloropropane and form the isopropyl carbocation:

$$H_3C-\underset{\displaystyle Cl}{\overset{\displaystyle H}{C}}-CH_3 + AlCl_3 \rightleftharpoons H_3C-\underset{+}{\overset{\displaystyle H}{C}}-CH_3 + AlCl_4^-$$

The mechanism for the electrophilic aromatic substitution reaction of benzene would be

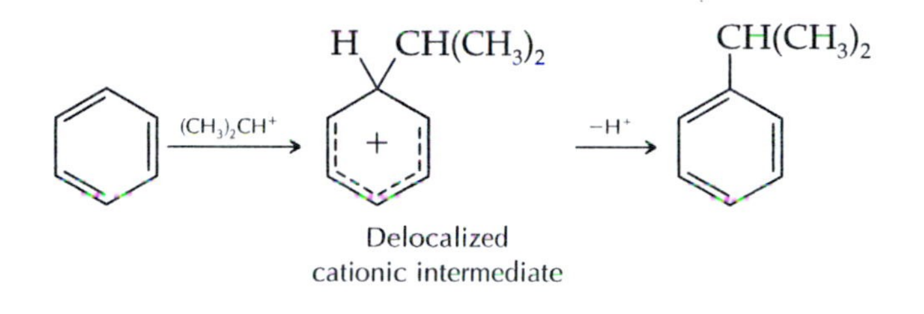

The delocalized cationic intermediate corresponds to three localized resonance forms:

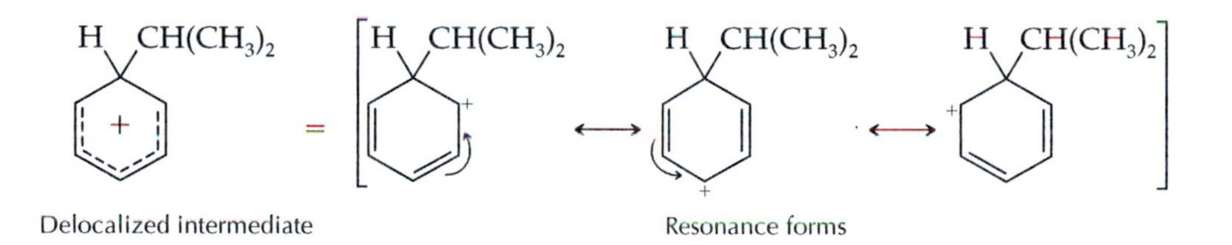

The delocalization of the positive charge over a large portion of the ring system stabilizes the cationic intermediate. In addition, electron-donating groups on the benzene ring speed up the substitution process by stabilizing the carbocation intermediate that is produced in the slow step of the reaction. When a substituent can donate electron density into the ring, it will be an ortho, para director.

The question in Experiment 23 is: In the alkylation of chlorobenzene with 2-chloropropane and $AlCl_3$, what are the percentages of ortho, meta, and para isomers in the product mixture and how can we understand the factors that control the product distribution?

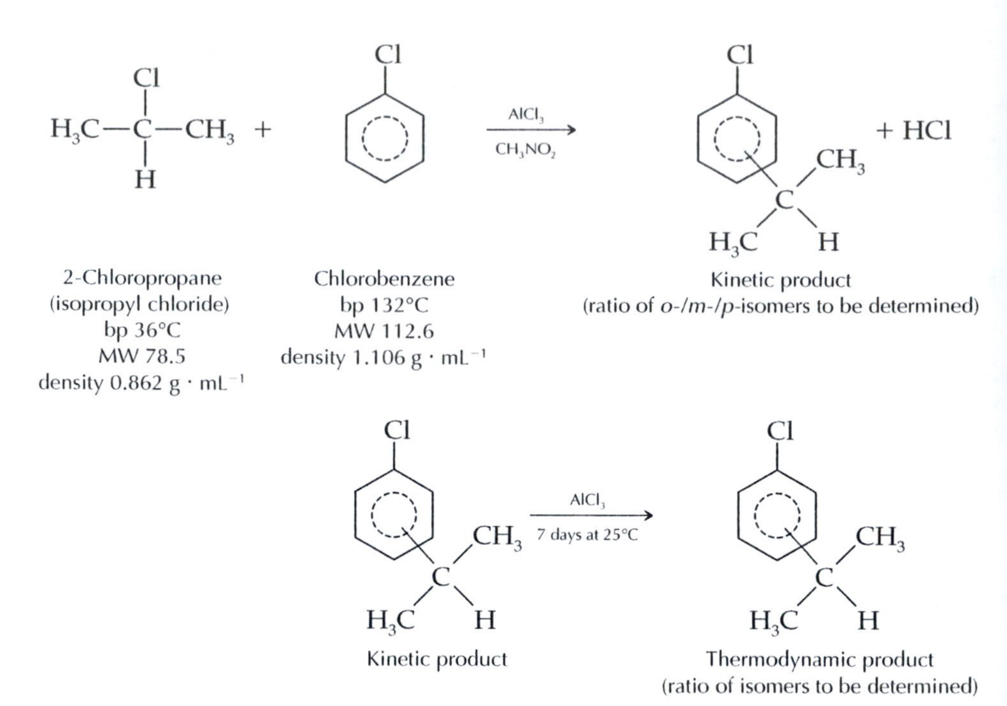

You will use a nitromethane/aluminum chloride mixture in the first experiment to temper the reactivity of the Lewis acid catalyst and form the product mixture that reflects kinetic control in the reaction:

$$\underset{\text{Nitromethane}}{CH_3{-}NO_2} + AlCl_3 \longrightarrow \underset{\text{Lewis acid-base complex}}{[CH_3{-}NO_2 \cdots\cdots AlCl_3]}$$

In the second reaction you will omit the nitromethane and use a much longer reaction time; under these conditions, thermodynamic control is expected.

MICROSCALE PROCEDURE*

PRELABORATORY ASSIGNMENT: Based on your understanding of the electronic directing effects of a chlorine substitutent on the regiochemistry of this Friedel-Crafts alkylation reaction, should Cl be an ortho, para director or a meta directing group? Why?

TEAMWORK: Work in a team of three students to plan the two optional experiments. One student should carry out the standard experiment and each of the other team members should carry out one of the experiments your team has designed.

*This experiment requires one laboratory period to prepare the kinetic product, a 7-day reaction period for conversion of the kinetic product to the thermodynamic product, and approximately 1 h of another laboratory period to work up the thermodynamic product and carry out gas chromatographic analysis of the product.

Techniques Microscale Extraction: Technique 8.6b
Gas Chromatography: Technique 16

SAFETY INFORMATION

Wear neoprene gloves while carrying out the experimental procedures.

Syringe needles are a puncture hazard.

Both **chlorobenzene** and **2-chloropropane** (isopropyl chloride) are flammable and are irritants. Avoid contact with skin, eyes, and clothing.

When not in use, the bottle of $AlCl_3$ should be tightly capped and stored in a desiccator.

Aluminum chloride is corrosive and moisture sensitive. It reacts with atmospheric moisture to form HCl.

Nitromethane is flammable, moisture sensitive, and toxic.

Kinetic Product

Place 1.40 mL of a solution containing 0.35 mmol · mL^{-1} of aluminum chloride and 0.70 mmol · mL^{-1} of nitromethane in chlorobenzene in a 15-mL centrifuge tube. Put a magnetic stirring bar into the tube and cap it with a fold-over rubber septum. Insert a syringe needle to act as a vent. Clamp the tube in an ice-water bath and stir the solution for several minutes. Prepare a solution containing 0.15 mL of 2-chloropropane and 0.80 mL of dry chlorobenzene in a small test tube or vial. Draw this solution into a 1-mL syringe.

When the solution in the tube is thoroughly chilled, insert the syringe through the septum and add the 2-chloropropane solution dropwise over a period of 5 min. Remove the syringe and the venting needle. Continue stirring the reaction mixture for 40 min while the ice melts and the cooling bath warms to room temperature.

Open the tube and add 2.0 mL of water. Stir the mixture for 1–2 min, then cap the tube and shake it. Allow the layers to separate. Transfer the lower organic phase to another centrifuge tube or test tube with a Pasteur pipet fitted with a syringe or a Pasteur filter pipet [see Technique 8.6b].

The densities of the two phases are similar, with the organic layer being slightly more dense; this similarity in densities may cause the cloudy organic layer to be partially suspended in the aqueous phase rather than completely separated. Should incomplete separation occur, cap the centrifuge tube and spin it in a centrifuge for 2–3 min.

SAFETY PRECAUTION

Some CO_2 will form during the Na_2CO_3 wash; open the cap slowly to vent the centrifuge tube.

Wash the organic phase, now in the second centrifuge tube, with 1.0 mL of 5% sodium carbonate solution. After this washing, transfer the organic phase to a dry 13 × 75 mm test tube. Add approximately 150–200 mg of anhydrous potassium carbonate to the test tube and allow the mixture to stand for 5 min. With a dry Pasteur filter pipet, transfer the product solution to another test tube; rinse the drying agent with 0.5 mL of chlorobenzene and transfer the rinse to the test tube containing the product. Dry the product a second time with potassium carbonate for 10 min.

Drying the product mixture twice with small amounts of drying agent is more effective than using a larger amount initially.

Separate the product from the drying agent with a clean Pasteur filter pipet, transferring it to a small screw-capped vial. Remove a sample of your product for GC analysis, then use the product solution remaining in the vial to prepare the thermodynamic product.

CLEANUP: Combine the aqueous solutions from the extractions and neutralize the solution with solid sodium carbonate **(Caution: Foaming.)** before washing it down the sink or placing it in the container for aqueous inorganic waste. The spent potassium carbonate drying agent is coated with halogenated organic compounds; place it in the container for hazardous solid waste.

Thermodynamic Product

Weigh a sample of aluminum chloride between 40 and 50 mg as quickly as possible and add it to the kinetic product in the screw-capped vial. Cap the vial tightly and swirl it occasionally for a period of 15 min. Place the vial upright in your drawer for 7 days.

After 7 days, add 2.0 mL of water to the product mixture, swirl the vial to dissolve the aluminum chloride, and transfer the entire contents of the screw-capped vial to a 15-mL centrifuge tube. Carry out the separation, washing, and drying of the organic phase by the same procedure used for the kinetic product. Analyze the product mixture by gas chromatography. Submit any remaining product mixture to your instructor.

If the volume of your product is less than 1 mL, add 1.0 mL of diethyl ether to the mixture in the centrifuge tube.

CLEANUP: Follow the same cleanup procedure you used for the kinetic product.

Gas Chromatographic Analysis of the Product Solutions

For a packed column gas chromatograph, use a 15% polyphenyl ether on Chromasorb W (acid-washed, 60/80 mesh) column at 110°C or a nonpolar column such as OV-101 at 100°C (Ref. 1). The product solution can be analyzed without further dilution. Consult your instructor about the appropriate injection volume for the packed column instruments in your laboratory. For a capillary column chromatograph, use a nonpolar column such as polydimethylsiloxane (SE-30 or OV-1) and a column temperature of 100°C. Prepare a solution for GC analysis by combining 2 drops of your product mixture and 0.5 mL of diethyl ether; inject 1 μL of this ether solution into the chromatograph.

In analyzing your chromatograms you will find a peak for chlorobenzene and peaks for the ortho, meta, and para isomers of chloro(isopropyl)benzene. Identify the chlorobenzene peak by the peak enhancement method [see Technique 16.6]. The chloro(isopropyl)-benzenes elute in close succession in the order of ortho, meta, and para isomers.

Interpretation of Experimental Data

Isomeric compounds usually have virtually identical molar response factors.

Calculate the relative percentages of the three isomers in each product mixture using the integration data from the recorder or the peak areas themselves [see Technique 16.7]. Explain why the ratio of products observed for the kinetic control experiment changed in the thermodynamic control experiment.

OPTIONAL EXPERIMENTS Role of Nitromethane

After consulting with your instructor, carry out the experiments you (or your team) design in answer to Questions 4 and 5, which follow. Analyze the results of each experiment and compare them to the product ratio observed in the standard experiment. Explain any differences between the results of the optional experiments and those observed in the standard experiment.

Reference

1. Kolb, K. E.; Standard, J. M.; Field, K. W. *J. Chem. Educ.* **1988,** *65,* 367.

Questions

1. Why is the product mixture washed with 5% sodium carbonate solution? Write a balanced equation for the reaction that occurs.
2. Which isomers form the kinetic product in the alkylation of chlorobenzene? Why do these isomers form faster?
3. How can the product of thermodynamic control be more stable but form more slowly?
4. Why must the aluminum chloride catalyst be deactivated with nitromethane to form the kinetic product? Outline an experiment to test your answer.
5. Outline an experiment designed to test whether the isomerization of the kinetic product to the thermodynamic product occurs only with nondeactivated aluminum chloride catalyst.

EXPERIMENT

24

REDUCTION REACTIONS OF 3-NITROACETOPHENONE

QUESTION: What is the regioselectivity when two different functional groups are possible sites for reduction?

> Working in a team of two people, you will discover in Experiment 24 which of the two functional groups of 3-nitroacetophenone is reduced by $NaBH_4$ and by Sn/HCl, using melting points and IR spectroscopy to find the answer.

Oxidation-reduction reactions are extremely diverse and proceed by a variety of mechanisms. The unifying factor is that one substrate is oxidized and another is reduced. In the inorganic chemistry of metal cations, this process can be thought of most easily as an electron transfer phenomenon; that is, oxidation is a loss of electrons and reduction is a gain of electrons. Even though this approach also applies to covalent organic compounds, it is awkward to use. It is often easier to think of reduction as the gain of hydrogen (the equivalent of H_2) or loss of oxygen atoms.

Increasing state of oxidation ⟶

RCH_3	RCH_2OH	$R(C{=}O)H$	$R(C{=}O)OH$
Alkane	Alcohol	Aldehyde	Carboxylic acid

Inorganic Reducing Agents

When reduction of an organic substrate is needed, the reducing agent is often an inorganic compound. Sodium borohydride ($NaBH_4$) has become a popular mild reducing agent because of its functional group selectivity and ease of handling. When $NaBH_4$ is used as a reducing agent, the borohydride anion acts as the source of nucleophilic hydride ions.

All hydride reducing agents, which make up a large family of reagents, are not only reducing agents but also bases, which can react with Brønsted acids to produce H_2 gas. Sodium borohydride is a relatively safe reducing reagent. Unlike the more hazardous lithium aluminum hydride ($LiAlH_4$), which reacts with water to form H_2 with explosive force, $NaBH_4$ reacts only slowly with water. In fact, water and alcohols can be used as solvents for sodium borohydride reductions. When BH_4^- reduces an organic compound in an alcohol (ROH) solvent, it is oxidized to $B(OR)_4^-$. After the reaction is finished and water is added, $B(OR)_4^-$ is hydrolyzed, producing salts of boric acid [$B(OH)_3$].

Another major class of reduction reactions, known as dissolving metal reactions, uses metals as the inorganic reducing agents, often in the presence of acid. Sn, Zn, and Fe are common choices for the metal. In Experiment 24.2 you will use tin (Sn) metal as the reducing agent in the presence of 6 M hydrochloric acid. Tin provides the necessary electrons to reduce the organic substrate by a number of one-electron steps involving free-radical intermediates. When excess tin metal is present in the reaction, it becomes oxidized to Sn^{2+}. Under your experimental conditions, Sn will largely be converted to Sn^{2+} mixed with a small amount of Sn^{4+}. After the reduction is complete and the solution made basic, the soluble Sn^{2+} is converted to insoluble tin oxide (SnO).

Reduction Sites of 3-Nitroacetophenone

3-Nitroacetophenone

The substrate in this experiment, 3-nitroacetophenone, has two potential sites for reduction. The nitro (NO_2) group will reduce to an amino (NH_2) group and the ketone carbonyl (C=O) group will reduce to a secondary alcohol (OH). Certain reaction conditions favor reduction of the nitro group, whereas others lead to reduction of the ketone. Your team will use $NaBH_4$ in one experiment and tin and HCl in the other experiment to determine which reagent reduces the nitro group and which one reduces the carbonyl group.

Infrared Spectral Interpretation

To determine the outcome of the reduction reactions by infrared spectroscopy, it is important to look for the bands in the infrared spectrum corresponding to the functional groups of both the starting material (Figure 24.1) and the products. Here we are fortunate to find that all four functional groups of interest (C=O, OH, NO_2, and NH_2) have polar bonds that produce reasonably intense IR vibrational stretching bands [see Technique 18].

Consider first the reduction of the ketone C=O to the corresponding alcohol. Whereas the aromatic conjugated ketone displays a strong C=O stretching band at 1693 cm^{-1}, the O—H and C—O stretching vibrations of its reduction product will appear in quite different regions of the IR spectrum [see Table 18.2]. IR sampling techniques such as mulls normally give rise to the broad "associated" O—H stretching band in the 3550–3300 cm^{-1} region.

Reduction of a nitro group to an amino group can also be conveniently monitored by IR spectroscopy. The N—O bonds of nitro groups have two strong IR stretching vibrational bands, one in the 1570–1500 cm^{-1} region and the other in the 1380–1300 cm^{-1} region. The higher-frequency band is due to the asymmetric stretching vibration, the lower-frequency one to the symmetric stretching vibration. Reduction of the nitro group to an amino group causes the disappearance of these two bands. They are replaced by a

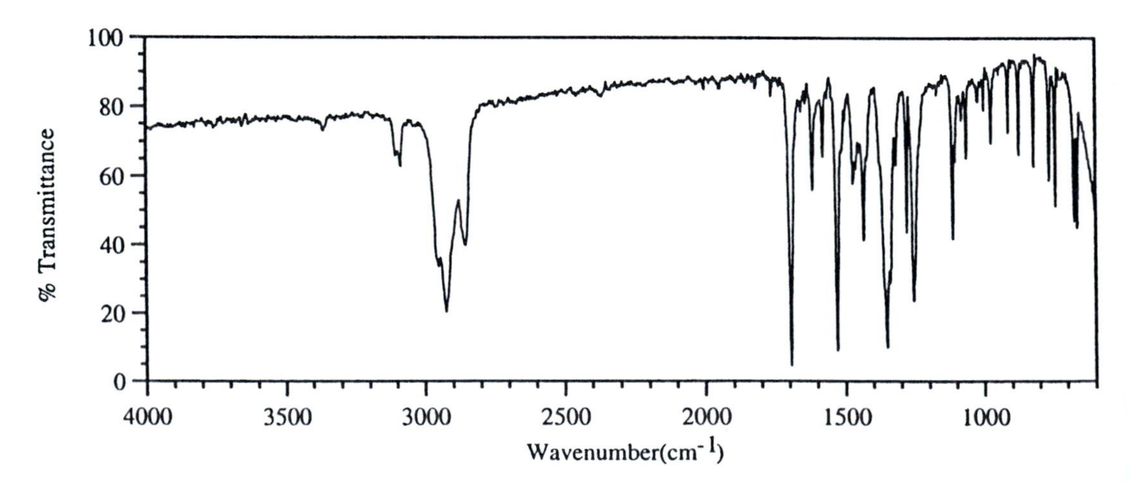

FIGURE 24.1 IR spectrum of 3-nitroacetophenone (Nujol mull).

pair of bands of moderate intensity due to stretching vibrations of the amino group. In practice, inadequate instrument resolution or poor sampling technique may cause a primary amine to show from one to three IR bands in the N—H stretching region.

24.1 Reduction of 3-Nitroacetophenone Using Sodium Borohydride

QUESTION: Which functional group of 3-nitroacetophenone is reduced by $NaBH_4$?

PRELABORATORY ASSIGNMENT: What vibrational bands would be prominent in the IR spectra of 3-aminoacetophenone and 1-(3-nitrophenyl)ethanol? At approximately what frequency would you expect to see each of these bands?

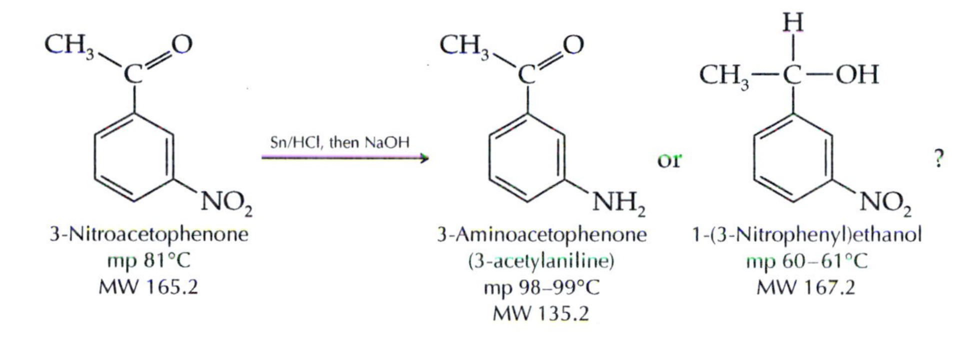

3-Nitroacetophenone
mp 81°C
MW 165.2

3-Aminoacetophenone
(3-acetylaniline)
mp 98–99°C
MW 135.2

1-(3-Nitrophenyl)ethanol
mp 60–61°C
MW 167.2

Techniques Microscale Extraction: Technique 8.6b
Microscale Separation of Drying Agent: Technique 8.8
IR Spectroscopy: Technique 18

SAFETY INFORMATION

3-Nitroacetophenone is an irritant. Wear gloves and avoid contact with skin, eyes, and clothing.

Sodium borohydride is harmful if swallowed, inhaled, or absorbed through the skin. Avoid breathing the dust and contact with skin, eyes, and clothing. It decomposes to flammable, explosive hydrogen gas.

Ethanol is flammable.

Dichloromethane is toxic, an irritant, absorbed through the skin, and harmful if inhaled. Use it only in a hood, and wear neoprene gloves while doing the extractions and evaporation.

Sodium borohydride is moisture sensitive; keep the bottle tightly closed and store it in a desiccator.

MICROSCALE PROCEDURE

Place 200 mg of 3-nitroacetophenone and 2.5 mL of absolute ethanol in a 10-mL Erlenmeyer flask. Dissolve the solid by stirring with a short stirring rod and warming the flask on a hot plate set at the lowest setting.

The color of the solution may change from yellow to light brown during the addition of $NaBH_4$.

Allow the solution to cool at room temperature for 2–3 min; a fine suspension of solid may form. Stir the suspension at room temperature while adding 70 mg of sodium borohydride powder in three portions over a 2- to 3-min period.

Add 2.0 mL of water and heat the contents of the flask to boiling on a hot plate set at the lowest setting. Cool the flask slightly and note whether the evolution of hydrogen gas from the decomposition of excess sodium borohydride has ceased. If bubbling continues when the solution has cooled, add 3-M hydrochloric acid dropwise and stir the solution to complete the decomposition process. Cool the reaction solution for 10 min in an ice-water bath to produce a suspension of the product.

You will need either a Pasteur pipet fitted with a syringe or two Pasteur pipets in the following extraction procedure—one with cotton in the tip for separation and transfers and one without cotton for mixing the phases [see Technique 8.5]. Pour the product mixture into a 15-mL centrifuge tube. Rinse the flask with 2 mL of dichloromethane and add the rinse to the centrifuge tube. Mix the two phases by drawing the mixture into a Pasteur pipet and expelling it back into the centrifuge tube four or five times. Allow the phases to separate and transfer the lower organic phase to a second centrifuge tube (or small test tube) [see Technique 8.6b]. Repeat the extraction procedure with two additional 2-mL portions of dichloromethane. Wash the combined dichloromethane solution with 2 mL of water; transfer the lower dichloromethane layer to a dry 10-mL Erlenmeyer flask. Dry the dichloromethane solution with anhydrous magnesium sulfate for about 10 min.

Filter half of the solution into a tared 13 $\times$ 150 mm test tube, using a Pasteur pipet as a funnel with cotton packed at the top of the tapered portion [see Technique 8.8, Figure 8.16b]. Use a clean Pasteur filter pipet to transfer the solution. Evaporate the dichloromethane with a stream of nitrogen or air in a hood. When most of the solvent is gone, filter the other half of the product solution into the test tube and complete the evaporation of the solvent. Alternatively, if nitrogen or air is unavailable, collect the filtered product solution on a large tared watch glass and evaporate the dichloromethane in the front of a hood with the sash drawn nearly shut. The product usually forms a white solid, but it may remain as a yellow oil after the removal of the solvent. The product is sufficiently pure for IR analysis without recrystallization.

Determine the mass of product, its melting point (if the product solidified), and the percent yield. Obtain an IR spectrum as directed by your instructor [see Technique 18.4].

Interpretation of Experimental Data

Analyze your IR spectrum for the functional groups shown by the strong vibrational bands [see Techniques 18.5 and 18.6 and Table 18.2]. Use your melting point and IR spectrum to determine whether your product is 3-aminoacetophenone or 1-(3-nitrophenyl)ethanol.

CLEANUP: The aqueous solution remaining from the extraction may be washed down the sink or placed in the container for aqueous inorganic waste. Place the magnesium sulfate drying agent in the container for non-hazardous waste or inorganic waste.

24.2 Reduction of 3-Nitroacetophenone Using Tin and Hydrochloric Acid

QUESTION: Which functional group of 3-nitroacetophenone is reduced by Sn/HCl?

PRELABORATORY ASSIGNMENT: What vibrational bands would be prominent in the IR spectra of 3-aminoacetophenone and 1-(3-nitrophenyl)ethanol? At approximately what frequency would you expect to see each of these bands?

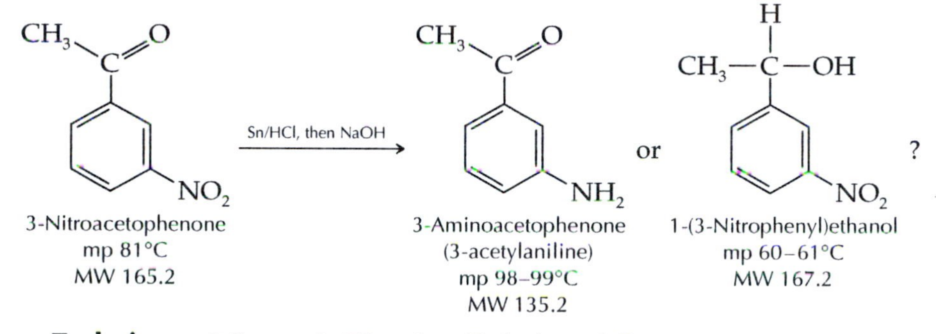

Techniques Microscale Filtration: Technique 9.7a
IR Spectroscopy: Technique 18

SAFETY INFORMATION

3-Nitroacetophenone is an irritant. Wear gloves and avoid contact with skin, eyes, and clothing.

Hydrochloric acid is corrosive and irritates the skin, eyes, and mucous membranes.

30% sodium hydroxide solution is corrosive and causes burns. Solutions as dilute as 9% (2.5 M) can cause severe eye injury.

MICROSCALE PROCEDURE

Place 200 mg of 3-nitroacetophenone and 400 mg of granular tin (20 mesh) in a 25-mL Erlenmeyer flask and add 4.0 mL of 6 M hydrochloric acid. Working in a hood, place the flask on a steam bath or in a boiling-water bath for 25–30 min or until most of the tin has dissolved and no brown oil remains. Stopper the flask loosely with a cork to minimize evaporation during the heating period. The reaction time may vary, depending on the particle size of the tin metal.

At the end of the reaction period, cool the flask in an ice-water bath. Add 30% (10 M) sodium hydroxide solution dropwise until the pH reaches 10 when checked with pHydrion or other pH paper. A thick yellow paste of tin salts and product will form; stir this mixture thoroughly while you are adding the NaOH solution. Heat the reaction flask for 10 min on a steam bath or in a boiling-water bath.

Assemble a vacuum filtration apparatus using a 5-cm Buchner funnel, prewetted filter paper, and a 25- or 50-mL filter flask. You will also need two small flasks of boiling water—one flask with approximately 5 mL of water, the other flask with approximately 12 mL of water. Do not let the water in these flasks boil for an extended time or an appreciable volume will vaporize. When you are ready to filter the heated reaction mixture, carefully pour the 12-mL portion of boiling water on the Buchner funnel with the vacuum turned on; this process will warm the Buchner funnel. Disconnect the vacuum and, holding the hot filter flask carefully with a towel, pour the water out of the filter flask. Quickly reassemble the filtration apparatus and immediately pour the hot reaction mixture on the Buchner funnel. Wash the precipitate (SnO) onto the Buchner funnel with the 5-mL portion of boiling water.

Cool the yellow filtrate to room temperature and then in an ice-water bath before collecting the crystalline product by vacuum filtration on a Hirsch funnel. Wash the crystallized product four times with 0.5-mL portions of ice-cold water. The product is sufficiently pure for analysis without recrystallization. Allow the product to dry completely before determining its mass, its melting point (if the product solidified), and percent yield. Obtain an IR spectrum of the product as directed by your instructor [see Technique 18.4].

Interpretation of Experimental Data

Analyze your IR spectrum for the functional groups represented by the strong vibrational bands [see Techniques 18.5 and 18.6 and Table 18.2]. Use your melting point and IR spectrum to determine whether your product is 3-aminoacetophenone or 1-(3-nitrophenyl)ethanol.

CLEANUP: Place the filter paper containing the SnO in the container for hazardous metals (or inorganic) waste. Neutralize the filtrate from the crystallized product with dropwise addition of 6 M hydrochloric acid until the pH reaches 6–7 before washing the solution down the sink or pouring it into the container for aqueous inorganic waste.

Questions

1. After you have determined the reduction product of 3-nitroacetophenone using Sn/HCl, write the balanced chemical equation for the reaction. Assume that Sn^{2+} is the oxidation product.
2. When butanoic acid is treated with sodium borohydride, 1-butanol is not obtained. There are, however, definite signs of reaction (bubbling, heat). What might the reaction be?
3. Suggest an appropriate reducing reagent for the following substrates and predict the product of the reaction: (a) 2-hexanone; (b) ethyl 4-nitrobenzoate (reduction of the nitro group); (c) hexanal.

EXPERIMENT

25

HORNER-EMMONS-WITTIG SYNTHESIS OF METHYL *E*-4-METHOXYCINNAMATE

PURPOSE: To make a conjugated *E*-alkene using modern synthetic methodology.

In Experiment 25 you will make methyl *E*-4-methoxycinnamate using the reaction of 4-methoxybenzaldehyde and a Wittig reagent synthesized from trimethylphosphonoacetate and sodium methoxide in methanol. You will evaluate the purity of your product from its melting point and NMR spectrum.

The Wittig reaction has proved to be a very useful synthetic method for preparing alkenes. As a result of his discovery of this reaction and exploration of its related chemistry, the German chemist Georg Wittig shared the 1979 Nobel prize in chemistry. In the reaction that bears his name, an aldehyde or ketone reacts with a Wittig reagent to produce an alkene:

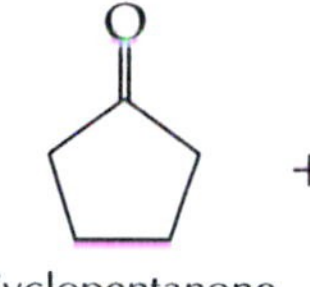

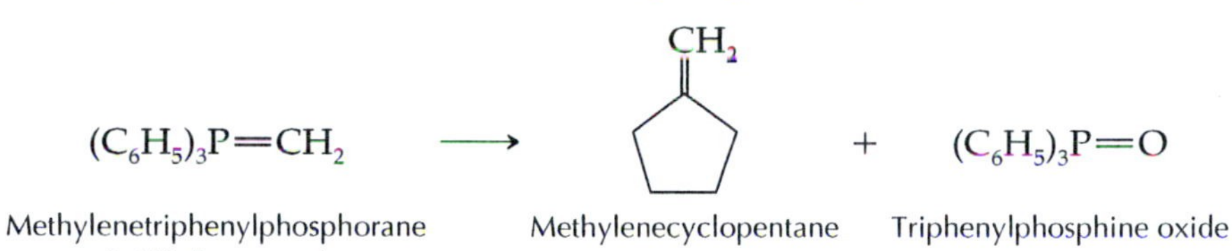

The value of the Wittig method becomes clear when we consider that alternative methods for preparing alkenes often give rise to complex product mixtures. Consider, for example, the synthesis of methylenecyclopentane by the dehydrohalogenation of an alkyl halide, a traditional method for alkene synthesis. Reaction of the tertiary alkyl bromide, 1-bromo-1-methylcyclopentane, with base gives a good yield of alkene, but the major product is 1-methylcyclopentene:

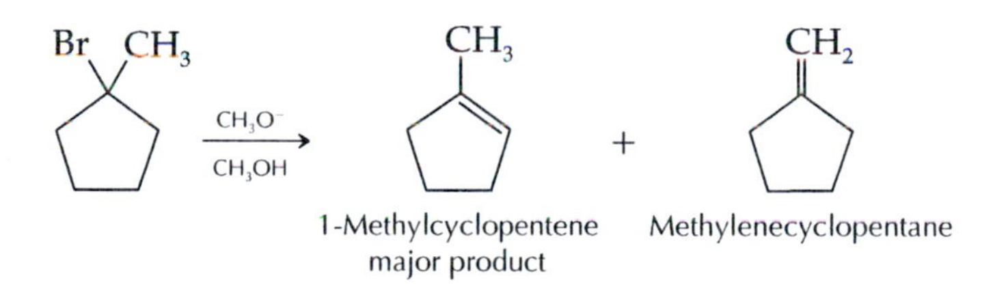

If one started with the primary alkyl halide, the synthesis of methylenecylopentane would involve multiple steps, and a competing S_N2 reaction forming an ether would lower the yield of the desired alkene:

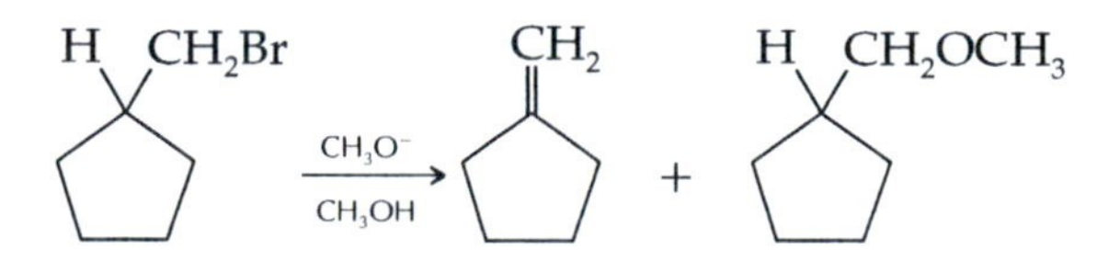

The Wittig reaction starts with readily available reactants and gives high yields of alkenes. To understand its chemistry, let us first examine how a Wittig reagent is formed. When alkyl halides are treated with triphenylphosphine, a strong nucleophile, they undergo S_N2 reactions to form phosphonium salts:

$$(C_6H_5)_3P\!: \quad CH_3\!-\!Br \longrightarrow (C_6H_5)_3\overset{+}{P}CH_3\,Br^-$$

Triphenylphosphine Bromomethane Methyltriphenylphosphonium bromide

When phosphonium salts are treated with strong bases (butyllithium is commonly used), the Wittig reagent is formed:

$$B\!:^- \quad H_3C\overset{+}{P}(C_6H_5)_3\,Br^- \longrightarrow {}^-\!:\!CH_2\!-\!\overset{+}{P}(C_6H_5)_3 + BH + Br^-$$

Base Wittig reagent

Wittig reagents have two major resonance forms. The first is merely the dipolar form we would expect as a result of the loss of a proton. The second has an expanded octet of 10 valence electrons around phosphorus and no formal charge separation. This behavior is not uncommon for third-row elements bearing d-orbitals that can participate in bonding:

$${}^-\!:\!CH_2\!-\!\overset{+}{P}(C_6H_5)_3 \longleftrightarrow CH_2\!=\!P(C_6H_5)_3$$

Wittig reagent

The mechanism of the Wittig reaction is complex, and we will discuss only a few major details. The first step is attack by the Wittig reagent at the carbon atom of the C=O to form a charged intermediate, which can form a four-membered ring by formation of a phosphorus-oxygen bond. This ring then opens to form the alkene product and triphenylphosphine oxide as a side product:

$$(C_6H_5)_3P\!=\!CH_2 + R_2C\!=\!\ddot{O}\!: \longrightarrow (C_6H_5)_3\overset{+}{P}\!-\!CH_2\!-\!CR_2\!-\!\ddot{O}\!:^- \longrightarrow (C_6H_5)_3P\!-\!\ddot{O}\!:\ (\text{ring with } CH_2\!-\!CR_2) \longrightarrow (C_6H_5)_3P\!=\!\ddot{O}\!: + CH_2\!=\!CR_2$$

You will use a modified phosphorus-containing substrate called the Horner-Emmons-Wittig reagent. This reagent is prepared from trimethyl phosphonoacetate, a commercially available compound that is much more acidic than simple phosphonium salts. The enhanced acidity means that sodium methoxide is a strong enough base to produce the

Horner-Emmons-Wittig reagent. Perhaps even more important, the final phosphorus-containing side product is water-soluble, a great advantage during isolation and purification of the product. Presumably, the mechanism of the Horner-Emmons-Wittig reaction parallels that for the simple Wittig reaction.

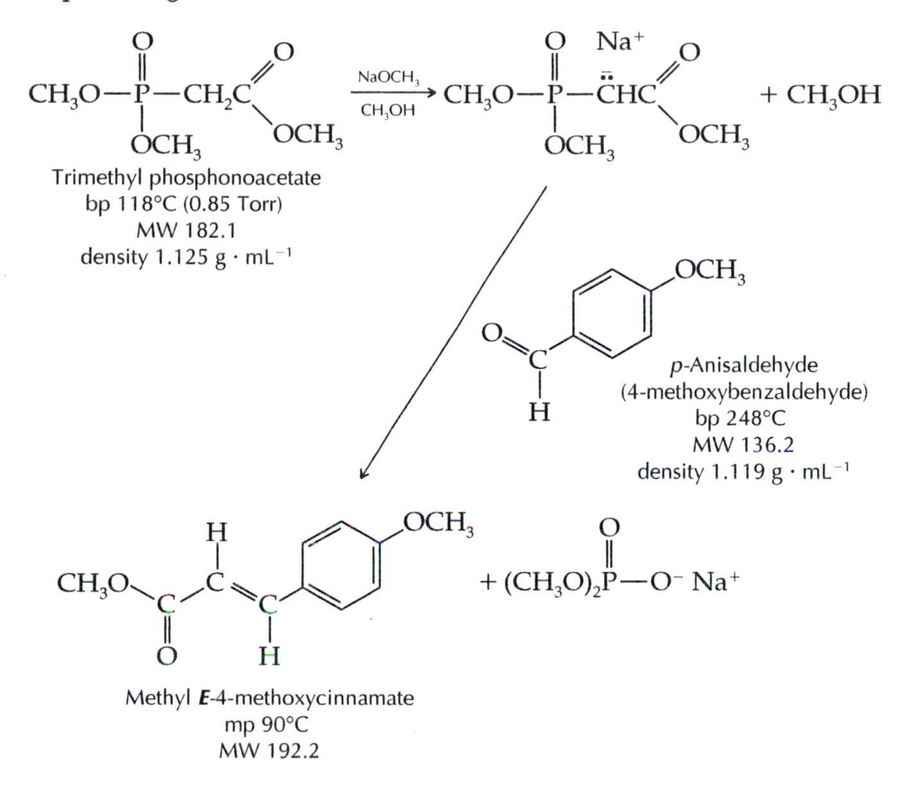

The stereochemistry of the Wittig reaction is also important. Not only is it quite specific for the synthesis of alkenes from aldehydes and ketones, but the configuration of the C=C bond also forms with substantial stereoselectivity. The original Wittig reagent produces largely ***Z*** or cis alkenes, whereas the Horner-Emmons-Wittig reagent that you will use produces the ***E*** or trans alkene. The reasons for this difference come about from subtle changes in the reaction pathway and conformational effects in the cyclic reaction intermediate.

As the name suggests, *p*-anisaldehyde (4-methoxybenzaldehyde) has an odor reminiscent of aniseed, the aromatic seed of the anise plant, whose oil is used in cooking and liqueurs for its licorice-like flavor. Your methyl 4-methoxycinnamate product gets its name from cinnamic acid, the oxidation product of cinnamaldehyde, which is a major flavor component of cinnamon. An analogue of your product, 2-ethylhexyl *E*-4-methoxycinnamate, is used in sunscreen products.

MINISCALE PROCEDURE

Techniques Mixed Solvent Recrystallization: Technique 9.2
Miniscale Recrystallization: Technique 9.5
NMR Spectroscopy: Techniques 19.2, 19.4–19.9
IR Spectroscopy: Techniques 18.4, 18.5

SAFETY INFORMATION

> Syringe needles are a puncture hazard.
>
> **Sodium methoxide in methanol solution** is flammable, corrosive, and moisture sensitive. Wear gloves and avoid contact with skin, eyes, and clothing. Keep the container tightly closed and store it in a desiccator.
>
> **Trimethyl phosphonoacetate** may be absorbed through the skin. Wear gloves and avoid contact with skin, eyes, and clothing.
>
> ***p*-Anisaldehyde (4-methoxybenzaldehyde)** is an irritant. Wear gloves and avoid contact with skin, eyes, and clothing.
>
> **Methanol** is flammable and toxic.
>
> **Ethanol** is flammable.

Place 4.0 mL of anhydrous methanol, 1.60 mL of 25 wt % sodium methoxide/methanol solution (4.4 mmol/mL), and 1.75 mL of trimethyl phosphonoacetate in a 50-mL Erlenmeyer flask containing a magnetic stirring bar. Quickly close the flask with a rubber septum and insert a syringe needle through the septum to act as a vent. Set the flask on a magnetic stirrer and clamp it to a ring stand.

In a small, dry test tube, prepare a solution of 0.80 mL of *p*-anisaldehyde and 2.0 mL of anhydrous methanol. Draw this solution into a syringe. Insert the syringe through the rubber septum and add the solution slowly to the reaction mixture over a period of 10 min. Remove the syringe. Continue to stir the reaction mixture at room temperature for 1 h. During this time, precipitation occurs and the solution may become light brown.

At the end of the reaction period, remove the venting needle and rubber septum. Add 8.0 mL of water and stir thoroughly. Collect the crude product by vacuum filtration on a Buchner funnel. Rinse the flask with 1–2 mL of water and add this rinse to the funnel.

Recrystallize the product in 95% ethanol [see Technique 9.5]. When the crystals have dissolved completely, add water dropwise to the hot solution until cloudiness occurs, then add ethanol dropwise until the solution clears [see Technique 9.2]. Allow the solution to cool slowly to room temperature before cooling it in an ice-water bath. Collect the purified product by vacuum filtration. Allow the product to dry thoroughly by drawing air over the crystals for 30 min with the vacuum source turned on or by leaving them in your desk overnight. Determine the melting point and the percent yield.

Spectroscopic Analysis of the Product

Obtain the NMR spectrum of your dried product in $CDCl_3$ as directed by your instructor [see Technique 19.2]. Analyze the spectrum and assign all the peaks [see Techniques 19.4 and 19.9 and Tables 19.2 and

19.6]. How would the spectrum of *p*-anisaldehyde differ from that of your product? Alternatively, the IR spectrum of the product can be obtained and analyzed [see Techniques 18.4 and 18.5].

CLEANUP: Dilute the aqueous filtrate from the crude product with water and neutralize it with 5% HCl solution before washing it down the sink or pouring it into the container for aqueous inorganic waste. Pour the filtrate from the recrystallization into the container for flammable (organic) waste.

MICROSCALE PROCEDURE

Techniques Mixed Solvent Recrystallization: Technique 9.2
Microscale Recrystallization: Technique 9.7a
NMR Spectroscopy: Techniques 19.2, 19.4–19.9
IR Spectroscopy: Techniques 18.4, 18.5

SAFETY INFORMATION

Syringe needles are a puncture hazard.

Sodium methoxide in methanol solution is flammable, corrosive, and moisture sensitive. Wear gloves and avoid contact with skin, eyes, and clothing. Keep the container tightly closed and store it in a desiccator.

Trimethyl phosphonoacetate may be absorbed through the skin. Wear gloves and avoid contact with skin, eyes, and clothing.

***p*-Anisaldehyde (4-methoxybenzaldehyde)** is an irritant. Wear gloves and avoid contact with skin, eyes, and clothing.

Methanol is flammable and toxic.

Ethanol is flammable.

Place 1.0 mL of anhydrous methanol, 0.40 mL of 25 wt % sodium methoxide/methanol solution (4.4 mmol/mL), and 0.43 mL of trimethyl phosphonoacetate in a 5-mL conical vial containing a magnetic spin vane. Quickly close the vial with a screwcap and septum, then insert a syringe needle through the septum to act as a vent. Clamp the vial so that it is positioned on a magnetic stirrer.

In a small, dry test tube, prepare a solution of 0.20 mL of *p*-anisaldehyde and 0.50 mL of anhydrous methanol. Draw this solution into a syringe. Insert the syringe through the septum and add the solution slowly to the reaction mixture over a period of 10 min. Remove the syringe. Continue to stir the reaction mixture at room temperature for 1 h. During this time, precipitation occurs and the solution may become light brown.

At the end of the reaction period, remove the venting needle and open the vial. Add 2.0 mL of water, cap the vial, and shake it to mix the contents thoroughly. Collect the crude product by vacuum filtration on a Hirsch funnel [see Technique 9.7a]. Rinse the flask with a few drops of water and add this rinse to the Hirsch funnel.

Recrystallize the product in 95% ethanol in a reaction tube using a boiling stick to prevent bumping. Use a water bath or an aluminum block heated to 90°C as the heat source. When the crystals have dissolved completely, add water dropwise to the hot solution until cloudiness occurs, then add ethanol dropwise until the solution clears. Cool the tube to room temperature before cooling it in an ice-water bath. Collect the product by vacuum filtration on a Hirsch funnel. Allow the product to dry thoroughly before determining the melting point and the percent yield. Carry out an NMR analysis of your product as directed in the miniscale procedure.

CLEANUP: Dilute the aqueous filtrate from the crude product with water and neutralize it with 5% HCl solution before washing it down the sink or pouring it into the container for aqueous inorganic waste. Pour the liquid remaining in the filter flask from the recrystallization into the container for flammable (organic) waste.

References

1. Crandall, J. K.; Mayer, C. F. *J. Org. Chem.* **1970,** *35,* 3049–3053.
2. Bottin-Strzalko, T. *Tetrahedron* **1973,** *29,* 4199–4204.

Questions

1. Suggest a reason why the phosphorus-containing side product from the Horner-Emmons-Wittig reaction is more water-soluble than the triphenylphosphine oxide side product obtained from the standard Wittig reaction.
2. What advantage does an ethanol/water solution have over anhydrous ethanol as the recrystallization solvent in this experiment?
3. Sodium methoxide is a weaker base than butyllithium. Suggest a reason why it is not surprising to find that a weaker base is sufficient to prepare the Horner-Emmons-Wittig reagent.
4. If the solution of sodium methoxide in methanol were not provided from your stockroom, what chemical reaction would you use to make it, starting from methanol?
5. The *E*-configuration of methyl 4-methoxycinnamate might very well be established in the cyclic intermediate in the reaction mechanism. Write out the reaction mechanism for the synthesis of methyl *E*-4-methoxycinnamate from *p*-anisaldehyde. Suggest a reason why the cyclic intermediate might have two important groups in a trans relationship that could favor the *E*-alkene.

EXPERIMENT

26

SYNTHESIS OF TRIPHENYLMETHANE DYES

Crystal Violet and Malachite Green

PURPOSE: To use Grignard reactions to synthesize dyes and to test the dyes on various fabrics.

In Experiment 26 you will work in a two-person team to make a Grignard reagent from 4-bromo-*N*,*N*-dimethylaminobenzene and react it with diethyl carbonate or methyl benzoate to form two dyes, evaluating them by their dyeing properties.

The dyes you will study in this experiment, crystal violet and malachite green, are highly colored compounds. Both dyes are salts, which renders them soluble in polar solvents, including water. Crystal violet and malachite green are green solids, but their colors in water solution differ—crystal violet is a deep purple color and malachite green has a bluish-green color. In addition to being used as dyes, both these compounds have also been used as stains and antiseptics.

Grignard Reagents

Crystal violet and malachite green can be synthesized by the reaction of an ester and a Grignard reagent. ***Grignard reagents,*** which are alkyl- and arylmagnesium halides, are versatile intermediates in the synthesis of many organic compounds. They are substances containing carbon-magnesium bonds that readily attack the carbonyl (C=O) groups of aldehydes, ketones, and esters.

$$\mathrm{R{-}X} + \mathrm{Mg} \xrightarrow{\mathrm{THF}} \overset{\delta-}{\mathrm{R}}{-}\overset{\delta+}{\mathrm{Mg}}{-}\overset{\delta-}{\mathrm{X}}$$

Formation of a Grignard reagent takes place in a heterogeneous reaction at the surface of the magnesium metal, and the surface area and reactivity of the magnesium are crucial factors in the rate of the reaction. It is thought that the alkyl or aryl halide reacts with the surface of the metal to produce a carbon-free radical and a magnesium-halogen radical, which combine to form the Grignard reagent. This gives the magnesium atom two covalent bonds, but it must somehow acquire two more since magnesium has a coordination number of 4. In an ether solvent, such as tetrahydrofuran (THF), the ether molecules occupy the other two coordination sites. The oxygen atom of tetrahydrofuran provides both electrons in the magnesium-ether bond:

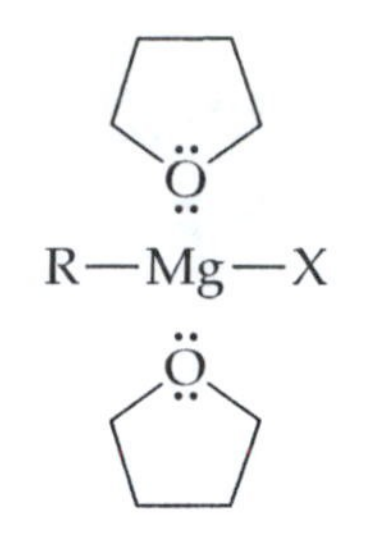

Tetrahydrofuran has excellent solvent properties and a higher boiling point (66°C) than diethyl ether. The Grignard reagent is quite soluble in THF. In its absence, the reaction of magnesium and an aryl halide soon stops because the surface of the metal becomes coated with polymeric Grignard reagent. In the presence of THF, the surface of the metal is kept clean and the reaction proceeds until the limiting reagent is entirely consumed.

Not only are Grignard reagents strong nucleophiles, they are also strong bases. They react quickly and easily with acids, even relatively weak acids such as water. This reaction destroys the Grignard reagent before the carbonyl compound is added and therefore reduces the yield of the desired product. It is important that the reaction conditions be completely anhydrous. The THF solvent, which is miscible with water, must be dried before use. All glassware and reagents must also be thoroughly dry before beginning a Grignard experiment.

Treatment of a solution of 4-bromo-*N*,*N*-dimethylaminobenzene in tetrahydrofuran with magnesium metal results in the formation of the Grignard reagent, *p*-dimethylaminophenylmagnesium bromide. Insertion of the magnesium between carbon and bromine greatly enhances the reactivity of the aromatic structure and reverses the polarity of the carbon formerly bonded to bromine. In the Grignard reagent, the carbon-metal bonds have a high degree of ionic character, with a good deal of negative charge on the carbon atom, making it a strong nucleophile.

Reaction of Grignard Reagents with Esters

The mechanism of the Grignard reaction with a carboxylate ester is actually quite complex, but it can easily be rationalized as a nucleophilic addition-elimination reaction. The intermediate ketone cannot be isolated because it rapidly undergoes a second nucleophilic addition reaction:

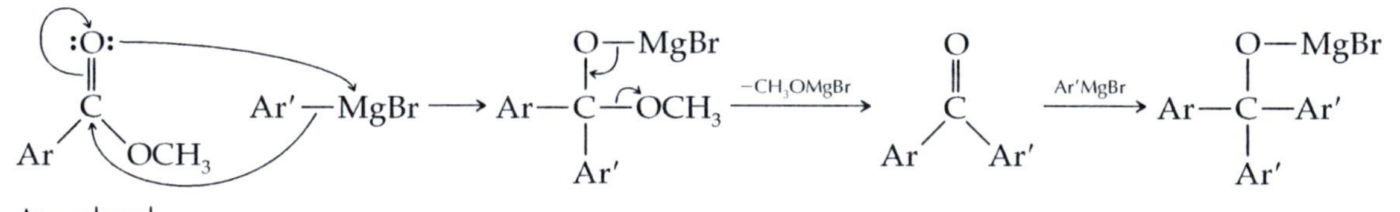

Ar = phenyl
Ar′ = 4-*N*,*N*-dimethylaminophenyl

Crystal violet is formed by the reaction of three moles of *p*-dimethylaminophenylmagnesium bromide and one mole of diethyl carbonate, whereas malachite green is formed by reaction of two moles of the Grignard reagent and one mole of methyl benzoate:

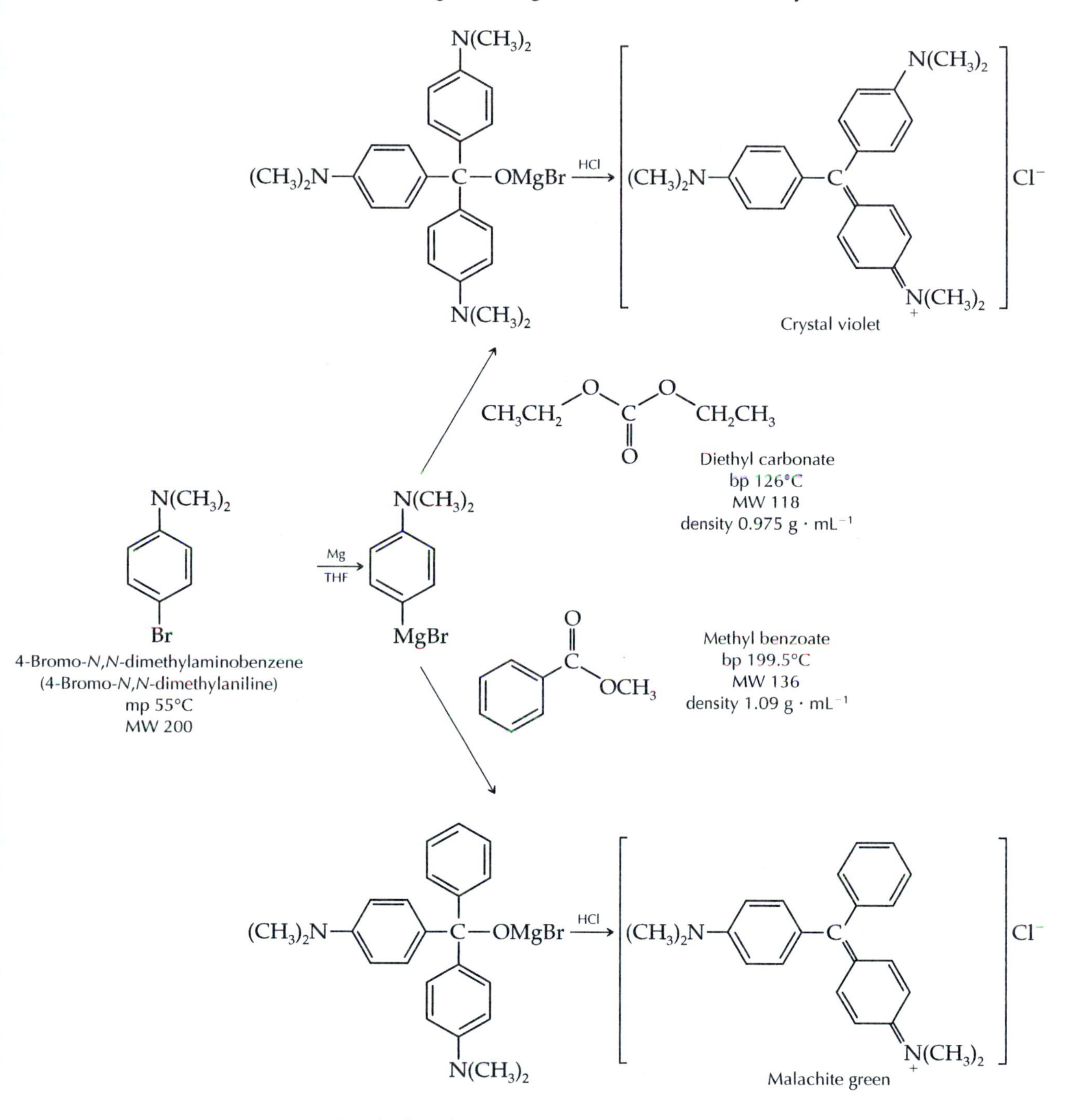

Diethyl carbonate, a diester of carbonic acid, reacts with the Grignard reagent in two nucleophilic addition-elimination reactions to form a diaryl ketone. This ketone cannot be isolated because it rapidly undergoes addition of a third mole of the arylmagnesium halide to form a magnesium salt, which when acidified gives the tertiary alcohol. This alcohol loses a molecule of water in the presence of hydrochloric acid and forms a highly stable colored cation:

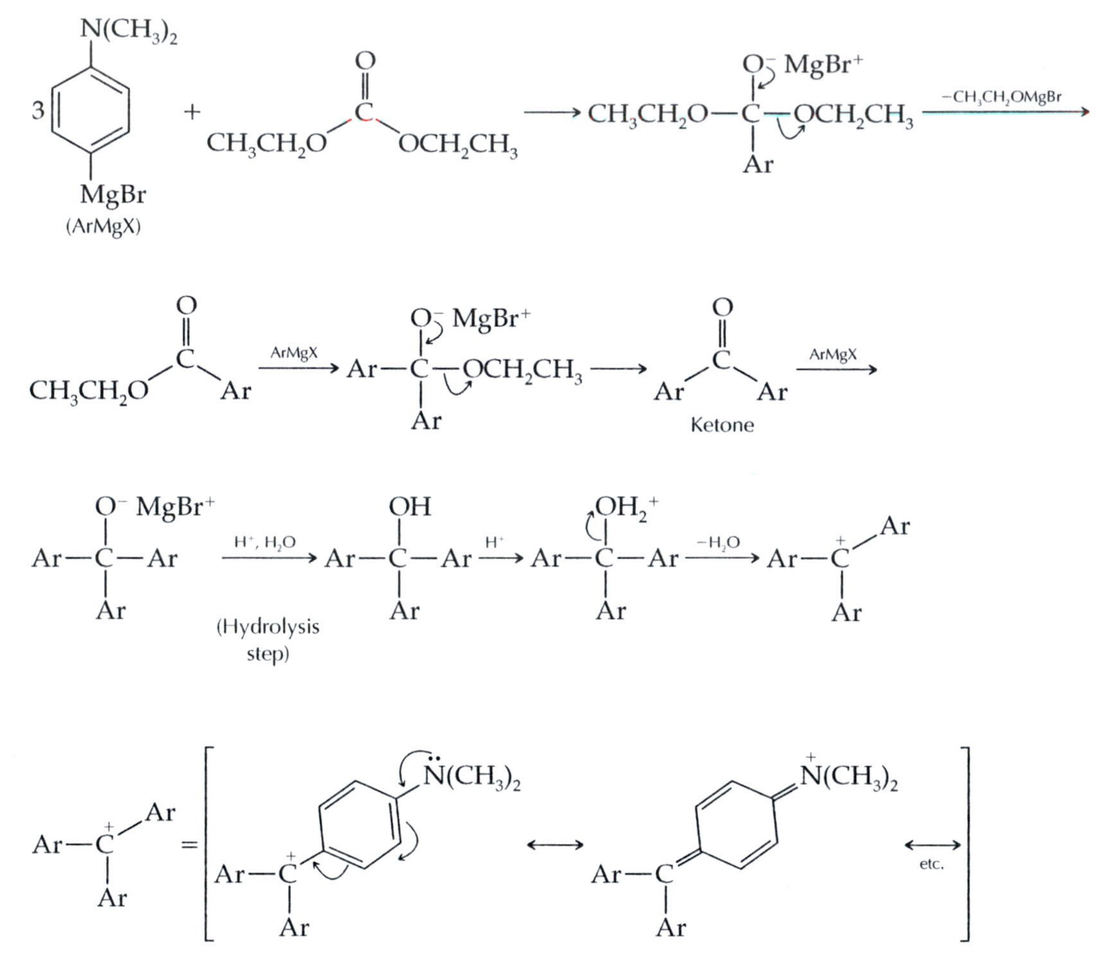

Methyl benzoate, an ester of benzoic acid, reacts with *p*-dimethylaminophenylmagnesium bromide in a nucleophilic addition-elimination reaction to form a different diaryl ketone, where one of the aromatic rings is a phenyl substituent. Again, this ketone cannot be isolated because it rapidly undergoes addition of a second mole of the arylmagnesium halide to form a magnesium salt, which when acidified gives the tertiary alcohol. This alcohol also loses a molecule of water in the presence of hydrochloric acid and forms a highly stable colored cation.

An important aspect of both malachite green and crystal violet is their extensive conjugation, which is responsible for their colors. The positive charge is extensively delocalized due to the overlap of the orbitals on the cationic center with the conjugated π-system of the aromatic rings and even the para dimethylamino groups.

MINISCALE PROCEDURE

TEAMWORK: You and your teammate will coordinate which of the syntheses you will do. Both syntheses start with the preparation of the Grignard reagent from 4-bromo-*N*,*N*-dimethylaminobenzene.

Technique Anhydrous Reaction Conditions: Technique 7.3.

Keep the reagent bottle of THF and your flask containing THF tightly closed. THF rapidly absorbs water from the atmosphere.

SAFETY INFORMATION

4-Bromo-*N,N*-dimethylaniline is toxic and an irritant. Wear gloves while weighing it.

Tetrahydrofuran (THF) is very flammable.

Diethyl carbonate and **methyl benzoate** are irritants. Avoid contact with skin, eyes, and clothing.

If possible, conduct the experiment in a hood.

The purity of the dyes made in the experiment is not suitable for clothing; only test fabrics should be dyed with it.

The dyes made in this experiment are toxic and cause stains on skin, clothing, and anything else they contact. **Wear old clothes to lab.** Wear gloves while dyeing the fabric samples and cleaning the glassware that contained the dye.

The Grignard reagent is moisture sensitive. All glassware used for this experiment must be dry.

Synthesis of the Grignard Reagent

Prepare a drying tube filled with anhydrous calcium chloride [see Technique 7.3]. Obtain the following reagents before setting up the reaction: 2.0 g of 4-bromo-*N,N*-dimethylaminobenzene (4-bromo-*N,N*-dimethylaniline), 0.25 g of magnesium turnings, 25 mL of dry tetrahydrofuran (THF), and one or two small crystals of iodine.

The reagents need to be added quickly and the apparatus assembled rapidly to minimize exposure to atmospheric moisture.

If the glassware has not been oven- or flame-dried, rinse a 50-mL round-bottomed flask with 2–3 mL of THF, pour 2–3 mL of THF through the condenser, and discard these rinses in the flammable waste container. Clamp the round-bottomed flask in place, and add the 4-bromo-*N,N*-dimethylaminobenzene, the remaining 19–20 mL of THF, the magnesium turnings, and the iodine crystal. With a clean, dry stirring rod, gently rub one or two magnesium turnings to expose a fresh surface, which helps initiate the reaction.

SAFETY PRECAUTION

Hold the round-bottomed flask firmly by the neck and be careful not to press so hard on a magnesium turning that you crack the flask.

Fit the condenser to the flask and swirl the mixture. Adjust the water so that it flows gently through condenser, and place the drying tube at the top of the condenser. Heat the reaction with a 70–75°C water bath. Maintain a gentle reflux for 30 min; swirl the flask every 5 min during the heating period. The initial dark color fades and is replaced by the grayish mixture typical of Grignard reagents. Cool the reaction flask in a beaker of tap water until it reaches room temperature.

Synthesis of the Dye

Choose the dye you wish to make.

For malachite green: Weigh 0.68 g of methyl benzoate into a small vial (set in a 30-mL beaker so that it will not tip over). Add 3.0 mL of THF to the vial. **For crystal violet:** Weigh 0.39 g of diethyl carbonate into a small vial (set in a 30-mL beaker so that it will not tip over). Add 3.0 mL of THF to the vial.

Remove the condenser. Using a Pasteur pipet, add the ester solution to the reaction flask dropwise, swirling after each drop. After the addition is complete, replace the condenser and heat the reaction mixture under reflux for 5 min. Swirl the flask occasionally while heating it. Cool the flask to room temperature.

Dyeing Test Samples

Put on gloves before starting the dye synthesis. Pour the reaction mixture into a 150-mL beaker. Slowly add 10 mL of 5% HCl solution to the beaker with stirring; some bubbling will occur as the residual magnesium reacts with the acid.

A variety of fabric samples will be available to you and your teammate. Dip each sample in the dye solution using a pair of forceps and leave it at least 1 min. Remove the sample with forceps, rinse it with tap water into a beaker, and blot it dry. The dye solution is very concentrated, so intense color should be produced in the fabrics, depending on how well the particular type of fiber accepts the dye. Record the types of fabrics you tested and describe any variations in the intensity of the color observed for different fibers. Compare your results with those of your teammate who synthesized the other dye.

Test fabric strips 10 cm wide, containing a variety of fibers, are available from TestFabrics, Inc., PO Box 26, West Pittston, PA 18643 (phone: 570-603-432) or from Kontes (800-223-7150).

If 10- to 12-inch samples of cotton jersey fabric (for example, cut from old T-shirts) are available, you can experiment with tie-dyeing effects by tying or knotting the fabric sample before dyeing it.

Allow the samples to dry and attach them to your report. Keep them in a small plastic bag or wrapped in paper because the dye may rub off on anything the samples touch.

CLEANUP: The dye solution should be poured in the waste container labeled "Dye Solution." Dye stains on glassware can be removed with a few milliliters of 6 M HCl, then washing with water. Neutralize the acid washings with sodium carbonate before pouring them down the sink or into the container for aqueous inorganic waste.

Reference

1. Taber, D. F.; Meagley, R. P.; Supplee, D. *J. Chem. Educ.* **1996,** *73,* 259–260.

Questions

1. What precipitate formed when the ester was added to the solution of your Grignard reagent?
2. What advantage(s) does tetrahydrofuran (THF) have over diethyl ether as a solvent in this experiment?
3. What would be the correlation between the color of your dye solution and the wavelength of its maximum absorbance in the visible spectrum of the solution?

EXPERIMENT

27 ENOLATE CHEMISTRY

Synthesis of *trans*-1,2-Dibenzoylcyclopropane

PURPOSE: To synthesize a cyclopropane-ring compound using enolate chemistry.

In Experiment 27 you will carry out a base-catalyzed halogenation of 1,3-dibenzoylpropane using I_2; then you will make a three-membered ring by the intramolecular S_N2 reaction of an enolate anion, all in one reaction vessel. You will characterize *trans*-1,2-dibenzoylcyclopropane by melting point and NMR spectroscopy.

Enolate Chemistry

The nucleophilic reactions of enolate anions have tremendous utility in the synthesis of carbon-carbon bonds. These reactions include the aldol and Claisen condensations and Michael addition reactions. Enolate anions are also effective nucleophiles in S_N2 reactions.

The α-hydrogen atoms of carbonyl compounds have significant acidity. The pKa of an α-proton of a ketone or aldehyde is approximately 19, so OH^- and CH_3O^- are strong enough bases to produce a small but effective concentration (about 0.1%) of the conjugate base of the carbonyl compound. This acid-base reaction produces an enolate anion, which behaves as a strong nucleophile in reaction with an alkyl halide.

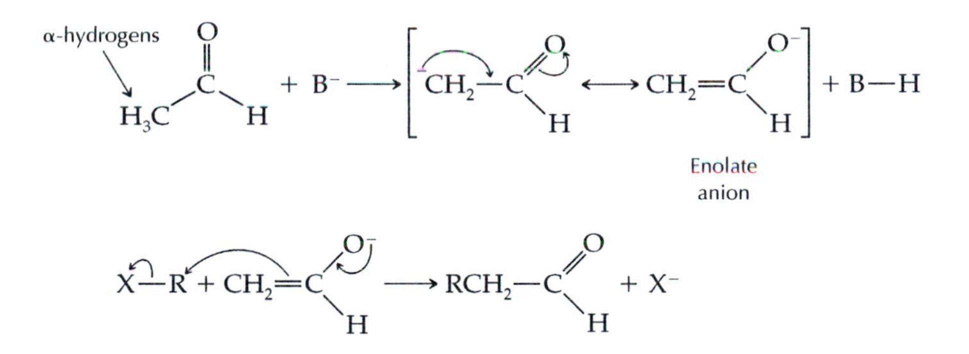

This experiment will use the intramolecular reaction of an enolate anion in an S_N2 reaction to make a cyclopropane ring. One of the α-protons of the symmetrical diketone substrate is removed by OH^- or CH_3O^-, forming an enolate anion, which then undergoes an S_N2-like reaction on molecular iodine to form an α-iodoketone:

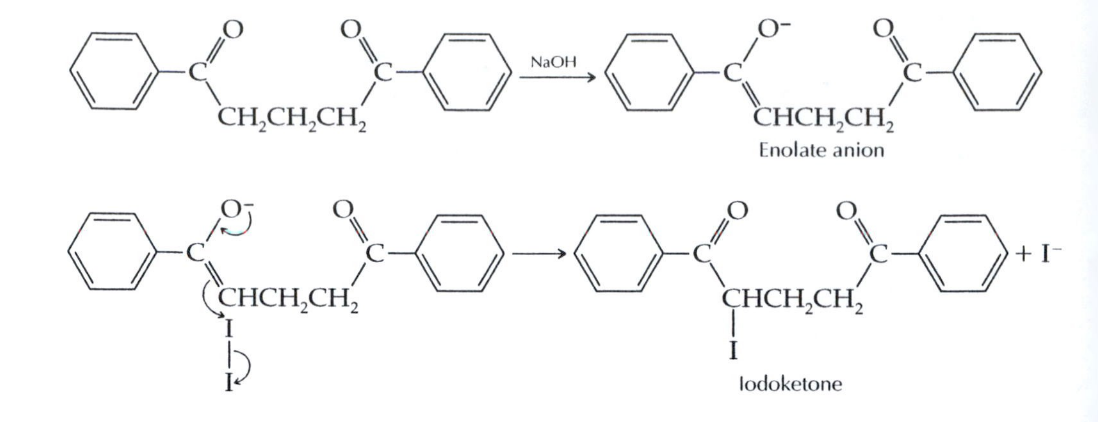

Abstraction of a second proton by OH^- or CH_3O^- transforms the intermediate iodoketone to another enolate anion, which then undergoes ring closure by an S_N2 reaction to form *trans*-1,2-dibenzoylcyclopropane:

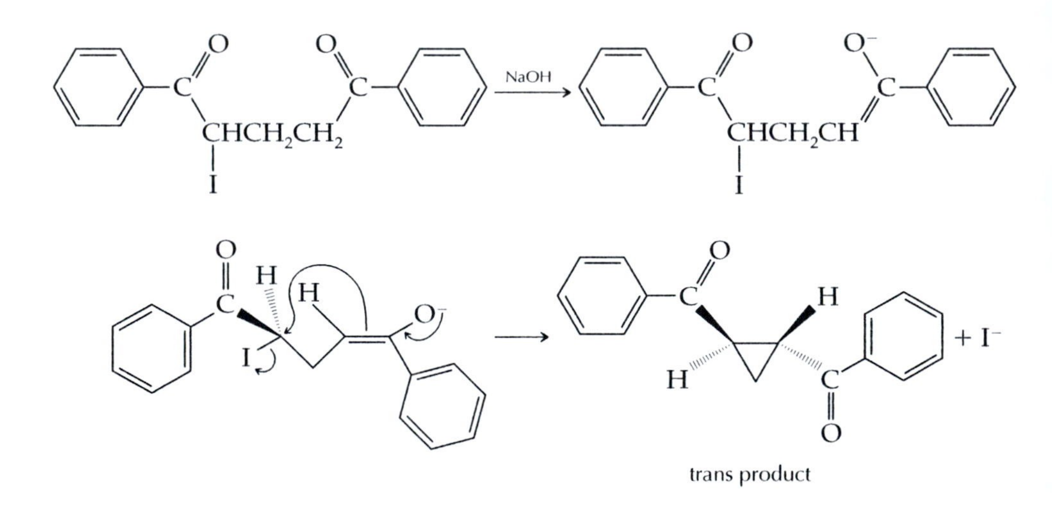

Cyclopropane Rings

The synthesis of ring compounds involves two major concepts, ring strain and ease of ring formation based on entropy effects. Small rings of three or four carbon atoms are destabilized because they have bond angles smaller than the ideal tetrahedral bond angle of 109.5°. Bond-angle strain is at a maximum for the flat cyclopropane ring, and each carbon-carbon bond of a cyclopropane ring is actually bent out of the line connecting the two carbon atoms (Figure 27.1).

Ring strain factors are to some degree counterbalanced by entropy effects in the formation of three-membered rings. First of all, bimolecular substitution reactions have an unfavorable entropy of activation because two molecules must come together in the transition state. This condition

FIGURE 27.1
Molecular-orbital picture of the bent bonds in cyclopropane.

imposes a highly ordered state, which costs energy to produce. An intramolecular S_N2 reaction has less entropic cost because both the nucleophile and the attackable carbon atom are within the same molecule. Even so, an improbable conformation must often be involved when a C—C bond of a large ring is formed.

The formation of a cyclopropane ring has an entropic advantage in forming the S_N2 transition state because the increase in order needed to form the ring is at a minimum. The nucleophilic center is directly adjacent to the carbon atom bearing the leaving group. Once the S_N2 reaction has taken place, there is no effective pathway for reopening the cyclopropane ring under the reaction conditions even though there is a good deal of ring strain. The rate of the reverse reaction is very slow because the enolate anion is a poor leaving group.

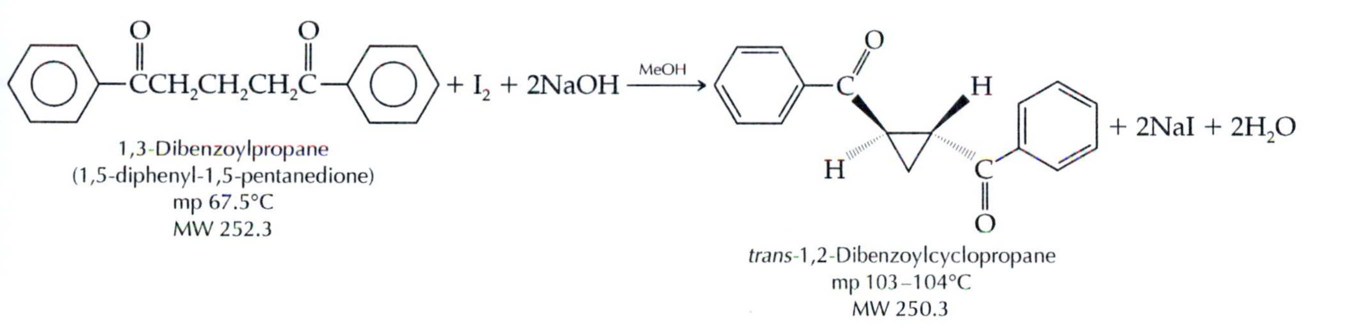

1,3-Dibenzoylpropane
(1,5-diphenyl-1,5-pentanedione)
mp 67.5°C
MW 252.3

trans-1,2-Dibenzoylcyclopropane
mp 103–104°C
MW 250.3

Chemical shifts of methylene protons:

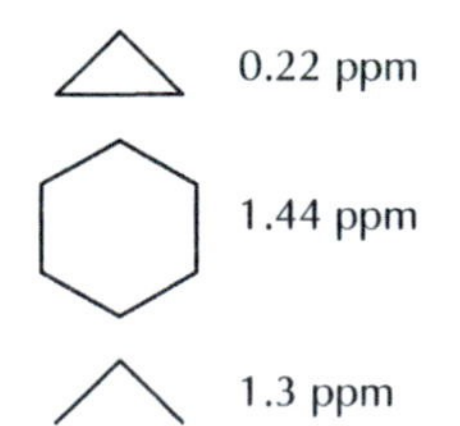

The ^1H NMR spectrum of *trans*-1,2-dibenzoylcyclopropane has some interesting features. The electron-withdrawing carbonyl (C═O) group has a deshielding effect on the α-protons and on the *ortho*-protons of the benzene ring. The cyclopropane protons are somewhat upfield (shielded) relative to the usual chemical shifts of alkyl protons. This shielding is an anisotropic effect [see Technique 19.7], which arises from the quasi π-bonding system of the bent bonds in the cyclopropane ring.

MINISCALE PROCEDURE

Techniques Recrystallization: Technique 9.5
NMR Spectroscopy: Techniques 19.2, 19.4–19-9

SAFETY INFORMATION

> **Methanol** is volatile, toxic, and flammable. **Iodine** is toxic and corrosive. Wear gloves while working with the **iodine/methanol** solution and pour the solution in a hood.
>
> Avoid contact with the **sodium hydroxide/methanol** solution, which is corrosive.
>
> **1,3-Dibenzoylpropane** and the product require no special precautions.

Place 1.0 g of 1,3-dibenzoylpropane and 12 mL of 0.67 M sodium hydroxide in methanol solution in a 50-mL round-bottomed flask containing a magnetic stirring bar. Assemble a Claisen adapter, separatory funnel, and water-cooled condenser as shown in Technique 7, Figure 7.3a, omitting the drying tubes at the top of both the separatory funnel and the condenser. Pour 6.0 mL of 0.67 M iodine in methanol solution into the separatory funnel. Put a glass stopper into the separatory funnel with a small strip of paper inserted between the stopper and the neck of the funnel to prevent an airlock.

Control the rate of iodine addition so that the brown color of the iodine disappears before the next drop is added.

Place a 40°C water bath under the round-bottomed flask and begin stirring the mixture. When the solid dissolves, begin dropwise addition of the iodine solution. After the addition is complete, continue to heat the reaction mixture for 45–60 min, until precipitation of white solid ceases.

Dismantle the reaction apparatus, cool the flask to room temperature, and then cool it further in an ice-water bath. Collect the crude product by vacuum filtration. Pour the methanol filtrate out of the filter flask into a 50-mL beaker before washing the product, which remains on the Buchner funnel. Wash the crude product on the funnel three times with 5-mL portions of water.

You may collect a second crop of crude product by evaporating the methanol filtrate on a steam bath or 80°C water bath in the hood. Stir the resulting reddish residue with 5 mL of 10% sodium bisulfite solution to reduce any iodine present. Again collect the solid by vacuum filtration and wash it three times with 5-mL portions of water.

If time permits, you may wish to recrystallize the second crop of crude product separately from the first because the second crop usually contains more impurities than the first.

Recrystallize the combined crude product from methanol, using a steam bath or 80°C water bath as the heat source [see Technique 9.5]. Allow the recrystallized product to dry before determining the melting point and percent yield.

NMR Analysis of the Product

Obtain the NMR spectrum of your dried product in $CDCl_3$ as directed by your instructor [see Technique 19.2]. Analyze the spectrum and assign all the peaks [see Techniques 19.4 and 19.9 and Tables 19.3 and 19.4]. How would the spectrum of the 1,3-dibenzoylpropane differ from that of your product?

CLEANUP: If you did not collect a second crop from the methanol filtrate, pour this solution into the container for halogenated waste. Pour the filtrate from the recrystallization into the container for flammable waste. The aqueous filtrate from washing the crude product can be washed down the sink or placed in the aqueous waste container, as directed by your instructor.

MICROSCALE PROCEDURE

Techniques Microscale Recrystallization: Technique 9.7b
NMR Spectroscopy: Techniques 19.2, 19.4–19.9

SAFETY INFORMATION

Methanol is volatile, toxic, and flammable. **Iodine** is toxic and corrosive. Wear gloves while working with the **iodine/methanol** solution and pour the solution in a hood.

Avoid contact with the **sodium hydroxide/methanol** solution, which is corrosive.

1,3-Dibenzoylpropane and the product require no special precautions.

Syringe needles are puncture hazards.

Control the rate of iodine addition so that the brown color of the iodine disappears before the next drop is added.

Place 200 mg of 1,3-dibenzoylpropane and 2.4 mL of 0.67 M sodium hydroxide in methanol solution in a 10-mL round-bottomed flask containing a magnetic stirring bar. Assemble a Claisen adapter, septum cap, and water-cooled condenser as shown in Technique 7, Figure 7.4a, omitting the drying tube at the top of the condenser. Measure 1.2 mL of 0.67 M iodine in methanol solution into a small screw-capped vial. Draw this iodine solution into a syringe and insert the syringe needle through the septum cap.

Place a 40°C water bath under the round-bottomed flask and begin stirring the mixture. When the solid dissolves, begin dropwise addition of the iodine solution. After the addition is complete, continue to heat the reaction mixture for 35–40 min. A white precipitate may form.

Dismantle the reaction apparatus, cool the flask to room temperature, and then cool it further in an ice-water bath. Collect the crude product by vacuum filtration on a Hirsch funnel. Wash the product three times with 0.75-mL portions of water. A second crop of crude product will precipitate in the filtrate. This second crop should also be collected and washed the same way.

Transfer the combined crude product to a Craig tube, insert a boiling stick, and recrystallize the product from a minimum amount of methanol, using a steam bath as the heat source [see Technique 9.7b]. After the product has been dissolved in the methanol, add water dropwise (only a few drops are needed) until the hot solution is cloudy, then warm the solution briefly to boiling and add methanol dropwise until the cloudiness clears. Place the plug in the tube and set the Craig apparatus in a 25-mL Erlenmeyer flask to cool slowly. When crystallization is well under way and the Craig tube has cooled to room temerature, place it in an ice-water bath for a few minutes to complete crystallization. Remove the mother liquor from the recrystallized product by centrifugation [see Technique 9.7b, Figure 9.9]. Allow the product to dry before determining the melting point and percent yield. Carry out NMR spectroscopic analysis of your product as described in the miniscale procedure.

CLEANUP: Pour the filtrates from the crude product and the recrystallization into the container for flammable waste.

Reference

1. Colon, I.; Griffin, G. W.; O'Connell, Jr., E. J. *Organic Syntheses;* Wiley: New York, 1988; Collect. Vol. VI, pp. 401–402.

Questions

1. Sodium bisulfite is used to reduce any residual iodine remaining in the second crop of crude product. Write a balanced equation for the reaction of sodium bisulfite with I_2.
2. How would the ^{1}H NMR spectrum of 1,3-dibenzoylpropane differ from that of the product, *trans*-1,2-dibenzoylcyclopropane? Would the chemical shifts of the protons α and β to the carbonyl groups be appreciably different? Explain.
3. In this experiment you are asked to heat the reaction mixture until the formation of a white solid ceases. What do you think the white solid is?
4. Rationalize the formation of *trans*-1,2-dibenzoylcyclopropane (rather than *cis*) by using molecular models to examine the S_N2 transition state for the ring closure. What difficulties arise in the closure to the cis compound?

EXPERIMENT

28

AMIDE CHEMISTRY

PURPOSE: To carry out syntheses of amides from amines and benzoyl chloride.

In Experiment 28.1 a team of three students will use benzoyl chloride to convert the amino acids glycine, alanine, and valine to their benzoylamino acid derivatives and compare the results. Each team will evaluate the structure and purity of its amide products by melting points and IR and NMR spectroscopy. In an optional experiment, you can synthesize and compare the properties of the benzoylamino acid derivatives of optically active alanine and valine using polarimetry and spectroscopy.

Experiment 28.2 is a two-week experiment in which you will use the reaction of thionyl chloride or bis(trichloromethyl) carbonate with 3-methylbenzoic acid (*meta*-toluic acid) to synthesize its acid chloride; then you will react diethylamine with *m*-toluoyl chloride to synthesize the amide, *N,N*-diethyl-*m*-toluamide (DEET), a commonly used insect repellent. You will purify DEET by column chromatography and assess its structure and purity by IR and NMR spectroscopy.

Benzoylamino Acids

Generally, it is necessary to carry out a two-step reaction sequence for the synthesis of an amide. The first step usually involves initial conversion of the carboxylic acid to an acid chloride, which is far more reactive toward nucleophilic attack at the carbonyl carbon. Acid chlorides are readily converted to amides by treatment with ammonia or the appropriate amine:

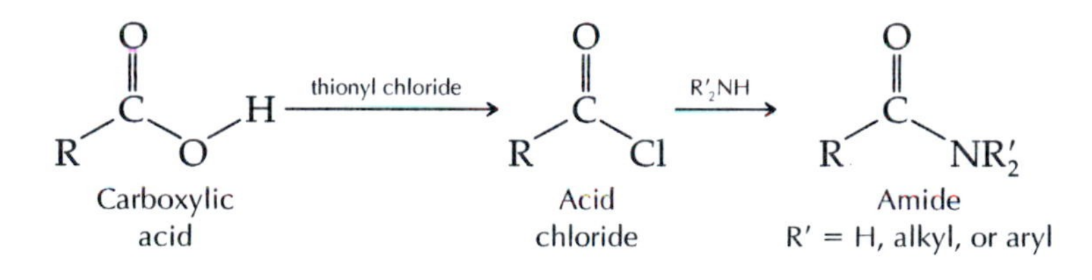

Your synthesis of benzoylamino acids starts with benzoyl chloride and one of three amino acids glycine, alanine, and valine, amino acids that are commonly found linked together by amide bonds in proteins:

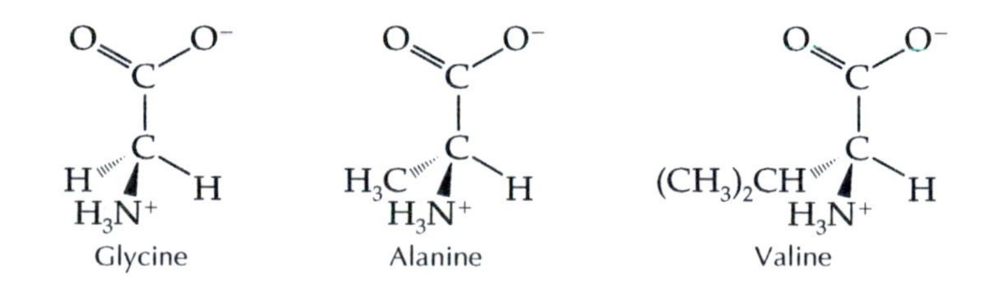

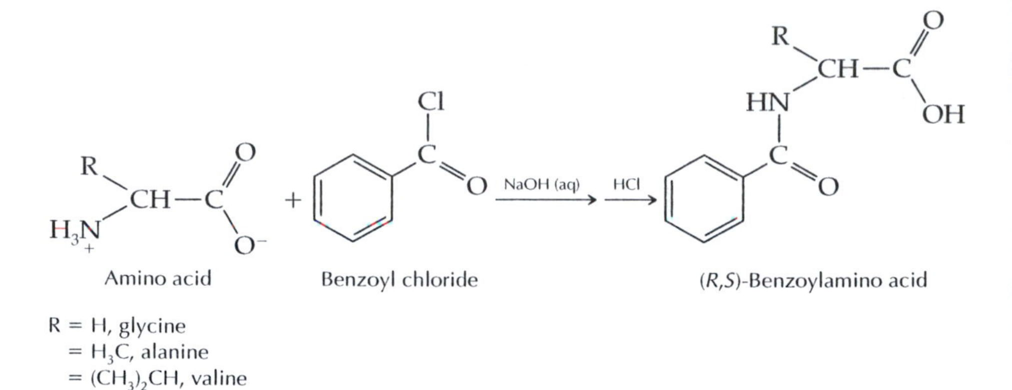

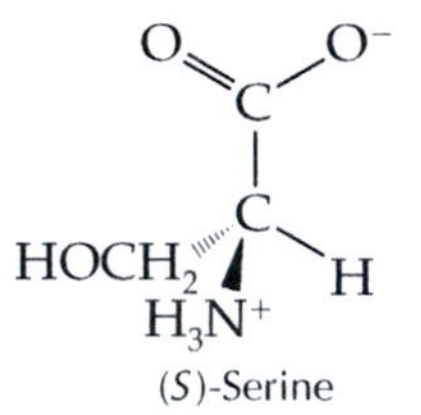
(*S*)-Serine

You may have noticed that unlike glycine, both alanine and valine have stereocenters. There are two different notations used for the configurations of amino acids and their derivatives. One type is the familiar *R,S*-notation for absolute configuration. The other is the D,L system, which is based on the configuration of the stereocenter relative to L-serine, another simple amino acid. Naturally occurring serine has the L-configuration, which is identical to the *S*-configuration. You will be using the *R,S*-amino acid racemic mixtures in this experiment.

The reaction of the amino acid glycine with benzoyl chloride in the presence of an alkaline catalyst yields the amide benzoylglycine. When glycine is in the zwitterionic (dipolar) form, as it is in water solution, the nitrogen has no nonbonding electron pair and is not nucleophilic. Treatment of glycine with OH^- converts the amino acid to a nucleophilic form, and the amino group bearing the electron pair readily attacks benzoyl chloride to initiate amide formation:

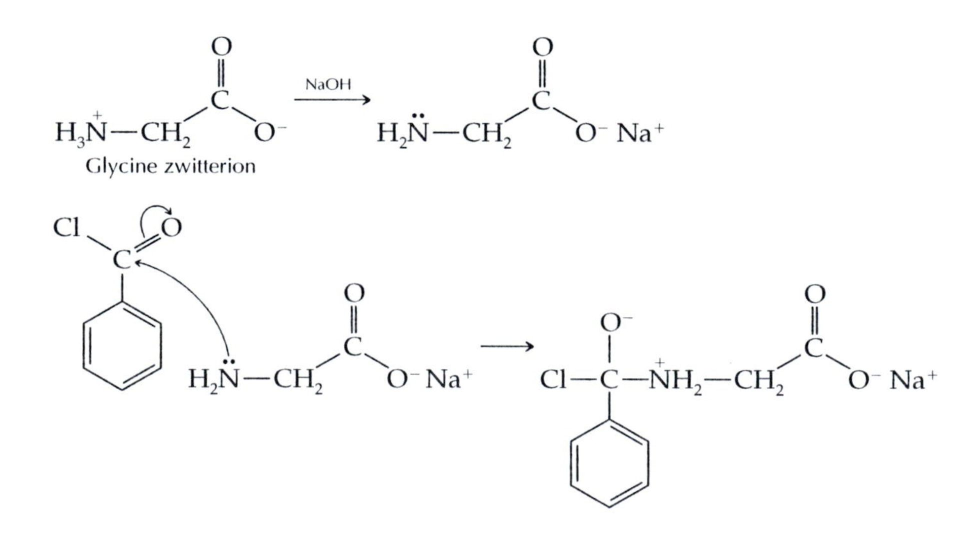

Loss of the elements of HCl yields a carboxylate salt, which when acidified produces the carboxylic acid

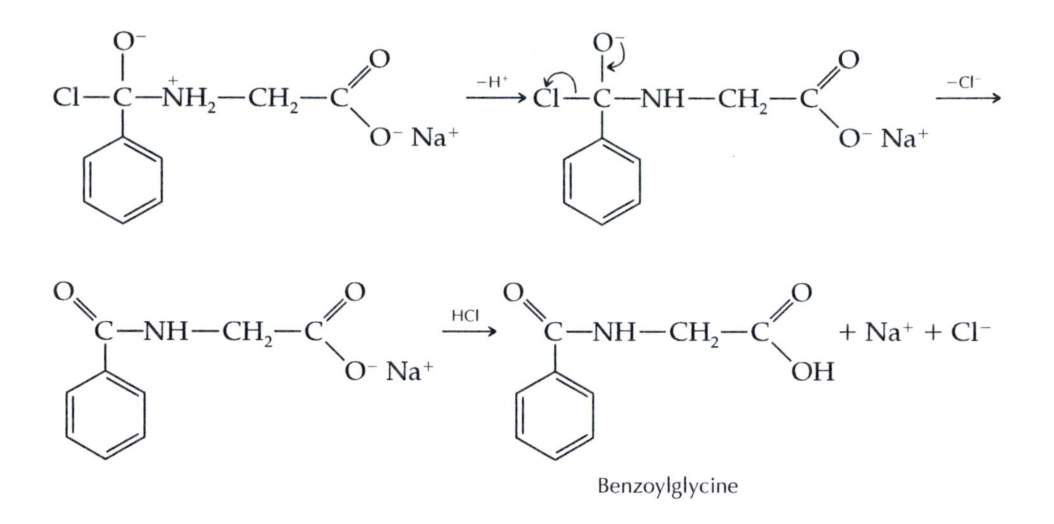

The great strength of the bond between the carbonyl carbon and the amide nitrogen is one of the main reasons why amides are among the most stable and least reactive derivatives of carboxylic acids. The stability of amides can be at least partially understood by the existence of a significant degree of double-bond character in the carbon-nitrogen bond induced by resonance:

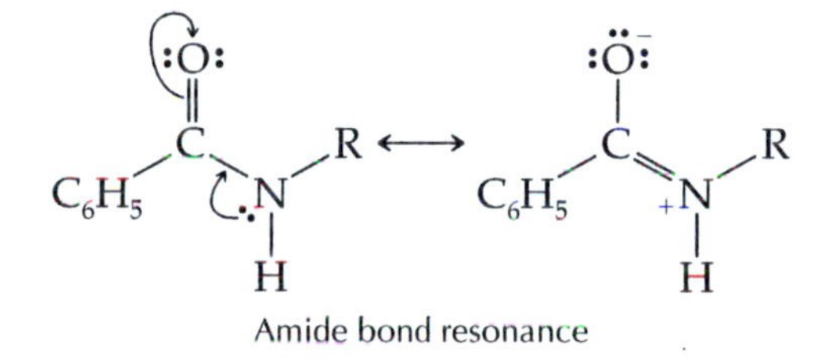
Amide bond resonance

The large amount of double-bond character in the carbon-nitrogn bond of amides produces a planar geometry and restricted rotation about the bond, just as one sees in the structures of alkenes. On the NMR time scale, two methyl groups on the amide nitrogen have distinctly different chemical shifts, because one methyl group is cis to the carbonyl oxygen while the other one is cis to the carbon substituent. The planarity of the amide bond is a crucial benchmark in the use of X-ray diffraction to decipher the three-dimensional structures of proteins. The planarity induced by resonance in the peptide bond of proteins is shown here:

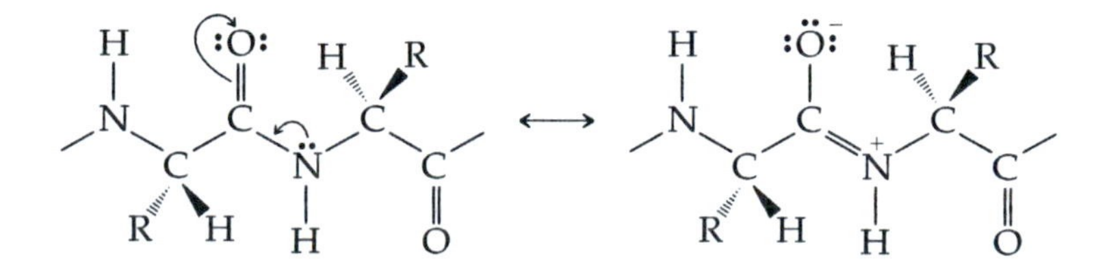

Synthesis of N,N-Diethyl-meta-toluamide (DEET)

Two compounds have been extensively used as insect repellents: *N,N*-diethyl-*meta*-toluamide (DEET) and 2-ethyl-1,3-hexanediol. In Experiment 28.2 you will synthesize DEET using a two-step procedure. In the first step of the miniscale synthesis you will convert 3-methylbenzoic acid (*m*-toluic acid) to *m*-toluoyl chloride using thionyl chloride ($SOCl_2$). In the second step you will react *m*-toluoyl chloride with two moles of diethylamine to produce *N,N*-diethyl-*m*-toluamide.

The formation of the acid chloride probably involves the formation of a sulfur-containing intermediate, which is very reactive to attack by chloride ion:

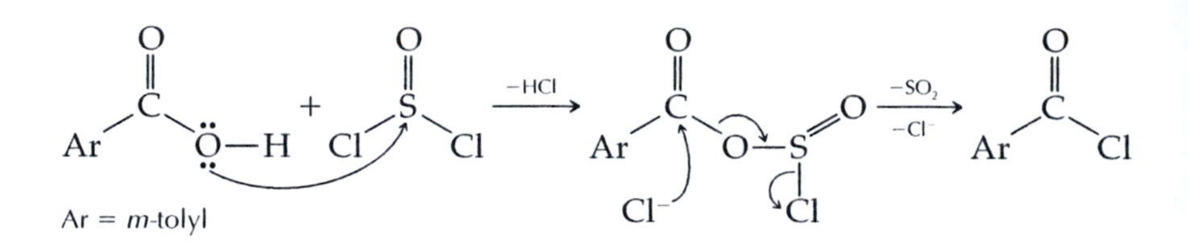

It is common to use the acid chloride produced in this way with no further purification. Although synthetic intermediates are often purified before carrying out subsequent steps, the simpler approach works here for two reasons. First, the side products, sulfur dioxide and hydrogen chloride, are gases that are readily lost from the reaction mixture. Second, because the acid chloride is highly reactive, the amine undergoes reaction with it more quickly than with impurities in the reaction mixture.

In the second step the nucleophilic amine attacks the acid chloride to form a tetrahedral intermediate that readily loses HCl to form the amide (DEET) product:

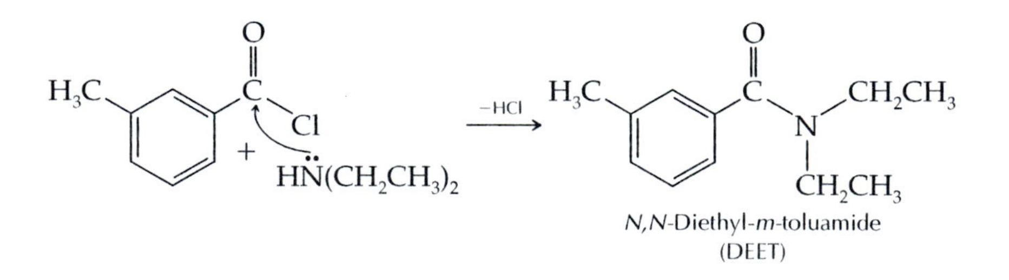

The second mole of diethylamine serves as a base to neutralize the side product HCl as it forms:

$$(CH_3CH_2)_2NH + HCl \longrightarrow (CH_3CH_2)_2NH_2^+ \, Cl^-$$

Diethylammonium chloride

Because both diethylamine and hydrogen chloride are volatile, they often produce a significant portion of the diethylammonium chloride in the air above the reaction mixture, causing it to appear white and "smoky."

In the microscale synthesis of DEET, you will use bis(trichloromethyl) carbonate (triphosgene), rather than thionyl chloride, to produce *m*-toluoyl chloride from *m*-toluic acid:

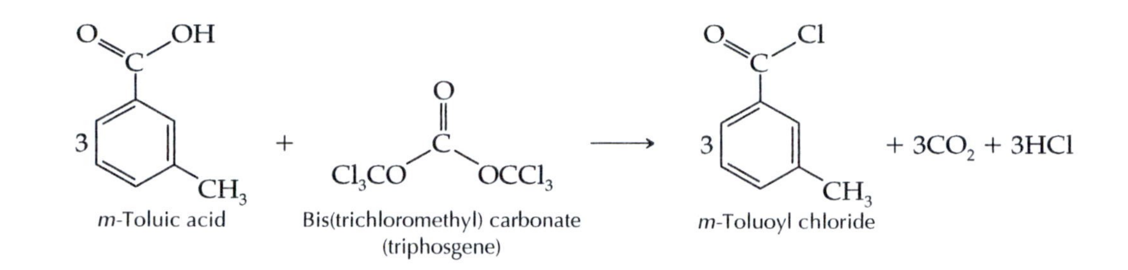

Although the mechanism of this reaction is quite complex, it is known how *N,N*-dimethylformamide (DMF) acts as a catalyst to produce the actual chlorinating species from triphosgene:

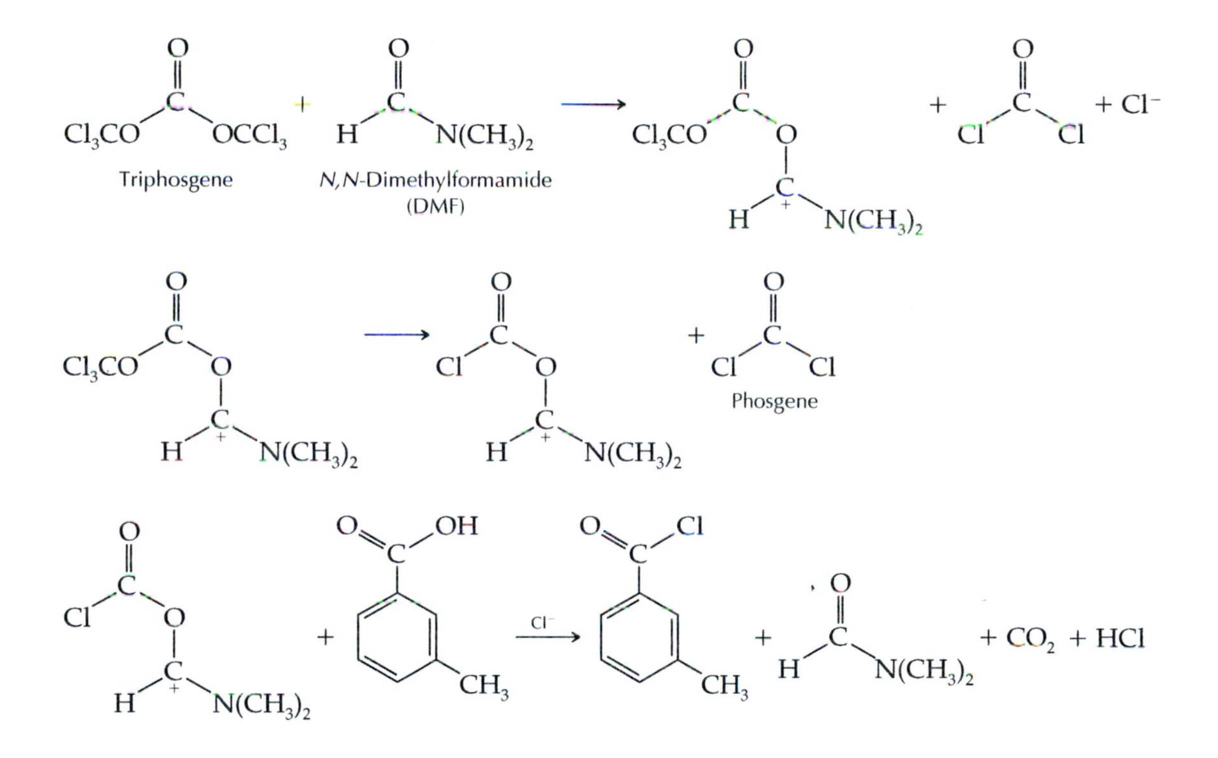

The two molecules of phosgene ($COCl_2$) that are made in the first two steps also react with *m*-toluic acid to produce two additional molecules of *m*-toluoyl chloride, in a mechanism analogous to the one given earlier for thionyl chloride.

Figure 28.1 shows the ^1H NMR spectrum of *N,N*-diethyl-*m*-toluamide. Of particular interest in the spectrum is the magnetic inequality of the two ethyl groups attached to nitrogen because of the planarity of the amide bond and its restricted rotation.

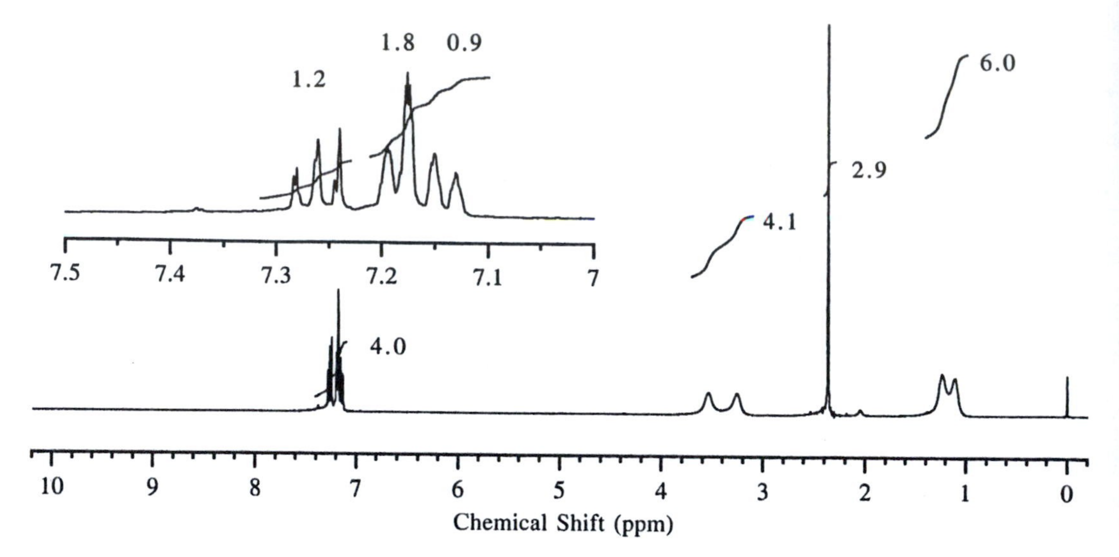

FIGURE 28.1 ^{1}H NMR spectrum (300 MHz) of *N,N*-diethyl-*m*-toluamide (in $CDCl_3$). Assignments: 1.05, 1.2 ppm: two broadened singlets, 3H each (methyl protons of the ethyl groups). At elevated temperatures these two signals coalesce into a 6H triplet. 2.35 ppm: singlet, 3H (ring methyl). 3.2, 3.5 ppm: two broadened singlets, 2H each (methylene protons of the ethyl groups). At elevated temperatures these two signals coalesce into a 4H quartet. 7.0–7.3 ppm: complex multiplet, 4H (aromatic protons).

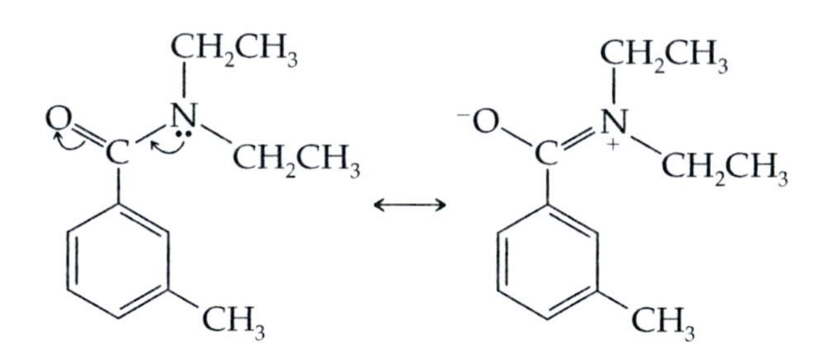

The signals for the ethyl protons are broadened because of the magnetic properties of the nitrogen nucleus, and the normal doublet/triplet splitting pattern of each ethyl group is not readily apparent in the spectrum.

28.1 Synthesis of Benzoylamino Acids from Glycine, Alanine, and Valine

PURPOSE: To convert the amino acids glycine, alanine, and valine into their benzoylamino acid derivatives and compare the results.

TEAMWORK: Work in teams of three students with each student using one of the three amino acids for the reaction. Share your data with the other members of the team.

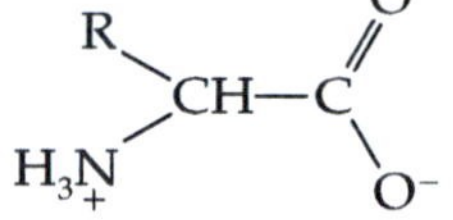

Amino acid

R = H, glycine MW 75.1
= H_3C, alanine MW 89.1
= $(CH_3)_2CH$, valine MW 117.2

+ Benzoyl chloride

bp 197°C
MW 140.6
density 1.21 g · mL^{-1}

NaOH (aq) → HCl →

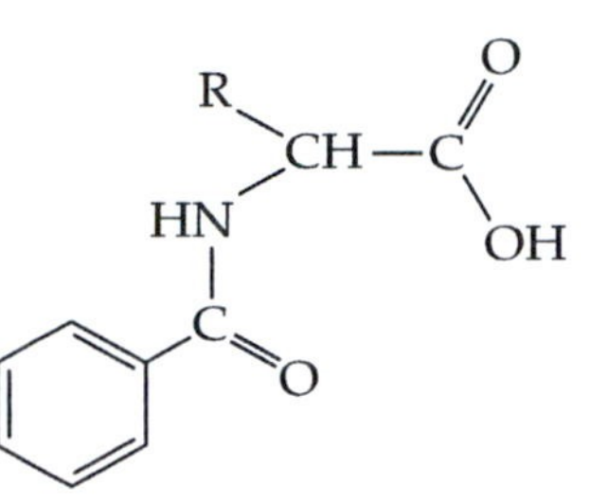

(*R*,*S*)-benzoylamino acid

	MW	mp,°C
(*R*,*S*)-benzoylglycine	179.2	190–193
(*R*,*S*)-benzoylalanine	193.2	164–166
(*R*,*S*)-benzoylvaline	221.3	129–130

MINISCALE PROCEDURE

Techniques IR Spectroscopy: Technique 18
NMR Spectroscopy: Technique 19

SAFETY INFORMATION

Benzoyl chloride is a lachrymator (it causes tears) and a skin irritant. Wear gloves and dispense it only in a hood.

Aqueous **sodium hydroxide** solutions are corrosive and cause burns. Solutions as dilute as 9% (2.5 M) can cause severe eye injury.

Diethyl ether is extremely flammable. Be certain that no flames are used in the laboratory and that hot electrical devices are not in the vicinity where ether is being used.

Concentrated hydrochloric acid is corrosive and causes burns. The vapor is extremely irritating to mucous membranes and the upper respiratory tract. Measure it in a hood and avoid contact with skin, eyes, and clothing.

Place 0.028 mol of crystalline glycine, alanine, or valine in a 125-mL Erlenmeyer flask that contains a mixture of 22 mL of distilled water and 3.0 mL of 6 M sodium hydroxide solution. Swirl the mixture until the solution clears.

Wearing gloves and working in a hood, measure 3.7 mL of benzoyl chloride in a dry graduated cylinder and quickly add it to the amino acid solution. Still working in the hood, rinse the graduated cylinder with 3.0 mL of 6 M NaOH solution and add this rinse to the reaction flask. Stopper the flask tightly with a solid rubber stopper.

Shake the Erlenmeyer flask until the layers are completely mixed. The reaction mixture will become warm, so vent it cautiously in the hood by removing the stopper. If the reaction mixture becomes hot, cool it in a

Monitor the pH by dipping the tip of a stirring rod into the solution and then touching it to a piece of pH test paper.

water bath. Monitor the pH, then replace the stopper immediately. When the mixture becomes acidic, add another 6.0 mL of 6 M NaOH solution over a 2- to 3-min period, shaking the flask after each addition. It may be necessary to add another 1–2 mL of 6 M NaOH toward the end of the reaction to maintain an alkaline pH.

The reaction is complete when it forms a clear yellow solution and no oil droplets remain at the bottom of the flask. If any precipitate is present at this point, filter the mixture using vacuum filtration. Save the filtrate (your product) and discard any solids on the filter paper.

Place approximately 15 g of ice in a 250-mL beaker and carefully pour 15 mL of concentrated hydrochloric acid over the ice. Pour your reaction solution into the acid/ice mixture with stirring. Collect the solid product by vacuum filtration. Remove the filtrate from the filter flask and set it aside in a beaker for treatment later.

To remove any benzoic acid formed by hydrolysis of excess benzoyl chloride, turn off the vacuum source and cover the crystals with 6 mL of ether. **(Caution: Flammable.)** Remove the ether by reapplying the suction. Repeat this ether washing two more times.

Sometimes a gummy white product forms when the reaction mixture is poured into the cold hydrochloric acid. If this occurs in your reaction, cool the mixture in an ice-water bath while you continue to stir and rub the gummy material. A solid that can be filtered should result. After filtration and before the ether washing, the solid should be ground to a fine powder; use a mortar and pestle if it is very lumpy. Thin-layer chromatography can be used to assay the purity of your product.

Normally, the benzoylamino acid does not need to be recrystallized. If you choose to recrystallize it, you can use a 1:1 ethanol/water mixture [see Technique 9.2 on recrystallization from a mixed solvent]. The product should be recrystallized if colored impurities are present (consult your instructor). Use a steam bath or a hot plate as the heat source. Add decolorizing carbon (2–3 spatulas full) and swirl the hot mixture [see Technique 9.5]. Remove the decolorizing carbon by gravity filtration of the hot solution into an Erlenmeyer flask through a fluted filter paper placed in a heated stemless funnel or plastic filling funnel. Also heat the receiving flask during the filtration as a way of minimizing crystallization in the funnel. On cooling the filtrate, the benzoylamino acid will crystallize and can be collected by vacuum filtration. Allow the benzoylamino acid to dry before weighing it and determining the melting point. Calculate your percent yield.

Spectroscopic Analysis of the Product

Obtain an IR spectrum of your benzoylamino acid [see Technique 18.4] and an NMR spectrum in acetone-d_6 [see Technique 19.2], as directed by your instructor. Analyze your spectra. What IR bands [see Techniques

18.5 and 18.6, including Tables 18.2 and 18.3] indicate that you have the desired product? Does your NMR spectrum fit your prediction of the spectrum for the desired product [see Techniques 19.4 and 19.9, including Tables 19.2 and 19.3]?

CLEANUP: Neutralize the aqueous filtrate from the crude product with solid sodium carbonate **(Caution: Foaming.)** before washing the solution down the sink or pouring it into the container for aqueous inorganic waste. Pour the ether filtrate into the container for flammable (organic) waste. The aqueous ethanol filtrate from the recrystallization can be washed down the sink or placed in the flammable (organic) waste container. Place the filter paper containing the decolorizing carbon in the container for nonhazardous solid waste.

MICROSCALE PROCEDURE

Techniques Microscale Recrystallization: Technique 9.7a
IR Spectroscopy: Technique 18
NMR Spectroscopy: Technique 19

SAFETY INFORMATION

Benzoyl chloride is a lachrymator (it causes tears) and a skin irritant. Wear gloves and dispense it only in a hood.

Aqueous **sodium hydroxide** solutions are corrosive and cause burns. Solutions as dilute as 9% (2.5 M) can cause severe eye injury.

Diethyl ether is extremely flammable. Be certain that no flames are used in the laboratory and that hot electrical devices are not in the vicinity where ether is being used.

Concentrated hydrochloric acid is corrosive and causes burns. The vapor is extremely irritating to mucous membranes and the upper respiratory tract. Measure it in a hood and avoid contact with skin, eyes, and clothing.

Set the conical vial in a 30-mL beaker so that it will not tip over.

Place 2.0 mmol of crystalline glycine, alanine, or valine, 1.6 mL of water, and 0.4 mL of 6 M sodium hydroxide solution in a 5-mL conical vial. Close the vial with a screw cap and plastic septum; shake it until the amino acid dissolves.

Monitor the pH by putting the tip of a stirring rod into the solution, and then touching it to a piece of pH test paper.

Wearing gloves and working in a hood, measure 0.26 mL of benzoyl chloride with a graduated pipet and quickly add it to the amino acid solution. Cap the vial and shake it until the layers are completely mixed. The reaction mixture may become warm, so vent it cautiously in the hood by loosening the screw cap. Monitor the pH, then replace the septum immediately. When the mixture becomes acidic, add another 0.40 mL of 6 M NaOH solution. Continue shaking the vial. It may be necessary to add another 0.10 mL of 6 M NaOH toward the end of the reaction to maintain an alkaline pH.

The reaction is complete when it forms a clear yellow solution and there are no oil droplets at the bottom of the vial. If any precipitate is present at this point, filter the mixture using a Hirsch funnel and vacuum filtration. Save the filtrate (your product).

Place 0.40 mL of water in a 30-mL beaker, cool the beaker in an ice-water bath, and add 0.40 mL of concentrated hydrochloric acid. Pour your reaction solution (or the filtrate, if you had to filter out precipitate) into the acid/water solution while stirring. Collect the solid product by vacuum filtration on a Hirsch funnel [see Technique 9.7a]. Remove the filtrate from the filter flask and set it aside in a beaker for treatment later.

To remove any benzoic acid formed by hydrolysis of excess benzoyl chloride, turn off the vacuum source and cover the crystals with approximately 1 mL of diethyl ether. **(Caution: Flammable.)** Remove the ether by reapplying the suction. Repeat this ether washing two more times. Allow the benzoylamino acid to dry before weighing it and determining the melting point. Calculate your percent yield.

If colored impurities are present, recrystallize the crude product using 1:1 (v/v) ethanol/water solution as the solvent. Place the crystals and a boiling stick in a 13 × 100 mm test tube; use a steam bath or a beaker of boiling water as the heat source. Add 10–15 Norit pellets (decolorizing carbon) and heat the mixture for another 1–2 min [see Technique 9.5]. Transfer the hot solution to another test tube with a Pasteur pipet. Cool the tube until crystallization is complete and collect the product by vacuum filtration on a Hirsch funnel.

Determine the melting point, mass, and percent yield of your product after it has dried at least overnight. Carry out spectroscopic analysis of your product as directed in the miniscale procedure.

CLEANUP: Neutralize the aqueous filtrate from the crude product with solid sodium carbonate before washing the solution down the sink or pouring it into the container for aqueous inorganic waste. Pour the ether filtrate into the container for flammable (organic) waste. The aqueous ethanol filtrate from the recrystallization can be washed down the sink or placed in the container for flammable (organic) waste. Put the Norit into the container for nonhazardous solid waste.

OPTIONAL EXPERIMENT Benzoylamino acids from L-Alanine and L-Valine

PURPOSE: To synthesize and study the properties of the benzoylamino acids of optically active alanine and valine using polarimetry and IR and NMR spectroscopy.

Prepare the benzoylamino acid of either L-alanine or L-valine by the same procedure you used for the racemic amino acid. Determine the observed

TABLE 28.1 Properties of *N*-benzoylamino acids

	mp (°C)	$[\alpha]_D$, Solvent
N-benzoyl-L-alanine	152–154	+35°, 1M KOH (Ref. 1)
N-benzoyl-L-valine	127	+22°, ethanol (Ref. 2)

rotation of your optically active benzoylamino in the solvent indicated in Table 28.1 [see Technique 14]. Calculate the specific rotation. Determine the melting point of your product. Carry out spectroscopic analysis of your product as directed in the miniscale procedure.

Explain any differences in the melting points of the racemic compound and the optically active one. What could make optically active benzoylalanine have a different melting point than racemic benzoylalanine? Compare the IR and NMR spectra of the racemic and optically active benzoyl amino acids. Explain similarities and/or differences in the spectra for the racemic compound versus the optically active one.

References

1. Daffe, V.; Fastrez, J. *J. Am. Chem. Soc.* **1980,** *110,* 3601–3605.
2. Dean, B. M.; et al. *J. Chem. Soc.* **1961,** 3394–3400.

Questions

1. Why is the benzoylamino acid product soluble in the aqueous reaction solution until hydrochloric acid is added?
2. Why must sodium hydroxide be present for the reaction between an amino acid and benzoyl chloride to take place at a reasonable rate?
3. A possible alternative procedure would be to add all the 6 M NaOH solution at the beginning of the reaction. What disadvantage could this have?

28.2 Synthesis of a Mosquito Repellant: *N,N*-Diethyl-*meta*-Toluamide (DEET)

PURPOSE: To synthesize DEET by a two-step process and purify it by column chromatography.

MINISCALE PROCEDURE

Techniques Removal of Noxious Gases: Technique 7.4
Extraction: Technique 8.4
Liquid Chromatography: Technique 17
IR Spectroscopy: Technique 18
NMR Spectroscopy: Technique 19

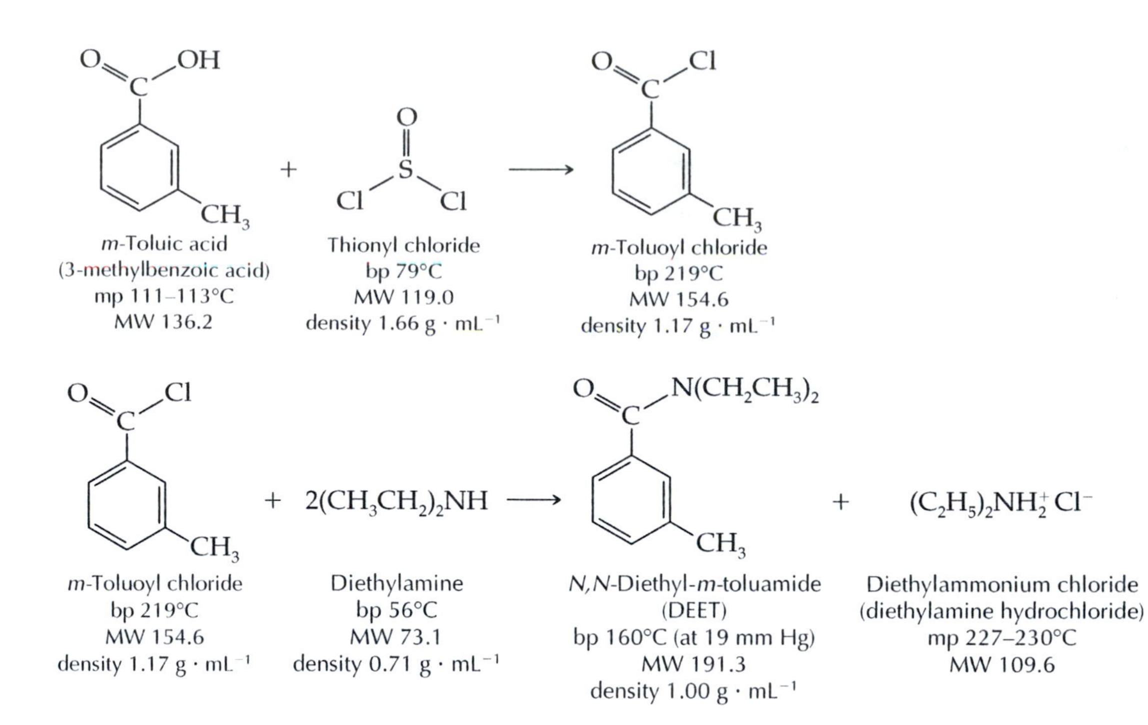

SAFETY INFORMATION

Wear gloves throughout the procedure and conduct the experiment in a hood.

Thionyl chloride is a lachrymator (it causes tears) and causes serious burns. It is also volatile and undergoes reaction with atmospheric moisture, forming hydrogen chloride and sulfur dioxide. Wear gloves while handling it and dispense it only in a hood.

Diethylamine is flammable, toxic, and has a noxious odor. Wear gloves while handling it and dispense it only in a hood.

Diethyl ether is very flammable. Use it only in a hood and keep it away from flames or hot electrical heating devices.

Alumina is an eye and respiratory irritant. Avoid breathing any fine particles while handling it.

Heptane is very flammable.

The product, ***N,N*-diethyl-*m*-toluamide,** is a skin irritant in high concentrations.

Hydrogen chloride and sulfur dioxide, produced as side products in this synthesis, are noxious, toxic gases that must be trapped by dissolution in water.

Step 1: Synthesis of m-Toluoyl Chloride

If your laboratory is not equipped with water aspirators, assemble the gas trap shown in Technique 7.4, Figure 7.5a, using a thermometer adapter, a 125-mL filter flask, and a piece of U-shaped glass tubing with a one-hole rubber stopper that fits the filter flask on the long arm of the tubing. If your laboratory is equipped with water aspirators, use the vacuum adapter and a piece of heavy-walled rubber tubing, as shown in Technique 7.4, Figure 7.5b, to trap the noxious vapors.

Both thionyl chloride ($SOCl_2$) and the acid chloride formed in this procedure react violently with water, so your apparatus and all reagents must be completely dry.

Use a 100-mL or 250-mL three-necked round-bottomed flask or a single-necked flask fitted with a Claisen adapter as the reaction vessel. Close the other necks of the flask or the Claisen adapter with glass stoppers.

Place 2.5 g of *m*-toluic acid in the round-bottomed flask. Add one or two boiling stones and 2.0 mL of thionyl chloride to the reaction flask. Heat the mixture in a water bath at 70°C, submerging the flask just enough to keep the mixture bubbling gently. Continue to heat the reaction until the bubbling slows and the reaction mixture is a clear liquid. The entire heating period should take about 15 min.

The reaction will turn from a creamy mixture to a light brown solution as it goes to completion.

Step 2: Synthesis of N,N-Diethyl-m-Toluamide

The diethylamine should be a colorless liquid, uncontaminated by colored oxidation products; otherwise your final product may be highly colored.

Cool the reaction mixture to room temperature, then pour 30 mL of anhydrous ether into the reaction flask and replace one glass stopper with a separatory funnel. Be sure that the stopcock of the separatory funnel is closed. The rest of the apparatus remains unchanged. Pour 10 mL of anhydrous diethyl ether into the separatory funnel and add 5.0 mL of diethylamine to the ether in the funnel.

Add the diethylamine/ether solution dropwise over a period of about 10 min. Large amounts of white smoke will evolve, possibly accompanied by sizzling noises. The reaction is quite dramatic but not dangerous. After the addition is complete and the reaction has ceased, swirl the flask vigorously to ensure thorough mixing. Add 15 mL of 2.5 M sodium hydroxide solution and swirl the flask to dissolve the solid.

Transfer the reaction mixture to a separatory funnel and remove the lower aqueous layer; set the aqueous layer aside in a separate flask. Wash the ether layer with 15 mL of cold 3-M hydrochloric acid and then with 15 mL of water [see Technique 8.4].

SAFETY PRECAUTION

The reaction of hydrochloric acid with the residual sodium hydroxide in the separatory funnel produces a considerable amount of heat, which may vaporize the diethyl ether. Before you stopper the funnel, swirl the ether and hydrochloric acid layers together to complete most of the acid-base reaction. If the solution becomes warm, add a few pieces of crushed ice. After you stopper the funnel, immediately invert it and open the stopcock to relieve any pressure buildup.

After removing the final aqueous solution from the separatory funnel, transfer the upper ether layer to a clean Erlenmeyer flask and dry it over anhydrous magnesium sulfate for at least 10 min [see Technique 8.7]. Decant the ether solution through a fluted filter paper into a tared 100-mL round-bottomed flask. Rinse the magnesium sulfate remaining in the Erlenmeyer flask with a few milliliters of ether and pour the rinse through the filter paper. Remove the ether by simple distillation [see Technique 11.3], using a steam bath as the heat source, or with a rotary evaporator [see Technique 8.9]. You should have a viscous tan liquid left after the ether is removed.

Step 3: Column Chromatography

Do not allow the solvent level to fall below the top of the adsorbent during the entire chromatography process.

Your product can be purified by column chromatography [see Technique 17]. Following the procedure in Technique 17.4, pack a 1.6-cm interior-diameter column with 20–25 g of alumina using heptane as the solvent. After the column has been prepared, allow the heptane to drip from the column until the liquid level is just at the top of the sand above the adsorbent. Close the stopcock and transfer your product to the column with a Pasteur pipet. Rinse the round-bottomed flask with 2–3 mL of heptane and also transfer this rinse into the column. Start the column flowing again and, when the solution of your product is just level with the sand, slowly add about 60 mL of heptane to the column. This eluent can be the same heptane that you used to make your column.

Elute the product from the column with heptane and collect the faintly yellow amide fraction. You can add a few milliliters more heptane to the top of the column if you have not added enough to elute all the *N,N*-diethyl-*m*-toluamide. If you have doubts as to when all the amide has been eluted, use a watch glass to collect a few drops of the liquid draining from the column. Stop the flow of the column and evaporate the liquid on a steam bath in a hood. If the amide is still eluting, a viscous residue will remain on the watch glass after the heptane has evaporated.

Remove the heptane from your product by simple distillation [see Technique 11.3] or with a rotary evaporator [see Technique 8.9]. When the distillation slows as the removal of heptane nears completion, pour the heptane out of the receiving flask, replace the receiving flask, and apply full water-aspirator suction at the vacuum adapter nipple to complete the removal of heptane from your product [see Technique 11.6]. If your laboratory is not equipped with water aspirators, work in a hood and remove the residual heptane by blowing a stream of dry nitrogen over the product until the flask and product have reached a constant mass (consecutive weights within 0.05 g). The viscous, high-boiling *N,N*-diethyl-*m*-toluamide will remain as a colorless or light tan liquid. Weigh your product and calculate the percent yield.

Spectroscopic Analysis of the Product

Obtain the NMR spectrum of your product in $CDCl_3$ [see Technique 19.2], as directed by your instructor, as well as its IR spectrum [see Technique 18.4]. Verify the structure of your product from the two spectra. Assign each peak in the NMR spectrum [see Techniques 19.4 and 19.9, including Tables 19.2–19.5] and account for its chemical shift and spin-spin splitting pattern. Are the peak areas consistent with your assignments of the peaks? What vibrational bands in the IR spectrum [see Techniques 18.5 and 18.6, including Tables 18.2 and 18.3] indicate that your synthesis was successful?

CLEANUP: The aqueous phase initially separated from the reaction mixture contains diethylamine; pour this into the container for flammable (organic) waste. Combine the HCl and water washes remaining from the extractions with the acid solution from the gas trap in an 800-mL beaker. Neutralize the solution with solid sodium carbonate **(Caution: Foaming.)** before washing it down the sink or pouring it into the container for aqueous inorganic waste. Place the recovered ether and recovered heptane into the bottle for flammable (organic) waste. Place the spent drying agent (magnesium sulfate) in the container for inorganic or nonhazardous solid waste, as directed by your instructor.

MICROSCALE PROCEDURE

Techniques Anhydrous Reaction Conditions: Technique 7.3
Removal of Noxious Vapors: Technique 7.4
Microscale Extraction: Technique 8.6b
Flash Chromatography: Technique 17.8
Thin-Layer Chromatography: Technique 15
IR Spectroscopy: Technique 18
NMR Spectroscopy: Technique 19

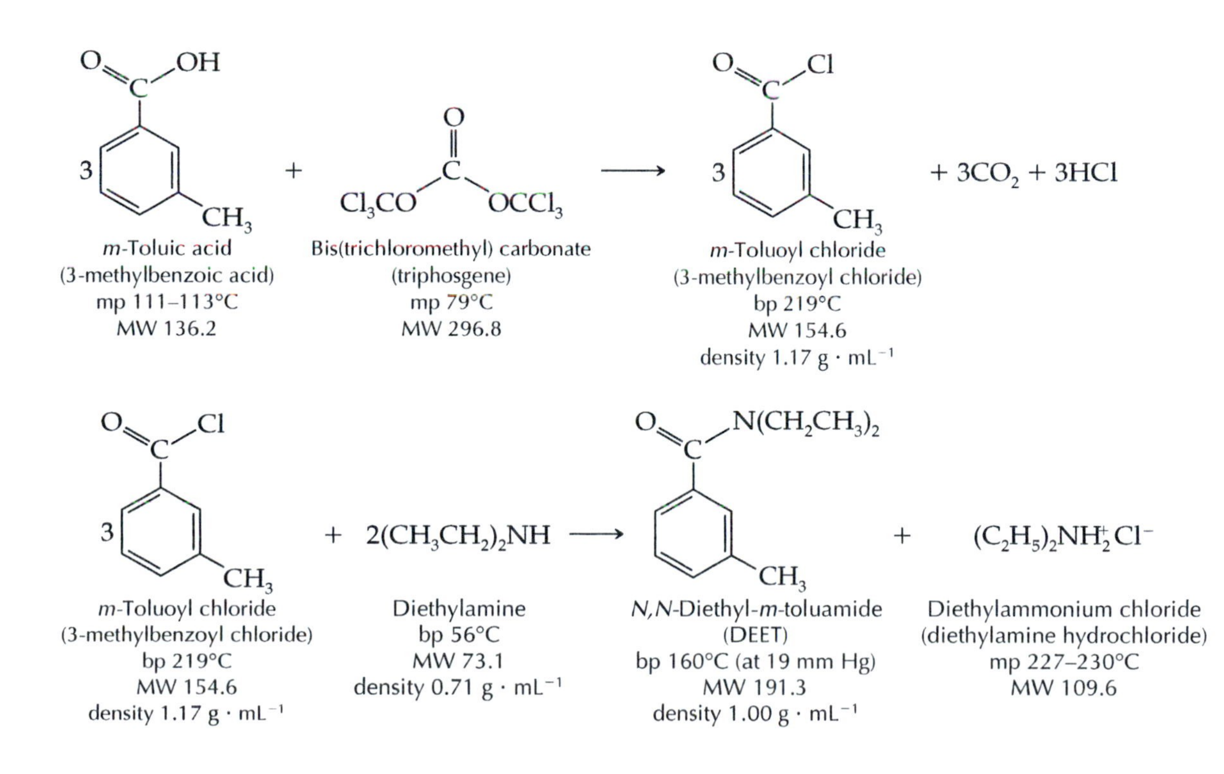

SAFETY INFORMATION

Wear gloves and conduct this synthesis in a hood.

3-Methylbenzoic acid (*m*-toluic acid) presents no particular hazards.

Bis(trichloromethyl) carbonate (triphosgene) is a lachrymator (it causes tears). Use it in a hood. The compound is moisture sensitive; keep it in a desiccator.

Diethylamine is flammable and corrosive. Wear gloves while working with it.

Dichloromethane is toxic, an irritant, absorbed through the skin, and harmful if inhaled. Use it only in a hood and wear neoprene gloves while doing the extractions.

Aqueous **sodium hydroxide** solutions are corrosive and cause burns. Solutions as dilute as 9% (2.5 M) can cause severe eye injury.

Synthesis of m-Toluoyl Chloride

Pack a microscale drying tube with anhydrous calcium chloride, as shown in Technique 7, Figure 7.4a. Put a water-jacketed condenser, a Claisen adapter, the drying tube filled with $CaCl_2$, and a 5-mL conical vial containing a spin vane in a 250-mL beaker and dry this glassware in a 125°C oven for 20 min. Cool the glassware in a desiccator to room temperature.

While the glassware is heating, prepare a gas trap. Thread one end of an 18- to 20-inch piece of Teflon tubing (1/16 inch in diameter) through a fold-over septum that fits the end of the drying tube and the other end through a septum that fits the top of a 25-mL filter flask [see Technique 7.4, Figure 7.7]. Put approximately 10 mL of 5% sodium hydroxide solution in the filter flask and close the flask with the rubber septum; adjust the tubing inside the flask so that the open end is just above the surface of the liquid.

Place 136 mg of *m*-toluic acid in the 5-mL conical vial and add 1.0 mL of dichloromethane. Stir the mixture until the acid dissolves, then add 100 mg of bis(trichloromethyl) carbonate (triphosgene) and one drop of dry dimethylformamide (DMF). Assemble the apparatus as shown in Technique 7, Figure 7.4a, omitting the syringe. Place the small septum on the gas-trap tubing on the end of the drying tube.

Heat and stir the reaction mixture in an 80°C water bath for 1 h. At the end of the healing period, lift the apparatus out of water bath. The acid chloride is used immediately in the next step.

Synthesis of N,N-Diethyl-m-Toluamide

During the heating period, mix 0.50 mL of dichloromethane and 0.22 mL of freshly distilled diethylamine in a dry 1-dram vial. Cap the vial and cool it in an ice-water bath.

Cool the reaction vial (with the condenser and drying tube still in place) for 5 min at room temperature, then for 10 min in an ice-water bath. Draw the dichloromethane/diethylamine solution into a dry 1-mL syringe and insert the syringe needle through the septum in the Claisen adapter. Add the diethylamine solution dropwise over 2–3 min to the cooled reaction mixture. A white precipitate of diethylamine hydrochloride forms immediately. After all the diethylamine has been added, remove the Claisen adapter and gas trap assembly. Then withdraw a drop of the reaction mixture with a Pasteur pipet and test it for basicity with wetted litmus paper or pH test paper. If the mixture is still acidic, add diethylamine dropwise until the solution is basic. Remove the conical vial from the ice-water bath and allow it to warm to room temperature.

Add 1 mL of dichloromethane and 1 mL of 10% NaOH to the conical vial and stir until the precipitate dissolves. Transfer the mixture to a 15-mL centrifuge tube with a Pasteur filter pipet or a Pasteur pipet fitted with a syringe [see Technique 8.5]. Rinse the flask with 2 mL of dichloromethane and transfer the rinse to the centrifuge tube. Cap the tube and shake it briefly, allow the phases to separate, then transfer the lower organic layer to another centrifuge tube [see Technique 8.6b]. Wash the organic layer with 1 mL of 3 M HCl solution and finally with 1 mL of water. After the water wash, transfer the organic layer to a dry 13 × 100 mm test tube, cork the tube, and dry the solution for 10 min with 200 mg of anhydrous sodium sulfate.

Transfer the dried solution to a tared 10-mL Erlenmeyer flask. Rinse the Na_2SO_4 with 0.5 mL of fresh dichloromethane and add this rinse to the flask. Evaporate the dichloromethane by leaving the flask open in a hood until the next laboratory period. Alternatively, evaporate the solvent with a stream of nitrogen in a hood. Weigh the flask and determine the percent yield of *N,N*-diethyl-*m*-toluamide.

CLEANUP: Combine the aqueous washes in the beaker with the NaOH from the gas trap and determine the pH with pH test paper. Adjust the pH to approximately 7 with either solid sodium carbonate or a few drops of 3 M HCl. Wash the neutralized solution down the sink or pour it into the container for aqueous inorganic waste. Allow the solvent to evaporate before putting the solid Na_2SO_4 in the container for inorganic waste.

A good stopping point.

Purification of N,N-Diethyl-m-Toluamide by Flash Chromatography

SAFETY INFORMATION

Ethyl acetate and **petroleum ether** are very volatile and flammable.

Obtain 250 mL of 2:3 (v/v) ethyl acetate/petroleum ether (bp 30–60°C). Dissolve your crude *N,N*-diethyl-*m*-toluamide by adding 1 mL of the ethyl acetate/petroleum ether solvent.

Use a 30 cm × 7 mm interior-diameter chromatography column. Prepare the column by the procedure described in Technique 17.4, using approximately 15 g of silica gel* and the ethyl acetate/petroleum ether solution as solvent. Fill the column almost to the top with solvent (be sure the stopcock is closed).

Place a number-2 rubber stopper—fitted with a glass Y-joint in the hole of the stopper—at the top of the column as a simple airflow controller. Connect one branch of the Y to an air or nitrogen line with a piece of Tygon or rubber tubing. Open the stopcock again, turn on the air valve slightly, and use your index finger on the other branch of the Y to manipulate the pressure so that the silica gel packs tightly, forcing all entrapped air out the bottom. Control the pressure so that solvent flows through the column at a rate of ~5 cm per min. Never let the solvent level fall below the top of the column; add more solvent, if necessary. When the column is packed tightly, allow the solvent level to drop almost to the top of the sand, close the stopcock, release the pressure, and add the solution of your product to the column, using a Pasteur pipet for the transfer. Allow the solution to sink into the sand at the top of the column.

Fill the column with solvent and adjust the air pressure so that the solvent flows through the column at a rate of ~5 cm per minute. Collect 20 fractions of 10 mL each in labeled test tubes. You will need to stop and refill the column several times with solvent while you are collecting the fractions.

A good stopping point. Tightly cork the test tubes containing the fractions.

Using thin-layer chromatography on silica gel plates impregnated with fluorescent indicator, analyze the fractions to determine which fractions contain *N,N*-diethyl-*m*-toluamide [see Technique 15]. Develop the TLC plates in the same solvent used for the liquid chromatography; visualize the chromatograms with UV light [see Technique 15.6]. Combine the fractions containing *N,N*-diethyl-*m*-toluamide and evaporate the solvent using a rotary evaporator [see Technique 8.9] or by a simple distillation [see Technique 11.3], using a steam bath or 80–90°C water bath as the heat source to yield pure *N,N*-diethyl-*m*-toluamide as a colorless oil. Weigh your product and calculate the percent yield.

Carry out the spectroscopic analysis of your product as described in the miniscale procedure.

CLEANUP: Pour the recovered ethyl acetate/petroleum ether solution into the container for flammable (organic) waste. Place the silica gel from the column in the container for solid inorganic waste.

*Silica Gel 60, 200-400 mesh, E. Merck, # 9385

References

1. Wang, B. J.-S. *J. Chem. Educ.* **1974,** *51,* 631.
2. LeFevre, J. W. *J. Chem. Educ.* **1990,** *67,* A278–A279.
3. Echkert, H.; Forster, B. *Ang. Chem. Int. Ed. Engl.* **1987,** *26,* 894–895.

Questions

1. Write the balanced chemical equation that describes the reaction of *m*-toluoyl chloride with water.
2. Diethyl ether is used as a solvent for the reaction of diethylamine with *m*-toluoyl chloride in the miniscale experiment. Why is a solvent necessary for this reaction?
3. What is the white solid that precipitates from the reaction mixture as diethylamine is added to *m*-toluoyl chloride in the miniscale experiment? Why would this compound not be soluble in ether? Write an equation for this reaction.
4. Why is it crucial that a chromatography column never go dry after preparation or during use?
5. Arrange the following compounds in order of their decreasing ease of elution from a silica gel chromatographic column: (a) 2-octanol; (b) *m*-dichlorobenzene; (c) *tert*-butylcyclohexane; (d) benzoic acid.

PART

2

Projects

PROJECT

1

IDENTIFICATION OF A WHITE SOLID

A Team Approach

QUESTION: Who else has my white solid and what is its identity?

> In this two-week project you will first determine the melting point, thin-layer chromatographic behavior, and solubility of a white solid. Then you will look for other students in your lab class with similar data. In the second week you will work with your team to prove that you all have the same compound and to identify it.

In the process of answering the question posed in Project 1, you will learn several basic techniques used by chemists in the identification of organic compounds: melting-point determination, solubility tests, and thin-layer chromatography. You will be given a small amount of a white organic compound, and you will collect data about its melting point; solubility in water, acetone, and 2.5 M NaOH; and how it moves on a thin-layer chromatographic plate. Your first goal for the project is to find the other students in your laboratory section who have the same compound as you do. Potential matches of different samples can be made by comparing your data with the data of each student in your laboratory section. If the comparisons are ambiguous, you can (1) spot the samples side by side or cospot the samples on a TLC plate and (2) use a mixture melting-point determination.

Once a group of students has shown that all its members have the same compound and that no other student in the lab section has this compound, your second goal is to determine the identity of your group's compound. The class will be given a list of compounds with which to compare the observed properties of their unknown samples. Samples of the possible compounds will be available for making actual comparisons.

MICROSCALE PROCEDURE

PRELABORATORY ASSIGNMENT: List the factors that determine the R_f of a compound on a silica gel TLC plate.

Techniques Melting Points: Technique 10
Thin-Layer Chromatography: Technique 15

SAFETY INFORMATION

Wear gloves when handling the unknown compounds.

Acetone is flammable and an irritant. Avoid contact with skin, eyes, and clothing.

Aqueous **sodium hydroxide** solutions are corrosive and cause burns. Solutions as dilute as 2.5 M can cause severe eye injury. Avoid contact with skin, eyes, and clothing.

First Week: Individual Work

Obtain an unknown sample. Determine its melting point until you have two determinations that agree within 1°C [see Technique 10.3].

Carry out thin-layer chromatographic analysis on your unknown sample to determine its R_f value in each of the following developing solvents: dichloromethane, hexane, and ethyl acetate [see Technique 15]. Prepare a spotting solution by dissolving 10–15 mg of your unknown compound in 1.0 mL of diethyl ether.

Determine the solubility of your unknown in water, acetone, and 2.5 M NaOH solution. Weigh 15 mg (0.015 g) of your unknown on glassine weighing paper and transfer the sample to a small test tube. Add 0.5 mL of the solvent being tested and shake the test tube vigorously or agitate its contents on a vortex mixer. Record your observations.

Summarize your data in tabular form and compare it to that of all other students in your lab section. Identify all students with data similar to yours.

CLEANUP: Pour any remaining dichloromethane from the TLC analysis into the container for halogenated organic waste. Pour the TLC spotting solution, any remaining acetone and ethyl acetate from the TLC analysis, and the acetone mixture from the solubility test into the container for flammable (organic) waste. Pour the aqueous mixtures from the solubility tests into the container for aqueous inorganic waste.

Second Week: Team Work

QUESTION: Who else has my white solid and what is its identity?

Meet with the other members of your team either before or during your lab period, as directed by your instructor. Your tasks this week are to prove that the group's samples are identical and to determine the identity of the group's compound using the reference compounds available in the laboratory. Address the following questions during your group discussion and work out a strategy for proving that your samples are the identical compound and for determining its identity.

1. What are the properties of my substance?
2. How do these properties compare with the properties reported by others in the lab?
3. What constitutes a valid comparison?
4. How will we use the techniques we employed last week to prove that our samples are identical and to determine the identity of our unknown substance?
5. How does the group's compound compare with the known reference compounds?

Record both your data and the discussions with the others in your group. Divide the work so that each team member contributes equally in collecting the team's data.

CLEANUP: Follow the protocol specified for the first week of the project.

Interpretation of Experimental Data

Your team has collected data about the melting point, solubility and TLC behavior of your compound. You have also compared its properties to those of one or more known compounds. How did you use the team's data to decide that each team member has the same substance? How did you use the data to determine the identity of your team's substance?

OPTIONAL EXPERIMENT **IR or NMR Spectroscopy**

During the first week of the project, obtain the infrared (IR) spectrum [see Technique 18.4] or nuclear magnetic resonance (NMR) spectrum [see Technique 19.2] of your white solid, as directed by your instructor. Include a comparison of your spectrum to that of others in your lab section when you are comparing data during the first week of the project. How can you use the spectral data in identifying your group's compound?

Reference

1. Coppola, B. P.; Lawton, R. G. *J. Chem. Educ.* **1995**, *72*, 1120–1122.

Questions

1. Why did you observe different R_f values for the three development solvents tested in the TLC analysis of your compound? Explain.
2. Why is the R_f of your compound greater in ethyl acetate than in hexane?
3. Explain why your compound is more soluble in acetone than in water.
4. Explain why a compound that is insoluble in water might dissolve in aqueous 2.5 M NaOH solution.
5. A student performed two melting-point determinations on a crystalline compound. In one determination, the capillary tube contained a sample about 1–2 mm in height, and the melting range was found to be 141–142°C. In the other determination, the sample height was 4–5 mm, and the melting range was found to be 141–145°C. The reported melting point for the compound is 143°C. Explain what may have caused the broader melting range observed for the second sample.
6. Another student reported 136–138°C for the melting range of the compound in Question 5 and recorded that the rate of heating was about 12° per minute. Further analysis of this student's compound did not reveal any impurities. Explain the low melting range.

PROJECT

2 USING EXTRACTION TO SEPARATE A MIXTURE

QUESTION: What are the identities of the individual compounds in a mixture of two organic chemicals?

In this two-week project you will use extraction to separate a mixture containing an acid and a neutral compound. The recovered solids will be purified by recrystallization and assayed by thin-layer chromatography, and their identities will be determined by melting point and infrared (IR) spectroscopy.

Read Technique 8 before starting this project.

The separation of a mixture containing a carboxylic acid and a neutral organic compound can be accomplished by extraction with an aqueous solution of inorganic base. The mixture is dissolved in an organic solvent that is immiscible with water. The extraction with dilute aqueous sodium hydroxide solution deprotonates the carboxylic acid creating a carboxylate anion.

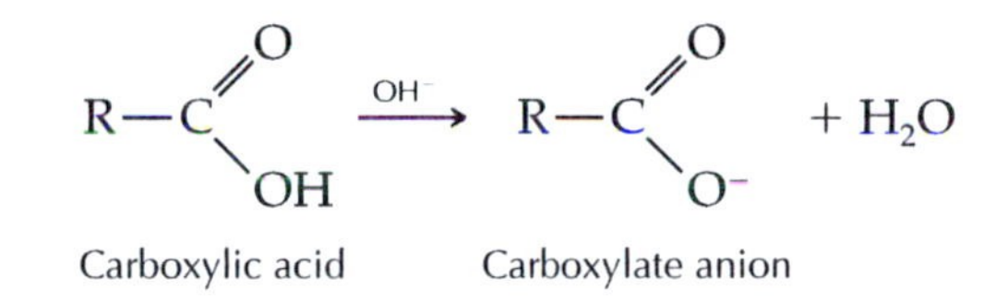

Because the carboxylate anion is an ionic species, it is soluble in the aqueous base solution and not soluble in most organic solvents. This solubility effectively separates the carboxylate anion from the organic solution containing the rest of the mixture. The carboxylic acid is recovered by neutralizing the aqueous solution containing the carboxylate anion with hydrochloric acid.

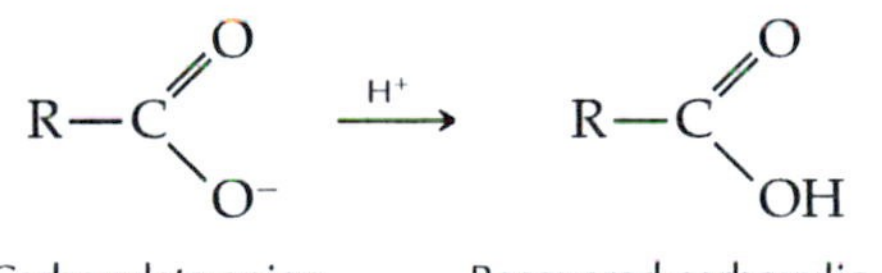

Following the extraction with aqueous base, only the neutral compound remains in the original organic solution. The organic solution is washed with water to remove traces of aqueous NaOH solution and dried with an anhydrous inorganic salt. The neutral compound is recovered by evaporating the organic solvent.

The recovered compounds are purified by recrystallization and their purity assayed by thin-layer chromatography. Their identities are determined by melting point and IR spectroscopy.

2.1 Separation and Purification of the Compounds in the Unknown Mixture: First Week

MINISCALE PROCEDURE

PRELABORATORY ASSIGNMENT: Explain the factors that make most neutral organic compounds more soluble in organic solvents than in water.

Techniques Extraction: Techniques 8.1–8.4
Drying Organic Liquids/Recovery of Product: Techniques 8.7–8.9
Recrystallization: Techniques 9.1–9.5

SAFETY INFORMATION

Aqueous **hydrochloric acid** solution (6 M) is a skin irritant. Avoid contact with skin, eyes, and clothing.

Diethyl ether, commonly called "ether," is extremely volatile and flammable. Use it in a hood, if possible. Be certain that hot electrical devices are not in the vicinity where it is being used.

Ethanol is flammable.

Extraction and Isolation of the Acid

Place your unknown mixture (0.60 g) in a 50-mL Erlenmeyer flask and add 30 mL of diethyl ether. Stir the contents of the flask until the solid is dissolved. Using a conical funnel, pour the solution into a 125-mL separatory funnel. Add 20 mL of 1.5 M sodium hydroxide solution to the separatory funnel. Place the stopper in the top; shake and vent the funnel [see Technique 8.2]. Allow the phases to separate. Drain the lower aqueous phase into a labeled 100-mL beaker. Wash the organic phase remaining in the separatory funnel with 10 mL of water. Allow the phases to separate, and drain the water layer into the beaker holding the first aqueous phase.

Pour the upper organic phase into a dry, labeled Erlenmeyer flask; add anhydrous sodium sulfate to dry the solution [see Technique 8.7]. Cork the flask and allow it to stand for at least 10 min.

Obtain 5 mL of 6 M hydrochloric acid solution. Add the acid in small portions with a dropper to the aqueous solution in the beaker until precipitation of the unknown acid is complete and the solution is acidic (blue) to Congo red test paper.

Congo red turns blue at pH 3.

Cool the beaker containing the precipitated acid in an ice-water bath. Collect the crystals by vacuum filtration on a Buchner funnel [see Technique 9.5, Figure 9.4]. Transfer the acid crystals to a labeled 50-mL Erlenmeyer flask and set the flask aside while you isolate the neutral compound.

Isolation of the Neutral Compound

Filter the dried ether solution of the neutral compound into a dry 50-mL Erlenmeyer flask [see Technique 8.8]. Place a boiling stick or boiling stone in the Erlenmeyer flask and, working in a hood, evaporate the ether using a steam bath or a hot-water bath until only a solid residue (your recovered neutral compound) remains in the flask [see Technique 8.9]. Alternatively, the ether can be removed with a rotary evaporator [see Technique 8.9]. The recovered neutral compound is now ready for recrystallization.

Recrystallization of the Recovered Compounds

Follow the recrystallization procedure given in Technique 9.5. Recrystallize the acid from water and the neutral compound from ethanol.

When the compound has dissolved completely in its boiling solvent, set the flask aside on the bench top to cool and allow crystallization to begin slowly. After crystals have formed, cool each flask in an ice-water bath; then collect each compound on a Buchner funnel by vacuum filtration [see Technique 9.5]. Determine the tare mass of two labeled watch glasses. Place each compound on its watch glass to dry and store them until the next laboratory period.

CLEANUP: Neutralize the aqueous filtrate from the isolation of the acid with solid sodium carbonate. Wash the neutralized solution down the sink or place it in the container for inorganic waste. Pour the filtrates from the two recrystallizations into the container for flammable waste.

2.2 Melting Points, Purity, and Identification of the Compounds in the Mixture: Second Week

MINISCALE PROCEDURE

PRELABORATORY ASSIGNMENT: Consider the following data: A compound melts at 120–122°C on one apparatus and at 128–129°C on another. Unfortunately, neither apparatus is calibrated. Explain how you might check the identity of your sample without calibrating either apparatus.

Techniques Melting Points: Technique 10
Thin-Layer Chromatography: Technique 15
IR Spectroscopy: Technique 18

Melting Points

Determine the melting ranges of the two unknown compounds that you isolated and recrystallized in Project 2.1 [see Technique 10.3]. Determine each melting point a second time or until you have two determinations that agree within 1°C or less. Compare your data to the list of known compounds available in the laboratory. Samples of the known compounds are also available for mixture melting points [see Technique 10.4].

Thin-Layer Chromatography

Prepare a separate TLC test solution for each recrystallized compound in a labeled small test tube or vial by dissolving 5–10 mg of the solid in

0.50 mL of acetone. Prepare a TLC development jar using 0.5% acetic acid in ethyl acetate (v/v) as the developing solvent [see Technique 15.4]. Use silica gel TLC plates impregnated with a fluorescent indicator.* Prepare the plate and spot both compounds on the same plate [see Technique 15.3]. Visualize the chromatogram with a UV lamp and circle all spots with a pencil. Calculate the R_f values for all spots. Does each compound show only one spot or is there evidence of the other compound in the original mixture still present in your recrystallized substances?

Infrared (IR) Spectrum

Obtain an IR spectrum for each of your recrystallized compounds by the thin-film method, or another method specified by your instructor [see Technique 18.4]. Analyze each spectrum and assign the principal vibrational bands using Table 18.2 in Technique 18.

CLEANUP: Pour the remaining TLC solutions and developing solvent into the container for flammable organic waste. Place the used TLC plates in the container for nonhazardous solid waste.

Interpretation of Experimental Data

What did the TLC analysis indicate about the purity of your recrystallized compounds? Why might the neutral compound show traces of the acid as an impurity whereas the acid might not show any traces of the neutral compound? Do the functional groups and structural features identified in each IR spectrum support the identification of the two unknowns made from the melting point data? What have your mixture melting points indicated about the identity of the acid and the neutral compound present in your unknown mixture?

Questions

1. Why doesn't the neutral organic compound dissolve in 1.5 M sodium hydroxide solution?
2. Why is diethyl ether a good choice for the organic solvent in this experiment?
3. What experimental difficulty would you encounter if you had neglected to include a drying step before evaporating the ether solution of the neutral organic compound?
4. Why are the two organic compounds recrystallized before their melting points are determined?
5. What infrared (IR) bands are most useful in distinguishing a carboxylic acid from a neutral organic compound?
6. Describe the characteristics of a good recrystallization solvent.

*EM Science Silica Gel 60 F-254, No. 5554-7, TLC plates were used in developing the procedure.

PROJECT

3 HYDROLYSIS OF AN UNKNOWN ESTER

QUESTION: How can you find the structure of an unknown ester by using the identities of its hydrolysis products?

In this three-week project you will hydrolyze an unknown ester, a reaction that yields an alcohol and a carboxylic acid as the products. You will learn how to carry out an organic reaction and use the techniques of distillation, extraction, and recrystallization, to separate and purify the products. Finally, you will identify the products using boiling point, refractive index, and melting point.

Hydrolysis

All derivatives of carboxylic acids including esters can be split apart by the action of water, a process called ***hydrolysis.*** Hydrolysis of an ester yields a molecule of an alcohol and a molecule of a carboxylic acid:

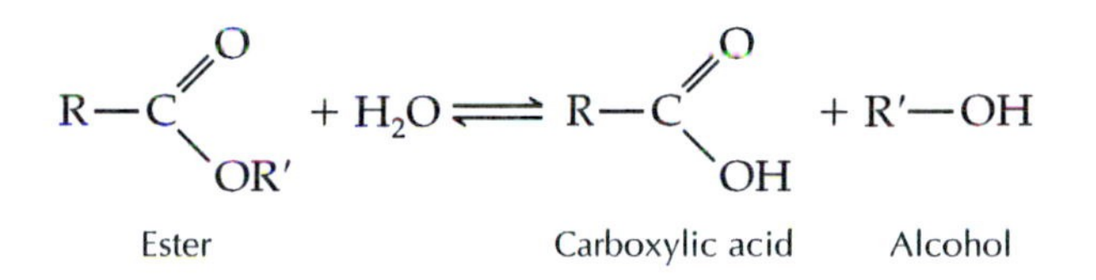

If you can determine the identities of the carboxylic acid and the alcohol, you can piece together what the structure of the original ester must have been.

Normally, aqueous base is used to catalyze the hydrolysis of an ester, a reaction called saponification (literally, "soap making"). The attack of hydroxide forms the carboxylate anion and the alcohol in a two-step reaction:

$$R{-}C(=O){-}OR' \xrightarrow[H_2O]{OH^-} R{-}C(=O){-}O^- + R'{-}OH$$

Carboxylate anion

Although hydrolysis of an ester involves a functional group that you may not yet have studied in organic chemistry, you can think of the reaction in a simple way and not be too far off. Reactions of two chemical species are often dominated by charge complementarity; positive attracts negative and negative attracts positive. The ester and the hydroxide ion are the two reagents in the key step of ester hydrolysis. The negative hydroxide ion reacts at the carbonyl carbon atom, the positive site of the ester. In the presence of strong base, the carboxylic acid produced is then converted to the carboxylate anion:

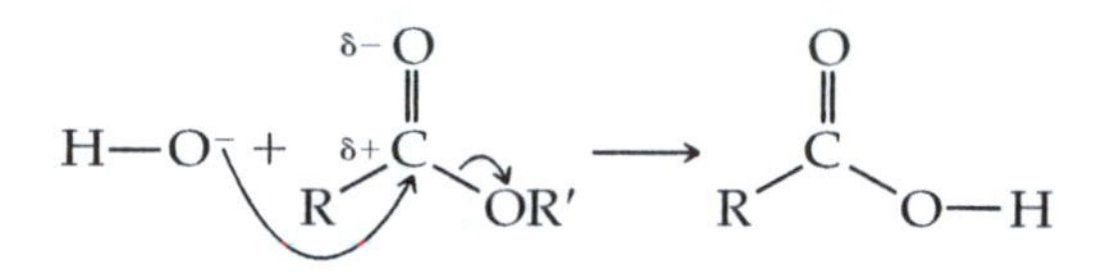

The reaction solvent will be a mixture of water and ethylene glycol, which will dissolve both the organic ester and sodium hydroxide, a polar inorganic compound. Ethylene glycol, widely used as antifreeze in automobile radiators, also has a high boiling point (197°C). Using a solution of ethylene glycol and water allows the hydrolysis to be carried out at a higher reflux temperature than with water as the solvent. The higher reflux temperature produces a faster rate of reaction.

Isolation of the Products

Separation of unknown alcohol. When the hydrolysis is complete, the reaction flask contains the sodium salt of the unknown carboxylic acid, the unknown alcohol, excess sodium hydroxide, water, and ethylene glycol. The alcohol product is much more volatile than ethylene glycol or the carboxylate product, which is an ionic salt. Therefore, the alcohol can be removed from the reaction mixture by codistillation with water, an example of azeotropic distillation. The steps for the hydrolysis of the unknown ester and separation of the alcohol from the reaction mixture are shown in the following flowchart:

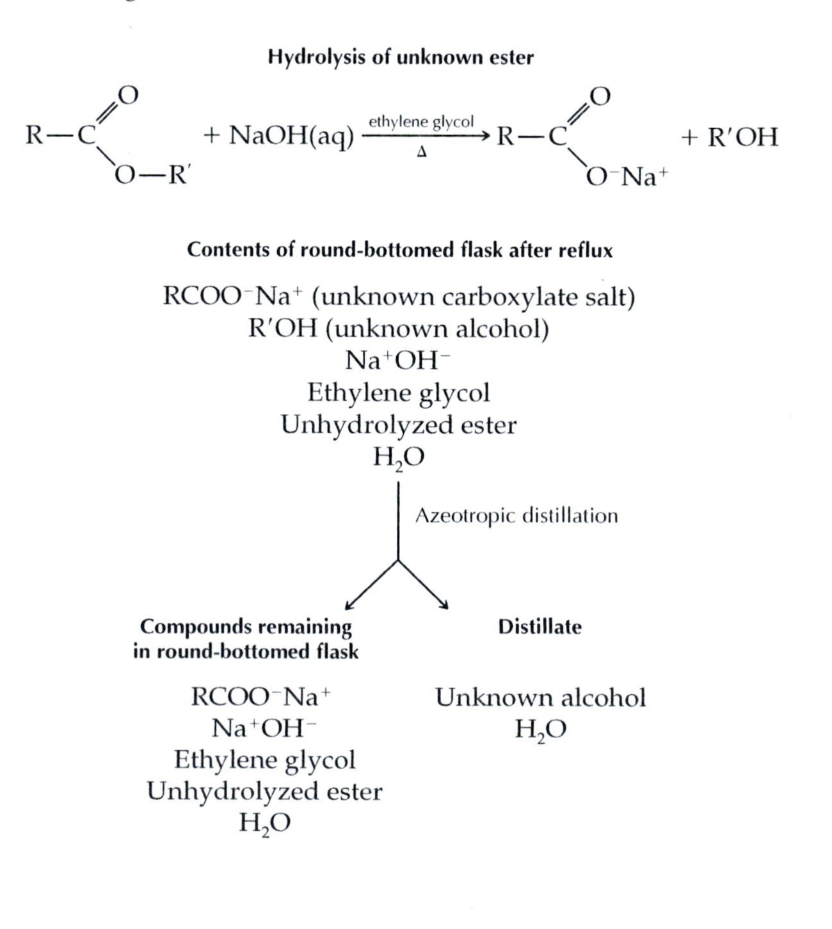

Isolation of unknown carboxyl acid. The mixture remaining in the round-bottomed flask contains an aqueous solution of the sodium salt of the carboxylic acid, excess sodium hydroxide, ethylene glycol, and perhaps a small amount of unhydrolyzed ester. To recover the pure carboxylic acid, you will first remove any unhydrolyzed ester by extraction with diethyl ether. All the other compounds remain dissolved in the aqueous phase. The addition of excess hydrochloric acid to this aqueous solution converts the carboxylate salt to the carboxylic acid, which is insoluble and precipitates:

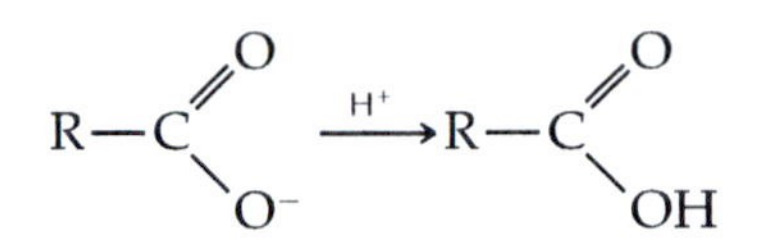

The hydrochloric acid also neutralizes the excess sodium hydroxide. The ethylene glycol remains dissolved in the aqueous solution. The precipitated unknown acid is recovered by vacuum filtration and recrystallized. The steps for recovery of the unknown acid, RCOOH, are summarized in the following flowchart:

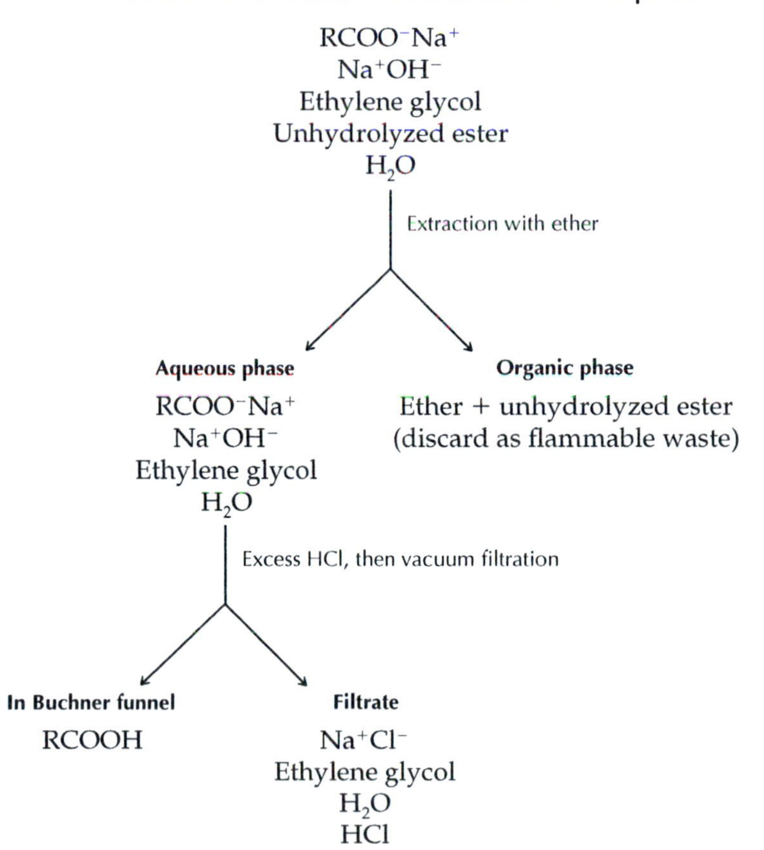

Extraction and drying of unknown alcohol. You will use an ether extraction to separate the unknown alcohol from the water with which it codistilled in the azeotropic distillation. After the ether solution has been dried with anhydrous sodium sulfate, the ether is removed by distillation. The flowchart shows the steps for extraction and drying of the unknown alcohol.

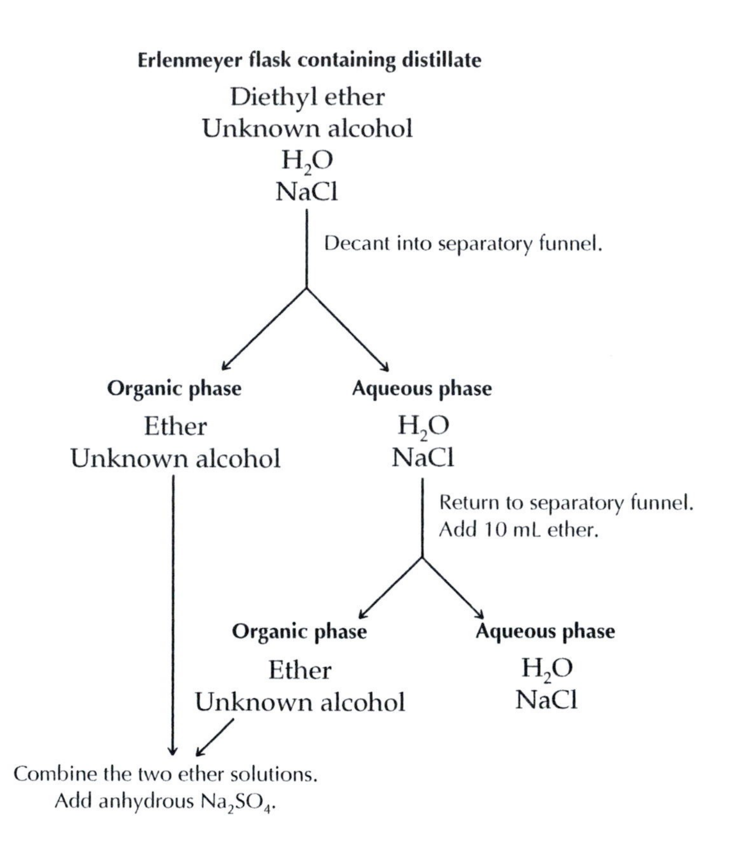

3.1 Hydrolysis and Azeotropic Distillation: First Week

MINISCALE PROCEDURE

PRELABORATORY ASSIGNMENT: Explain why it is easy to separate the unknown alcohol from ethylene glycol by azeotropic distillation.

Techniques Boiling Stones: Technique 6.1
Reflux: Technique 7.1
Azeotropic Distillation: Technique 11.5
Simple Distillation: Technique 11.3

Solid sodium hydroxide is hygroscopic and rapidly absorbs water from the atmosphere; keep the reagent bottle tightly closed.

SAFETY INFORMATION

Sodium hydroxide solutions are corrosive and cause severe burns. The solution prepared in this experiment is very concentrated. Wear gloves while weighing and transferring NaOH pellets. Avoid contact with skin, eyes, and clothing. Notify the instructor if any solid is spilled.

Ethylene glycol and the **unknown ester** may cause skin irritation. Avoid contact with skin, eyes, and clothing. Wash hands thoroughly after handling.

Hydrolysis

Clamp a 100-mL round-bottomed flask in an upright position. Place 7.0 g of sodium hydroxide pellets in the flask and add, in order, 2 mL of water, 20 mL of ethylene glycol, 15 mL of the unknown ester, and two boiling stones. Fit a water-cooled condenser above the flask [see Technique 7.1] and position a heating mantle under the flask. Connect the heating mantle to the variable transformer and plug the transformer into the electrical outlet [see Technique 6.2, Figure 6.1]. Alternatively, use a sand bath on a hot plate as the heat source. Have the instructor check your apparatus before you begin heating. After the reaction mixture begins to boil, continue heating under reflux for 30 min. Adjust the heat input, if necessary, to maintain gentle boiling.

A cake of white solid forms (see Question 2) during the reflux period, and the upper layer of liquid ester should disappear.

Azeotropic Distillation

At the end of the reflux period, lower the heating mantle from the flask or carefully raise the apparatus out of the sand bath by moving the clamp **(Caution: Hot.)** Allow the flask to cool for 10 min and add 35 mL of water through the top of the condenser. Add two more boiling stones.

Set up a simple distillation apparatus [see Technique 11.3] using a 50-mL Erlenmeyer flask as the receiving flask. Keep a slow stream of cold water running through the condenser. Fit the heat source under the flask. Have the instructor check your apparatus before you begin heating. The unknown alcohol forms an azeotrope with water that boils below 100°C [see Technique 11.5]; when the azeotrope has completely distilled, the boiling point will rise to that of water. Adjust the rate of heating so that the distillate drops from the condenser at a rapid rate of 1 drop every 2–3 s. Record the temperature of the vapor when distillate begins to collect in the receiving flask. Continue the distillation until the temperature reaches 100°C or until about 20 mL of distillate have been collected. Stop the heating process by turning off the heat source and lowering the heating mantle from the flask or carefully raising the distillation apparatus out of the sand bath. **(Caution: Hot.)**

Organic compounds have lower solubility in water saturated with sodium chloride than in water without dissolved salt.

Add 4.0 g of sodium chloride to the flask containing the distillate that you collected. Cork the flask and swirl it to dissolve most of the NaCl. Label the flask and store it in your drawer until the next laboratory period. Allow the mixture remaining in the round-bottomed flask to cool. Then label the flask, cork it tightly, and place it in a beaker so that it will not tip over. Store the round-bottomed flask in your drawer until the next laboratory period.

3.2 Recrystallization and Extraction: Second Week

PRELABORATORY ASSIGNMENT: What characteristics must water have to be an effective recrystallization solvent for this experiment?

MINISCALE PROCEDURE

Techniques Extraction: Technique 8.2
Drying Agents: Techniques 8.7–8.9
Recrystallization: Technique 9.5

SAFETY INFORMATION

Diethyl ether is extremely volatile and flammable. Be certain that there are no flames in the laboratory. Also be sure that there are no hot electrical devices in the vicinity while you are pouring ether or performing the extractions. Diethyl ether should be used in a hood, if possible.

Recovery of the Carboxylic Acid

If the mixture in the round-bottomed flask contains any solid, add 10–20 mL of water and stir to dissolve it. Working in a hood, if possible, pour the aqueous mixture from the round-bottomed flask into a 125-mL separatory funnel; use a conical funnel in the neck of the separatory funnel to prevent spills. Rinse the flask with 5–10 mL of water and also add this rinse to the separatory funnel. Then add 10 mL of diethyl ether to the separatory funnel and place the stopper in it. Alternately shake the separatory funnel gently to mix the layers and vent the funnel [see Technique 8.2]. Allow the two phases to separate and drain the aqueous phase into a clean 125-mL Erlenmeyer flask.

What is the density of ether? Will the ether layer be the upper or the lower one in the extraction?

Add 30 mL of 6 M hydrochloric acid to the aqueous solution in the Erlenmeyer flask and stir. This amount of acid should be enough to neutralize both the excess sodium hydroxide and the sodium salt of the acid, giving a copious white precipitate. Check the pH using pH test paper such as pHydrion paper. Add another 1 or 2 mL of acid if the pH is greater than 1. Cool the mixture in an ice-water bath before collecting the crystals by vacuum filtration [see Technique 9.5, Figure 9.4].

To test the pH, dip the tip of a stirring rod in the aqueous mixture and then touch the tip to the pH test paper.

CLEANUP: Discard the ether solution remaining in the separatory funnel in the container for flammable (organic) waste. The aqueous solution in the filter flask should be neutralized with solid sodium carbonate **(Caution: Foaming.)** before washing it down the sink or pouring it into the container for aqueous inorganic waste.

Recrystallization

Dissolution is not an instantaneous process, so let the solution boil a minute or two before concluding that more water is needed.

On a hot plate, heat about 200 mL of water to boiling in an Erlenmeyer flask. Place the recovered acid in a 250-mL Erlenmeyer flask and add a boiling stick or boiling stone. Carefully pour about 75 mL of the hot water over the solid, using flask tongs to hold the hot flask. Bring the mixture to a boil. Continue adding hot water in 10- to 15-mL increments (estimate this volume; it does not need to be known exactly), allowing the mixture to boil after each addition, until the solid and the oily layer in the bottom of the flask just dissolve. When all the solid and oil have dissolved, add an excess of about 20–25 mL of hot water. Estimate the total volume of water used to dissolve the solid.

Set the flask on the desk top to cool slowly; cooling its contents to room temperature requires at least 20 min. Proceed with the extraction of the alcohol while you are waiting for the carboxylic acid solution to cool.

When the flask has cooled to room temperature and crystals have formed throughout the solution, cool the flask in an ice-water bath for at least 10 min. Collect the crystals by vacuum filtration and place them on a watch glass to dry in your drawer until the next laboratory period.

CLEANUP: The filtrate remaining in the filter flask can be washed down the sink or poured into the container for aqueous inorganic waste.

Extraction and Drying of the Alcohol

SAFETY PRECAUTION

Diethyl ether vapors pose a fire hazard. The hot plate used for the recrystallization may still be very hot; remove it from your work area before beginning the extraction.

Sodium chloride, added to the alcohol/water mixture from the previous laboratory period, dissolves in the water and reduces the solubility of the unknown alcohol in water.

Working in a hood, if possible, add 15 mL of diethyl ether to the mixture of unknown alcohol, water, and NaCl saved from the previous laboratory period. Cork the flask again and swirl it. Carefully decant the ether/water mixture into a separatory funnel so that any undissolved NaCl remains in the flask. Stopper the separatory funnel. Gently shake it to mix the layers and allow the phases to separate [see Technique 8.2]. Drain the aqueous layer into a clean, labeled flask and pour the ether layer out of the top of the funnel into another clean, labeled flask. Return the aqueous phase to the separatory funnel and extract it again with a 10-mL portion of ether. Separate the layers and drain the aqueous layer into the same flask you previously used for it; then pour the ether phase into the flask that contains the first portion of ether. Set the flask containing the aqueous solution aside.

Dry the ether solution with anhydrous sodium sulfate. Add sodium sulfate in small scoops one at a time, swirling the flask between additions. When some of the sodium sulfate moves freely, that is, when it is not clumped together or stuck to the glass, you have added enough [see Technique 8.7]. Cork the flask tightly and store the solution until the next laboratory period. Because ether fumes may be emitted, consult your instructor about whether to store the flask in your drawer or in some other designated place in the laboratory.

CLEANUP: The aqueous solution remaining from the extraction may be washed down the sink or poured into the container for aqueous inorganic waste.

3.3 Distillation, Boiling Points, and Melting Points: Third Week

PRELABORATORY ASSIGNMENT: Explain why the placement of the thermometer bulb is so important in measuring an accurate boiling point in a distillation.

Techniques Simple Distillation: Technique 11.3
Melting Points: Technique 10.3
Refractive Index: Technique 13

Simple Distillation of the Alcohol

SAFETY INFORMATION

Diethyl ether vapors pose a fire hazard. Perform the evaporation in a well-ventilated hood and use a steam bath or hot-water bath as the heat source.

Place a small (pea size) plug of cotton in the bottom of a small funnel [see Technique 8.8, Figure 8.16] and set the funnel into a 50-mL round-bottomed flask that is clamped firmly. Decant the ether/alcohol solution into the funnel, rinse the sodium sulfate remaining in the Erlenmeyer flask with 5 mL of ether, and add the rinse to the funnel. Put a boiling stone in the round-bottomed flask and evaporate the solution on a steam bath or

80–90°C water bath in a hood until the volume is reduced by approximately two-thirds. Alternatively, the ether can be removed on a rotary evaporator [see Technique 8.9]. Then set up a simple distillation apparatus [see Technique 11.3, Figure 11.6] adding another boiling stone and using a 25-mL round-bottomed flask cooled in an ice-water bath for the receiver. Collect the first fraction until the boiling point reaches 100–110°C; this fraction is mainly residual ether. Change the receiver to a tared (weighed) 50-mL Erlenmeyer flask or wide-mouthed vial. This receiver does not need to be cooled in an ice-water bath. Collect the unknown alcohol fraction. The temperature will stabilize while most of the alcohol distills—this temperature is the boiling point of the unknown alcohol.

Stop the distillation just before the boiling flask reaches dryness by removing the heat source. Determine the weight of the alcohol and its refractive index [see Technique 13.3].

SAFETY PRECAUTION

Be certain to stop the distillation before the boiling flask reaches dryness. Diethyl ether may form peroxides. If the flask goes to dryness, the temperature inside it rises rapidly, and the peroxides could explode.

CLEANUP: Pour the residual ether fraction and residue remaining in the boiling flask into the container for flammable (organic) waste. Place the sodium sulfate in the container for nonhazardous waste or inorganic waste.

Melting Point of the Carboxylic Acid

Weigh the recrystallized unknown acid. Determine the melting-point range [see Technique 10.3]. Ideally this range will be 2° or less. Allow the melting-point apparatus to cool at least 20° below your observed melting point, prepare a new sample, and do a second determination of the melting point to verify your first measurement. The two determinations should agree within 1°; if not, continue making determinations until you have two melting-point ranges that do agree within this limit.

3.4 Identification of the Unknown Ester: Third Week

Identification of the Alcohol

Compare your experimental data for the unknown alcohol with the information listed in Table 3.1. What alcohol has properties that most closely match your data?

TABLE 3.1 Possible compounds for the unknown alcohol

Compound	Boiling point, °C	n_D^{20}	MW
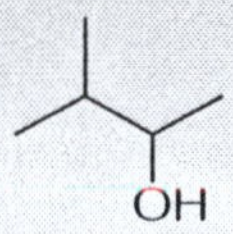 3-Methyl-2-butanol	112.9	1.4089	88
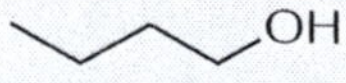 1-Butanol	117.6	1.3993	74
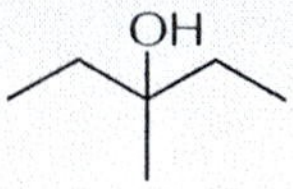 3-Methyl-3-pentanol	122.4	1.4186	102

Refractive index is a temperature-dependent property, and the values given in Table 3.1 were reported for 20°C. If you measured the refractive index at a temperature other than 20°C, you will need to calculate the temperature correction [see Technique 13.4] and use your corrected refractive index for comparison with the values in Table 3.1. The refractive index of diethyl ether is 1.3555 at 15°C and that of water is 1.3330 at 20°C; if either substance remains in your product, its presence will lower the observed refractive index.

Identification of the Carboxylic Acid

Compare your data with the melting points given in Table 3.2. What compound most closely matches your data? When you have completed the identification of your products, submit them to your instructor as directed.

Identification of the Unknown Ester

Now that you have identified the hydrolysis products of your unknown ester, draw the structure of the ester itself. Write the balanced equation for its hydrolysis to the carboxylic acid and alcohol you identified.

Look up the density of the now-identified ester in a handbook. In some handbooks, esters are listed under the carboxylic acid of which they are a derivative. For example, the ester methyl acetate may appear under the listing *acetic acid, methyl ester.*

Calculate the theoretical yield and percent yield for both the carboxylic acid and the alcohol you identified as the hydrolysis products.

TABLE 3.2 Possible compounds for the unknown carboxylic acid

Compound	Melting point, °C	MW
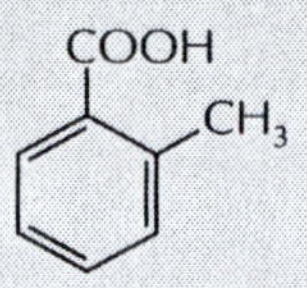 2-Methylbenzoic acid (2-toluic acid)	104–105	136
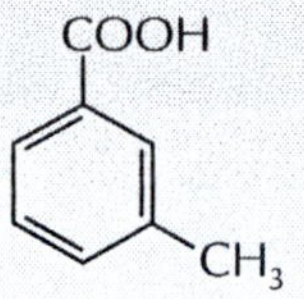 3-Methylbenzoic acid (3-toluic acid)	111	136
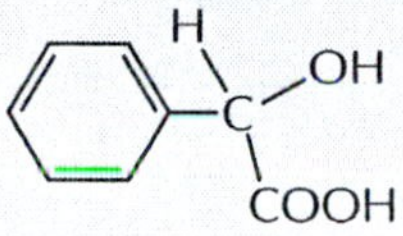 α-Hydroxyphenylacetic acid	118	152
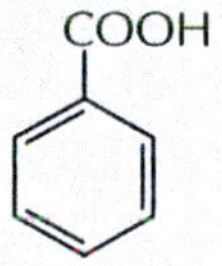 Benzoic acid	121–122	122
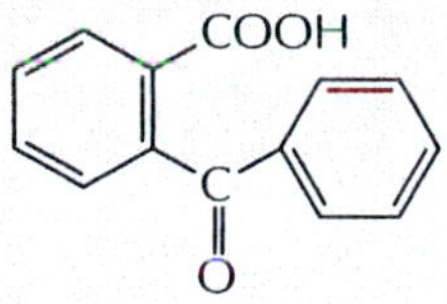 2-Benzoylbenzoic acid	127–129	226

Questions

1. Explain why a reaction mixture can be heated under reflux for an extended period of time without losing the solvent.
2. What is the white precipitate that forms while the reaction mixture is heated?
3. If you were doing this experiment in Denver, Colorado, the observed boiling point would never reach 100°C during the azeotropic distillation. Why not?
4. Explain why the unknown alcohol cannot be separated from water by simple distillation.
5. Show, by calculations, why 30 mL of 6 M HCl is enough to neutralize both the excess sodium hydroxide and the sodium salt of the carboxylic acid in this experiment.
6. Discuss the principles of recrystallization and how the process helps to purify a solid product.

7. Suppose the solubility of your unknown carboxylic acid is 6.8 g/100 mL water at 95°C and 0.17 g/100 mL water at 0°C. From the approximate volume of water you used for the recrystallization, calculate how much of the carboxylic acid would remain in your recrystallization solution at 0°C.

8. The solubility of a carboxylic acid is 67 g/100 mL in boiling ethanol and 43 g/100 mL in cold ethanol. Would ethanol be a good recrystallization solvent? Explain.

PROJECT

4

INTERCONVERSION OF 4-*tert*-BUTYLCYCLOHEXANOL AND 4-*tert*-BUTYLCYCLOHEXANONE

QUESTION: What is the stereoselectivity of the $NaBH_4$ reduction of 4-*tert*-butylcyclohexanone, and how does the product mixture of the *cis*- and *trans*-4-*tert*-butylcyclohexanols compare to the composition of the commercial mixture of these isomers?

In Project 4, a two- to three-week project, you will investigate two oxidation-reduction reactions involving the interconversion of a mixture of *cis*- and *trans*-4-*tert*-butylcyclohexanol and 4-*tert*-butylcyclohexanone. You will use sodium hypochlorite to oxidize the 4-*tert*-butylcyclohexanols in the first reaction; in the second reaction you will use sodium borohydride to reduce 4-*tert*-butylcyclohexanone. You will follow the progress of both reactions by TLC and IR spectroscopy and determine the ratio of cis and trans isomers formed in the reduction reaction by gas chromatography or NMR spectroscopy.

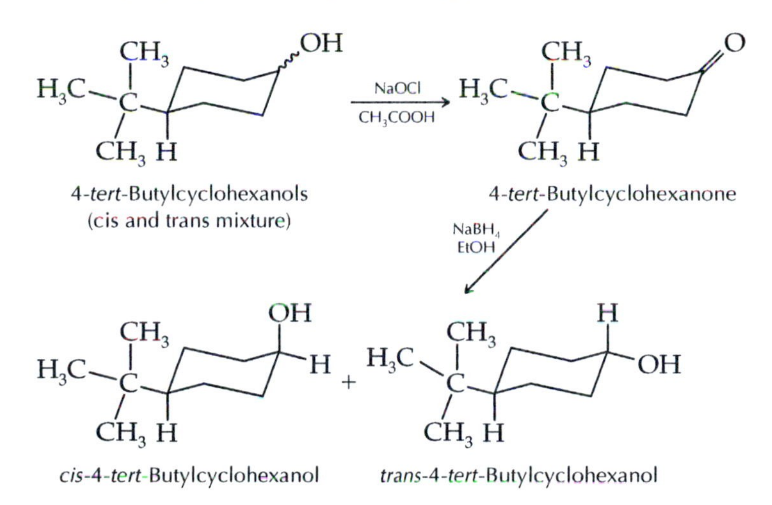

Oxidation-Reduction Reactions

Oxidation-reduction reactions are extremely diverse, and they proceed by a variety of mechanisms. The unifying factor is that one substrate is oxidized and another is reduced. In the inorganic chemistry of metal cations, this process can be thought of most easily as an electron transfer phenomenon; that is, oxidation is a loss of electrons and reduction is a gain of electrons. While this approach also applies to covalent organic systems, it is awkward to use. It is often easier to think of oxidation as the loss of hydrogen (the equivalent of H_2) or gain of oxygen atoms.

Increasing state of oxidation ⟶

RCH_3	RCH_2OH	R(C=O)H	R(C=O)OH
Alkane	Alcohol	Aldehyde	Carboxylic acid

In the oxidation of a secondary alcohol to a ketone, the α-hydrogen atom is lost from carbon and the hydrogen is also lost from the oxygen atom.

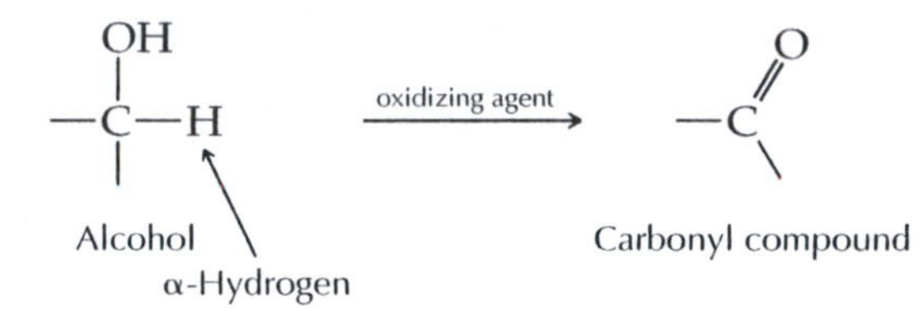

Many alcohols can be oxidized in this manner; primary alcohols can be oxidized to aldehydes and secondary alcohols to ketones. Tertiary alcohols, because they bear no α-hydrogens, do not easily undergo oxidation.

Oxidation with Chlorine Bleach

Project 4.1 uses household bleach in the presence of acetic acid as the oxidizing agent. The bleach is reduced to sodium chloride by the oxidation of a mixture of *cis*- and *trans*-4-*tert*-butylcyclohexanol with NaOCl, forming 4-*tert*-butylcyclohexanone.

Most bleach is an aqueous solution of sodium hypochlorite, prepared by adding Cl_2 gas to aqueous sodium hydroxide:

$$Cl_2 + 2NaOH \rightleftharpoons NaOCl + NaCl + H_2O$$

When sodium hypochlorite is added to acetic acid, the following acid-base reaction occurs:

$$NaOCl + CH_3{-}C(=O){-}OH \rightleftharpoons HOCl + CH_3{-}C(=O){-}O^-Na^+$$

In the acidic solution HOCl (hypochlorous acid) is in equilibrium with Cl_2:

$$HOCl + HCl \rightleftharpoons Cl_2 + H_2O$$

Cl_2 gas is toxic, but under the experimental conditions used in this project, Cl_2 is present in very low and safe concentrations in the atmosphere when one takes simple precautions.

Both Cl_2 and HOCl are sources of positive chlorine, a species that has two fewer electrons than the chloride anion does. A key step in reactions of positive chlorine reagents is the transfer of Cl^+ to the substrate. It is reasonable to expect that the first step in the oxidation of the 4-*tert*-butylcyclohexanols is exchange of positive chlorine with the hydroxyl proton. Subsequent E2 elimination of HCl from the resulting alkyl hypochlorite forms the ketone. If HOCl is not present, the first step cannot take place and the overall oxidation is very slow. It is important to have enough acid present in the reaction mixture to bring the pH of the aqueous solution to less than pH 7, so that the alkyl hypochlorite intermediate can form:

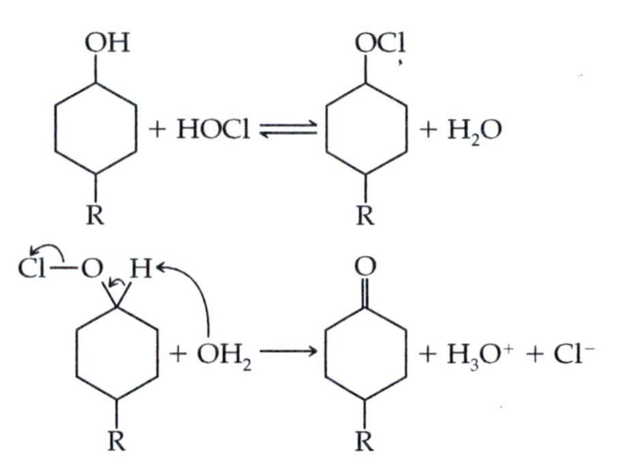

In the first reaction, Cl^+ is transferred to the substrate, and in the second reaction, Cl^- is lost. The change is a reduction by two electrons. The two electrons are provided by 4-*tert*-butylcyclohexanol as it is oxidized to 4-*tert*-butylcyclohexanone.

Reduction with Sodium Borohydride

In Project 4.2, 4-*tert*-butylcyclohexanone is reduced by $NaBH_4$ to give a mixture of the cis and trans isomers of 4-*tert*-butylcyclohexanol. When reduction of an organic substrate is needed, the reducing agent is often an inorganic compound. Sodium borohydride has become a popular mild reducing agent because of its functional group selectivity and ease of handling.

The probable reaction mechanism involves a nucleophilic hydride-ion transfer from boron to the electropositive carbonyl carbon of the ketone:

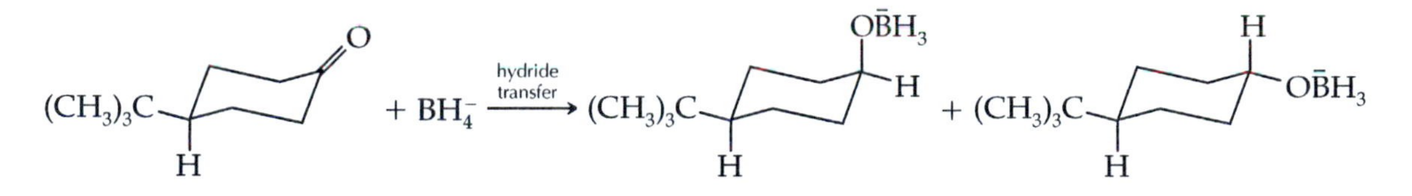

Substitution of an alkoxy group for a hydrogen atom on boron does not seriously alter the reducing ability of the borohydride. Thus, when BH_4^- reduces an organic compound in an alcohol (ROH) solvent, it is oxidized to $B(OR)_4^-$. The overall stoichiometry is

$$4R{-}\overset{\overset{\displaystyle O}{\|}}{C}{-}R + Na^+BH_4^- \longrightarrow (R_2CHO)_4B^-Na^+$$

After the reaction is finished and water is added, $B(OR)_4^-$ is hydrolyzed, producing salts of boric acid $[B(OH)_3]$.

$$(R_2CHO)_4B^-Na^+ + 2H_2O \longrightarrow 4R_2CHOH + NaBO_2$$

All hydride reducing agents, which constitute a large family of reagents, are not only reducing agents but also bases that can react with Brønsted acids to produce H_2 gas. Sodium borohydride is a relatively safe reducing reagent. Unlike the more hazardous lithium aluminum hydride ($LiAlH_4$), which reacts with water to form H_2 with explosive force, $NaBH_4$ reacts only slowly with water. In fact, water and alcohols can be used as solvents for sodium borohydride reductions.

Conformational Isomers of Cyclohexanes

The most stable conformational isomer of cyclohexane is a structure in the form of a chair. In the chair form, all the carbon-carbon bond angles are 109°, the stable tetrahedral angle.

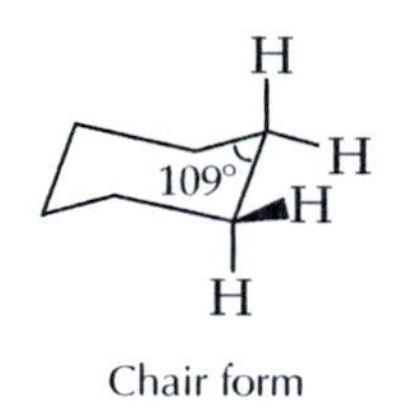

Chair form

Under ordinary conditions, a single chair form has a finite but short existence before it undergoes a ring-flipping process to another conformational isomer of equal stability:

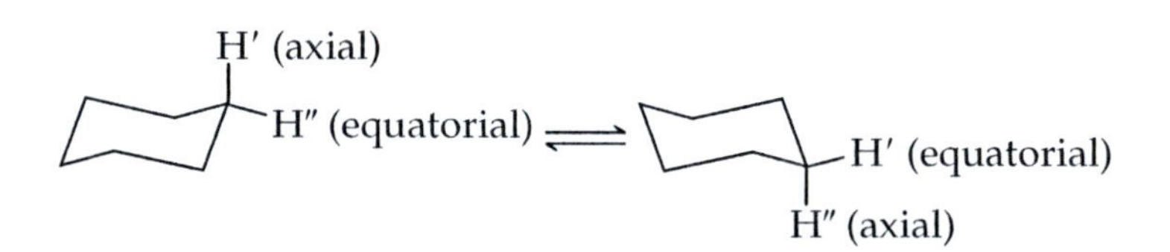

This ring flipping in cyclohexane converts axial hydrogens to equatorial ones and vice versa.

A large bulky group on a cyclohexane ring can effectively freeze the cyclohexane ring in a particular conformation. Steric interference between the bulky group and other atoms in the molecule, particularly 1,2-gauche interactions and 1,3-diaxial interactions, becomes so large that one conformer is almost completely favored over the other. Such a group is the *tert*-butyl group:

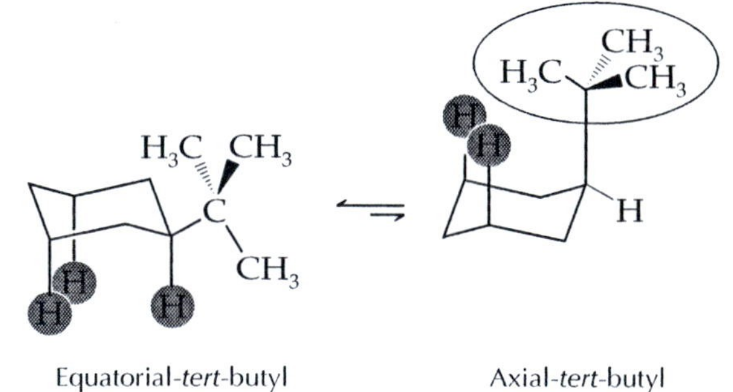

Equatorial-*tert*-butyl
No 1,3-diaxial interactions
>99.9%

Axial-*tert*-butyl
Severe 1,3-diaxial interactions
<0.1%

The large *tert*-butyl group freezes the conformation of the cyclohexane ring because it strongly prefers the equatorial position. The hydroxyl group in *cis*-4-*tert*-butylcyclohexanol is therefore held in the axial position, while both large groups are in equatorial positions in the more stable trans compound:

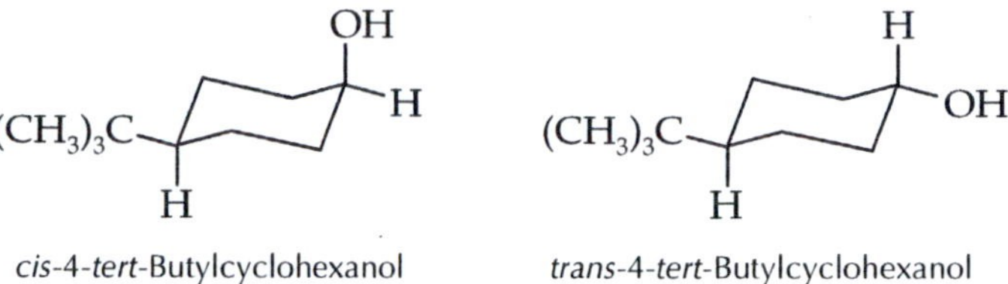

cis-4-*tert*-Butylcyclohexanol

Equatorial *tert*-butyl group
Axial hydroxyl group

trans-4-*tert*-Butylcyclohexanol

Equatorial *tert*-butyl group
Equatorial hydroxyl group

Oxidation of the mixture of cis and trans starting materials forms only one product because the alcohol functional group is oxidized to the trigonal carbonyl group:

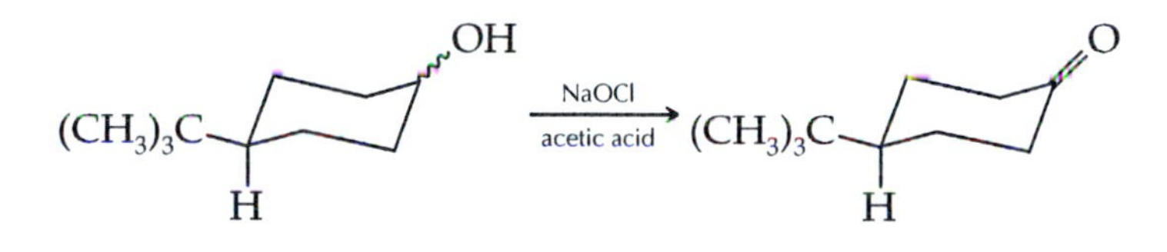

The question in this project is whether the mixture of *cis*- and *trans*-4-*tert*-butylcyclohexanols that you used in the oxidation reaction is reproduced by your sodium borohydride reduction of 4-*tert*-butylcyclohexanone. If the cis/trans ratios are the same, your initial starting material may have been synthesized by the $NaBH_4$ reduction of 4-*tert*-butylcyclohexanone. If the ratios are different, it is likely that it was synthesized by another route, such as the catalytic reduction of the benzene ring of 4-*tert*-butylphenol:

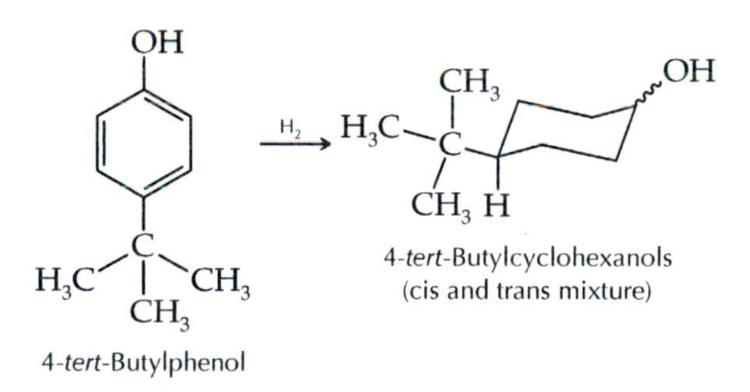

4-*tert*-Butylphenol

4-*tert*-Butylcyclohexanols
(cis and trans mixture)

NMR Analysis of the cis/trans Ratio

Although the overall ^1H NMR spectrum is rather complex, you can use NMR spectroscopy in a straightforward manner to analyze the cis/trans isomeric ratio of the starting material, 4-*tert*-butylcyclohexanol. The methine proton on the carbon bearing the hydroxyl group is shifted downfield from the other protons. In the cis isomer, this proton is equatorial and has a chemical shift of 4.03 ppm. In the trans isomer, this proton is axial, which makes it somewhat shielded by an anisotropic effect; the axial proton can be observed at 3.5 ppm. Because each isomer has only one of these protons, the integrated areas of these methine hydrogens can be used to determine the ratio of isomers:

$$\%\ \text{cis} = \frac{\text{area of 4.03 ppm signal}}{\text{sum of areas of 4.03 ppm and 3.5 ppm signals}} \times 100$$

You will follow the course of both the oxidation and reduction reactions by thin-layer chromatography and determine the ratio of cis to trans isomers formed in the reduction reaction by using either NMR or gas chromatographic analysis.

4.1 Green Chemistry: Sodium Hypochlorite Oxidation of 4-*tert*-Butylcyclohexanol

PRELABORATORY ASSIGNMENT: Explain why the R_f values for 4-*tert*-butylcyclohexanone and for *cis*- and *trans*-4-*tert*-butylcyclohexanol are so different.

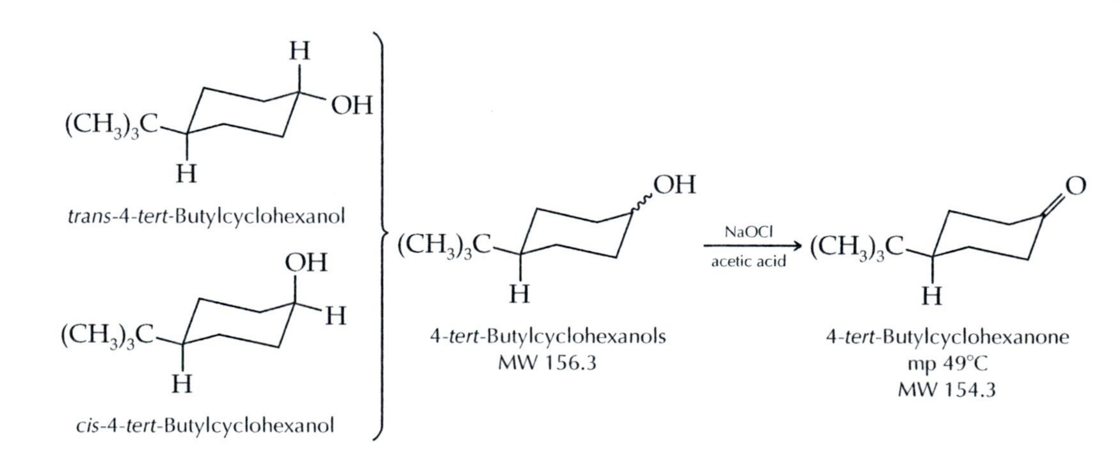

MINISCALE PROCEDURE

Techniques Thin-Layer Chromatography: Technique 15
Extraction: Technique 8.2
Drying Organic Liquids: Techniques 8.7–8.9
Simple Distillation: Technique 11.3
IR Spectroscopy: Technique 18

SAFETY INFORMATION

Conduct this experiment in a hood, if possible.

4-*tert*-Butylcyclohexanol and the product **4-*tert*-butylcyclohexanone** may be irritants. Wash thoroughly after handling either one.

Acetone is extremely flammable and volatile, it is a severe eye irritant, and it may cause skin irritation. Avoid contact with skin, eyes, and clothing.

Sodium hypochlorite solution emits chlorine gas, which is toxic and a respiratory and eye irritant. Use it in a hood, if possible.

Glacial acetic acid is a dehydrating agent and irritant, and it causes burns. Dispense it only in a hood. Wear gloves. Avoid contact with skin, eyes, and clothing.

Diethyl ether is extremely volatile and flammable. Be certain that there are no flames in the laboratory or hot electrical devices in the vicinity while you are pouring ether or performing the extractions. Use ether in a hood, if possible.

Monitoring the Reaction by Thin-Layer Chromatography

Note to instructor: Prepare p-anisaldehyde reagent solution by mixing together 5.1 mL of p-anisaldehyde, 2.1 mL of acetic acid, 6.9 mL of concentrated sulfuric acid, and 186 mL of 95% ethanol.

Before using TLC [Technique 15] to monitor the progress of the reaction, it will be necessary to determine the R_f values for 4-*tert*-butylcyclohexanone and for both isomers of 4-*tert*-butylcyclohexanol by spotting 2% standard solutions of 4-*tert*-butylcyclohexanol and 4-*tert*-butylcyclohexanone, developing the plates in diethyl ether, and using *p*-anisaldehyde reagent for visualization [see Technique 15.]. Dip the TLC plate in a jar containing the *p*-anisaldehyde and allow the excess to drip back into the jar. Wipe the back of the TLC plate with a tissue. In a hood, heat the TLC plate on a hot plate heated at a medium setting; the hot plate setting should produce colored spots within about 1 min without burning or excessively darkening the entire TLC plate. Use the same developing and visualizing procedure to monitor the reaction.

Oxidation Reaction

Use recently purchased chlorine household bleach, such as Clorox or a supermarket brand. Bleach solutions that have been stored for many months may have decreased concentrations of sodium hypochlorite.

Place 1.5 g of 4-*tert*-butylcyclohexanol, 1.20 mL of glacial acetic acid, and 20 mL of acetone in a three-necked 100-mL or 250-mL round-bottomed flask that is firmly clamped to a ring stand or metal support. Put a magnetic stirring bar in the flask and insert a water-cooled condenser in one neck, a separatory funnel in the second neck, and a glass stopper in the third neck. Pour 13 mL of 5.25% (0.75 M) sodium hypochlorite solution into the separatory funnel. Place the flask in a room-temperature water bath and begin stirring the solution. Add sodium hypochlorite solution dropwise over a period of approximately 15 min. A moderate to fast stirring rate is necessary to ensure thorough mixing of the two-phase system because the acetone solution of 4-*tert*-butylcyclohexanol is not totally miscible with aqueous hypochlorite solution.

Monitor the reaction by TLC at the midpoint of the hypochlorite addition and after the addition is complete. To do this, turn off the stirrer and allow the two phases to separate. Touch the top of the upper (acetone) layer with a capillary pipet suitable for spotting TLC plates. Spot a TLC plate with the reaction solution in one channel and 2% 4-*tert*-butylcyclohexanol solution in a second channel. Develop the TLC plate using the same procedure you used previously on the known solutions.

When the addition of hypochlorite solution is complete, test the lower aqueous layer for the presence of excess hypochlorite ion. Turn off the stirrer and let the two phases separate. Expel the air from a rubber bulb on a Pasteur pipet before inserting the pipet to the bottom of the reaction flask. Release the pressure only enough to draw a few drops of lower aqueous solution into the pipet and remove the pipet from the flask. Put a drop of the solution on a strip of wet starch-iodide test paper. (If the bleach is too concentrated, colorless iodate may form.) A bluish black color indicates excess hypochlorite ion. If no color appears, put an additional 2 mL of bleach solution into the separatory funnel and add the solution dropwise to the stirred reaction mixture.

Continue monitoring the reaction by TLC until complete conversion of 4-*tert*-butylcyclohexanol to 4-*tert*-butylcyclohexanone has occurred. When TLC analysis indicates that the reaction is complete, cool the reaction mixture to room temperature and add 2 mL of saturated sodium bisulfite solution to reduce any excess hypochlorite, stir the mixture briefly, and test as directed earlier for excess hypochlorite ion. If a blue color appears on the test paper, add an additional 2 mL of $NaHSO_3$ solution and repeat the hypochlorite test. If no color appears on the test paper, proceed to the next step.

In each extraction, remove the lower aqueous phase before adding the next aqueous solution; the ether solution remains in the funnel during all the washings.

When performing extractions, it is prudent to save all solutions until the procedure is complete.

Transfer the contents of the flask to a separatory funnel. Rinse the flask with 20 mL of diethyl ether and add the ether to the separatory funnel. Shake the funnel thoroughly to mix the layers [see Technique 8.2]. Remove the lower aqueous layer and wash (extract) the organic phase consecutively with 10 mL water, 10 mL of saturated sodium bicarbonate solution **(Caution: Foaming may occur.)**, if and 10 mL of distilled water. Drain off each aqueous phase before adding the next aqueous solution. Pour the ether solution into a dry 50-mL Erlenmeyer flask and add anhydrous magnesium sulfate to dry the solution [see Technique 8.7]. Place a cork in the Erlenmeyer flask and allow the solution to stand with the drying agent for at least 10 min.

Place a fluted filter paper in a conical funnel and filter the dried solution through the filter paper into a dry 100-mL round-bottomed flask. Rinse the drying agent with approximately 2 mL of ether and add this ether rinse to the funnel. Remove most of the ether by simple distillation on a steam bath or 70–75°C water bath [see Technique 11.3] or with a rotary evaporator [see Technique 8.9]. Transfer the residue to a tared (weighed) 50-mL Erlenmeyer flask. Rinse the round-bottomed flask with approximately 2 mL of ether and add this rinse to the Erlenmeyer flask. Working in a hood, finish the evaporation of ether, using a stream of nitrogen or air until the product forms a dry white solid. Weigh the flask and calculate the percent yield. Obtain an IR spectrum of your product as directed by your instructor.

Tightly cork the flask containing your 4-*tert*-butylcyclohexanone and store it in your laboratory drawer until you are ready to perform the reduction procedure given in Project 4.2.

CLEANUP: The aqueous solutions remaining from the extractions should be combined and neutralized with sodium carbonate **(Caution: Foaming.)** before being washed down the sink or poured into the container for aqueous inorganic waste. Place the spent magnesium sulfate in the container for solid inorganic waste. Pour any ether remaining in the TLC developing jar into the container for flammable (organic) waste. Place glass-backed TLC plates in the appropriate container for glass waste, as directed by your instructor; aluminum-backed TLC plates should be placed in a container for metal waste.

4.2 Sodium Borohydride Reduction of 4-*tert*-Butylcyclohexanone

PRELABORATORY ASSIGNMENT: Which nonbonded interactions make *trans*-4-*tert*-butylcyclohexanol more stable than the cis isomer? Explain.

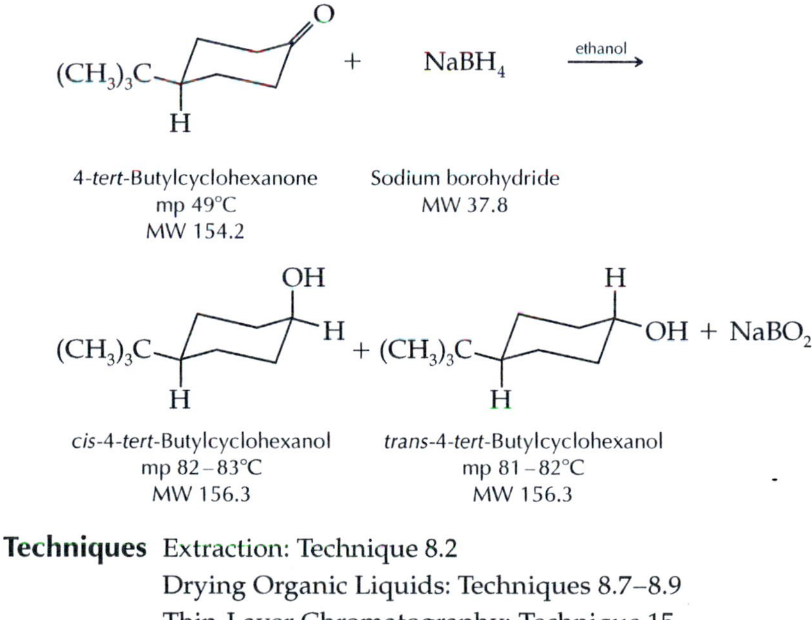

MINISCALE PROCEDURE

Techniques Extraction: Technique 8.2
Drying Organic Liquids: Techniques 8.7–8.9
Thin-Layer Chromatography: Technique 15
Gas Chromatography: Technique 16
IR Spectroscopy: Technique 18
NMR Spectroscopy: Technique 19

SAFETY INFORMATION

4-*tert*-Butylcyclohexanone and the product, **4-*tert*-butylcyclohexanol,** may be irritants. Avoid breathing the dust. Avoid contact with skin, eyes, and clothing.

Sodium borohydride is harmful if swallowed, inhaled, or absorbed through the skin. Avoid breathing the dust. Avoid contact with skin, eyes, and clothing. It decomposes to flammable, explosive hydrogen gas.

Diethyl ether is extremely volatile and flammable. Use it in a hood, if possible. Be sure that there are no flames in the laboratory and no hot electrical devices in the vicinity while you are pouring ether or performing the extractions.

$NaBH_4$ is moisture sensitive; keep the bottle tightly closed when not in use. Store the reagent in a desiccator.

Proportionally adjust the quantities of all reagents for the amount of ketone you synthesized in 4.1. Carry out the reaction in the Erlenmeyer flask containing your stored 4-*tert*-butylcyclohexanone. If you have less than 0.30 g of ketone, use the microscale procedure.

Reduction Reaction

Dissolve 0.80 g of 4-*tert*-butylcyclohexanone in 6.0 mL of 95% ethanol in a 50-mL Erlenmeyer flask. Slowly add 0.60 g of sodium borohydride. The mixture may become warm, so have a cold-water bath handy. Keep the reaction temperature between 25°C and 35°C. You may see a slow evolution of hydrogen gas during the reaction. After 15 min, check for completeness of reaction (indicated by the disappearance of the 4-*tert*-butylcyclohexanone spot) by the TLC method used in Project 4.1 [see Technique 15].

When the reaction is complete, add 10 mL of water and then carefully add 3 M hydrochloric acid solution in 1-mL portions (2–4 mL may be needed) until the evolution of hydrogen gas becomes very slow or ceases. Then heat the mixture to the boiling point. Cool the mixture to room temperature or slightly lower. Add 15 mL of diethyl ether and stir the mixture. If a white inorganic solid forms, decant the liquid away from the solid into a separatory funnel. If no solid is present, simply pour the mixture into a separatory funnel. Rinse the Erlenmeyer flask with 5 mL of ether and add this rinse to the separatory funnel.

Working in a hood, if possible, shake the separatory funnel to mix the phases and drain the lower aqueous layer into a labeled flask [see Technique 8.2]. Pour the ether layer into another labeled flask. Return the aqueous phase to the separatory funnel and extract it a second time with 10 mL of ether. Remove the lower aqueous phase and combine the two ether solutions in the separatory funnel. Wash the combined ether layers with two 10-mL portions of water to remove most of the ethanol that remains dissolved in the ether layer. Transfer the ether solution to a dry

Be sure to separate the first water wash before adding the second.

50-mL Erlenmeyer flask after the last water wash. Dry the ether/product solution with anhydrous magnesium sulfate [see Technique 8.7].

Filter the product solution into a tared (weighed) 50-mL Erlenmeyer flask using a fluted filter paper [see Technique 8.8], and evaporate the ether on a steam bath or in a beaker of 70–75°C water in a hood, using a boiling stick or stone to prevent bumping. Alternatively, the ether may be evaporated in a hood using a stream of nitrogen or air or evaporated using a rotary evaporator [see Technique 8.9]. Weigh the product after all the ether has been removed and calculate the percent yield.

Analysis of the Product

Obtain an IR spectrum of your product as directed by your instructor [see Technique 18.4].

Analyze the product composition by quantitative GC analysis [see Technique 16.7]. Use a nonpolar column at 120–125°C. Consult your instructor about sample preparation and sample size for a packed column chromatograph. For a capillary column instrument, prepare a solution containing 5 mg of your product in 1.0 mL of ether; inject a 1-μL sample of this solution into the chromatograph. On both a packed Carbowax column and a nonpolar capillary column, such as SE-30 or OV-101, *cis*-4-*tert*-butylcyclohexanol has a shorter retention time than does *trans*-4-*tert*-butylcyclohexanol, but on the packed column, 4-*tert*-butylcyclohexanone and the cis alcohol may have very similar retention times. On a nonpolar capillary column, the compounds elute in the following order: *cis*-4-*tert*-butylcyclohexanol, *trans*-4-*tert*-butylcyclohexanol, 4-*tert*-butylcyclohexanone. If you have two peaks on your gas chromatogram, demonstrate that one of them is or is not 4-*tert*-butylcyclohexanone by using the peak enhancement method [see Technique 16.6].

The product may also be analyzed by NMR spectroscopy. Prepare an NMR sample as directed by your instructor, using $CDCl_3$ as the solvent [see Technique 19.2].

Analysis of the 4-tert-Butylcyclohexanol used in Project 4.1

The 4-*tert*-butylcyclohexanol available commercially is an unspecified mixture of cis and trans isomers. Determine its composition by quantitative GC analysis under the same chromatographic conditions that you use for the product.

Alternatively, the isomeric mixture in commercial 4-*tert*-butylcyclohexanol may be analyzed by NMR spectroscopy. Prepare an NMR sample as directed by your instructor using $CDCl_3$ as the solvent.

CLEANUP: The aqueous solutions remaining from the extractions may be washed down the sink or poured into the container for aqueous inorganic waste. Put any white solid formed in the reaction mixture into the container for inorganic waste. Place the magnesium sulfate drying agent in the container for nonhazardous waste or inorganic waste. Place the glass-backed TLC plates in the appropriate glass waste container, as directed by your instructor. Place aluminum-backed TLC plates in the metal waste container. If you prepared ether solutions for the GC analysis, pour any remaining solutions into the container for flammable (organic) waste.

Interpretation of Experimental Data

Analyze the IR spectra of the 4-*tert*-butylcyclohexanone you prepared in Project 4.1 and the 4-*tert*-butylcyclohexanol you prepared in Project 4.2 and identify the principal vibrational bands in each spectrum [see Techniques 18.5 and 18.6]. What vibrational band(s) would be present in the spectrum of 4-*tert*-butylcyclohexanone if the oxidation were not complete? Similarly, what vibrational band(s) would be present in the spectrum of 4-*tert*-butylcyclohexanol if the reduction were not complete?

Analyze the chromatograms for your 4-*tert*-butylcyclohexanol prepared by $NaBH_4$ reduction and for the commercially available 4-*tert*-butylcyclohexanol. Calculate the ratio of cis and trans isomers present in each sample. Is there evidence of unreacted 4-*tert*-butylcyclohexanone in your product?

If you used NMR analysis to compare the 4-*tert*-butylcyclohexanol prepared by $NaBH_4$ reduction and the commercially available 4-*tert*-butylcyclohexanol, determine the ratio of cis and trans isomers present in each spectrum from the integrated areas of the methine hydrogens on the carbon bearing the —OH group.

What does the observed ratio of cis- and trans-isomeric alcohols indicate about the stereoselectivity of the $NaBH_4$ reduction of 4-*tert*-butylcyclohexanone?

Compare the cis- and trans-isomeric ratio of 4-*tert*-butylcyclohexanol that you prepared with that of the commercially available material. What does this comparison suggest about the method used to produce commercial 4-*tert*-butylcyclohexanol?

MICROSCALE PROCEDURE

Techniques Microscale Extraction: Technique 8.6c
Drying Organic Solutions: Techniques 8.7–8.9
Thin-Layer Chromatography: Technique 15
Gas Chromatography: Technique 16
IR Spectroscopy: Technique 18
NMR Spectroscopy: Technique 19

$NaBH_4$ is moisture sensitive; keep the bottle tightly closed when not in use. Store the reagent in a desiccator.

SAFETY INFORMATION

4-*tert*-Butylcyclohexanone and the **4-*tert*-butylcyclohexanol** product may be irritants. Avoid breathing the dust. Avoid contact with skin, eyes, and clothing.

Sodium borohydride is harmful if swallowed, inhaled, or absorbed through the skin. Avoid breathing the dust. Avoid contact with skin, eyes, and clothing. It decomposes to flammable, explosive hydrogen gas.

Diethyl ether is extremely volatile and flammable. Use it in a hood, if possible. Be sure that there are no flames in the laboratory and no hot electrical devices in the vicinity while you are pouring ether or performing the extractions.

Use this procedure if you have less than 300 mg of 4-*tert*-butylcyclohexanone. Proportionally adjust the quantities of all reagents for your amount of ketone.

Note to instructor: Prepare the sodium borohydride solution just prior to use in a quantity sufficient for the entire class. The solution will be cloudy and will start to decompose slowly; it should be used as soon as possible after preparation.

Place 200 mg of 4-*tert*-butylcyclohexanone and a magnetic stirring bar in a conical vial. Add 1.0 mL of 95% ethanol. Set the vial in a 50-mL beaker of tap water and begin stirring the mixture to dissolve the solid. Pipet 0.80 mL of a freshly prepared sodium borohydride solution that contains 150 mg of $NaBH_4$ per milliliter of 95% ethanol into a 3-mL conical vial and cap the vial. Make slow dropwise additions of the sodium borohydride solution to the reaction vial containing the 4-*tert*-butylcyclohexanone solution. The initial reaction may be vigorous. After the addition is complete, remove the water bath and continue to stir the reaction. After 15 min, check for completeness of reaction (indicated by the disappearance of the 4-*tert*-butylcyclohexanone spot) by the TLC method used in Project 4.1.

When the reaction is complete, add 1.0 mL of water to the reaction mixture, then slowly add 0.25 mL of 3 M hydrochloric acid. **(Caution: If you add the HCl too quickly, it may cause the reaction mixture to bubble out of the vial.)** When the evolution of hydrogen gas ceases, add 1.0 mL of ether to the vial. Cap the vial and shake it to mix the layers thoroughly. Open the cap cautiously to vent the vial. Remove the lower aqueous layer with a Pasteur pipet fitted with a syringe or a rubber bulb [see Technique 8.5]. Transfer the aqueous layer to a clean 5-mL conical vial. Repeat the extraction of the aqueous layer using another 1.0 mL of ether. Remove the lower aqueous phase and transfer it to a small test tube. Transfer the first ether extract to the conical vial containing the second ether extract. Then wash the combined ether layers successively with two 1.0-mL portions of water. Dry the ether solution with 100 mg of anhydrous magnesium sulfate [see Technique 8.7].

The water washes remove most of the ethanol that is dissolved in the ether solution.

Transfer the dried ether solution to a tared (weighed) 10 × 75 mm test tube or 3-mL conical vial, using a Pasteur filter pipet [see Technique 8.5]. Under a hood, evaporate all the ether from the product mixture, using a gentle stream of nitrogen or air. Alternatively, transfer the dried ether solution to a tared watch glass. Set the watch glass in the front of a hood with the sash nearly closed until the ether evaporates. Weigh the dried product and determine the percent yield.

Analyze the product composition by IR spectroscopy and by quantitative GC or NMR spectroscopy, and complete the interpretation of experimental data as directed in the miniscale procedure.

CLEANUP: The aqueous solutions remaining from the extractions may be poured down the sink or into the container for aqueous inorganic waste. Place the magnesium sulfate drying agent in the container for nonhazardous solid inorganic waste. Place the glass-backed TLC plates in the appropriate container for glass waste, as directed by your instructor;

place aluminum-backed TLC plates in the container for metal waste. If you prepared ether solutions for the GC analysis, pour any remaining solutions into the container for flammable (organic) waste.

4.3 Computational Chemistry Experiment

Build a molecule of cyclohexanone. Optimize the structure first by using a molecular mechanics package and then save the structure to use as a template for building molecules of axial-4-*tert*-butylcyclohexanone and equatorial-4-*tert*-butylcyclohexanone. Optimize these structures using molecular mechanics and record their steric energies. Which molecule is lower in energy?

Using the difference in the steric energies, calculate the equilibrium constant (K_{eq}) for the interconversion of the two structures. Assume that the change in enthalpy ($\Delta H°$) can be approximated by the difference in steric energies. Use the following equations in your calculation:

$$\Delta G° = \Delta H° - T\Delta S°$$
$$\Delta G° = -RT \ln K_{eq} = -2.303\ RT \log K_{eq}$$
$$\Delta G° = -1.36 \log K_{eq} \text{ at } 25°C\ (298\ K)$$

Assume that the entropy change, $\Delta S°$, is close to zero; thus, $\Delta G°$, the Gibbs standard free energy change, is approximately the same as the change in the steric energy, $\Delta H°$.

Hydride reductions are often highly stereoselective, with delivery of hydrogen from the less hindered side of the substrate molecule. Using the lowest-energy molecule of 4-*tert*-butylcyclohexanone, measure the distances between the carbon atom of the carbonyl group and the axial hydrogen atoms at C2 and C3. Which is the less hindered side of the molecule? Using the space-filling rendering mode, view the structure. What stereoisomer will be produced as the result of hydride attack from the less hindered side of the molecule?

Reference

1. Mohrig, J. R.; Nienhuis, D. M.; Linck, C. F.; Van Zoeren, C.; Fox, B. G.; Mahaffy, P. G. *J. Chem. Educ.* **1985,** *62,* 519–521.

Questions

1. Which IR band is most likely to be observed for unreacted starting material in the oxidation product?
2. Reference 1 suggests that 2-propanol rather than sodium bisulfite ($NaHSO_3$) can be used to destroy the excess sodium hypochlorite. What organic compound would be formed and why would it not be a contamination problem in this reaction?
3. Balance the equation for the oxidation-reduction reaction that occurs between bisulfite and hypochlorite ions to give sulfate and chloride ions.
4. Predict the product that will be obtained if *cis*-2-methylcyclohexanol is oxidized with NaOCl.
5. How might you minimize the hydrogen evolution when using $NaBH_4$ as a reducing agent in water?

PROJECT

5

SYNTHESIS OF 1-BROMOBUTANE AND THE GRIGNARD SYNTHESIS OF AN ALCOHOL

PURPOSE: To synthesize 1-bromobutane and use it as a substrate in a Grignard synthesis that you will design.

> In this three-week project you will first prepare 1-bromobutane by an acid-catalyzed S_N2 reaction of 1-butanol. Then you will convert it to a Grignard reagent by reaction with magnesium metal in diethyl ether and add the Grignard reagent to an aldehyde or ketone of your choice. The final product will be a secondary or tertiary alcohol. You will characterize it by boiling point, GC analysis, and IR and NMR spectroscopy. In an optional experiment, you can synthesize a different alkyl halide to use in the Grignard synthesis.

S_N2 Chemistry

Acid-catalyzed synthesis of a primary alkyl bromide from a primary alcohol is the first synthetic step in this project. The reaction uses a mixture of sodium bromide and the strong mineral acid sulfuric acid. The acid is essential because it turns R—OH into $R—OH_2^+$, which converts a very poor leaving group into a reasonably good one. Because alcohols are weak bases, a strong acid is required to protonate them. Sodium bromide provides the bromide ion, an excellent nucleophile, which directly displaces the protonated hydroxyl group in an S_N2 reaction.

$$R—OH + H^+ \rightleftharpoons R—OH_2^+$$

$$R—OH_2^+ + Br^- \longrightarrow R—Br + H_2O$$

Rate-determining step

Examination of the S_N2 transition state shows that for an instant there are effectively five bonds to the central carbon. This cluttered state is achievable with primary alcohols where two of the three groups originally bonded to the primary carbon are hydrogen atoms. Because the bonds are made and broken in a concerted fashion, there is no opportunity for any rearrangement to occur. This S_N2 reaction produces a good yield of 1-bromobutane.

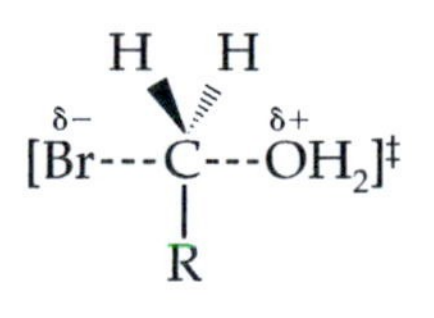

S_N2 transition state

Grignard Syntheses

Among the most important organometallic reagents are the alkyl- and arylmagnesium halides, which are almost universally called ***Grignard reagents*** after the French chemist Victor Grignard, who first realized their tremendous potential in organic synthesis; Grignard received the 1912 Nobel prize in chemistry. Grignard reagents readily attack the C═O groups of aldehydes and ketones and produce the synthesis of highly specific carbon-carbon bonds in excellent yields.

Grignard reagents contain carbon-metal bonds. Their synthesis requires the reaction of an alkyl or aryl halide with magnesium metal in the presence of an ether solvent:

$$R{-}X + Mg \xrightarrow{\text{ether}} \overset{\delta-}{R}{-}\overset{\delta+}{Mg}{-}\overset{\delta-}{X}$$

Grignard reagent

Formation of a Grignard reagent takes place in a heterogeneous reaction at the surface of the solid magnesium metal, and the surface area and reactivity of the magnesium are crucial factors in the rate of the reaction. It is thought that the alkyl halide reacts with the surface of the metal to produce a carbon free radical and a magnesium-halogen radical, which combine to form the Grignard reagent. This reaction gives the magnesium atom two covalent bonds, but it must somehow acquire two more because magnesium has a coordination number of 4. When an ether is the solvent, ether molecules occupy the other two coordination sites. The oxygen atom of an ether molecule provides both electrons in the magnesium-ether bond:

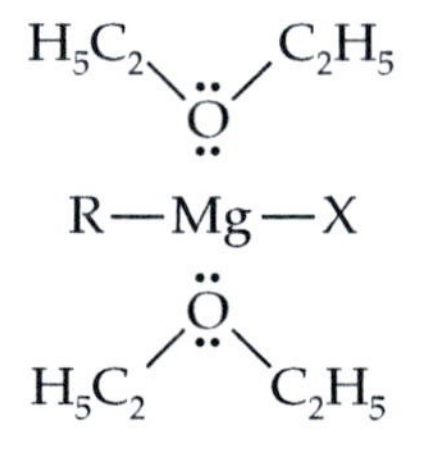

The Grignard-ether complexes are quite soluble in ether. In the absence of an ether solvent, the reaction of magnesium and the alkyl halide takes place rapidly but soon stops because the surface of the metal becomes coated with polymeric Grignard reagent. In the presence of ether, the surface of the metal is kept clean and the reaction proceeds until the limiting reagent is entirely consumed.

Treatment of an ether solution of 1-bromobutane with magnesium metal results in the formation of the Grignard reagent, butylmagnesium bromide. Insertion of the magnesium between carbon and bromine reverses the polarity of the carbon formerly bonded to bromine. In the Grignard reagent, the carbon-metal bonds have a high degree of ionic character with a good deal of negative charge on the carbon atom, making it a strong nucleophile:

$$CH_3CH_2CH_2{-}\overset{\delta+}{C}H_2{-}\overset{\delta-}{Br} \xrightarrow[\text{polarity reversal}]{Mg} CH_3CH_2CH_2{-}\overset{\delta-}{C}H_2{-}\overset{\delta+}{MgBr}$$

The mechanism of the Grignard reaction with carbonyl compounds is actually quite complex, but it can easily be rationalized as a simple nucleophilic addition reaction:

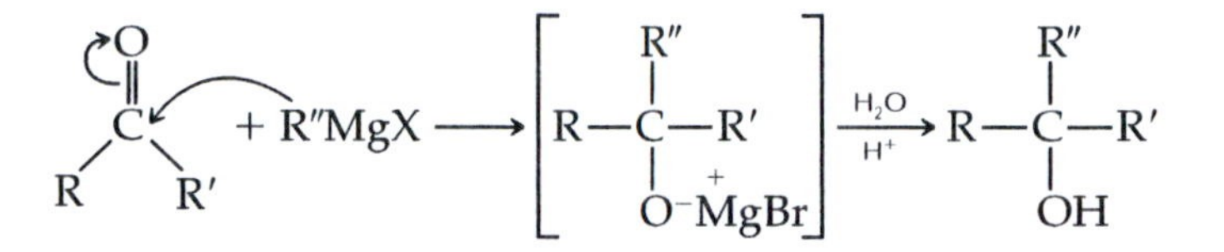

The Grignard reaction produces a magnesium salt of the organic product. These magnesium salts are readily hydrolyzed to the organic product by reaction with acidic water solution. It is common to use an aqueous mineral acid, such as sulfuric or hydrochloric acid, to expedite hydrolysis. Not only does this cause the protonation to occur more readily, but Mg(II) is converted from insoluble salts to water-soluble sulfates or chlorides.

Unwanted Side Reactions

Because Grignard reagents have such tremendous reactivity as nucleophiles, it is not surprising that they are also strong bases. They react quickly and easily with acids, even relatively weak acids such as water and alcohols. This reaction is a nuisance because it destroys the Grignard reagent before the carbonyl compound is added and therefore reduces the yield of the desired product.

$$\mathrm{RX} \xrightarrow{\mathrm{Mg}} \mathrm{RMgX} \xrightarrow{\mathrm{H_2O}} \mathrm{RH} + \mathrm{HOMgX}$$

The presence of water or other acids also inhibits the formation of the Grignard reagent. Thus, it is important that the reaction conditions be completely anhydrous. All glassware and reagents must be thoroughly dry before beginning a Grignard experiment.

Other reactions that may interfere with Grignard syntheses include Grignard coupling and reaction with molecular oxygen:

$$2\mathrm{RX} + \mathrm{Mg} \longrightarrow \mathrm{R{-}R} + \mathrm{MgX_2}$$

$$\mathrm{RMgX} + \mathrm{O_2} \longrightarrow \mathrm{ROOMgX} \xrightarrow{\mathrm{RMgX}} 2\,\mathrm{ROMgX}$$

The coupling reaction takes place at the surface of the solid magnesium when two free radicals react to form a stable hydrocarbon dimer. This side reaction is favored by a high concentration of the 1-bromobutane, where the concentration of free radicals is greater. The combination of the Grignard reagent with molecular oxygen is a side reaction when a synthesis is run in the presence of air. Low-boiling-point diethyl ether (bp 35°C) helps to exclude air near the surface of the reaction solution because of its great volatility.

Designing a Grignard Synthesis

In Project 5.2 you will react butylmagnesium bromide with one of five aldehydes or ketones. You may be assigned an aldehyde or ketone from the following list or you may be asked to select one. The final product needs to have eight or fewer carbons to have a boiling point under 200°C, thus allowing it to be distilled at atmospheric pressure without decomposition.

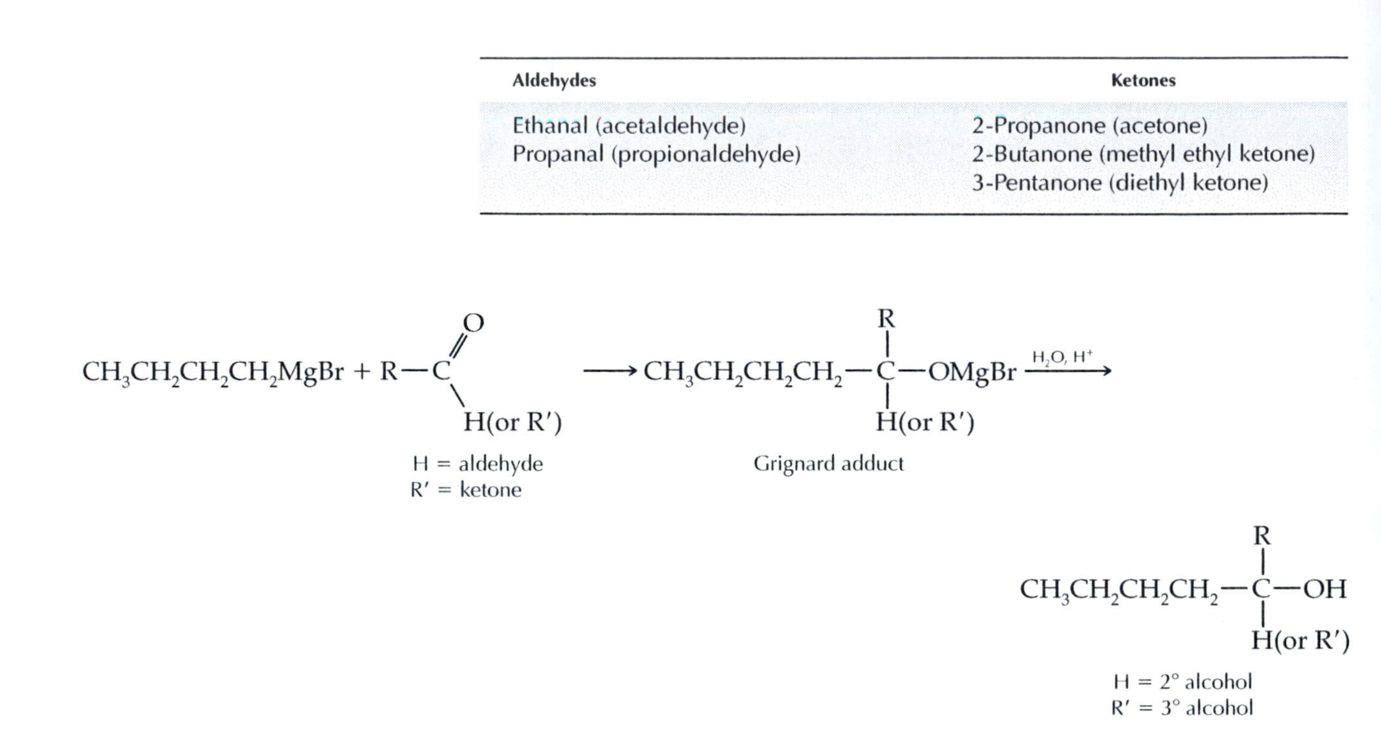

Aldehydes	Ketones
Ethanal (acetaldehyde)	2-Propanone (acetone)
Propanal (propionaldehyde)	2-Butanone (methyl ethyl ketone)
	3-Pentanone (diethyl ketone)

5.1 Synthesis of 1-Bromobutane from 1-Butanol

PRELABORATORY ASSIGNMENT: Explain why 1-bromobutane and water codistill at a temperature lower than the boiling point of either one of them.

$$CH_3CH_2CH_2CH_2OH + NaBr + H_2SO_4 \longrightarrow CH_3CH_2CH_2CH_2Br + H_2O + NaHSO_4$$

1-Butanol	Sodium bromide	1-Bromobutane
bp 117.2°C	MW 102.9	bp 101.3°C
MW 74.1		MW 137.0
density 0.810 g·mL^{-1}		density 1.27 g·mL^{-1}

MINISCALE PROCEDURE

Techniques Reflux: Technique 7.1
Steam Distillation: Technique 11.7
Extraction: Technique 8.2
Drying Agents: Techniques 8.7–8.9
Simple Distillation: Technique 11.3
IR Spectroscopy: Technique 18
NMR Spectroscopy: Technique 19

SAFETY INFORMATION

The **1-butanol** is flammable and an irritant to skin and eyes. Avoid contact with skin, eyes, and clothing.

The **1-bromobutane** synthesized is harmful if inhaled, ingested, or absorbed through the skin. Wear neoprene gloves when handling it and work in a hood, if possible.

Sodium bromide is a skin irritant.

Sulfuric acid is corrosive and causes severe burns.

Concentrated hydrochloric acid is corrosive, causes burns, and emits HCl vapors. Use it in a hood.

Wear gloves while adding the sulfuric acid.

Pour 70 mL of 8.7 M sulfuric acid into a 125-mL Erlenmeyer flask and cool the flask in an ice-water bath. Using a conical funnel, pour 18.2 mL of 1-butanol into a 250-mL round-bottomed flask that is firmly clamped at the neck to a rack or ring stand. Using a powder funnel, add 40.8 g of sodium bromide to the flask. Add one or two boiling stones or use magnetic stirring. Attach a water-cooled condenser to the flask, as shown for a reflux apparatus [see Technique 7.1]. Place a conical funnel in the top of the condenser and pour the chilled sulfuric acid solution down the condenser in three portions. Gently swirl the flask and its contents between each addition.

Place a heating mantle or sand bath under the flask and bring the mixture to a boil. Heat the reaction mixture under reflux for 45 min; the flask contents should be boiling vigorously so that the two phases mix well. Remove the heat source and let the flask cool for 5 min.

The addition of water allows a steam distillation to be carried out [see Technique 11.7].

Test for completeness of the distillation by removing the receiving flask and collecting several drops of distillate in a test tube containing about 1 mL of water.

Pour 60 mL of water down the condenser. Add another boiling stone and set up the apparatus for a simple distillation [see Technique 11.3]. Use a 125-mL Erlenmeyer flask as the receiver. Collect 40–45 mL of distillate, or collect distillate until it no longer contains water-insoluble droplets. When the distillation is complete, remove the heat source from the round-bottomed flask. During this steam distillation, place 32 mL of concentrated hydrochloric acid in a corked flask and chill the acid in an ice-water bath until you need it during the extraction procedure.

After you finish the distillation, wash the distillation head, thermometer adapter, condenser, and vacuum adapter. Place them in a drying oven for 10–15 min, and allow them to cool to room temperature before assembling the final distillation apparatus. Alternatively, the glassware can be dried by rinsing it with a small amount of acetone and placing the glassware in a hood until the acetone evaporates. **(Caution: If you use acetone, do not put the glassware in an oven.)**

SAFETY PRECAUTIONS

Wear gloves while performing the extractions and work in a hood to keep hydrochloric acid vapors out of the laboratory.

Foaming from CO_2 will occur in the extraction with sodium bicarbonate. Vent the funnel frequently to relieve the gas pressure.

Review the extraction procedure in Technique 8.2 before you begin this part of the procedure.

Transfer the distillate, which contains water, 1-bromobutane, and any 1-butanol that did not react, to a separatory funnel using a conical funnel. Allow the phases to separate and drain the lower organic phase into an Erlenmeyer flask labeled "Organic phase." Pour the upper aqueous phase out of the top of the separatory funnel into a flask labeled "Aqueous phase." Return the organic phase to the separatory funnel. With each extraction, it is necessary to repeat this process of removing both layers from the funnel and returning the organic phase for the next extraction. The same flask can be used to hold the organic phase after each separation. Wash the organic phase with two 16-mL portions of chilled, concentrated hydrochloric acid, then with 15 mL of cold water. Finally, wash the organic phase with 12 mL of saturated sodium bicarbonate solution. **(Caution: Pressure buildup. Vent the funnel frequently.)** After the phases separate, drain the lower organic phase into a clean, dry 50-mL Erlenmeyer flask. Add anhydrous potassium carbonate and dry the product for at least 10 min [see Technique 4.6].

Any unreacted 1-butanol would be difficult to remove from your product by distillation. Using hydrochloric acid extracts it into the aqueous phase.

$$ROH + H_3O^+ \longrightarrow R\overset{+}{O}H_2 + H_2O$$

Filter the dried 1-bromobutane through a dry conical funnel containing a small cotton plug into a clean, dry 50-mL round-bottomed flask [see Technique 8.8, Figure 8.15]. Add two boiling stones and assemble the apparatus for simple distillation [see Technique 11.3]. Collect the product fraction from 97°–102°C in a tared (weighed) 20-mL screw-capped vial. **Be sure to remove the heat source before the distillation flask reaches dryness.** Weigh your product and determine the percent yield.

Analysis of the Product

Obtain an NMR spectrum in $CDCl_3$ as directed by your instructor [see Technique 19.2]. Does your NMR spectrum show evidence of impurities? Alternatively, you can analyze your product by IR spectroscopy [see Technique 18.4]. What vibrational band(s) would you expect to see if unreacted alcohol were present in your product?

After determining the mass of the product and its NMR or IR spectrum, add just enough anhydrous potassium carbonate to cover the bottom of the vial. Cap the vial tightly and store it upright in a small beaker in your desk until you are ready to do the next step of the project.

The extra treatment with drying agent ensures that no water will be present in your 1-bromobutane when you use it in the Grignard reaction.

CLEANUP: The residue remaining in the distillation flask following the first distillation contains sulfuric and hydrobromic acids. Carefully pour it into a large beaker containing 300 mL of water. Add the hydrochloric acid and other aqueous washes from the extractions to the beaker. Add sodium carbonate in small portions **(Caution: Foaming.)** until the acid is neutralized. The solution can then be washed down the sink or poured into the container for aqueous inorganic waste. The residue remaining in the distillation flask after the final distillation should be poured into the container for halogenated waste. The potassium carbonate used as the drying agent is saturated with 1-bromobutane; discard it in the container for hazardous solid waste.

OPTIONAL SYNTHESES OF ALKYL HALIDES

Substitute 0.20 mol of 1-pentanol, 3-methyl-1-butanol, or 1-propanol for 1-butanol; calculate the number of milliliters that you will need to use for the alcohol you select. The rest of the procedure remains the same. Each of these compounds is a primary alcohol that will undergo a S_N2 reaction with Br^- in the presence of a strong acid. You need to find the density of the alcohol you will use and the boiling point of the corresponding alkyl bromide product in *The Aldrich Catalog* or the *CRC Handbook of Chemistry and Physics* as part of your prelaboratory preparation.

5.2 Grignard Synthesis of Secondary and Tertiary Alcohols

PRELABORATORY ASSIGNMENT: Write a balanced equation for the Grignard synthesis of the secondary or tertiary alcohol that you will make from the reaction of your aldehyde or ketone with butylmagnesium bromide (or other alkylmagnesium bromide that you have synthesized). Also, in the *CRC Handbook of Chemistry and Physics* or the *Dictionary of Organic Compounds,* look up the boiling points of your alcohol product and the coupled alkane that may arise from your bromoalkane in the Grignard synthesis.

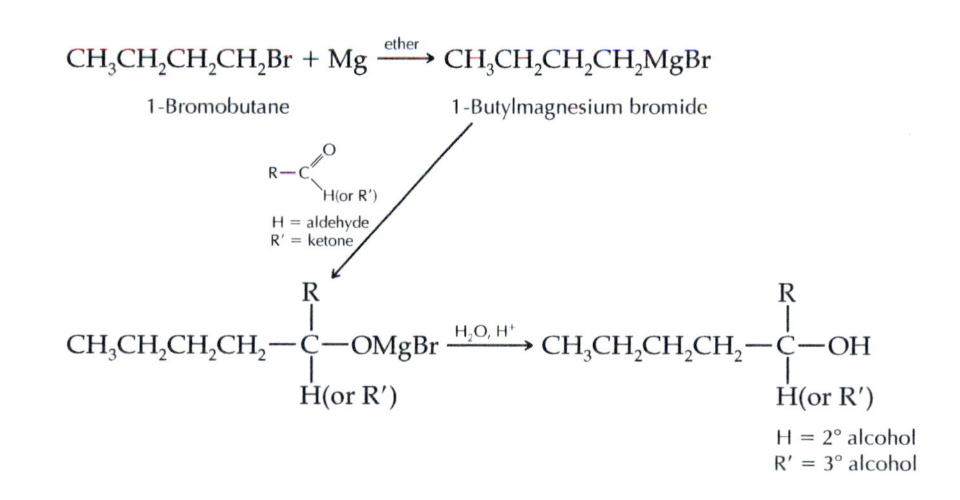

MINISCALE PROCEDURE

Techniques Reflux under Anhydrous Conditions: Technique 7.3
Extraction: Technique 4.2
Drying Agents: Techniques 8.7–8.9
Simple Distillation: Technique 11.3
Gas Chromatography: Technique 16
IR Spectroscopy: Technique 18

SAFETY INFORMATION

Diethyl ether is extremely flammable. Make sure that there are no flames in the laboratory or hot electrical devices nearby, and use ether only in a hood.

The **alkyl halides** used in this experiment are harmful if inhaled, ingested, or absorbed through the skin. Wear neoprene gloves when handling them and use them in a hood, if possible. They are also flammable.

The **aldehydes** are flammable and corrosive. Wear gloves when handling them and dispense them in a hood. **Acetaldehyde** is also a lachrymator and a suspected carcinogen and may be an immune-response sensitizer.

The **ketones** are flammable.

All glassware and reagents need to be thoroughly dry before beginning the experiment.

This synthesis is based on 0.10 mol of bromoalkane. If you have more or less than 0.10 mol of your 1-bromobutane (or other alkyl halide), you must adjust proportionally the amounts of all other reagents. Consult your instructor before you begin the procedure.

Synthesis of the Grignard Reagent

Fit a 250-mL three-necked round-bottomed flask with a dropping funnel (or a separatory funnel) and a condenser; close the third opening with a glass stopper. If you are using a one-necked flask, fit it with the Claisen adapter and insert the dropping funnel into one neck of the adapter and the condenser into the other. Place a drying tube filled with calcium chloride in the top of the condenser and another in the top of the dropping funnel [Technique 7.3, Figure 7.3].

Weigh 2.4 g of magnesium turnings; crush several pieces with a glass stirring rod to expose a fresh surface. Place the magnesium turnings in the round-bottomed flask. Obtain 75 mL of anhydrous diethyl ether in a corked graduated cylinder. Cover the magnesium turnings with approximately 35 mL of anhydrous ether. Put a magnetic stirring bar into the flask.

Keep the ether can tightly closed as much as possible. Do not let your ether stand in an open container because water from the air will dissolve into it.

Measure 0.10 mol of your alkyl bromide (check the quantity with your instructor) and pour it into the dropping funnel (be sure the stopcock is closed). Pour the remaining 40 mL of ether into the dropping funnel, stopper the funnel, and shake it thoroughly to produce a homogeneous solution; put the drying tube filled with calcium chloride back into the top of the separatory funnel. Add about 3 mL of the ether/alkyl bromide solution to the reaction flask. Do not stir the contents of the flask; initiation of the reaction occurs more rapidly if there is a high concentration of halide in the vicinity of the magnesium. Warm the flask for a few minutes in a water bath at 45–50°C. The ether will begin to boil. Within 5–10 min, you should see tiny bubbles forming at the magnesium surface, and a small amount of cloudy precipitate may appear.

The precipitate is probably magnesium hydroxide generated from any moisture in the flask or the reagents.

Once the reaction is well under way, remove the warm-water bath and begin stirring the reaction mixture slowly. Add the alkyl bromide solution at a rate that allows for moderate reflux to take place. The condensing ether vapor should fill only the lower half of the condenser. Have a beaker of ice water available to cool the flask by briefly immersing it; if the reaction becomes too vigorous, too much cooling will stop the reaction.

When all the alkyl bromide has been added, continue stirring the reaction mixture until it ceases to reflux. Then heat the mixture at reflux for 15 min, using a water bath at 45–50°C.

Addition of the Carbonyl Compound

Pour 0.098 mol of the aldehyde (or ketone) that you selected (check the quantity with your instructor) into the dropping funnel, rinse the graduated cylinder used for the aldehyde (or ketone) with 10 mL of anhydrous ether, and also add this ether rinse to the dropping funnel. Place an ice-water bath under the reaction flask and run the stirrer at a moderate speed. Add the aldehyde (or ketone) solution slowly over a period of 15–20 min. This reaction may be vigorous. Use a rate of addition that maintains the condensing ether vapors in the lower half of the condenser. After the addition is complete and the reaction is no longer refluxing of its own accord, heat it under reflux in a water bath at 45–50°C for 15 min. Cool the reaction mixture to room temperature before proceeding with the hydrolysis.

A good stopping point. Instead of heating the reaction mixture, allow it to stand until the next laboratory period; stopper the flask tightly with a cork and store it as directed by your instructor.

Hydrolysis of the Grignard Adduct

SAFETY PRECAUTION

Ether vapors will be present during the hydrolysis and extractions. Be sure that there are no flames in the laboratory or hot electrical heating devices in your work area. Work in a hood, if possible, and wear gloves.

In the hood, pour the reaction mixture slowly into a 250-mL Erlenmeyer flask containing 50 g of crushed ice and 50 mL of 10% sulfuric acid. Stir or swirl the mixture until the precipitate dissolves. If the ether begins to boil, add more ice.

When the residual magnesium has all dissolved, transfer the mixture to a separatory funnel and separate the layers [Technique 8.2]. Keep both layers. Return the aqueous layer to the funnel and extract it twice with 25-mL portions of ether. Combine the two ether extracts with the ether layer from the first separation and dry the combined ether solution with anhydrous potassium carbonate. Set the aqueous layer aside for later neutralization.

Anhydrous ether is not necessary for the extractions because you are adding it to an aqueous solution.

Alternatively, the ether may be removed with a rotary evaporator [see Technique 8.9].

Filter part of the dried ether solution into a 100-mL round-bottomed flask, filling the flask no more than two-thirds full [Technique 8.8]. Remove the ether by simple distillation on a steam bath or 80–90°C water bath [Technique 11.3]; place the receiving flask in an ice-water bath. Stop the distillation and filter the remaining ether solution into the round-bottomed flask; add another boiling stone. Continue the distillation until the ether stops distilling. Pour the distilled ether into the recovered ether bottle in the hood. Transfer the liquid remaining in the boiling flask to a 50-mL round-bottomed flask, rinse the larger flask with 1–2 mL of ether, and add the rinse to the 50-mL flask. Set up the apparatus for simple distillation, using a heating mantle or sand bath as the heat source.

Collect as a first fraction the residual ether. Use the boiling points of your alcohol and the coupled alkane to determine the temperature range for collecting the product fraction. Collect your product fraction in a tared (weighed) Erlenmeyer flask. Stop the distillation when only a small amount of liquid remains in the boiling flask but be very careful not to let the flask boil dry. Record the boiling-point range for the product fraction. Determine the mass of your product and calculate the percent yield.

SAFETY PRECAUTION

Ether may form peroxides that can explode if the flask goes dry at the end of the distillation.

CLEANUP: Pour the residue remaining in the distillation flask into the container for flammable (organic) waste. Place the recovered ether in the "recovered ether" container or the container for flammable (organic) waste, as directed by your instructor. Neutralize the aqueous solution remaining from the extractions with sodium carbonate **(Caution: Foaming.)** before washing it down the sink or pouring it into the container for aqueous inorganic waste.

Assessing the Purity of the Product

You will determine the purity of your alcohol product by gas chromatography. The competing reaction that produces the coupled alkane may occur to a significant degree in this Grignard reaction. The boiling point of the coupled alkane lies within 30° of the alcohol product's boiling point for some of the possible combinations of alkyl bromide and carbonyl compounds. This small difference in boiling points means that you may not have been able to separate the alkane from the alcohol in your distillation.

Determine the purity of your product by GC analysis using a nonpolar column such as SE-30 or OV-1. For capillary GC, dissolve 1 drop of your product in 0.5 mL of ether and inject 1 μL of the solution into the chromatograph. Consult your instructor about sample preparation for a packed column GC. The alkane and alcohol elute in order of increasing boiling point from a nonpolar column such as SE-1 or OV-1. Analyze the chromatogram of your product. What does it show about the purity of your product?

Obtain an IR spectrum of your product [see Technique 18.4]. Does the spectrum show the strong vibrational bands that you would predict?

Questions

1. Treatment of 1-butanol with phosphoric acid (H_3PO_4) mixed with sodium chloride should result in formation of 1-chlorobutane when done at elevated temperatures. Write a mechanism for this reaction.
2. Comment on the viability of each of the following reactions for its ability to lead to 1-bromobutane:
   (a) 1-butanol + $NaBr/H_2O$
   (b) 1-butanol + $NaBr/H_3PO_4$
   (c) 1-butanol + KBr/H_2SO_4
   (d) 1-butanol + NaBr/acetic acid
3. Why was the product washed with aqueous sodium bicarbonate solution in the 1-bromobutane synthesis?
4. What is the copious white precipitate that forms when the carbonyl compound is added to an ether solution of butylmagnesium bromide?
5. If water is present in the Grignard reaction mixture, what organic product will form?
6. Which of the following compounds will undergo reaction with a Grignard reagent? If a reaction occurs, what will the products be?
   (a) 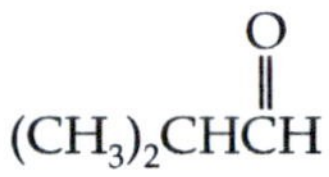 $(CH_3)_2CHCH{=}O$ (O double-bonded to the terminal C)
   (b) 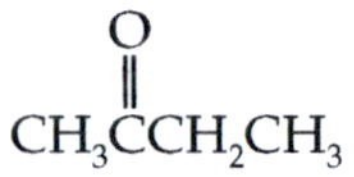 $CH_3C(=O)CH_2CH_3$
   (c) CH_3NH_2
   (d) $CH_3CH_2CH_2CH_2CH_3$
   (e) CH_3CH_2Cl
7. Suggest Grignard syntheses for the following compounds.
   (a) 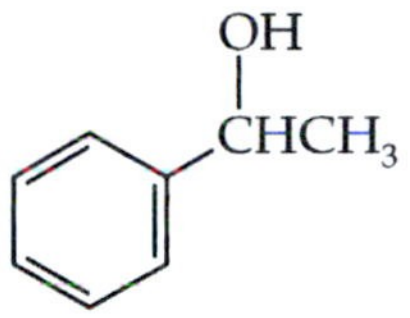 $C_6H_5CH(OH)CH_3$ (benzene ring bearing CHCH₃ with OH on the CH)
   (b) $CH_3CH(OH)CH_3$
   (c) $CH_3CH(CH_3)CH_2CH_2CH_2OH$

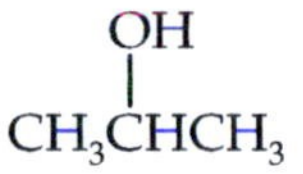

PROJECT

6 E1/E2 ELIMINATION REACTIONS

QUESTION: How does the mixture of isomeric alkenes produced by an acid-catalyzed dehydration reaction compare with the alkene mixture from a base-catalyzed dehydrochlorination reaction? Are either or both sets of products influenced primarily by product stability?

This is a two-week team project. In Project 6.1 you will carry out the dehydration of 2-methyl-2-butanol with H_2SO_4 and determine the product composition by gas chromatography. Meanwhile, in Project 6.2 your partner will synthesize 2-chloro-2-methylbutane from 2-methyl-2-butanol and HCl. Then the following week in Project 6.3, each of you will dehydrochlorinate 2-chloro-2-methylbutane by treatment with KOH and together compare the ratio of alkenes produced in the base-catalyzed dehydrochlorination to the ratio of alkenes produced by the acid-catalyzed dehydration of 2-methyl-2-butanol.

Acid-Catalyzed Dehydration of 2-Methyl-2-butanol

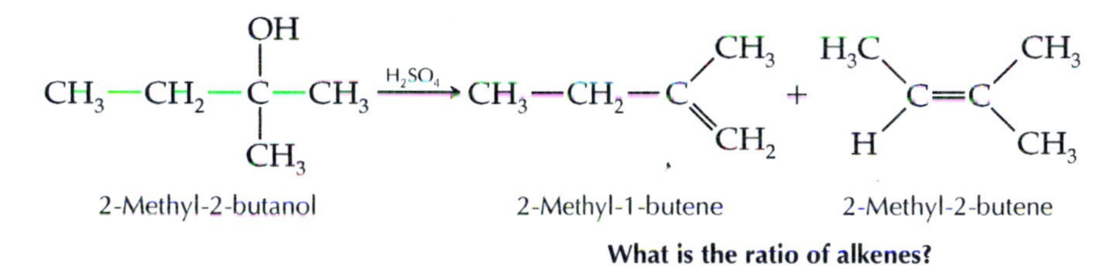

When 2-methyl-2-butanol is heated in the presence of sulfuric acid, the major product is a mixture of alkenes. The 1,2-elimination of a water molecule from an alcohol is called a dehydration reaction:

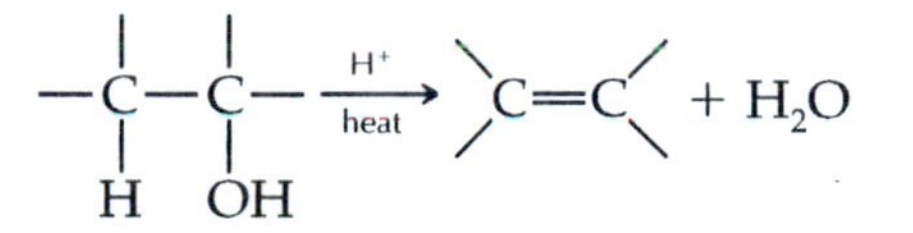

Elimination reactions require a leaving group that departs with its bonding electrons. The other group, which is lost from the adjacent carbon atom, is a proton. The acid is essential because it converts R—OH into R—OH_2^+, thereby providing a reasonably good leaving group. When tertiary alcohols are used, protonation and subsequent loss of water form a carbocation:

$$\mathrm{H{-}\overset{|}{\underset{|}{C}}{-}\overset{|}{\underset{OH}{C}}{-}} + H_2SO_4 \rightleftharpoons HSO_4^- + \mathrm{H{-}\overset{|}{\underset{|}{C}}{-}\overset{|}{\underset{{}^+OH_2}{C}}{-}} \rightleftharpoons \mathrm{H{-}\overset{|}{\underset{|}{C}}{-}C^+\!\!<} + H_2O$$

Carbocation

The rate of formation of a carbocation depends heavily on the stability of the carbocation being formed. Carbocation stability increases markedly with an increase in the number of alkyl or aryl substituents attached to the carbon atom bearing the positive charge. These substituents allow the charge to be dispersed throughout a larger space and thereby stabilize it.

In all but the most highly acidic environments, even tertiary carbocations are too unstable to persist. They react with nucleophiles to give S_N1 substitution products or lose a proton to give E1 elimination products. The term "E1" simply means unimolecular elimination. The reaction is said to be unimolecular because only one molecule is involved in the slowest or rate-determining step, which is the loss of a water molecule from the protonated alcohol. Carbocations are very strong acids, and the loss of a proton to form an alkene is fast.

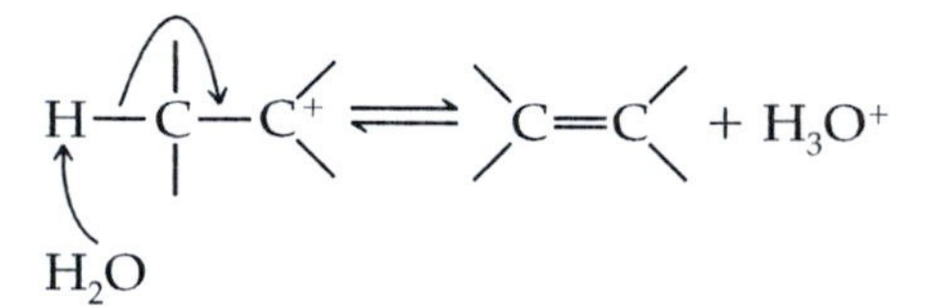

Sulfuric acid is the chosen catalyst for the dehydration of 2-methyl-2-butanol, in part because its conjugate base is a very poor nucleophile and does not lead to the formation of S_N1 products. Naturally, when a catalyst is selected for the synthesis of an alkene, we want to minimize yield-reducing substitution reactions.

The entire pathway for the E1 elimination of water from an alcohol in the presence of strong acid is

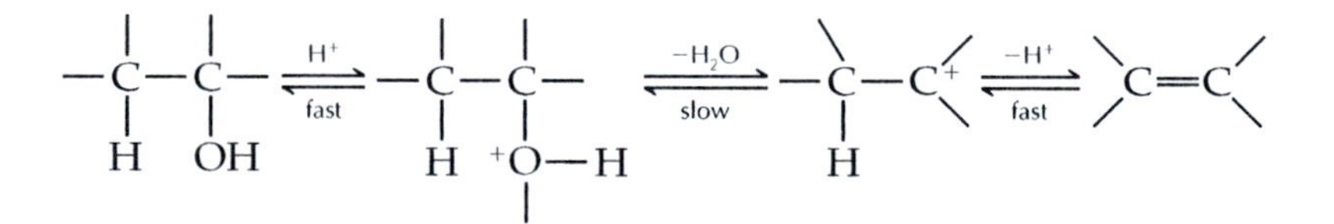

All the reaction steps in the E1 mechanism are reversible. Thus, the dehydration of alcohols is just the reverse of the mechanism for hydration of alkenes. This reversibility also means that alkenes formed under these conditions can revert to alcohols unless proper experimental conditions are used. To drive the elimination to completion, the alkene is distilled from the reaction mixture as it is formed. This strategy allows the equilibrium to be shifted continually in favor of the product.

The two possible products from the dehydration of 2-methyl-2-butanol in Project 6.1 are 2-methyl-1-butene and 2-methyl-2-butene. Because 2-methyl-2-butene has three alkyl groups attached to the C=C, it is more stable than 2-methyl-1-butene, which has only two alkyl groups attached to the C=C. If the more stable product forms more quickly, the product mixture should be composed mainly of 2-methyl-2-butene. On the other hand, it is possible that the factor controlling the ratio of the two alkenes is not product stability. It could be a statistical factor in the step where a proton is lost to form the alkene. 2-Methyl-1-butene might form faster because there are six methyl protons, any of which could be lost to form 2-methyl-1-butene. By comparison, there are only two methylene protons that could be lost in the formation of 2-methyl-2-butene.

The situation is made more complex by the possibility that the two alkenes might isomerize to one another in the presence of strong acid. The ratio of products will then reflect their relative stabilities. If this is the case, the relative rates of proton loss would not be a determining factor in the product ratio. You will be able to discover which factors are more important by using gas chromatography to determine the ratio of the two alkene products.

Synthesis of 2-Chloro-2-methylbutane

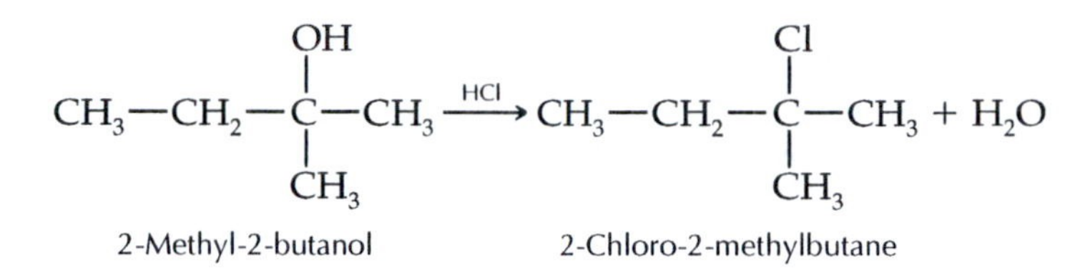

The synthesis of a tertiary alkyl halide from a tertiary alcohol and HX is an S_N1 reaction, a thoroughly studied and well-established organic chemistry reaction mechanism. The slowest or rate-determining step of this first-order nucleophilic substitution reaction is the formation of a carbocation intermediate by loss of H_2O from the conjugate acid of the alcohol substrate. The acid is essential because it turns R—OH into $R{-}OH_2^+$, converting a very poor leaving group into a reasonably good one. Because alcohols are weak bases, a strong acid is required to protonate them.

$$R{-}OH + H^+ \longrightarrow R{-}OH_2^+$$

Rate-determining step $$R{-}OH_2^+ \longrightarrow R^+ + H_2O$$

$$R^+ + Cl^- \longrightarrow R{-}Cl$$

It is only when the carbocation is reasonably stable that a water molecule can be lost quickly enough for an S_N1 reaction to take place. Otherwise, competing reaction pathways win out. Tertiary carbocations are the most stable alkyl carbocations, so this S_N1 pathway is viable for 2-methyl-2-butanol. Although the chloride anion is not a very strong nucleophile, it is nucleophilic enough so that the carbocation quickly combines with Cl^- to form 2-chloro-2-methylbutane.

Base-Catalyzed Dehydrochlorination of 2-Chloro-2-methylbutane

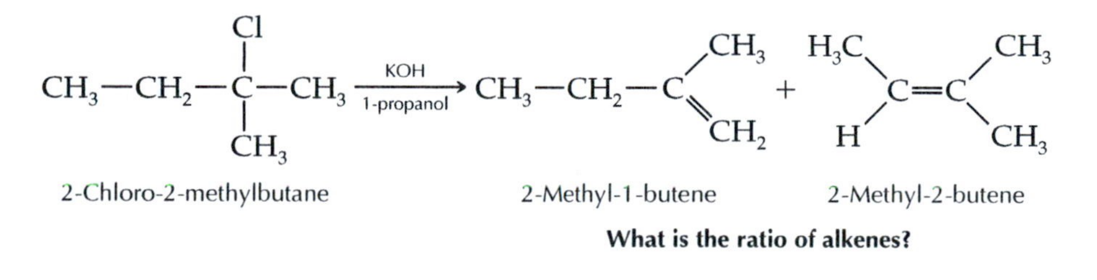

What is the ratio of alkenes?

The reaction of a strong base and a tertiary alkyl halide produces an alkene by an E2 elimination process. A tertiary alkyl halide has too much steric hindrance for an S_N2 substitution reaction to compete effectively.

When the substrate is an alkyl chloride, the loss of HCl is called a dehydrochlorination reaction. It is a 1,2-elimination reaction, in that when the base abstracts a proton, the elimination of chloride occurs from the adjacent carbon atom. The reaction is called E2 because the rate-determining step is bimolecular. The reaction is first order in base and first order in alkyl chloride. The E2 reaction occurs in a single step. As the proton is removed by the base, the C=C double bond forms and Cl^- is simultaneously eliminated. Once an alkene has formed in the presence of OH^-, it is stable and does not isomerize.

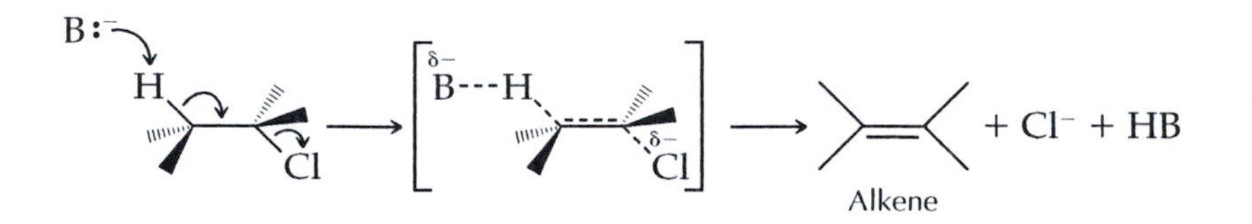

Again, as in the acid-catalyzed dehydration reaction, the question being addressed deals with the ratio of the two possible 1,2-elimination products, 2-methyl-1-butene and 2-methyl-2-butene. As discussed earlier, trisubstituted alkenes such as 2-methyl-2-butene are more stable than disubstituted C=C double bonds. If the E2 transition state has a large amount of double-bond character, the greater stability of the trisubstituted double bond predicts that 2-methyl-2-butene will be the major product. On the other hand, there are six methyl protons and only two methylene protons that could be lost in a 1,2-elimination reaction of 2-chloro-2-methylbutane. If the statistics dominate, 2-methyl-1-butene will form in a greater amount. In addition, the base is also part of the process of abstracting a β-proton in the E2 transition state. There may be steric factors in play. The methyl protons are more out in the open than the methylene protons. Perhaps, even though trisubstituted-alkene products are more stable, they might have a difficult time forming for steric reasons. In any case, the product ratio will reflect the relative rates of the two E2 elimination reactions that lead to 2-methyl-1-butene and 2-methyl-2-butene.

In Project 6.3 you will use gas chromatography to discover whether the ratio of the two alkene isomers obtained in the dehydrohalogenation reaction reflects the product stability. Does the thermodynamically preferred product predominate or are statistical and steric effects the important factors? Your experimental data will tell.

Opportunities for Teamwork

One attractive option for teamwork is for students to work in pairs on this project. During the first week one student carries out the dehydration of 2-methyl-2-butanol (Project 6.1) and obtains a gas chromatogram of the product while the other student synthesizes 2-chloro-2-methylbutane (Project 6.2). During the second week both students do the dehydrohalogenation (Project 6.3) individually, each using half of the alkyl chloride synthesized the previous week. As a team, two students analyze both dehydrohalogenation products by gas chromatography and then compare their results with each other and with the results obtained from the acid-catalyzed dehydration reaction.

6.1 Acid-Catalyzed Dehydration of 2-Methyl-2-butanol

QUESTION: Does product stability or do statistical factors determine the ratio of 2-methyl-2-butene to 2-methyl-1-butene?

PRELABORATORY ASSIGNMENT: Given the information on the pathway for this reaction, draw a likely potential energy diagram (a reaction profile) for the acid-catalyzed dehydration of 2-methyl-2-butanol.

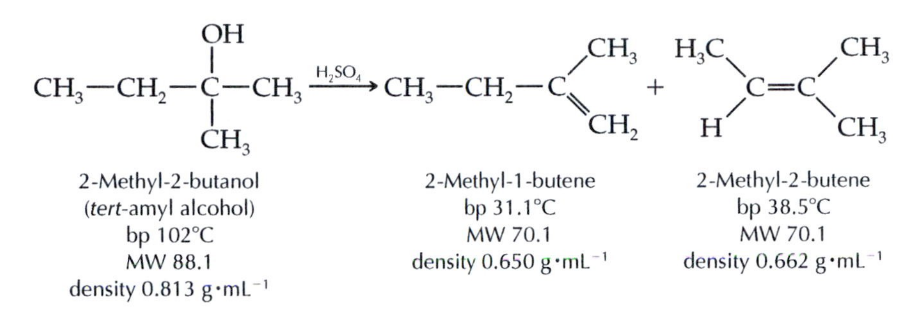

MINISCALE PROCEDURE

Technique Fractional Distillation: Technique 11.4

SAFETY INFORMATION

Wear gloves while dispensing and transferring the reagents.

2-Methyl-2-butanol is flammable and toxic. Avoid contact with skin, eyes, and clothing.

Sulfuric acid solution (6 M) is corrosive and causes burns. Notify the instructor if any acid is spilled.

In this experiment, the product is removed from the reaction mixture by fractional distillation as the reaction proceeds.

Place 3.0 mL (2.4 g) of 2-methyl-2-butanol in a 25-mL round-bottomed flask containing a magnetic stirring bar. Carefully add 15 mL of 6 M sulfuric acid to the material in the flask. Fit an air condenser to the flask and complete the apparatus for fractional distillation as shown in Technique 11.4, Figure 11.15. Use a tared 25-mL round-bottomed flask or screw-capped vial immersed in an ice-water bath as the receiving vessel. Begin stirring the reaction mixture.

Gently heat the reaction mixture in a 85–90°C water bath. When distillation begins, regulate the heating so that the temperature of the vapor reaching the thermometer does not exceed 45°C. The distillation rate should be about 1–2 drops per second. Collect all distillate that boils below 45°C.

GC Analysis of the Product

Analyze your product by gas chromatography using a nonpolar column, such as polydimethylsiloxane, heated no higher than 50°C. If you use a capillary GC, dissolve 2 drops of the product in 0.5 mL of heptane and inject 0.5–1.0 μL of the heptane solution into the chromatograph. The alkenes elute in order of increasing boiling point, with the solvent peak (heptane) occurring *after* the product peaks rather than before, as is usually the case.

Identification of the isomeric butenes can be done by the peak enhancement method, because both products are commercially available [see Technique 16.6]. Submit the remaining product to your instructor.

CLEANUP: Pour the residue remaining in the distilling flask into a 250-mL beaker containing about 20 mL of water. Rinse the flask with a few milliliters of water and add the rinse to the beaker. Add solid sodium carbonate in small portions until the acid is neutralized. **(Caution: Foaming.)** Wash the neutralized solution down the sink or pour it into the container for aqueous inorganic waste. Pour the hexane solution remaining from the GC analysis into the container for flammable waste.

6.2 Synthesis of 2-Chloro-2-methylbutane

PURPOSE: To synthesize 2-chloro-2-methylbutane for use in a dehydrochlorination reaction.

PRELABORATORY ASSIGNMENT: Explain the advantage of using a short-path distillation apparatus in the purification of 2-chloro-2-methylbutane.

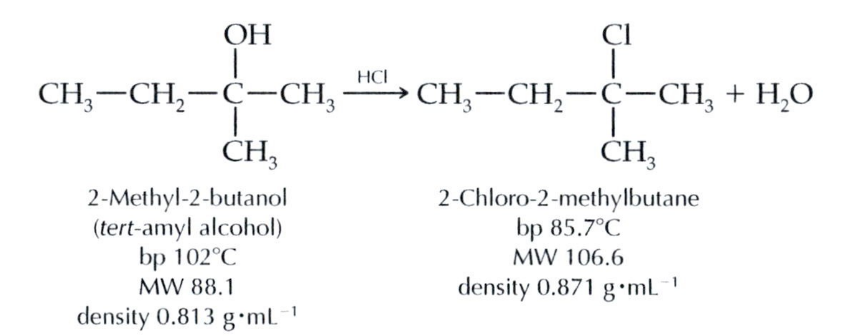

MINISCALE PROCEDURE

Techniques Extraction: Technique 8.2
Short-Path Distillation: Technique 11.3a

SAFETY INFORMATION

Wear gloves while conducting this experiment and carry it out in a hood.

2-Methyl-2-butanol is flammable and toxic. Avoid contact with skin, eyes, and clothing.

Concentrated hydrochloric acid (12 M) is corrosive and causes burns. The vapor is extremely irritating to mucous membranes and the upper respiratory tract. Wear gloves and measure it in a hood; avoid contact with skin, eyes, and clothing.

The product, **2-chloro-2-methylbutane,** is flammable and an irritant. Avoid contact with skin, eyes, and clothing.

Place 10 mL of 2-methyl-2-butanol and 25 mL of concentrated hydrochloric acid (12 M) in a 60- or 125-mL separatory funnel with its stopcock closed. Use a stirring rod to mix the two solutions. Then swirl the contents of the separatory funnel gently without the stopper in the funnel for 1 min. Place the stopper in the funnel, carefully invert it, and immediately open the stopcock to release any excess pressure. Shake the funnel for 4–5 min, periodically stopping to vent the funnel by inverting it and opening the stopcock. **(Caution: Point the separatory funnel toward the back of the hood while venting it.)** Allow the contents of the funnel to stand until the mixture has separated into two distinct layers.

Carefully drain the lower aqueous layer into a labeled 250-mL Erlenmeyer flask. Wash the upper organic layer with 20 mL of saturated sodium chloride solution. Separate the lower aqueous phase. Then wash the organic phase with 20 mL of saturated sodium bicarbonate solution. **(Caution: Vigorous evolution of CO_2 gas usually occurs.)** Stir the contents of the separatory funnel with a stirring rod until CO_2 evolution subsides. Stopper the funnel, invert it, and immediately open the stopcock. Shake and vent the funnel several times to ensure thorough mixing. Separate the aqueous phase, then wash the organic phase with 10 mL of water and then with 10 mL of saturated sodium chloride solution, separating the aqueous phase after each wash.

Transfer the organic phase to a clean, dry 25-mL Erlenmeyer flask and dry the crude product with anhydrous calcium chloride for at least 10 min. Transfer the dried product to a 25-mL round-bottomed flask with a Pasteur pipet. Assemble the apparatus for a short-path distillation [see Technique 11.3a]. Collect the product fraction boiling from 81–86°C in a tared (weighed) 25-mL round-bottomed flask or a tared (weighed) screw-capped vial immersed in an ice-water bath. Determine the percent yield of the product. Store the product for use in Project 6.3.

CLEANUP: Combine all the aqueous solutions remaining from the extractions in the Erlenmeyer flask, and add solid sodium carbonate until the solution is neutralized. **(Caution: Foaming.)** Wash the neutralized solution down the sink or pour it into the container for inorganic waste. Pour the residue remaining in the boiling flask into the container for halogenated waste.

6.3 Base-Catalyzed Dehydrochlorination of 2-Chloro-2-methylbutane

QUESTION: Does product stability or do statistical and steric factors determine the ratio of 2-methyl-2-butene to 2-methyl-1-butene?

PRELABORATORY ASSIGNMENT: Given the information on the pathway for this reaction, draw a likely potential energy diagram (a reaction profile) for the base-catalyzed dehydrochlorination of 2-chloro-2-methylbutane.

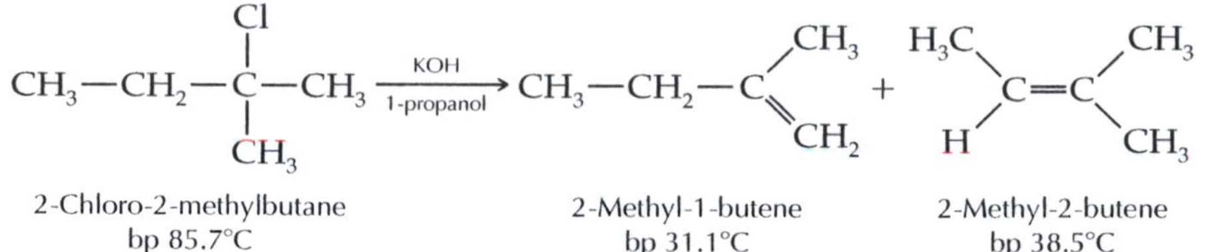

MINISCALE PROCEDURE*

Technique Fractional Distillation: Technique 7.4

SAFETY INFORMATION

2-Chloro-2-methylbutane and **1-propanol** are flammable and irritants. Avoid contact with skin, eyes and clothing.

Potassium hydroxide is corrosive and causes burns. Avoid contact with skin, eyes, and clothing. Solid potassium hydroxide takes up water from the air and after a time leaves a colorless concentrated KOH solution, which could cause burns to anyone who touches it unknowingly.

The quantity of KOH used in the procedure has been corrected for the fact that commercial potassium hydroxide contains approximately 15% by weight of water.

Place 3.0 g of potassium hydroxide and 23 mL of 1-propanol in a 50-mL round-bottomed flask containing a magnetic stirring bar. Warm and stir the mixture until the potassium hydroxide has dissolved. Cool the flask to room temperature or below in an ice-water bath.

Slowly pour 3.0 mL of 2-chloro-2-methylbutane into the flask containing the KOH/1-propanol solution. Assemble a fractional distillation apparatus using a Vigreux column or a jacketed-condenser as the fractionating column [see Technique 11.4, Figure 11.15]. Use a 25-mL tared, round-bottomed flask immersed in an ice-water bath as the receiving flask. Heat and stir the contents of the reaction flask for 1 h in a 75–80°C water bath; monitor the temperature closely so that it does not exceed 80°C during the reflux period. Replace the ice around the receiving flask as it melts. During the heating period a white solid precipitates from solution (see Question 2).

At the end of 1 hr, increase the temperature of the water bath to 90–95°C and distill the product mixture. Collect all distillate boiling below 45°C.

Analyze your product by gas chromatography as directed in Project 6.1.

*Experimental procedure corrected in second printing.

CLEANUP: Pour the solution remaining in the distillation flask into a 100-mL beaker and add 6 M hydrochloric acid dropwise until pH paper indicates a pH of 6–8. Collect the precipitated KCl by vacuum filtration. Pour the filtrate into the container for halogenated waste and place the KCl in the container for solid inorganic waste. Pour the solution remaining from the GC analysis into the container for flammable waste.

Interpretation of Experimental Data

Analyze the gas chromatograms of the products from both the dehydration and dehydrochlorination reactions. How many products are indicated by gas chromatography for each reaction? What are the relative amounts of butenes produced in each reaction? Compare the ratios of isomeric butenes obtained in the two reactions. Explain your results.

Questions

1. In Project 6.2, why is the product washed with sodium bicarbonate solution?
2. Why is saturated NaCl solution, rather than water, used to wash the organic layer in Project 6.2?
3. What is the white solid that forms in Project 6.3 while the dehydrochlorination reaction is undergoing reflux?
4. What is the difference in reaction conditions in Projects 6.1 and 6.2 that produces quite different products?
5. Diethyl ether is a commonly used solvent for GC analyses because of its low boiling point. Why was heptane used in this project instead of ether?
6. Acid-catalyzed dehydration of an alcohol is an equilibrium situation. How was the reaction forced to completion in Project 6.1?
7. Why do you suspect that 1-propanol is used as the reaction solvent in Project 6.3, rather than the more common ethanol?

PROJECT

7

SYNTHESIS AND HYDROBORATION-OXIDATION OF 1-PHENYLCYCLOHEXENE

QUESTION: What is the stereochemistry of water addition in the hydroboration-oxidation of 1-phenylcyclohexene?

In this three-week project you will synthesize 1-phenylcyclohexanol by a Grignard reaction and convert it to 1-phenylcyclohexene by acid-catalyzed dehydration. In the third reaction you will carry out a hydroboration-oxidation sequence to add the elements of water to the alkene, using NMR spectroscopy or gas chromatography to determine which isomer of 2-phenylcyclohexanol forms.

Grignard Syntheses

Among the most important organometallic reagents are the alkyl- and arylmagnesium halides, which are almost universally called ***Grignard reagents*** after the French chemist Victor Grignard, who first realized their tremendous potential in organic synthesis. Grignard received the 1912 Nobel prize in chemistry. Grignard reagents readily attack the C=O groups of aldehydes and ketones and allow the synthesis of highly specific carbon-carbon bonds in excellent yields.

Grignard reagents contain carbon-metal bonds. Their synthesis requires the reaction of an alkyl or aryl halide with magnesium metal in the presence of an ether solvent:

$$\mathrm{R{-}X + Mg \xrightarrow{\text{ether}} \overset{\delta-}{R}{-}\overset{\delta+}{Mg}{-}\overset{\delta-}{X}}$$

Formation of a Grignard reagent takes place in a heterogeneous reaction at the surface of the solid magnesium metal, and the surface area and reactivity of the magnesium are crucial factors in the rate of the reaction. It is thought that the alkyl halide reacts with the surface of the metal to produce a carbon free radical and a magnesium-halogen radical, which combine to form the Grignard reagent. This gives the magnesium atom two covalent bonds, but it must somehow acquire two more since magnesium has a coordination number of 4. In an ether solvent, ether molecules occupy the other two coordination sites. The oxygen atom of the ether molecule provides both electrons in the magnesium-ether bond:

H_5C_2 C_2H_5
Ö
R—Mg—X
Ö
H_5C_2 C_2H_5

These complexes are quite soluble in ether. In the absence of an ether solvent, the reaction of magnesium and the alkyl halide takes place rapidly but soon stops because the surface of the metal becomes coated with polymeric Grignard reagent. In the presence of ether, the surface of the metal is kept clean and the reaction proceeds until the limiting reagent is entirely consumed.

Treatment of an ether solution of bromobenzene with magnesium metal results in the formation of the Grignard reagent phenylmagnesium bromide. Insertion of the magnesium between carbon and bromine reverses the polarity of the carbon formerly bonded to bromine. In the Grignard reagent, the carbon-metal bonds have a high degree of ionic character with a good deal of negative charge on the carbon atom, making it a strong nucleophile:

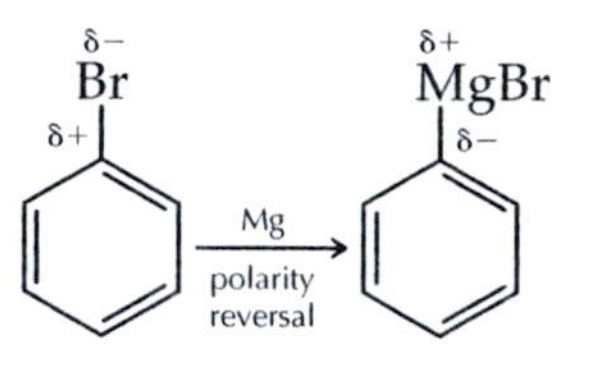

The mechanism of the Grignard reaction with carbonyl compounds is actually quite complex, but it can easily be rationalized as a simple nucleophilic addition reaction:

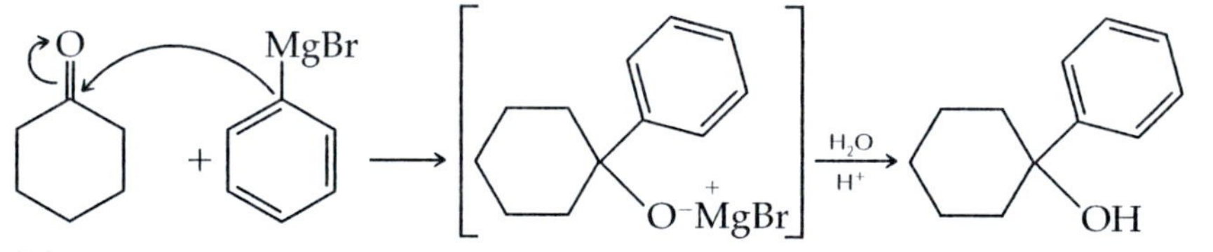

The Grignard reaction produces a magnesium salt of the organic product. These magnesium salts are readily hydrolyzed to the organic product by reaction with an acid—commonly an aqueous mineral acid, such as sulfuric or hydrochloric acid. Not only does a mineral acid cause the protonation to occur more readily, but Mg(II) is converted from insoluble salts to water-soluble sulfates or chlorides. For products that might react with strong acids, such as the tertiary alcohol prepared in this project, the weaker acid ammonium chloride is an excellent alternative. Strong acids may cause tertiary alcohols to dehydrate.

Unwanted Side Reactions in Grignard Syntheses

Because Grignard reagents have such tremendous reactivity as nucleophiles, it is not surprising that they are also strong bases. They react quickly and easily with acids, even relatively weak acids such as water and alcohols. This reaction can be a great nuisance because it destroys the Grignard reagent before the carbonyl compound is added and therefore reduces the yield of the desired product. The presence of water or other acids also inhibits the formation of the Grignard reagent. Thus, it is important that the reaction conditions be completely anhydrous. All glassware and reagents must be thoroughly dry before beginning a Grignard experiment.

Other reactions that may interfere with Grignard syntheses include Grignard coupling and reaction with molecular oxygen:

$$2\,RX + Mg \longrightarrow R{-}R + MgX_2$$
$$RMgX + O_2 \longrightarrow ROOMgX \xrightarrow{RMgX} 2\,ROMgX$$

The coupling reaction takes place at the surface of the solid magnesium when two free radicals react to form a stable hydrocarbon dimer. This side reaction is favored by a high concentration of an alkyl or aryl halide, where the concentration of free radicals is greater. Combination of the Grignard reagent with molecular oxygen is a side reaction when a synthesis is run in the presence of air. Low-boiling diethyl ether (bp 35°C) helps to exclude air near the surface of the reaction solution because its great volatility keeps a layer of ether vapor above the reaction mixture. Alternatively, Grignard reagents can be used under nitrogen or argon to avoid reactions with molecular oxygen.

Acid-Catalyzed Dehydration

When 1-phenylcyclohexanol is heated in the presence of sulfuric acid, the product is 1-phenylcyclohexene. The 1,2-elimination of a water molecule from the alcohol is called a dehydration reaction:

$$-\underset{H}{\overset{|}{\underset{|}{C}}}-\underset{OH}{\overset{|}{\underset{|}{C}}}- \xrightarrow[\text{heat}]{H^+} \;\gt C{=}C\lt\; + H_2O$$

Elimination reactions require a leaving group that departs with its bonding electrons. The other group, which is lost from the adjacent carbon atom, is a proton. The acid is essential because it converts R—OH into R—OH_2^+, thereby providing a reasonably good leaving group. When tertiary alcohols are used, protonation and subsequent loss of water result in a carbocation:

$$H-\overset{|}{\underset{|}{C}}-\underset{OH}{\overset{|}{\underset{|}{C}}}- + H_2SO_4 \rightleftharpoons HSO_4^- + H-\overset{|}{\underset{|}{C}}-\underset{{}^{+}O-H,\ H}{\overset{|}{\underset{|}{C}}}- \rightleftharpoons H-\overset{|}{\underset{|}{C}}-C^{+}\lt \; + H_2O$$

Carbocation

The rate of formation of a carbocation depends heavily on the stability of the carbocation being formed. Carbocation stability increases markedly with an increase in the number of alkyl or aryl substituents attached to the carbon atom bearing the positive charge. These substituents allow the charge to be dispersed throughout a larger space and thereby stabilize it. The 1-phenylcyclohexyl cation is quite a stable carbocation, and it forms easily.

In all but the most highly acidic environments, even tertiary carbocations are too unstable to persist. They react with nucleophiles to give S_N1 substitution products or lose a proton to give E1 elimination products. The rate-determining step is the loss of a water molecule from the protonated alcohol. Sulfuric acid is the chosen catalyst because its conjugate base is a very poor nucleophile and does not lead to the formation of S_N1 products. Carbocations are very strong acids, and the loss of a proton to form an alkene is fast. In this dehydration reaction, the product is the conjugated alkene.

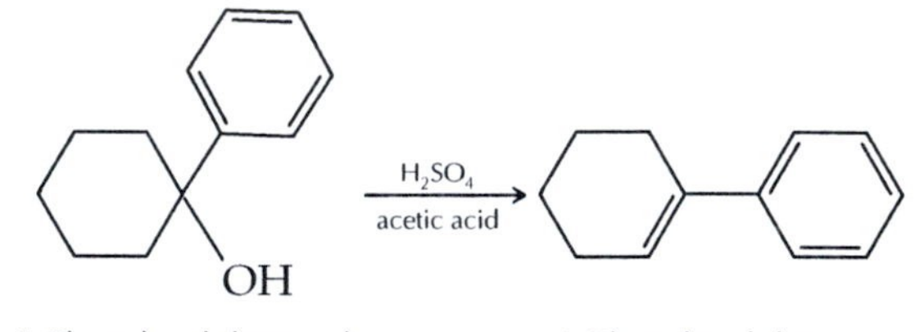

The entire pathway for the E1 elimination of water from an alcohol in the presence of strong acid is

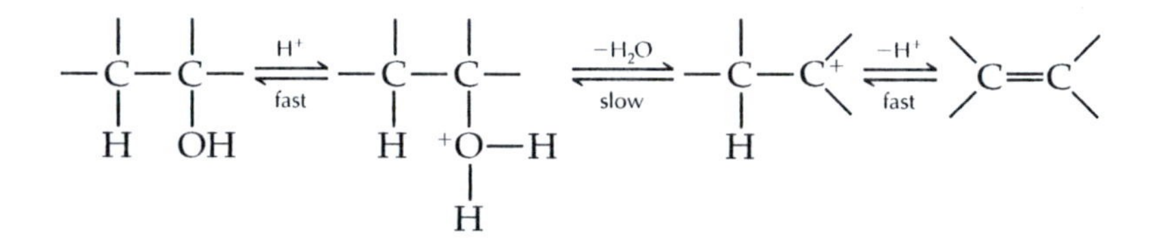

Hydroboration-Oxidation

The basic question in this project concerns the stereochemistry of the net addition of water in the hydroboration-oxidation of 1-phenylcyclohexene.

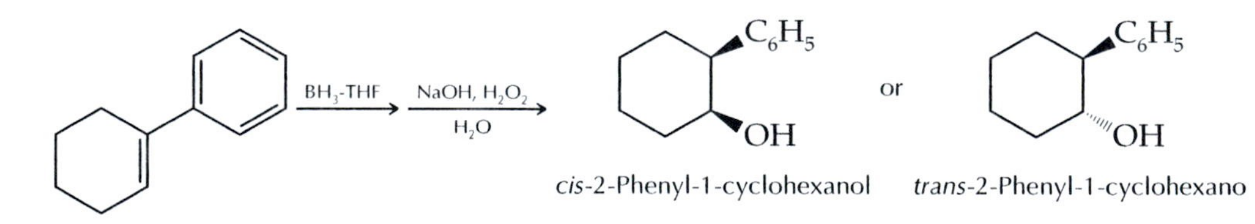

Borane, BH_3, would have only a sextet of valence electrons and is not a stable compound. In hydroboration reactions, its dimer B_2H_6 can be used, but more often the commercially available Lewis acid-base complex of BH_3 with tetrahydrofuran (THF) is used. Because borane reacts rapidly with water, it is common to handle the borane-THF solution using syringe techniques.

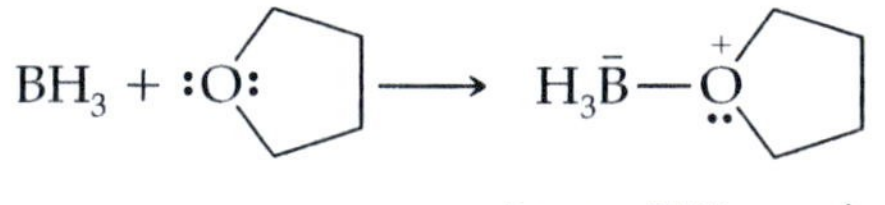

Borane-THF complex

The reaction of borane with alkenes was discovered a little over 50 years ago by the American chemist Herbert Brown. It is one of the most important reactions in the repertoire of synthetic organic chemists. Brown was awarded a Nobel prize in 1979 for discovering the potential of boron compounds in organic synthesis. The boron-hydrogen bond adds rapidly and quantitatively to carbon-carbon double bonds. With simple alkenes a trialkylborane is formed:

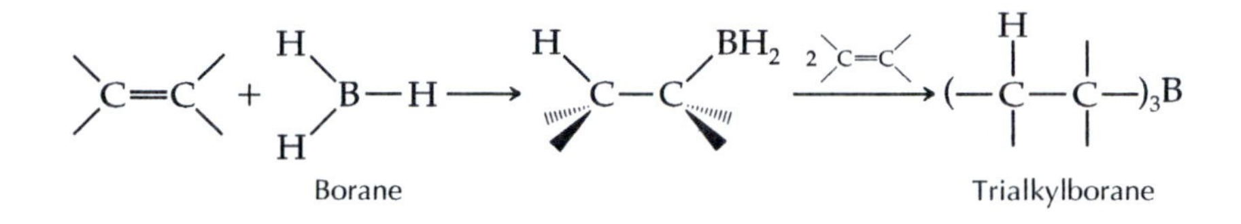

A great advantage of this reaction is that the carbon-boron bond can be oxidized to a carbon-oxygen bond in a straightforward manner with an alkaline solution of hydrogen peroxide:

$$3\,\mathrm{RCH{=}CHR} \xrightarrow{\mathrm{BH_3,\ THF}} (\mathrm{RCH_2{-}C(R)(H){-}})_3\mathrm{B} \xrightarrow[\mathrm{H_2O}]{\mathrm{H_2O_2,\ NaOH}} 3\,\mathrm{RCH_2{-}C(R)(H){-}OH}$$

This oxidation produces an alcohol in a stereospecific reaction. Alcohols, of course, are versatile intermediates in organic synthesis. The mechanism of the alkylborane oxidation involves nucleophilic attack of the hydroperoxy anion at boron, followed by migration of an alkyl group from boron to oxygen and loss of hydroxide:

$$-\mathrm{B}(\mathrm{R}) + {}^{-}{:}\ddot{\mathrm{O}}{-}\ddot{\mathrm{O}}\mathrm{H} \longrightarrow -\mathrm{B}^{-}(\mathrm{R}){-}\ddot{\mathrm{O}}{-}\ddot{\mathrm{O}}\mathrm{H} \xrightarrow{-\mathrm{OH}^{-}} -\mathrm{B}{-}\ddot{\mathrm{O}}{-}\mathrm{R}$$

This process is repeated until all three alkyl groups have migrated to oxygen atoms. This forms a trialkyl borate $(RO)_3B$, which subsequently hydrolyzes to three molecules of the alcohol and sodium borate:

$$(\mathrm{RO})_3\mathrm{B} + 3\,\mathrm{NaOH} \xrightarrow{\mathrm{H_2O}} \mathrm{Na_3BO_3} + 3\,\mathrm{ROH}$$

The regiochemistry of the initial addition of the B—H bond places the hydrogen atom at C1 and the boron atom at C2 of 1-phenylcyclohexene, so 2-phenylcyclohexanol is the product of the oxidation of the borane addition product. 2-Phenylcyclohexanol has cis and trans isomers, one produced by an *anti* 1,2-addition process and the other produced by *syn* 1,2-addition.

Conformational Isomers of Cyclohexanes

The most stable conformational isomer of cyclohexane is a structure in the form of a chair. In the chair form, all the carbon-carbon bond angles are 109°, the stable tetrahedral angle.

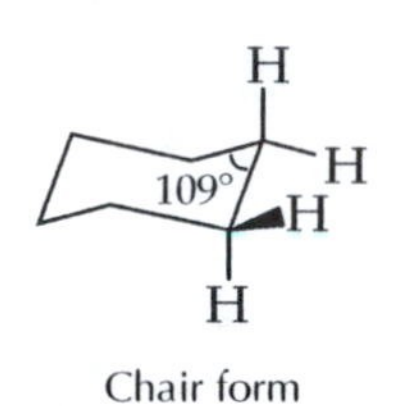

Chair form

Under ordinary conditions, a single chair form has a finite but short existence before it undergoes a ring-flipping process to another conformational isomer of equal stability:

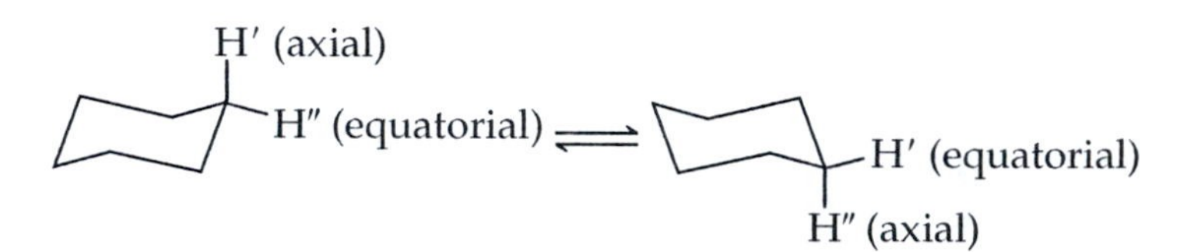

This ring flipping in cyclohexane interconverts axial and equatorial hydrogens.

A large bulky group on a cyclohexane ring has a much greater tendency to occupy an equatorial position on the ring. Steric interference between the bulky group and other atoms in the molecule, particularly 1,2-gauche interactions and 1,3-diaxial interactions, are smaller when the phenyl substituent occupies the equatorial position, and that conformer is strongly favored in the conformational equilibrium. The hydroxyl group occupies the axial position. In the more stable *trans*-2-phenylcyclohexanol both of the large groups can and do occupy equatorial positions, but in the cis isomer one of the large groups must occupy an axial position. The smaller of the two large groups, the hydroxyl, is in the axial position in the major conformer of *cis*-2-phenylcyclohexanol.

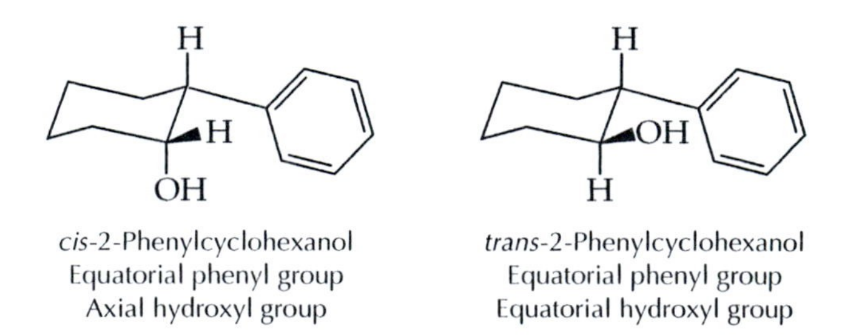

cis-2-Phenylcyclohexanol
Equatorial phenyl group
Axial hydroxyl group

trans-2-Phenylcyclohexanol
Equatorial phenyl group
Equatorial hydroxyl group

Although the overall ^1H NMR spectrum of your hydroboration-oxidation product will be complex, you can use NMR spectroscopy in a reasonably straightforward manner to determine its structure. The methine proton on the carbon atom bearing the hydroxyl group will be shifted substantially downfield, although not nearly so far as the aromatic protons. The methine proton on the carbon bearing the phenyl group will also be deshielded, but the phenyl group will not have as great a deshielding effect as the hydroxyl group. One additional point must be considered in analyzing the NMR spectrum, namely, the chemical shifts of equatorial and axial protons attached to cyclohexane rings. Axial protons are shielded through a diamagnetic anisotropic effect, and they are approximately 0.4 ppm upfield from similar equatorial protons [see Anisotropy in Technique 19.7]. For *trans*-2-phenylcyclohexanol the chemical shift of the methine proton at C1 appears at 3.64 ppm and for the cis isomer it is at 3.98 ppm in CDl_3.

7.1 Grignard Synthesis of 1-Phenylcyclohexanol

PURPOSE: To synthesize 1-phenylcyclohexanol for use in a subsequent dehydration reaction.

PRELABORATORY ASSIGNMENT: Anhydrous conditions are vital for a successful Grignard synthesis. To demonstrate this point, calculate the mass of water it would take to completely destroy the phenylmagnesium bromide that you will synthesize. What is the volume of this amount of water?

—Br + Mg $\xrightarrow{\text{dry ether}}$

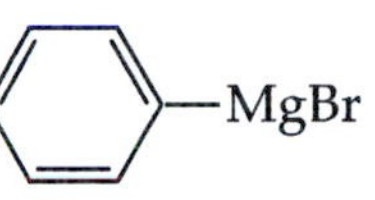

Bromobenzene
bp 156°C
MW 157.0
density 1.50 g/mL

Phenylmagnesium bromide
(intermediate to be used immediately)

+ O= $\xrightarrow{\text{dry ether}}$ OBrMg $\xrightarrow{\text{aq. } NH_4Cl}$ OH

Phenylmagnesium bromide

Cyclohexanone
bp 155°C
MW 108.2
density 0.947 g • mL^{-1}

1-Phenylcyclohexanol
mp 60–61°C
MW 176.3

MINISCALE PROCEDURE

Techniques Anhydrous Reaction Conditions: Technique 7.2
Removal of Solvents: Technique 8.9

SAFETY INFORMATION

Diethyl ether, commonly called ether, is extremely flammable. Be certain that no flames are used in the laboratory and that hot electrical devices are not in the vicinity where ether is being used. Work in a hood, if possible.

Bromobenzene is an irritant. Wear gloves. Avoid contact with skin, eyes, and clothing.

Cyclohexanone is corrosive and toxic. Wear gloves. Avoid contact with skin, eyes, and clothing.

Step 1. Grignard Synthesis of Phenylmagnesium Bromide

Make sure that all your glassware and reagents are thoroughly dry before beginning the experiment. Refer to p. 296 concerning the necessity for anhydrous conditions during a Grignard reaction.

Assemble the apparatus shown in Technique 7.2, Figure 7.3a or 7.3b, using a 100-mL round-bottomed flask, a water-jacketed condenser, and a 125-mL separatory funnel. For a one-necked flask, use a Claisen adapter to make two openings; for a three-necked flask, close the third opening with a glass stopper. Prepare two drying tubes containing anhydrous calcium chloride. Fit them to the top of the condenser and the separatory funnel with thermometer adapters or one-holed rubber stoppers.

Obtain 1.00 g of magnesium turnings and 8 mL of anhydrous ether. Open the round-bottomed flask and add the magnesium and the ether to the flask. With a dry stirring rod, gently break one or two magnesium pieces to expose a fresh surface. Place a magnetic stirring bar in the round-bottomed flask. Close the flask.

SAFETY PRECAUTION

Hold the round-bottomed flask firmly by the neck and be careful not to press so hard on a magnesium turning that you crack the flask.

Prepare a solution of bromobenzene in ether by dissolving 4.0 mL of bromobenzene in 10 mL of anhydrous ether, and pour it into the separatory (dropping) funnel with the stopcock closed. Add approximately one-third of this solution to the reaction flask and swirl the flask to mix the reagents thoroughly.

Place the flask in a warm-water bath (45–50°C) and swirl the flask from time to time. The ether will begin to boil. A small amount of cloudy precipitate may appear. Formation of the organometallic reagent is exothermic. Periodically remove the water bath and note whether the reaction has started. This process is indicated by refluxing of the ether, which results from the heat produced by the reaction.

The precipitate is probably magnesium hydroxide, which results from any moisture in the flask or the reagents.

When the reaction is well under way, remove the warm-water bath, add 10 mL of anhydrous ether through the condenser to dilute the solution, and begin stirring the reaction mixture. Add the rest of the bromobenzene/ether solution from the dropping funnel at a speed that maintains a moderate reflux rate. Have a cold-water bath available to cool the reaction *briefly* in case the condensation ring in the condenser is more than halfway up the condenser. Do not cool the reaction so much that it stops. When all the bromobenzene has been added, pour 3.0 mL of ether into the dropping funnel and add the ether to the reaction mixture. When the reaction stops refluxing spontaneously, heat the reaction in the warm-water bath (45–50°C) for 20 min. Cool the reaction flask in a cold-water bath before proceeding immediately to Step 2.

Step 2. Addition to Cyclohexanone

Remove the heating bath from the Grignard reagent mixture and cool the reaction flask in an ice-water bath for 5–10 min. Prepare a solution of 4.1 g of cyclohexanone in 20 mL of anhydrous ether. Place this solution in the dropping funnel and add the solution dropwise to the stirred Grignard reagent at such a rate that the reaction mixture gently refluxes. After all the cyclohexanone solution has been added, gently reflux the reaction mixture for another 15 min.

Cool the reaction mixture to room temperature, then pour 50 mL of 10% ammonium chloride solution into the flask while stirring the mixture. Stir the mixture until the magnesium salts have dissolved and the two phases are clear (but not necessarily colorless). Transfer the mixture to a separatory funnel. If unreacted bits of magnesium are present, filter the mixture through a fluted filter paper into the separatory funnel. Rinse the flask with a few milliliters of water, then with a few milliliters of ether; add the rinses to the separatory funnel.

Separate the phases and save the organic layer in a labeled flask. Return the aqueous phase to the separatory funnel and extract it twice with 20-mL portions of diethyl ether. Combine the second ether extract with the original ether layer. After the third extraction, remove the aqueous phase and pour the combined ether solution into the separatory funnel, which still contains the third ether extract. Wash the combined ether solution first with 30 mL of saturated sodium bicarbonate solution and then with 30 mL of saturated sodium chloride solution, separating the phases after each wash. Pour the ether solution into a clean, dry flask and dry the solution with anhydrous magnesium sulfate.

Filter the dried solution into a dry 125-mL Erlenmeyer flask. Remove the solvent with a stream of nitrogen in a hood or with a rotary evaporator. When the volume is reduced to about 10 mL, transfer the solution to a tared (weighed), dry 25-mL Erlenmeyer flask and complete the removal of ether. The final product may be a mixture of crystals and oil.

CLEANUP: Wash the aqueous solutions remaining from the extractions down the sink or pour them into the container for aqueous inorganic waste. Put the spent magnesium sulfate drying agent into the container for solid inorganic waste.

Product Analysis

Prepare a sample for NMR analysis in $CDCl_3$ as directed by your instructor [see Technique 19.2]. You can also assess the purity of your product by GC analysis on a 5% phenylmethyl silicone column, such as AT-5, OV-5, or SE-52, heated to 160°C. If your product crystallizes, determine its melting point.

7.2 Acid-Catalyzed Dehydration of 1-Phenylcyclohexanol

PURPOSE: To prepare 1-phenylcyclohexene for study of the stereochemistry of a hydroboration-oxidation reaction.

PRELABORATORY ASSIGNMENT: Describe the peaks and approximate chemical shifts you would expect to find in the NMR spectrum of 1-phenylcyclohexene.

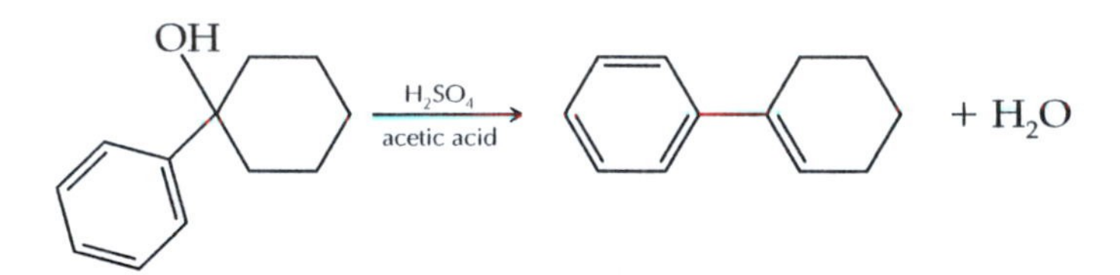

1-Phenylcyclohexanol
mp 60–61°C
MW 176.3

1-Phenylcyclohexene
bp 253°C (133°C/17 mm Hg)
MW 158.2

MINISCALE PROCEDURE

Techniques Extraction: Technique 8.2
IR Spectroscopy: Technique 18
NMR Spectroscopy: Technique 19

SAFETY INFORMATION

A solution of 20% (v/v) **sulfuric acid in acetic acid** is corrosive and causes burns. Wear gloves. Dispense it in a hood and avoid contact with skin, eyes, and clothing.

Diethyl ether is extremely volatile and flammable.

Reagent Quantities

Determine the mass of 1-phenylcyclohexanol in your remaining product from Project 7.1. Use up to 2.5 g of your alcohol as the starting material for Project 7.2. Proportionally adjust the amount of sulfuric acid/acetic acid solution for your amount of alcohol at a ratio of 2.3 mL of 20% (v/v) sulfuric acid in acetic acid per 1.0 g of 1-phenylcyclohexanol. Submit any additional 1-phenylcyclohexanol to your instructor.

Synthesis

Use a 25-mL Erlenmeyer flask as the reaction vessel. Add your 1-phenylcyclohexanol and the calculated volume of 20% sulfuric acid in acetic acid solution. Swirl the mixture for approximately 1 min and heat it on a steam bath in a hood for 5 min. Cool the reaction mixture to 10–15°C in an ice-water bath for 5 min. Add 10 mL of water to the flask and swirl the flask to mix the contents.

Carry out the following extractions in a hood. Obtain 25 mL of diethyl ether for your extractions. Pour the diluted reaction mixture into a separatory funnel. Rinse the flask with approximately 15 mL of water and add the rinse to the separatory funnel. Then rinse the flask twice, using approximately half of the ether for each rinse. Add the ether rinses to the separatory funnel. Stopper the funnel and mix the two phases, being sure to vent frequently [see Technique 8.2]. Separate the phases and wash the ether layer with 20 mL of water. Repeat the washing process with 15 mL of saturated sodium bicarbonate solution. **(Caution: Foaming.)** Transfer the ether solution to a clean, dry Erlenmeyer flask and dry the solution with anhydrous magnesium sulfate for 10–15 min.

Filter approximately 10 mL of the dried product solution into a tared 25-mL Erlenmeyer flask. Evaporate the ether using a stream of nitrogen or air. Repeat the filtration and evaporation process until all your product is in the tared flask. Weigh your product and determine the percent yield. Store your product for use in Project 7.3.

CLEANUP: Combine the aqueous solutions from the extractions and neutralize the combined solution with solid sodium carbonate before washing the solution down the sink or pouring it into the container for aqueous inorganic waste.

Product Analysis

Prepare a sample for NMR analysis in $CDCl_3$ as directed by your instructor [see Technique 19.2]. If the spectrum matches the expected spectrum of 1-phenylcyclohexene, it can be used for the subsequent hydroboration reaction without further purification. You can also obtain an IR spectrum of your product to assess whether the reaction has gone to completion. You can assess the purity of your product by GC analysis on a nonpolar methyl silicone column or a 5% phenylmethyl silicone column heated to 160°C.

7.3 Hydroboration-Oxidation of 1-Phenylcyclohexene

QUESTION: What is the stereochemistry of the net addition of water to 1-phenylcyclohexene?

PRELABORATORY ASSIGNMENT: Show the axial and equatorial substituents on carbons 1 and 2 of *cis*- and *trans*-2-phenylcyclohexanol and predict how the NMR spectra of *cis*-2-phenylcyclohexanol and *trans*-2-phenylcyclohexanol would differ.

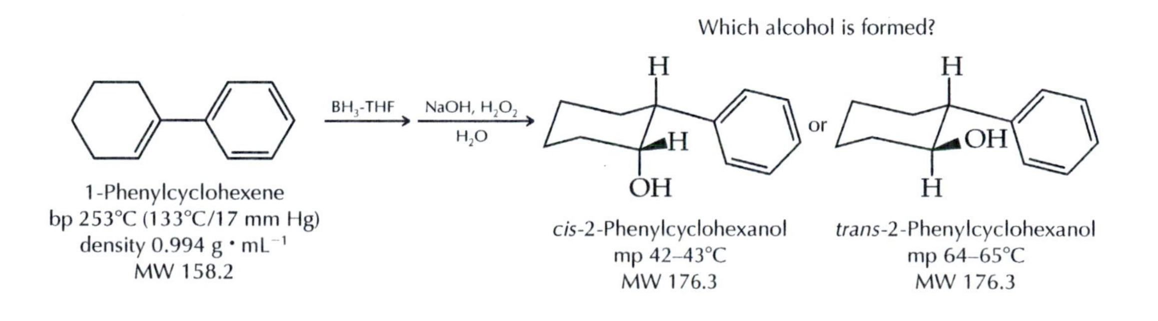

MICROSCALE PROCEDURE

Technique Microscale Extraction: Technique 8.6b

SAFETY INFORMATION

1-Phenylcyclohexene poses no particular hazards.

Borane-tetrahydrofuran complex is flammable and moisture sensitive.

Hydrogen peroxide solution (30%) is a strong oxidizing agent and causes serious burns. Wear gloves while handling it.

Aqueous **sodium hydroxide solution (3 M)** is corrosive and causes burns, and it can cause severe eye injury. Avoid contact with skin, eyes, and clothing.

Dichloromethane is toxic, an irritant, absorbed through the skin, and harmful if inhaled. Use it only in a hood and wear neoprene gloves while doing the extractions and evaporation.

Petroleum ether (pentanes) is extremely flammable. Be certain that no flames are used in the laboratory and that hot electrical devices are not in the vicinity where petroleum ether is being used. Work in a hood, if possible.

Hydroboration

Place 0.30 mL of the 1-phenylcyclohexene you prepared in Project 7.2 and a magnetic stirring bar in a 10-mL round-bottomed flask. Fit a water-jacketed condenser to the flask and clamp the apparatus above a magnetic stirrer. Close the top of the condenser with a screw cap and flat septum or with a fold-over rubber septum; insert a syringe needle through the septum to serve as a vent. Turn on the magnetic stirrer.

Obtain 2.0 mL of 1.0 M borane-tetrahydrofuran solution in a clean dry syringe. Insert the syringe into the septum and add the borane solution dropwise over a period of several minutes. After the addition is complete, stir the reaction mixture at room temperature for 1 h. Remove the septum and cap from the top of the condenser.

Oxidation

Add 1.0 mL of 3 M sodium hydroxide solution dropwise. Initially a vigorous reaction will occur with each drop; wait for the bubbling to subside before making a subsequent addition. Then add 1.0 mL of 30% hydrogen peroxide solution. Heat the reaction mixture at reflux for 45 min using a 75–80°C water bath.

It may be necessary to spin the centrifuge tube in a centrifuge to achieve a clean separation of the two phases.

At the end of the reflux period, transfer the reaction mixture to a centrifuge tube using a Pasteur pipet. After the mixture cools to room temperature, extract it with 3 mL of dichloromethane [see Technique 8.6b]. Allow the phases to separate. Remove the lower organic phase with a Pasteur pipet fitted with a syringe or a Pasteur filter pipet and set it aside in a second centrifuge tube. Repeat the extraction of the aqueous phase remaining in the first centrifuge tube two more times with 1-mL portions of dichloromethane. Combine the dichloromethane extracts in the second centrifuge tube. Wash the combined dichloromethane extracts with 2 mL of water. Transfer the organic phase to a clean, dry 10-mL Erlenmeyer flask and dry the solution with anhydrous potassium carbonate for at least 10 min.

Transfer the dried product solution to a tared, clean, dry 10-mL Erlenmeyer flask with a clean Pasteur filter pipet. Working in a hood, evaporate the solvent with a stream of air or nitrogen, or remove the solvent with a rotary evaporator. The crude product crystallizes as a white solid. If your product is an oil or a mixture of liquid and crystals, cooling the flask in a 2-propanol/dry ice bath should induce crystallization. Recrystallize the crude product from pentane or petroleum ether using a 50–60°C water bath as the heat source. If the product does not crystallize on cooling, further cooling in a 2-propanol/dry ice bath may be necessary. Collect the product on a Hirsch funnel by vacuum filtration. Use a few drops of chilled pentane or petroleum ether to aid in transferring the crystals and rinsing them. Allow the recrystallized product to air dry for at least 30 min before determining its melting point. Prepare a sample for NMR analysis as directed by your instructor using $CDCl_3$ as the solvent [see Technique 19.2].

CLEANUP: Neutralize the aqueous solutions from the extractions with 3 M hydrochloric acid until the pH is 6–8. Wash the neutralized solution down the sink or pour it into the container for aqueous waste. Place the spent drying agent in the container for hazardous solid waste.

Interpretation of Experimental Data

Analyze the NMR spectrum of your product to determine whether *cis*- or *trans*-2-phenylcyclohexanol was formed. Use your NMR analysis to discover the stereochemistry in the hydroboration-oxidation of 1-phenylcyclohexene.

7.4 Computational Chemistry Experiment

Technique Karplus Relationship and Splitting Trees: Technique 19.8

Build molecules representing the two chair cyclohexane conformers of *cis*-2-phenylcyclohexanol. Optimize the structures using molecular mechanics. Record the steric energies. Which conformer is lower in energy?

Next build molecules representing the two chair cyclohexane conformers of *trans*-2-phenylcyclohexanol. Optimize the structures using molecular mechanics. Record the steric energies. Which conformer is lower in energy?

The NMR signal of the hydrogen atom attached to C_1, the carbon bearing the hydroxyl group, appears in the region between 3 and 4 ppm in both isomers. However, the stereochemical relationship between this hydrogen atom and the hydrogen atoms attached to adjacent carbon atoms is different in the *cis* and *trans* isomers. Because coupling constants, J, vary with the dihedral angle, θ, we can expect that the appearance of the NMR signals in the two isomers will differ. The relationship between J and θ is given by the Karplus relationship [see Technique 19.8, Figure 19.17].

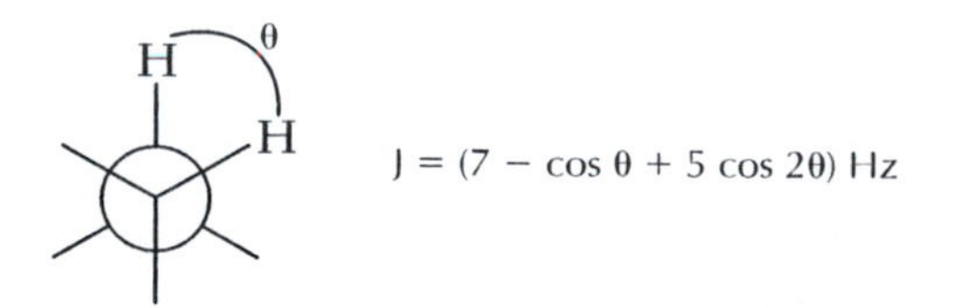

Using the lowest-energy conformers for the cis and trans diastereomers, measure the dihedral angles formed by the C_1—H_1 bond and the C—H bonds of adjacent carbon atoms. Estimate the coupling constants using the Karplus relationship. Prepare a table for each diastereomer similar to the following table and enter the dihedral angles and estimated coupling constants.

Dihedral angle	θ	J, Hz
H_1—C_1—C_2—H_2		
H_1—C_1—C_6—$H_{6(ax)}$		
H_1—C_1—C_6—$H_{6(eq)}$		

From the estimated values for the coupling constants, build a splitting tree [see Technique 19.8] to predict the appearance of the NMR signal due to H_1 in *cis*-2-phenylcyclohexanol and the appearance of the NMR signal due to H_1 in *trans*-2-phenylcyclohexanol. Which pattern is closer in appearance to the expanded NMR signal between 3 and 4 ppm in your spectrum of the product resulting from hydroboration-oxidation of 1-phenylcyclohexene?

Questions

1. Benzene is often produced as a by-product during the synthesis of phenylmagnesium bromide. How can its formation be explained? Write a balanced chemical equation for the formation of benzene.
2. If, by mistake, 100% ethanol rather than diethyl ether were used as the reaction solvent, would the Grignard synthesis still proceed as it should?
3. What might be the reason that bromobenzene is added to the reaction flask in small portions during the Grignard synthesis, rather than all at once?
4. Why was a sulfuric acid–acetic acid solution used for the dehydration in Project 7.2 instead of the more common reagent aqueous sulfuric acid?
5. In Project 7.2, why is the product washed with sodium bicarbonate solution?
6. The dehydration of 1-phenylcyclohexanol follows an E1 pathway. What produces the substantial stability of the intermediate carbocation?
7. Suggest what produces the vigorous reaction that occurs with bubbling when aqueous NaOH solution is added to the hydroboration mixture in Project 7.3.
8. What factors determine that 2-phenylcyclohexanol rather than 1-phenylcyclohexanol is the product of the hydroboration-oxidation reaction of 1-phenylcyclohexene?

PROJECT

8

ELECTROPHILIC AROMATIC SUBSTITUTION

The Diacetylation of Ferrocene

QUESTION: Which isomer of diacetylferrocene forms in the acylation of ferrocene?

> In this two-week project you and your teammate will investigate the acylation of ferrocene, a colored aromatic compound, using acetyl chloride and $AlCl_3$. Then you will use column chromatography to separate the reaction products and NMR spectroscopy or melting points to determine which isomer of diacetylferrocene forms.

Electrophilic Aromatic Substitution

The delocalized π-systems of aromatic compounds are much less vulnerable to attack by electrophiles (E^+) than the localized π-double bonds of alkenes. However, as the following equation shows, aromatic compounds do undergo electrophilic substitution reactions. Because the conjugated 6π-electron system of the aromatic ring is so stable, the carbocation intermediate loses a proton and regenerates the aromatic ring rather than undergoing reaction with a nucleophile:

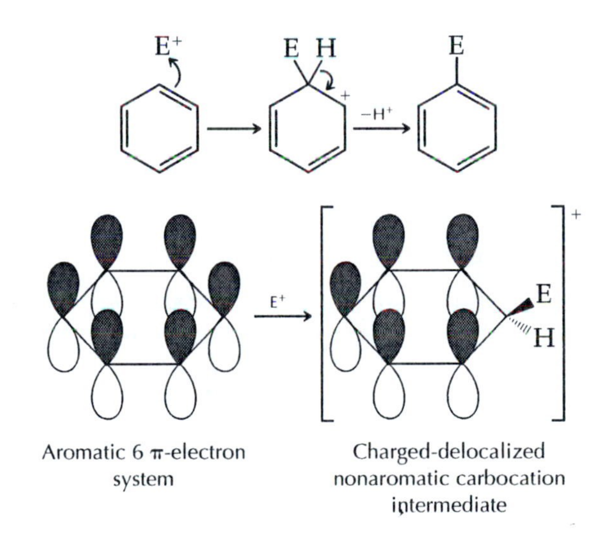

Aromatic 6 π-electron system

Charged-delocalized nonaromatic carbocation intermediate

The mechanism of electrophilic aromatic substitution has been thoroughly studied. Ring substituents strongly influence the rate and position of electrophilic attack. Electron-donating groups on the benzene ring speed up the substitution process by stabilizing the carbocation intermediate. If, however, the first substituent is an electron-withdrawing group, the aromatic substitution becomes slower because formation of the carbocation intermediate is more difficult. The electron-withdrawing group removes electron density from a species that is already positively charged and thus quite electron deficient. The more strongly a group is electron withdrawing, the more it destabilizes the carbocation intermediate and the more it slows the substitution process.

Friedel-Crafts Acylation Reactions

Friedel-Crafts acylation of aromatic compounds is an important example of electrophilic aromatic substitution reactions. Named after Charles Friedel and James Mason Crafts, the French and American chemists who discovered it over a hundred years ago, the Friedel-Crafts reaction has great utility in the synthesis of carbon-carbon bonds. Acyl groups can be substituted on aromatic rings by using acidic catalysts, such as $AlCl_3$, H_2SO_4, or H_3PO_4. In this type of electrophilic aromatic substitution reaction, an acyl chloride or acyl anhydride reacts with the aromatic compound in the presence of the acid catalyst. The product of an acylation reaction is an aromatic ketone:

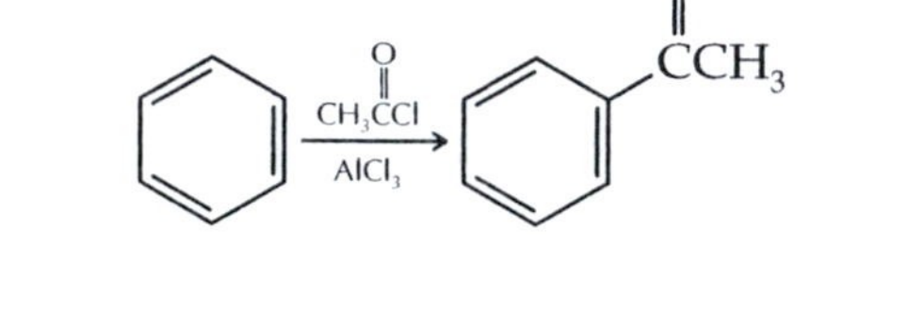

Diacylation of Ferrocene

Project 8 uses the unusual aromatic compound ferrocene, an organometallic compound that is composed of two negatively charged, planar five-membered rings that "sandwich" an iron ion:

Fe^{2+}

Ferrocene can be thought of as a compound formed by the bonding of an Fe^{2+} cation to two cyclopentadienide ligands, each bearing a negative charge. Each cyclopentadienyl anion is aromatic because it has six π-electrons in a completely conjugated ring:

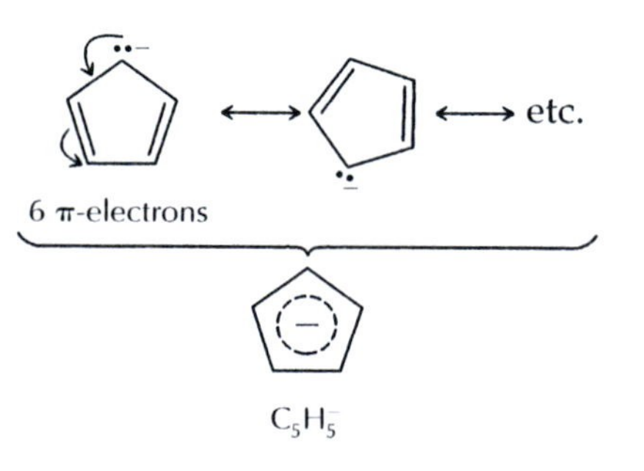

Friedel-Crafts acylation of ferrocene with acetyl chloride in the presence of aluminum chloride produces acetylferrocene.

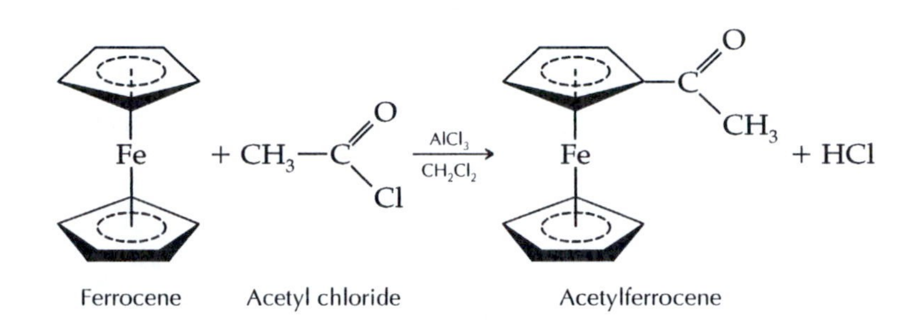

The catalyst, aluminum chloride, is an electron-deficient Lewis acid that can bond to acetyl chloride and produce an acetyl cation, which is a potent electrophile. The electrophile attacks the ring, resulting in substitution of the acetyl group for a ring proton.

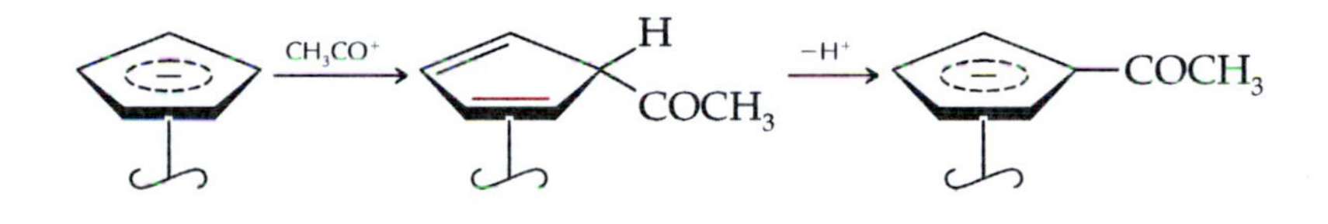

When acetylferrocene reacts with a second molecule of acetyl chloride, substitution can occur on either the unsubstituted cyclopentadienyl ring or on the ring already containing an acetyl group.

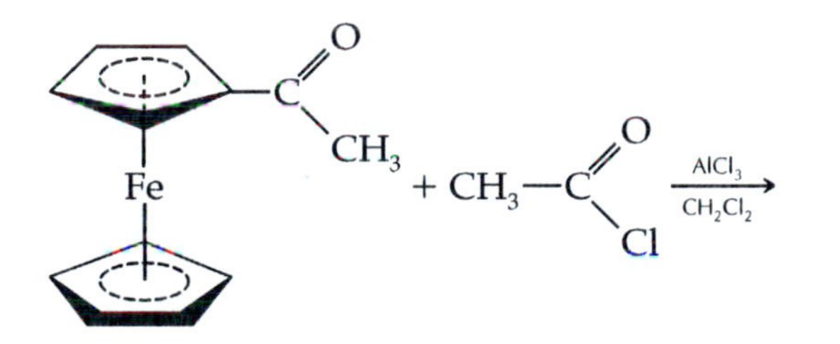

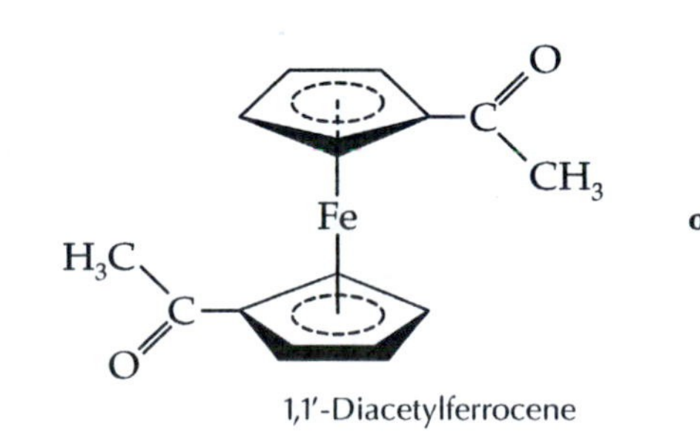

1,1′-Diacetylferrocene

or

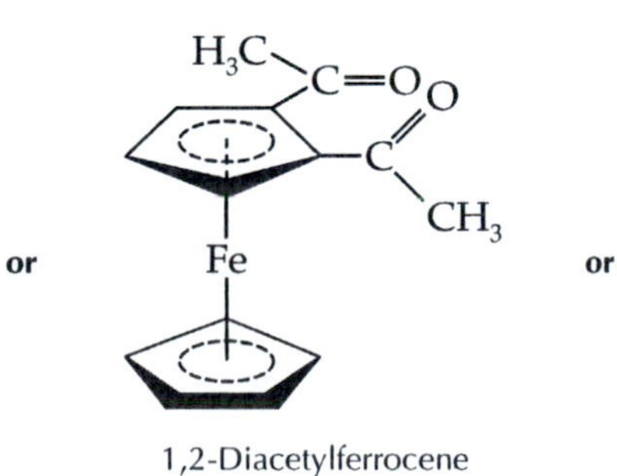

1,2-Diacetylferrocene

or

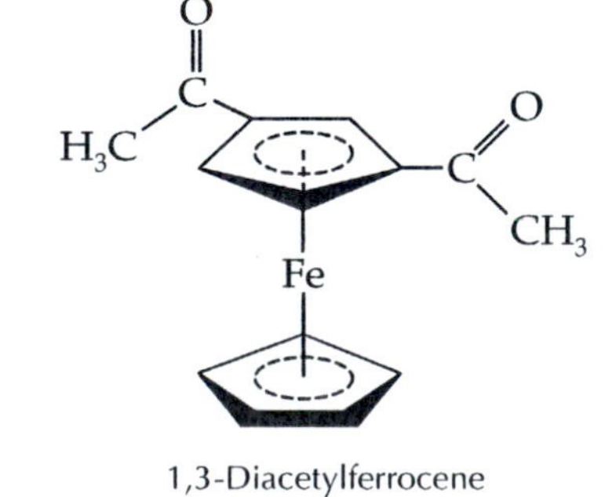

1,3-Diacetylferrocene

You will use column chromatography to separate diacetylferrocene from any acetylferrocene remaining after the reaction. The NMR spectra of acetylferrocenes are very helpful in determining their structures. In addition, the diacetylferrocenes have well-defined melting points after recrystallization from methanol. Using what you have already studied in the classroom about the theory of electrophilic aromatic substitution, you might be able to make a prediction about where substitution of the second acetyl group will occur. Or you may want to keep an open mind and see what answer your experimental data supports.

NMR Analysis of Substituted Ferrocenes

The ^{1}H NMR spectrum of ferrocene itself shows 10 equivalent aromatic protons as a singlet at about 4.2 ppm. The spectra of substituted ferrocenes are simplified because the two cyclopentadienyl rings are free to rotate much faster than the NMR experiment can distinguish rotational isomers.

You will probably isolate a mixture of acetylferrocene and diacetylferrocenes in your Friedel-Crafts reaction. Acetylferrocene can be characterized by examining its IR (Figure 8.1) and ^{1}H NMR (Figure 8.2) spectra. Note the intense carbonyl stretching vibration in the IR spectrum at about 1660 cm^{-1}. The ^{1}H NMR spectrum (see Figure 8.2) shows the methyl group of the acetyl substituent as a singlet (3H) at 2.4 ppm. The unsubstituted ring yields a singlet (5H) at 4.2 ppm, and the substituted ring reveals a pair of signals (2H each) as apparent triplets at 4.5 ppm and 4.8 ppm. The splitting patterns of substituted ferrocenes are actually quite complex because chemically equivalent protons on the substituted ring are often not magnetically equivalent. Chemical shifts and NMR integration are more reliable in interpreting the NMR spectra of substituted ferrocenes [see Techniques 19.5 and 19.6].

Three different diacetylferrocene isomers are possible. If the second acetyl group ends up on the formerly unsubstituted ring, the product is called 1,1′-diacetylferrocene. If the second acetyl group substitutes on the same cyclopentadienyl ring as the first acetyl group

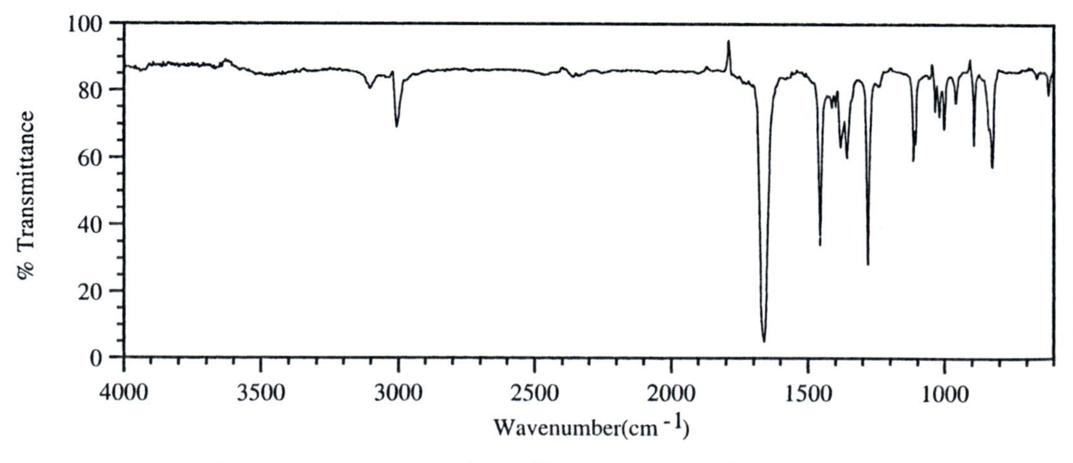

FIGURE 8.1 IR spectrum of acetylferrocene (in $CHCl_3$).

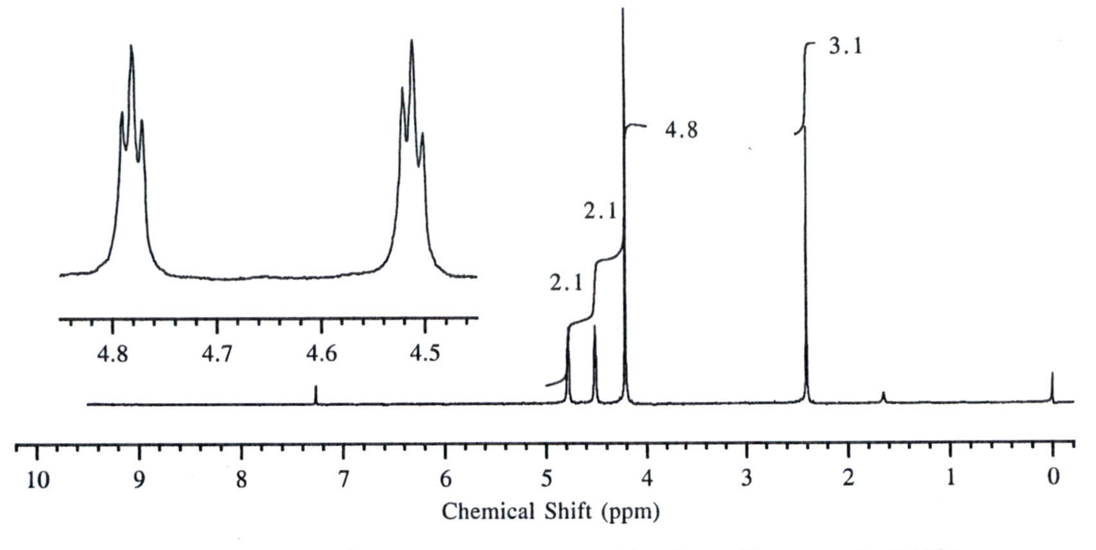

FIGURE 8.2 ^{1}H NMR spectrum (200 MHz) of acetylferrocene (in $CDCl_3$).

does, the product can be 1,2- or 1,3-diacetylferrocene. The chemical shifts and integration of the different types of protons on the cyclopentadienyl rings naturally differ. The first step in interpreting the NMR spectra is to identify the sets of protons that are different from each other on each of the three isomers. Protons on carbon atoms adjacent to an acetyl group will be more deshielded than protons farther away. Overall, the chemical shifts will be similar to those of acetylferrocene (see Figure 8.2).

If the chromatographic separation was not complete in your intermediate fractions, your spectra will reflect mixtures of products [see Technique 19.9]. In addition, the NMR samples may contain a small amount of hexane, ether, and even dichloromethane. If you used a steam bath to aid in the evaporation of your chromatographic samples, they may contain a water signal at 1.6 ppm.

8.1 Synthesis of Diacetylferrocene

PRELABORATORY ASSIGNMENT: Because it forms a stable complex with acetyl chloride, the $AlCl_3$ catalyst is used in a stoichiometric amount in the synthesis, rather than in a small catalytic amount. Draw a likely structure for the acetyl chloride-$AlCl_3$ complex and explain why its stability requires one to use a stoichiometric amount of $AlCl_3$.

TEAMWORK OPTION: The synthesis of diacetylferrocenes can be carried out individually or by a team of two students. If you plan to do the optional miniscale column chromatography in Project 8.3, it is better to work as a two-person team throughout the project.

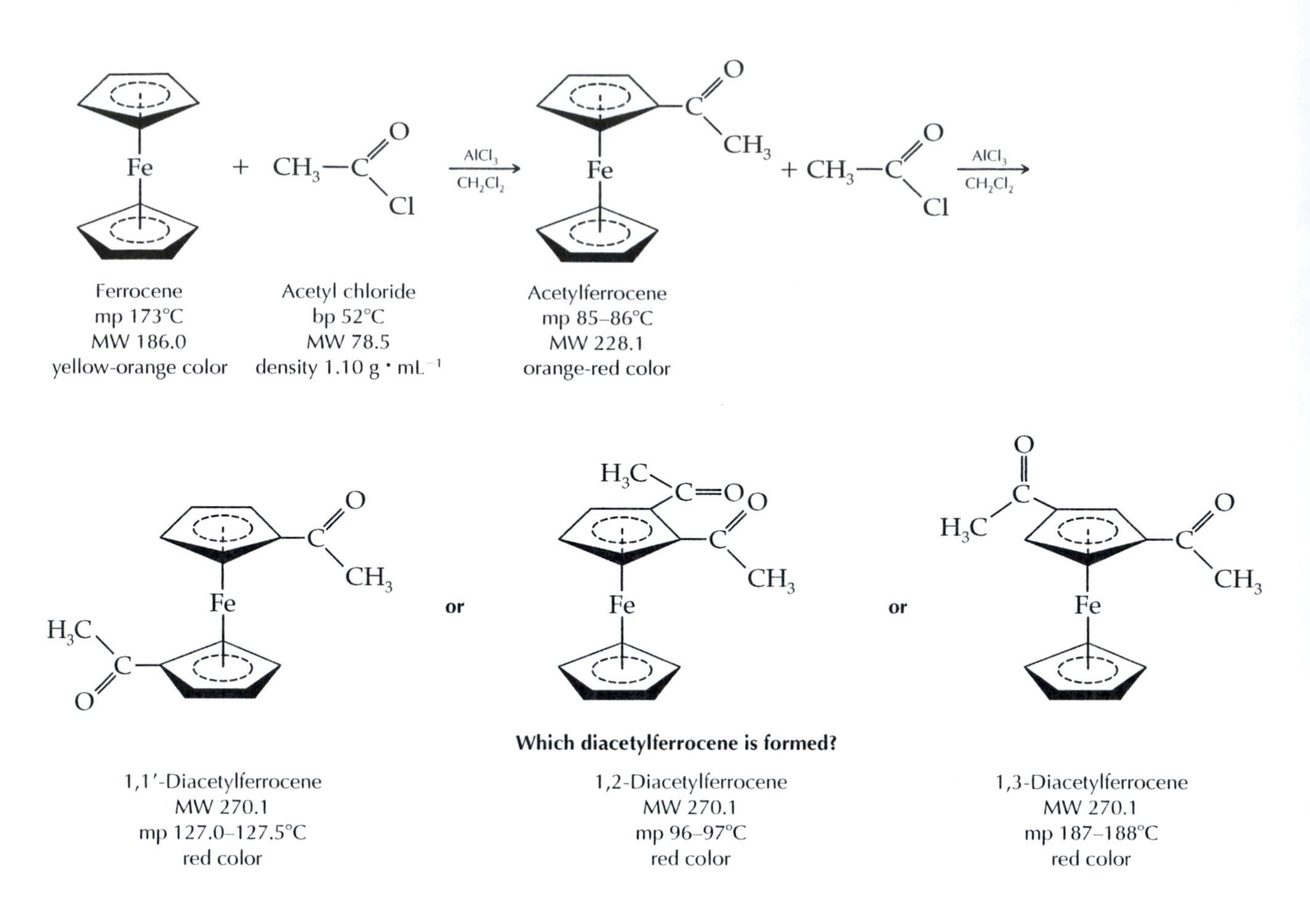

MICROSCALE PROCEDURE

Techniques Anhydrous Reaction Conditions: Technique 7.3
Microscale Extractions: Technique 8.6b
Thin-Layer Chromatography: Technique 15

SAFETY INFORMATION

Ferrocene is relatively nontoxic, but avoid contact with the skin.

Acetyl chloride is flammable and corrosive. Avoid contact with skin, eyes, and clothing. Dispense it in a hood.

Anhydrous aluminum chloride is corrosive and very moisture sensitive. It reacts with atmospheric moisture to form toxic, corrosive HCl.

Dichloromethane is toxic, an irritant, absorbed through the skin, and harmful if swallowed or inhaled. Use it only in a hood and wear neoprene gloves while handling it.

Diethyl ether is extremely volatile and flammable. Use it only in a hood.

Synthesis

Prepare a microscale drying tube using anhydrous calcium chloride as the desiccant. Fit the drying tube to the top of a water-jacketed condenser and attach the condenser to the curved arm of a Claisen adapter. Close the other opening of the Claisen adapter with a flat septum and screw cap or a fold-over rubber septum, as shown in Technique 7, Figure 7.4a; substitute a 10-mL round-bottomed flask for the conical vial.

Have a magnetic stirrer available. Remove the round-bottomed flask from the apparatus and set it in a 50-mL beaker. Place a magnetic stirring bar and 3.5 mL of dichloromethane in the flask. Weigh 300 mg ±15 mg of anhydrous aluminum chloride on weighing paper and quickly transfer it to the flask containing the dichloromethane. Immediately attach the flask to the Claisen adapter. Turn on the water flow through the condenser; begin stirring the contents of the flask.

Obtain 0.20 mL of acetyl chloride in a small vial. Draw the acetyl chloride into a 1-mL syringe. Insert the syringe needle through the septum at the top of the Claisen adapter and add the acetyl chloride dropwise. Continue stirring until the aluminum chloride is dissolved. (There may be some turbidity.) Prepare a solution of 150 mg of ferrocene and 3.5 mL of dichloromethane in a small vial. Draw the ferrocene solution into a syringe and add the solution dropwise over a period of 10 min while continuing to stir the reaction mixture. After addition of the ferrocene solution is complete, continue to stir the reaction mixture for 30 min.

Pour the reaction mixture into a 50-mL beaker containing 8–9 g of ice. Stir the mixture for 5 min on a magnetic stirrer, then pour the mixture into a 15-mL screw-capped centrifuge tube. Transfer the lower organic layer to another centrifuge tube with a Pasteur filter pipet or a Pasteur pipet fitted with a syringe [see Technique 8.5]. Extract the remaining aqueous phase twice with 3-mL portions of dichloromethane, combining each organic phase with the original organic layer. Wash the combined organic phase three times with 3-mL portions of water [see Technique 8.6b]. Transfer the organic phase to a clean, dry 25-mL Erlenmeyer flask after the third washing and dry the solution with anhydrous calcium chloride.

TLC Analysis of Product Mixture

Transfer the dried solution to a tared (weighed) 25-mL Erlenmeyer flask. Analyze the crude product solution by thin-layer chromatography on silica gel thin-layer plates impregnated with fluorescent indicator; use diethyl ether as the developing solvent [see Technique 15]. Also spot the plate with a reference solution of ferrocene. Calculate the R_f value for each spot observed on the developed TLC plate. The order of decreasing R_f values is ferrocene (yellow), acetylferrocene (orange), one of the diacetylferrocenes (orange), and the other diacetylferrocene (dark red).

Recovery of Product Mixture

Working in a hood, evaporate the solvent from the crude product with a stream of nitrogen or using a rotary evaporator until a constant mass is obtained. Record the mass of crude product and store it in the Erlenmeyer flask until your next laboratory period.

CLEANUP: Pour the aqueous waste remaining from the extractions into the container for hazardous inorganic waste. Place the spent drying agent in the container for hazardous solid waste. Pour the remaining TLC solvent into the container for flammable organic waste.

8.2 Purification by Column Chromatography

PRELABORATORY ASSIGNMENT: When the solvent level in a chromatography column is allowed to go below the top of the adsorbent, the adsorbent may form cracks. Explain why this must be avoided.

TEAMWORK: Work in teams of two students in carrying out the column chromatography. Each student should assemble all reagents and equipment needed for chromatographic separation of his/her own product. Once everything is at hand, undertake together the complete chromatographic process on one student's crude product of diacetylferrocene prepared in Project 8.1. When that procedure is completed, proceed to chromatograph the second student's sample.

MICROSCALE PROCEDURE

Techniques Microscale Column Chromatography: Technique 17.6
NMR Spectroscopy: Technique 19

SAFETY INFORMATION

Diethyl ether and **hexane** are extremely volatile and flammable. Use them only in a hood.

Column Chromatography

Assemble all the equipment and reagents that you will need for the entire chromatography procedure before you begin to prepare the column.

Read this procedure completely and review Technique 17.6 before you undertake this part of the project.

Obtain about 30 mL of hexane in a 50-mL Erlenmeyer flask fitted with a cork. Place 50 mg of your crude product in a small test tube and dissolve it in 0.5 mL of dichloromethane; cork the test tube. For collecting fractions from the column, label four 25-mL Erlenmeyer flasks "Fraction 1," "Fraction 2," and so on; record the mass of each flask to the nearest 0.001 g.

Prepare the other elution solvents and store them in labeled large test tubes or corked 50-mL Erlenmeyer flasks. You will need 20 mL of each solvent: 25:75 (v/v) ether/hexane, 50:50 (v/v) ether/hexane, and pure ether.

Obtain a large-volume Pasteur pipet* to use as the chromatography column and pack a small plug of glass wool down into the stem, using a thin stirring rod or a regular Pasteur pipet. Clamp the pipet in a vertical

*Large-volume Pasteur pipets, available from Fisher Scientific, catalog no. 13678-8, have a capacity of 4 mL.

position and place a 25-mL Erlenmeyer flask underneath it to collect the hexane that you will be adding to the column. Place approximately 1.7–1.8 g of silica gel (70–230 mesh) in a 25-mL Erlenmeyer flask; add approximately 15 mL of hexane to make a thin slurry. Transfer the silica gel slurry to the column, using a regular Pasteur pipet. Continue adding slurry until the column is one-half to two-thirds full of silica gel. Fill the column four or five times with hexane to pack it well. The eluted hexane can be reused for this purpose. **(Note: Do not let the hexane level fall below the top of the silica gel at any time.)** After the column is packed, add a 2–3 mm layer of sand by letting it settle through the hexane.

Be sure that the silica gel is covered with solvent throughout the chromatographic procedure.

If the column drains very slowly, place a pipet bulb or rubber ear syringe in the top of the column and press gently to exert a slight pressure. Lift the bulb out of the column before releasing the pressure on it so that the silica gel will not be disturbed by the vacuum inside the syringe as the syringe refills with air.

Allow the hexane level to fall almost to the top of the sand and then place the flask labeled "Fraction 1" under the column. Transfer your crude product solution to the top of the column, using a Pasteur pipet. Begin eluting the column with the appropriate solvent mixture only after the crude product solution is on the column and the solvent level has again reached almost to the top of the sand.

If your TLC analysis showed the presence of unreacted ferrocene, begin the elution with hexane until the leading yellow band of ferrocene has left the column. At this point, change the collection flask to the second fraction and begin eluting with 25:75 (v/v) ether/hexane solution until the second band has been collected.

If no ferrocene was found in the TLC analysis, begin the elution with 25:75 ether/hexane solution. Collect the eluent as Fraction 1 until the orange band (acetylferrocene) reaches the bottom of the column, then change collection flasks. Collect eluent as Fraction 2 until the acetylferrocene band is completely off the column.

Change to the Fraction 3 flask and begin elution with 50:50 ether/hexane until the orange band has left the column. Change to the Fraction 4 flask and elute the dark red band with pure ether.

TLC Analysis and Recovery of Product

Analyze all fractions by thin-layer chromatography. Develop the TLC plates as you did previously. Record the results and R_f values in your notebook. Has column chromatography purified the diacetylferrocene product?

Evaporate the solvent from each fraction with a stream of nitrogen in a hood or using a rotary evaporator. If you will analyze your product by NMR, do not heat the product extensively in the presence of air; O_2 can oxidize the Fe^{2+} of ferrocene to Fe^{3+}, which is paramagnetic. Even a small amount of this oxidation can cause broadening of the NMR peaks.

Weigh the flasks again to determine the amount of each product that was separated. Determine the relative ratio of the products and the percent yield of diacetylferrocene. One isomer of diacetylferrocene may be present in trace amounts. Prepare a sample of the diacetylferrocene present in the larger amount for NMR analysis as directed by your instructor, using $CDCl_3$ as the solvent [see Technique 19.2].

CLEANUP: Place the aqueous solutions remaining from the extractions in the container for aqueous inorganic waste. Pour any remaining TLC solvents into the container for flammable (organic) waste. Place the TLC plates and the silica gel from the column in the container for inorganic solid waste.

Interpretation of Data

Assign all major peaks in your NMR spectrum and identify the structure of the diacetylferrocene that is your major product. Compare your spectrum to that of your teammate. Describe how your experimental data show whether acetylferrocene is acetylated on the substituted or the unsubstituted ring.

8.3 Optional Column Chromatography for Isolation of the Minor Diacetylferrocene Isomer

PRELABORATORY ASSIGNMENT: When the solvent level in a chromatography column is allowed to go below the top of the adsorbent, the column may crack. Explain why this must be avoided.

TEAMWORK: Work in teams of two students in carrying out and analyzing the results of the column chromatography.

MINISCALE PROCEDURE

Technique Miniscale Column Chromatography: Technique 17

SAFETY INFORMATION

Dichloromethane is toxic, an irritant, absorbed through the skin, and harmful if swallowed or inhaled. Work in a hood and wear neoprene gloves while handling it.

Diethyl ether and **hexane** are extremely volatile and flammable. Use them only in a hood.

Preparing the Column

Read this procedure completely and review Techniques 17.4 and 17.5 before you undertake this part of the project.

Dissolve your crude product in 2.0 mL of dichloromethane; cork the Erlenmeyer flask. Label five 125-mL Erlenmeyer flasks to collect fractions from the column.

You will need 75–100 mL of each solvent: hexane, 25:75 (v/v) ether/hexane, 50:50 (v/v) ether/hexane, and pure ether for the chromatography. If you will be using a chromatography column with a stopcock at the bottom, you can obtain these solvents as you need them.

Obtain a 300 × 18 mm chromatography column and pack a small plug of glass wool down into the stem, using a piece of glass tubing. Clamp the column in a vertical position and place a 125-mL Erlenmeyer flask underneath it to collect the hexane that you will be adding to the column. Cover the glass wool plug with about 6 mm of white sand.

Place approximately 9–10 g of silica gel (70–230 mesh) in a 125-mL Erlenmeyer flask and add about 40 mL of hexane to make a thin slurry. Swirl the mixture vigorously to suspend the silica gel in the liquid, and quickly pour the slurry into the column through a powder funnel. Fill the column four or five times with hexane to pack it well. The eluted hexane can be reused for this purpose. Use a pipet bulb in the top of the column and press it to apply considerable pressure each time you pass hexane through the silica gel; this allows the adsorbent to pack tightly and avoids cracking. **(Note: Lift the bulb out of the column before releasing the pressure on it so that the silica gel will not be disturbed by the vacuum inside the bulb as the bulb refills with air. Also, do not let the hexane level fall below the top of the silica gel at any time.)** Alternatively, you can use compressed air or N_2 to apply the necessary pressure. After the column is packed, add a 2–3 mm layer of white sand by letting it settle through the hexane.

Be sure that the silica gel is covered with solvent throughout the chromatographic procedure.

Column Chromatography

Allow the hexane level to fall almost to the top of the sand and then place the flask labeled "Fraction 1" under the column. Transfer your crude product solution to the top of the column, using a Pasteur pipet. Try to add the solution at the center of the silica gel column; if you get some on the glass walls, wash them down with a small amount of dichloromethane (about 1 mL). Add a small amount of hexane to the column to ensure that all the acetylferrocenes have passed into the silica gel. Begin eluting the column with fresh hexane when the liquid level has again reached the top of the sand. If the column drains slowly, use a pipet bulb at the top to increase the solvent flow.

If your crude product mixture contains unreacted ferrocene, you will see a yellow band developing on the column; continue eluting the column with hexane until the ferrocene has eluted from the column. At this point, begin eluting with 25:75 (v/v) ether/hexane solution; continue collecting Fraction 1 until the orange acetylferrocene band reaches the bottom of the column. Then change the collection flask to the flask labeled "Fraction 2" and elute the orange acetylferrocene band as Fraction 2. Change to the Fraction 3 flask only when the acetylferrocene band is completely off the column. Begin elution with 50:50 ether/hexane and continue it until the early part of the paler orange band has left the column. Then change to the Fraction 4 flask and elute the second part of the orange band. Finally, collect as Fraction 5 the dark red band using pure ether as the solvent.

Running the chromatography faster, using pressure, is probably better than slower. Elution at 5 cm/min is often cited as the best speed.

The intermediate pale orange band contains the minor isomer of diacetylferrocene produced in the Friedel-Crafts reaction. It is possible that you won't even see this band, so you might have to make an arbitrary decision about when to switch from Fraction 3 to Fraction 4. Collecting two fractions at this stage of the chromatography gives you a greater chance of collecting the minor isomer of diacetylferrocene in reasonably pure form. The leading part of its band may co-elute partially with acetylferrocene and the trailing part may coelute with the major isomer of diacetylferrocene.

TLC Analysis of Product Fractions

Analyze Fractions 2–5 by thin-layer chromatography on a single TLC plate. It may be necessary to use overspotting (five times) with the intermediate fractions, which may contain only a small amount of your product. Develop the TLC plates as you did in Project 8.1. Record your TLC results and R_f values in your notebook. Has column chromatography purified the diacetylferrocene products?

Using a stream of nitrogen along with a steam bath in a hood, or using a rotary evaporator, evaporate the solvent from Fractions 2 and 5 and whichever intermediate fraction has the greater percentage of the minor isomer of diacetylferrocene. You need not evaporate the initial fraction containing the yellow ferrocene band or the intermediate fraction that has the smaller amount of the minor diacetylferrocene isomer. Do not heat the product solutions extensively in the presence of air, as O_2 can oxidize the Fe^{2+} of ferrocene to Fe^{3+}, which is paramagnetic. Even a small amount of this oxidation can cause broadening of the NMR peaks of your products.

NMR Analysis of Products

You will analyze by NMR spectroscopy the solids recovered from all three chromatographic fractions that you evaporated. After you have evaporated Fraction 5 to dryness, scrape out the major isomer of diacetylferrocene onto a weighing paper and record its mass. Determine the percent yield of the major diacetylferrocene isomer.

Prepare a sample of each of the three fractions you evaporated for NMR analysis as directed by your instructor, using $CDCl_3$ as the solvent [see Technique 19.2]. The minor isomer of diacetylferrocene may be present in trace amounts. Even so, there should be enough for an NMR spectrum if you wash the inside of its flask with 0.5 mL of $CDCl_3$ and use it for the bulk of its NMR sample.

CLEANUP: Place the aqueous solutions remaining from the extractions in the container for aqueous inorganic waste. Pour any remaining TLC solvents and the column chromatography fractions that you did not evaporate into the container for flammable (organic) waste. Place the TLC plates and the silica gel from the column in the container for inorganic solid waste.

Interpretation of Experimental Data

Assign all major peaks in your NMR spectra and identify the structures of both the major and minor isomers of diacetylferrocene in your product. Use the spectra of the samples from Fractions 2 and 5 to determine the exact chemical shifts of the protons in acetylferrocene and the major isomer of diacetylferrocene as reference points in the NMR spectral analysis of the intermediate fraction, which may contain a mixture of products. [see History of the NMR Sample: Mixtures of Compounds in Technique 19.9]. Describe how your experimental data show whether acetylferrocene is predominantly acetylated on the substituted or the unsubstituted ring and what the structure of the minor isomer of diacetylferrocene is.

Questions

1. In the column chromatography, diacetylferrocene elutes much more slowly than acetylferrocene. Ferrocene itself elutes sooner than either of the acetylferrocenes. Explain the order of elution of ferrocene, acetylferrocene, and diacetylferrocene on the chromatography column.
2. In the ^{1}H NMR spectrum of most aromatic compounds, the aromatic protons exhibit a chemical shift of 7–8 ppm. However, in ferrocene, the chemical shift of the aromatic protons is 4.2 ppm. Explain what factors might cause the upfield shift.
3. Explain how the ^{1}H NMR spectrum of ferrocene supports the assigned sandwich structure rather than a structure in which the iron atom is bound to only one carbon atom of each ring.
4. Use your knowledge of the mechanism and theory of electrophilic aromatic substitution to explain how your experimental results in this project can be understood.

PROJECT

9

STEREOCHEMISTRY OF ELECTROPHILIC ADDITION TO AN ALKENE

QUESTION: What is the stereochemistry for the addition of bromine to the carbon-carbon double bond of cyclohex-4-ene-*cis*-1,2-dicarboxylic acid? Does this stereochemistry support the existence of a bromonium ion intermediate in the reaction pathway?

In this three-week project you will carry out a four-step synthesis of the dimethyl ester of 4,5-dibromocyclohexane-*cis*-1,2-dicarboxylic acid, based on Diels-Alder methodology, and determine the stereochemistry of bromine addition by NMR analysis. The stereochemical data will provide insights on the mechanism of the electrophilic addition reaction.

Diels-Alder Reaction and Hydrolysis of the Product

In Project 9.1 you will carry out a Diels-Alder synthesis of cyclohex-4-ene-*cis*-1,2-dicarboxylic anhydride (step 1) and then hydrolyze the Diels-Alder product to cyclohex-4-ene-*cis*-1,2-dicarboxylic acid (step 2).

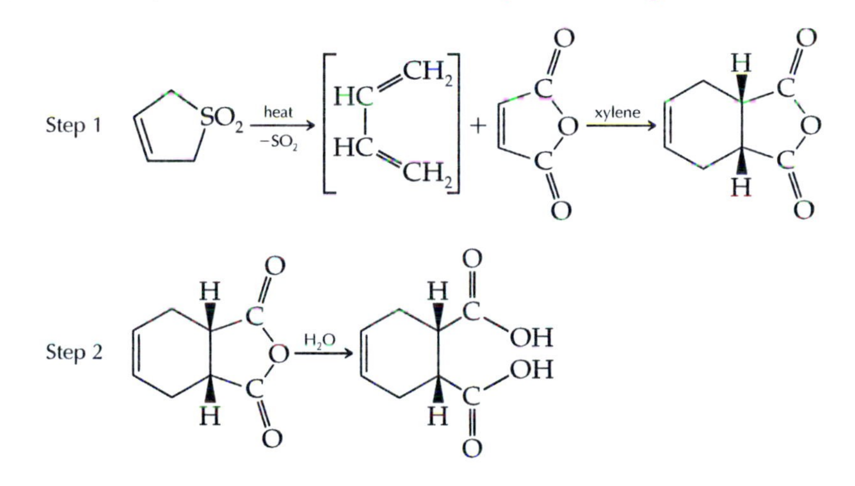

The cis configuration of the dienophile maleic anhydride and the stereospecificity of the Diels-Alder reaction determine the cis geometry of the Diels-Alder product in Project 9.1. The Diels-Alder reaction has a faster rate when the C=C double bond of the dienophile is electron-deficient. The two electron-withdrawing groups on maleic anhydride speed up the reaction significantly. The conjugated diene 1,3-butadiene is a gas at room temperature; thus it is inconvenient to handle. In Project 9.1 butadiene is formed in situ, within the reaction mixture, by the extrusion of sulfur dioxide from the solid reactant, butadiene sulfone. As butadiene forms, it reacts rapidly with maleic anhydride to form the Diels-Alder adduct before it can escape from the reaction mixture. Because a carbon-carbon double bond in a cyclohexene ring must have a cis configuration (*trans*-cyclohexene is unknown), the trans-like conformation of butadiene must rearrange to the cis-like conformation before reaction with maleic anhydride can proceed. The reaction is carried out at the boiling point of dimethylbenzene (xylene).

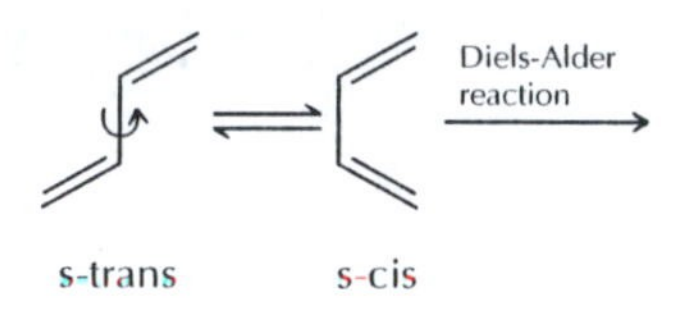

The Diels-Alder reaction has great utility in the chemical synthesis of six-membered ring compounds from simple cyclohexanes to complex steroids. Otto Diels and Kurt Alder received the Nobel prize in 1950 for discovering the reaction that bears their names. It is a [4 + 2] cycloaddition reaction in which two molecules add together in one step to form a cyclic compound. The conjugated diene provides four π-electrons and the dienophile provides the other two π-electrons, which allows a concerted reaction with a stabilized aromatic-like transition state having six π-electrons. In the transition state the p-orbitals of all six carbon atoms overlap simultaneously.

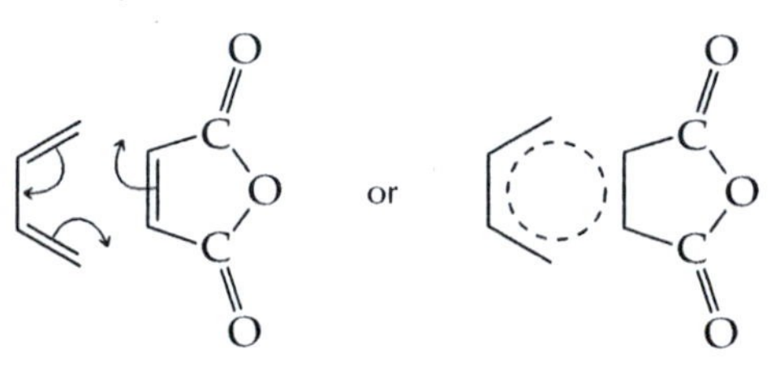

The second step of Project 9.1 is the hydrolysis of the bicyclic anhydride produced in the Diels-Alder reaction. This hydrolysis involves the addition of water to the carbonyl group and breaking of the anhydride linkage, giving the dicarboxylic acid. In water solution, the equilibrium lies on the side of the dicarboxylic acid product.

Bromine Addition

In Project 9.2 you will add molecular bromine across the carbon-carbon double bond of cyclohex-4-ene-*cis*-1,2-dicarboxylic acid.

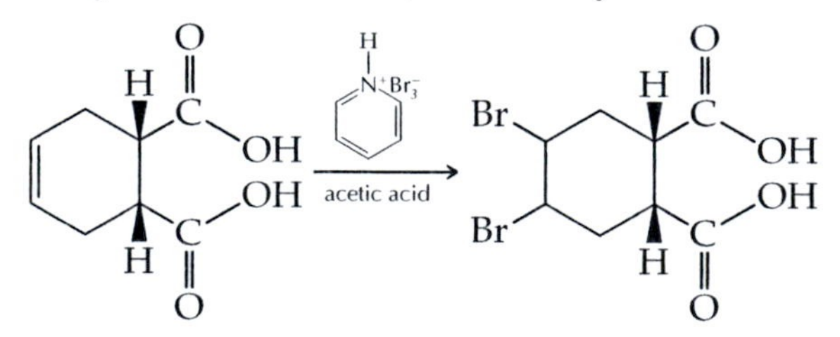

Pyridinium tribromide serves as the source of molecular bromine, Br_2. Because it is a crystalline solid, pyridinium tribromide is much safer to handle than Br_2, which is a volatile corrosive liquid. Pyridinium tribromide dissociates into molecular bromine and pyridinium bromide, and the bromine then adds to the carbon-carbon double bond of the alkene.

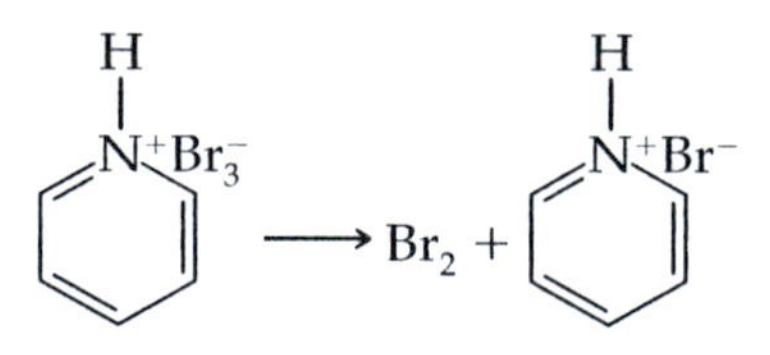

Synthesis of the Dimethyl Ester

In Project 9.3 you will synthesize the dimethyl ester of 4,5-dibromocyclohexane-*cis*-1,2-dicarboxylic acid and use NMR spectroscopy to determine the configuration of the product.

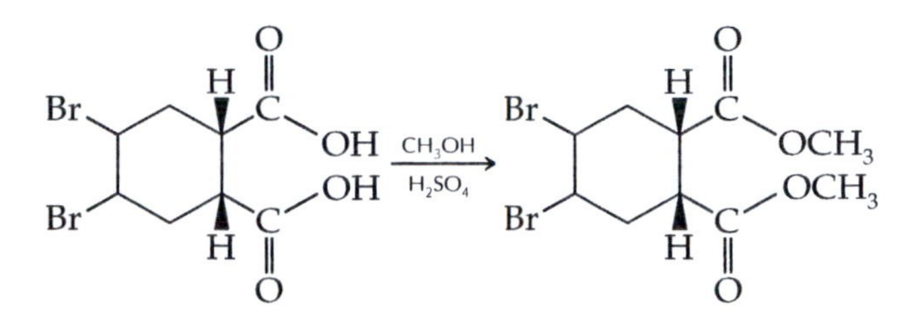

The mechanism of acid-catalyzed ester formation is well understood. The first step is protonation of the carboxylic acid by the sulfuric acid catalyst to produce the conjugate acid of the carboxylic acid:

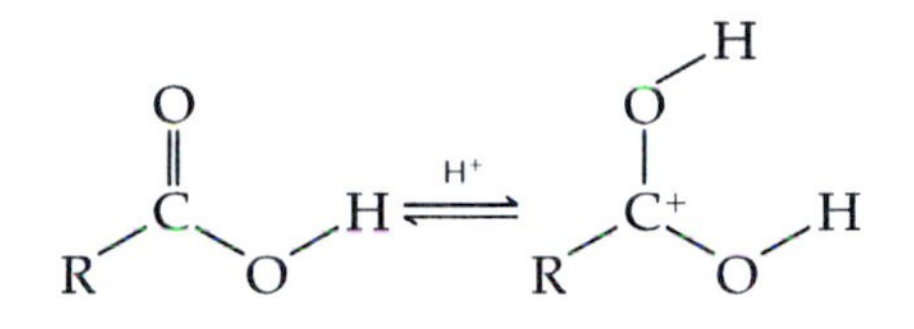

This protonated intermediate is very reactive and highly vulnerable to attack by an alcohol molecule:

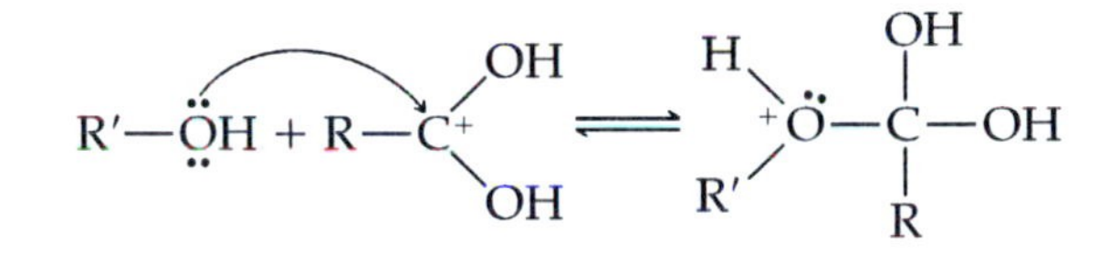

A proton shift then produces a new intermediate from which water is lost:

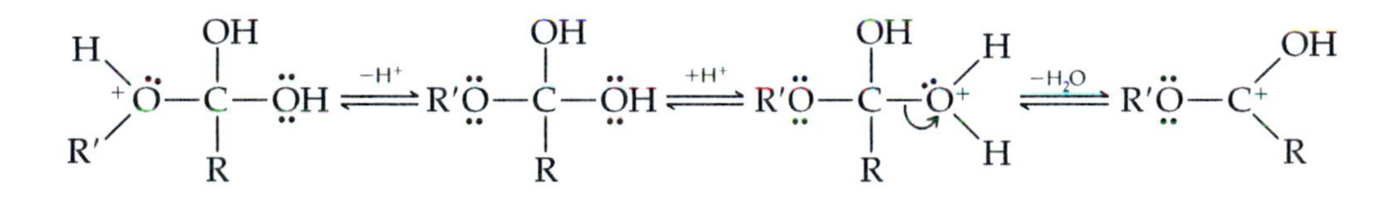

Finally, the carbocation loses a proton to form the ester:

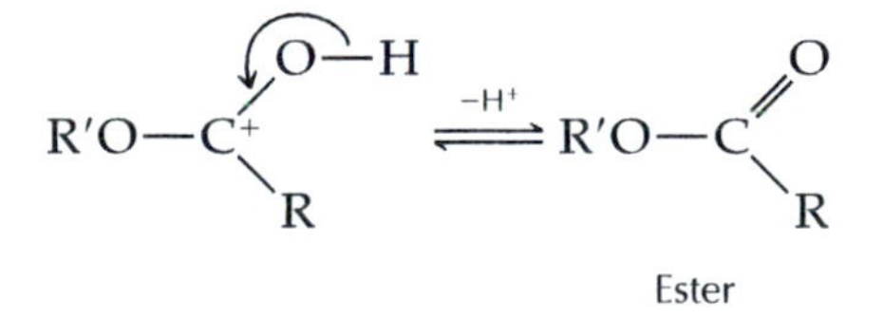

In this instance the reaction sequence repeats itself on the second carboxylic acid functional group, thereby producing the diester.

What Is the Mechanism of Bromine Addition to a Carbon-Carbon Double Bond?

The addition of Br_2 to a carbon-carbon double bond is a classic electrophilic addition reaction whose mechanism has been studied in depth. The generally accepted pathway goes through a bromonium ion intermediate. Bromine is a good electrophile in the step 1 of the reaction, and in step 2, the intermediate subsequently reacts with Br^-, an excellent nucleophile.

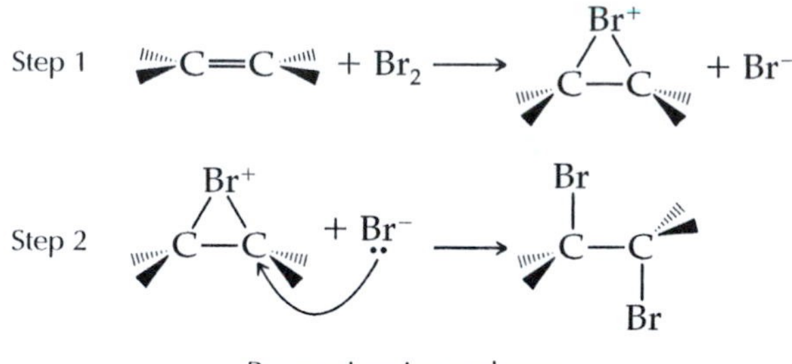

Bromonium ion pathway

How do we know whether this mechanism is the correct one for a particular alkene? What experimental data would support or disprove it? One crucial piece of evidence is the stereochemistry of the overall addition. If the bromonium ion pathway operates, the addition gives *anti* stereochemistry because step 2, which determines the configuration of the product, is an S_N2 reaction of Br^- with the bromonium ion. S_N2 reactions always proceed with inversion of configuration.

However, if the addition reaction proceeds by attack of Br^- on a carbocation rather than a bromonium ion, the stereochemistry will be more complex.

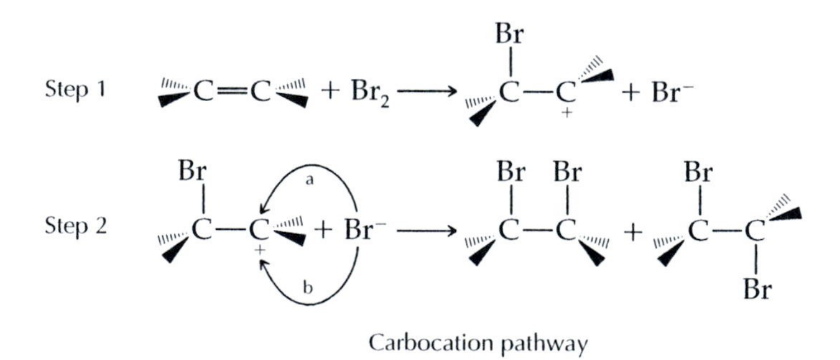

Carbocation pathway

The carbocation intermediate (step 1) can undergo reaction with Br^- from two sides (step 2). Attack on the top side (a), adjacent to the bromine atom already bound to carbon, gives a *syn* addition. Attack of Br^- on the bottom side (b) gives *anti* addition. Thus, a mixture of *syn* and *anti* addition products should result if a carbocation is the intermediate that leads to product formation.

When the alkene is a cyclic compound, *anti* addition produces a trans dibromide and *syn* addition gives cis dibromide products. In 9.3 you will discover the stereochemistry of the addition of Br_2 to cyclohex-4-ene-*cis*-1,2-dicarboxylic acid by determining the cis/trans ratio of the dibromide products.

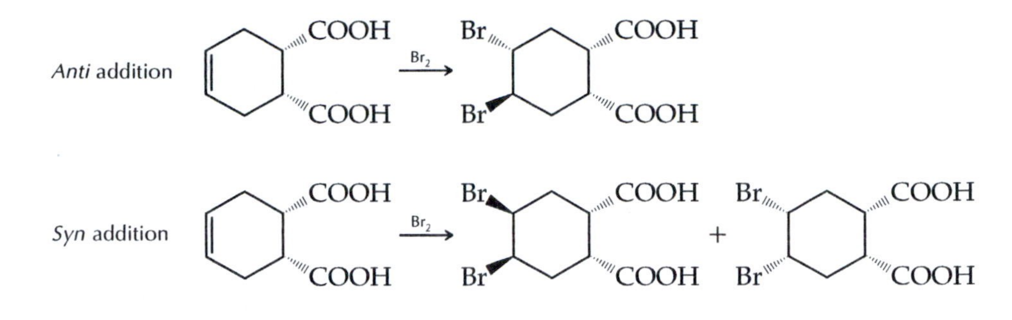

Using NMR to Probe Stereochemistry

NMR spectroscopy is often an excellent way to determine the stereochemistry of a chemical reaction, especially when molecular symmetry simplifies the spectrum. Let us first examine the spectrum of cyclohex-4-ene-*cis*-1,2-dicarboxylic acid (Figure 9.1). Because the dicarboxylic acid has a plane of symmetry, the spectrum is simplified by the equivalency of some groups of protons.

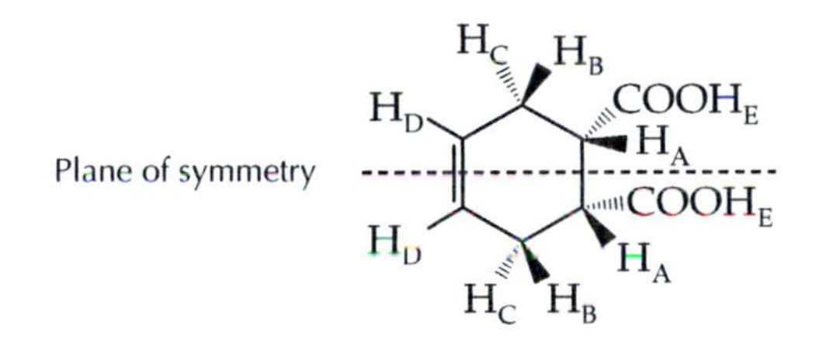

The NMR spectrum in Figure 9.1 shows four groups of peaks. The singlet at 5.7 ppm, due to the two highly deshielded vinyl protons (H_D) on the carbon-carbon double bond, has a relative integration of two protons.

The somewhat asymmetric triplet at 3.1 ppm is due to the two protons (H_A) that are in the α position relative to the carboxylic acid groups.

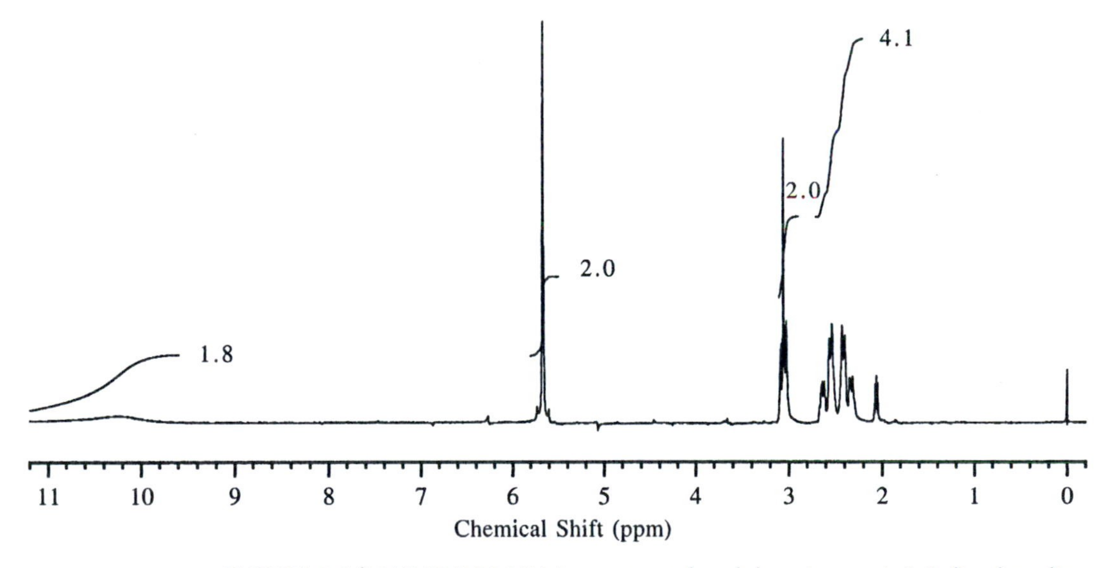

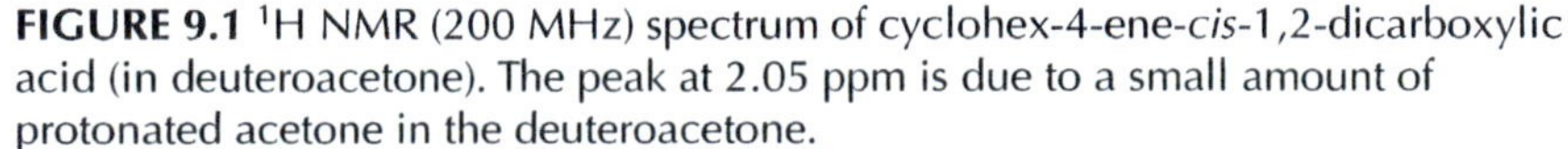

FIGURE 9.1 ^{1}H NMR (200 MHz) spectrum of cyclohex-4-ene-*cis*-1,2-dicarboxylic acid (in deuteroacetone). The peak at 2.05 ppm is due to a small amount of protonated acetone in the deuteroacetone.

They couple with the two adjacent protons of the methylene group. Centered at 2.5 ppm is a complex set of peaks due to the nonequivalent protons (H_B and H_C) of the allylic methylene groups [see "Diastereotopic Protons" in Technique 19.9,]. They couple with each other and with the protons that are α to the carboxylic acid groups. Overall, the peaks centered at 2.5 ppm have a relative integration of four protons.

At first glance you might not even notice the broad, low peak near 10 ppm that is produced by the protons (H_E) attached to the oxygen atoms of the carboxylic acid groups. Acidic protons are strongly H-bonded, which makes them highly deshielded and far downfield in the NMR spectrum. In addition, they experience a range of molecular environments, which can produce a very broad peak.

Finding the stereochemistry of electrophilic addition of Br_2 to the alkene is made much simpler by converting the dibromo-dicarboxylic acid to its dimethyl ester before NMR analysis. The NMR spectrum of dimethyl 4,5-dibromocyclohexane-*cis*-1,2-dicarboxylate, which you will synthesize in Project 9.3 from the diacid, methanol, and sulfuric acid, is still quite complex. However, the CH_3 peaks of the dimethyl ester appear at about 3.7 ppm, a region of the spectrum far from the complex peaks of the other protons in the molecule. It is the chemical shifts of these methyl signals that will help you answer the question on the stereochemistry of Br_2 addition.

***anti* Addition of Bromine**

First, let us consider the dimethyl ester that would form in a stereospecific *anti* addition.

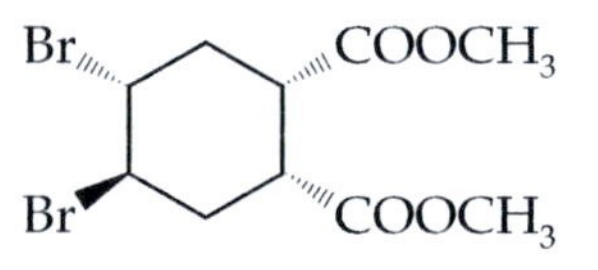

In the predominant chair conformation of the dimethyl dibromoester both bromine substituents are equatorial, while one of the two cis carbomethoxy groups is equatorial and the other is axial. Thus, three of the large substituents are equatorial and one is axial. The population of the less stable chair conformation is very small because it has three large axial substituents and only one equatorial group.

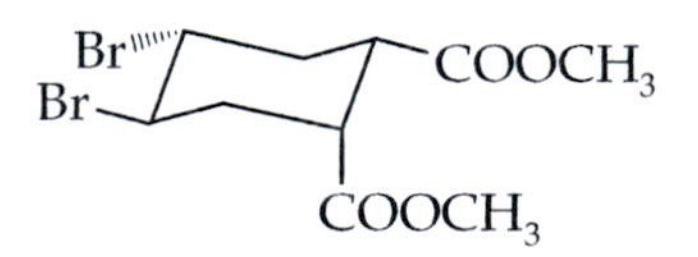

Dimethyl *trans*-4,5-dibromocyclohexane-*cis*-1,2-dicarboxylate has no plane of symmetry, so the environments of the two methyl groups on the ester are inherently different. The two methyl groups always experience somewhat different chemical shifts. To sum up, *anti* addition through a bromonium ion pathway would give two methyl singlets having equal NMR integrations.

***syn* Addition of Bromine**

The two possible products of *syn* addition are diastereomers of each other, so they would probably be formed in unequal amounts. Because each diastereomer has a plane of symmetry, each dimethyl ester would give a singlet NMR peak for the time-averaged identical methyl protons. Hence, for a *syn* addition one would expect two methyl singlets near 3.7 ppm, whose relative areas are unequal. If one of the diastereomers formed much faster than the other, only one singlet might actually appear.

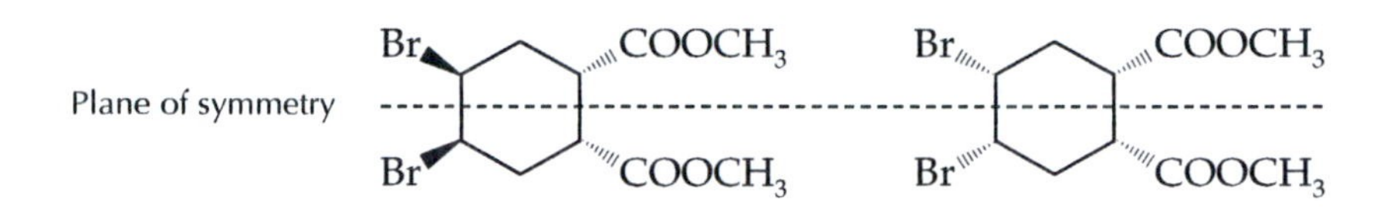

If the addition of Br_2 to the carbon-carbon double bond follows the carbocation mechanism, you would see both *syn* and *anti* addition products. It is likely that the NMR spectrum would show a complex pattern of four different methyl peaks of different integrations near 3.7 ppm.

Analysis of the NMR spectrum of your dimethyl ester product in Project 9.3 will determine whether the bromonium ion pathway or the carbocation pathway wins out in the electrophilic addition of Br_2 to cyclohex-4-ene-*cis*-1,2-dicarboxylic acid.

9.1 Synthesis of Cyclohex-4-ene-*cis*-1,2-dicarboxylic Acid

PURPOSE: To synthesize cyclohex-4-ene-*cis*-1,2-dicarboxylic acid, a necessary intermediate for studying the stereochemistry of electrophilic addition of Br_2 to a C=C double bond.

PRELABORATORY ASSIGNMENT: In what major ways will the IR spectrum of cyclohex-4-ene-*cis*-1,2-dicarboxylic anhydride differ from the IR spectrum of cyclohex-4-ene-*cis*-1,2-dicarboxylic acid?

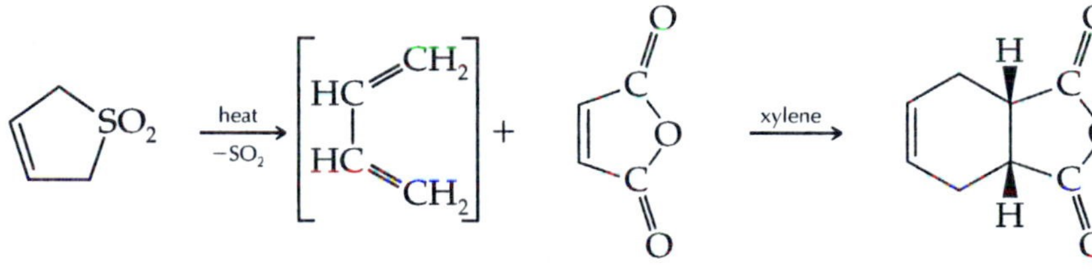

Butadiene sulfone
mp 65–66°C
MW 118.2

1,3-Butadiene

Maleic anhydride
mp 52°C
MW 98.1

Cyclohex-4-ene-*cis*-1,2-dicarboxylic anhydride
mp 105–106°C
MW 152.2

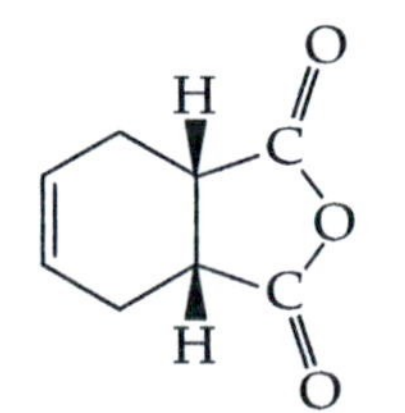

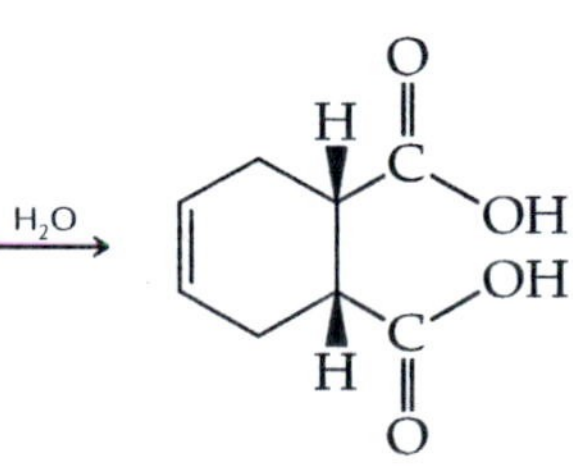

Cyclohex-4-ene-*cis*-1,2-dicarboxylic anhydride

Cyclohex-4-ene-*cis*-1,2-dicarboxylic acid
mp 168–169°C
MW 170.2

MINISCALE PROCEDURE

Techniques Removal of Noxious Vapors: Technique 7.4
IR Spectroscopy: Technique 18
NMR Spectroscopy: Technique 19

SAFETY INFORMATION

Conduct this experiment in a hood, if possible.

Butadiene sulfone (3-sulfolene) is an irritant. Wear gloves and avoid contact with skin, eyes, and clothing. This compound emits toxic corrosive sulfur dioxide when it is heated. Be sure that the gas trap is positioned before you begin heating the reaction mixture.

Maleic anhydride is toxic and corrosive. Avoid breathing the dust and avoid contact with skin, eyes, and clothing.

Xylene is flammable.

Synthesis of the Diels-Alder Adduct

Combine 2.0 g of butadiene sulfone, 1.2 g of finely ground maleic anhydride, 0.80 mL of xylene (a mixture of isomers is all right), and a boiling stone in a 25-mL round-bottomed flask. Fit a Claisen adapter to the flask, close one hole with a glass stopper, and put a water-cooled condenser in the other hole. Prepare a gas trap at the top of the condenser with a thermometer adapter, a piece of U-shaped glass tubing, and a 125-mL filter flask fitted with a one-hole rubber stopper and containing about 50 mL of water; position the U-tube so that its tip is *just above* the surface of the water [see Technique 7.4, Figure 7.5a]. Alternatively, in a laboratory equipped with water aspirators, place a vacuum adapter in the top of the condenser and attach its side arm to the side arm of the vacuum aspirator; turn on the water at a moderate flow [see Technique 7.4, Figure 7.5b].

Begin heating the mixture gently with a heating mantle or sand bath. After the solids dissolve, continue heating the mixture at a gentle reflux for 30 min. Remove the heat source and cool the reaction mixture for about 5 min before proceeding immediately with the hydrolysis of the anhydride product.

Hydrolysis of the Diels-Alder Adduct

Remove the gas trap from the top of the condenser. Pour 4 mL of water down the condenser, add another boiling stone, and heat the mixture under reflux for 30 min. Cool the solution to room temperature. If crystallization of the product does not occur, add 3 or 4 drops of concentrated sulfuric acid, stir the contents of the flask, and cool the resulting mixture in an ice-water bath for 5 min. Collect the product by vacuum filtration. Wash the crystals twice with 1-mL portions of ice-cold water. The dicarboxylic acid is usually quite pure without recrystallization. If time permits, the product can be recrystallized from water [see Technique 9.5]. Allow the product to dry overnight before determining the melting point and percent yield. Obtain an IR spectrum [see Technique 18.4] and an NMR spectrum in deuterated acetone [see Technique 19.2] as directed by your instructor.

CLEANUP: Pour the filtrate from the reaction mixture into the container for flammable organic waste. Pour the water in the gas trap into a large beaker. Neutralize the solution with solid sodium carbonate before washing it down the sink or pouring it into the container for aqueous inorganic waste.

9.2 Bromination of Cyclohex-4-ene-*cis*-1,2-dicarboxylic Acid

PURPOSE: To add Br_2 to cyclohex-4-ene-*cis*-1,2-dicarboxylic acid in preparation for studying the stereochemistry of this electrophilic addition reaction.

PRELABORATORY ASSIGNMENT: Draw a likely Lewis structure for the tribromide anion, Br_3^-.

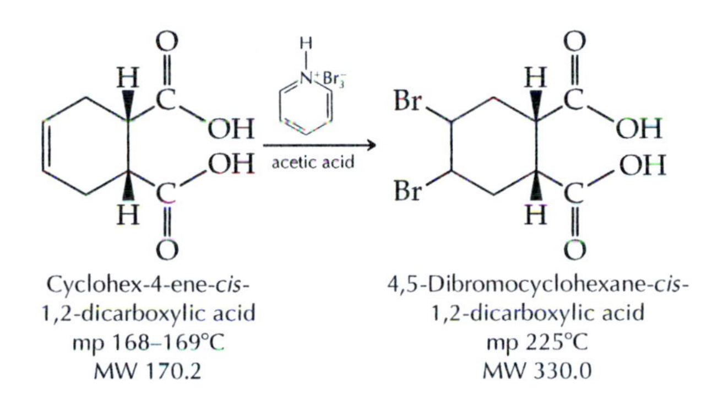

Cyclohex-4-ene-*cis*-1,2-dicarboxylic acid
mp 168–169°C
MW 170.2

4,5-Dibromocyclohexane-*cis*-1,2-dicarboxylic acid
mp 225°C
MW 330.0

MINISCALE PROCEDURE

SAFETY INFORMATION

Pyridinium tribromide is corrosive and a lachrymator (it causes tears). Wear gloves and avoid breathing its dust or contact with skin, eyes, and clothing.

Acetic acid is a dehydrating agent and an irritant, and it causes burns. Dispense it only in a hood. Wear gloves and avoid contact with skin, eyes, and clothing.

Pyridine is produced during the cleanup procedure. It is an irritant and has an unpleasant odor. Wear gloves and carry out the cleanup procedure in a hood.

If you are using the reagent quantities specified in the following procedure, submit your remaining product from Project 9.1 to your instructor. If your instructor asks you to use your entire product from Project 9.1 for this procedure, you will need to proportionally adjust the quantities of all reagents.

Place 0.66 g of cyclohex-4-ene-*cis*-1,2-dicarboxylic acid and 8.0 mL of glacial acetic acid in a 25-mL Erlenmeyer flask. Add a magnetic stirring bar and stir the mixture until the solid dissolves. Then add 1.30 g of pyridinium tribromide and continue stirring the reaction mixture. The red pyridinium tribromide will dissolve and the mixture should become light yellow. A solid will begin to precipitate. Continue stirring at room temperature for 45 min.

Pour the reaction mixture into a beaker containing about 20 g of ice and stir the resulting mixture with a magnetic stirrer until the ice has melted. Collect the crude product on a Buchner funnel by vacuum filtration. Disconnect the vacuum and wash the product by covering the crystals with cold water, then reconnect the vacuum to remove the water. Repeat the washing process two or three more times until the odor of acetic acid is no longer perceptible. Allow the crystals to air dry before determining their mass and melting point. The evolution of a gas will be visible at the melting point. Calculate the percent yield.

CLEANUP: Pour the filtrate from the reaction mixture into a beaker. Take the beaker to a hood. Reduce any remaining pyridinium tribromide by adding sodium bisulfite in small portions with a spatula until starch-iodide test paper no longer turns brown. Neutralize the solution with solid sodium carbonate. Pour the neutralized solution into a separatory funnel and extract the remaining pyridine with 5 mL of diethyl ether. The separated aqueous phase can be washed down the sink or placed in the container for aqueous inorganic waste. Place the ether solution in the container for flammable waste.

9.3 Synthesis of Dimethyl 4,5-Dibromocyclohexane-*cis*-1,2-dicarboxylate

QUESTION: What is the stereochemistry for the addition of Br_2 to the carbon-carbon double bond of cyclohex-4-ene-*cis*-1,2-dicarboxylic acid?

PRELABORATORY ASSIGNMENT: Draw the structures of the by-products that are removed from the desired dimethyl ester by using a saturated $NaHCO_3$ solution in the product workup.

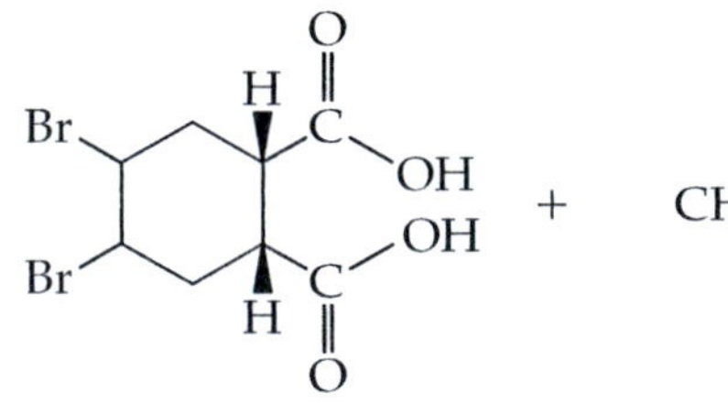

CH_3OH $\xrightarrow{H_2SO_4}$

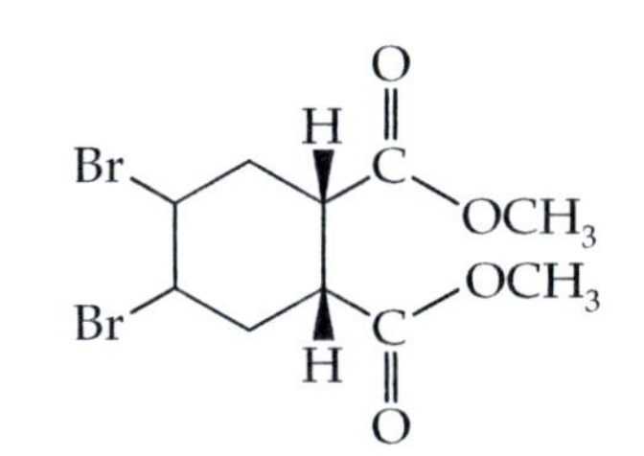

4,5-Dibromocyclohexane-*cis*-1,2-dicarboxylic acid

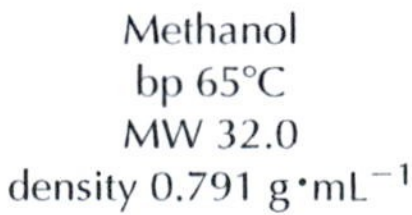

Methanol
bp 65°C
MW 32.0
density 0.791 g·mL^{-1}

Dimethyl 4,5-dibromocyclohexane-*cis*-1,2-dicarboxylate
MW 358.0

MINISCALE PROCEDURE

Techniques Extraction: Technique 8.2
NMR Spectroscopy: Technique 19

SAFETY INFORMATION

Methanol is volatile, toxic and flammable. Dispense it only in a hood.

Concentrated sulfuric acid is corrosive and causes severe burns. Wear gloves while measuring it and avoid contact with skin, eyes, and clothing.

Dichloromethane is toxic, an irritant, absorbed through the skin, and harmful if inhaled. Use it only in a hood and wear neoprene gloves while doing the extractions and evaporation.

Note: If you have more or less than 0.72 g of 4,5-dibromocyclohexane-*cis*-1,2-dicarboxylic acid from Project 9.2, proportionally adjust the reagent quantities for the amount of diacid you have.

Place 0.72 g of 4,5-dibromocyclohexane-*cis*-1,2-dicarboxylic acid, 4.0 mL of methanol, and 1.0 mL of concentrated sulfuric acid in a 10- or 25-mL round-bottomed flask. Add a magnetic stirring bar and fit a water-jacketed condenser to the flask. Stir and reflux the contents of the flask for 45 min at 100°C.

After cooling the flask to room temperature, *slowly* pour the reaction mixture into a 50-mL beaker containing 25 mL of saturated sodium bicarbonate solution. **(Caution: Foaming will occur.)** Stir the mixture until foaming ceases. If the mixture is not basic to litmus paper, add another 5 mL of bicarbonate solution.

Transfer the aqueous mixture to a separatory funnel. Before extracting the aqueous mixture three times with 10-mL portions of dichloromethane, rinse the beaker with the first 10-mL portion of dichloromethane, then add the dichloromethane rinse to the separatory funnel. Continue with the extraction by shaking and venting the funnel, allow the phases to separate, and drain the organic phase into a labeled Erlenmeyer flask. Repeat the extraction of the aqueous phase with the other two portions of dichloromethane. Combine the dichloromethane layers in the same flask. Set the aqueous phase aside.

Wash the combined dichloromethane extracts with 20 mL of saturated sodium chloride solution. Dry the organic phase with anhydrous magnesium sulfate for at least 10 min. Filter half the dried dichloromethane solution through a fluted filter paper into a tared (weighed) 25-mL Erlenmeyer flask. Add a boiling stick and evaporate most of the solvent on a steam bath. Filter the remaining dichloromethane solution into the flask and complete the evaporation of solvent. Alternatively, the solvent can be removed with a rotary evaporator. If crystallization does not occur, stir the residue in the flask with several milliliters of hexane to promote crystallization and collect the solid on a Hirsch funnel by vacuum filtration.

Allow your product to dry for 30 min in a hood. Prepare a sample for NMR analysis as directed by your instructor using $CDCl_3$ as the solvent [see Technique 19.2]. Determine the mass of your product and calculate the percent yield.

CLEANUP: The aqueous solutions remaining from the extractions can be washed down the sink or poured into the container for aqueous inorganic waste. The hexane filtrate should be poured into the container for halogenated waste.

Interpretation of Experimental Data

Analyze the NMR spectrum that you obtained for your final product. What does your spectrum indicate about the stereochemistry for the addition of Br_2 to the carbon-carbon double bond of cyclohex-4-ene-*cis*-1,2-dicarboxylic acid?

Reference

1. Sample, T. E.; Hatch, L. F. *J. Chem. Educ.* **1968,** *45,* 55–56.

Questions

1. How does using water as the reaction solvent in Project 9.1 ensure that the hydrolysis reaction goes to completion?
2. Write out a reaction mechanism for the hydrolysis of the dicarboxylic anhydride in Project 9.1.
3. What allows the pyridinium bromide byproduct in Project 9.2 to stay dissolved in the acetic acid/water mixture while the bromine addition product precipitates from solution?
4. Draw the most stable chair conformation for the product of *anti* addition of Br_2 to cyclohex-4-ene-*cis*-1,2-dicarboxylic acid.
5. Draw the most stable chair conformation for each of the two diastereomers that would be formed from the *syn* addition of bromine to cyclohex-4-ene-*cis*-1,2-dicarboxylic acid. Predict which diastereomer would have the lower energy.

PROJECT

10

DIELS-ALDER CYCLOADDITIONS OF DIENOPHILES TO 2,3-DIMETHYL-1,3-BUTADIENE

QUESTION: What are the relative rates of dienophiles and dienes in the Diels-Alder reaction?

In this three- to four-week project, with a choice of four separate options, you will be part of a two-person team investigating Diels-Alder reactions. In Project 10.1 you can compare the rates of two dienes, 2,3-dimethyl-1,3-butadiene and 1,3-cyclohexadiene with maleic anhydride, using GC analysis and reaction conditions in which the dienophiles directly compete with one another. Projects 10.2 and 10.3 use the same experimental approach to compare the reaction rates of 2,3-dimethyl-1,3-butadiene with different dienophiles, including cis and trans alkenes and an alkyne. In Project 10.4 you will be able to design a two-step synthesis of the aromatic compound dimethyl 4,5-dimethylbenzene-1,2-dicarboxylate by Diels-Alder and dehydrogenation chemistry. In each of these options there is an element of experimental design: the two-student team will choose experimental conditions for carrying out the reactions, the GC assays, and NMR spectroscopy.

Diels-Alder Reaction

The Diels-Alder reaction has great utility in the chemical synthesis of six-membered ring compounds, from simple cyclohexanes to complex steroids. Otto Diels and Kurt Alder received the Nobel prize in 1950 for discovering the reaction that bears their names. It is a [4 + 2] cycloaddition reaction in which two molecules add together in one step to form a cyclic compound. The conjugated diene provides 4 π-electrons and the dienophile, which is usually an alkene or an alkyne, provides the other 2 π-electrons. This allows a concerted reaction with a stabilized aromatic-like transition state having 6 π-electrons. In the transition state, *p*-orbitals of all six carbon atoms overlap simultaneously.

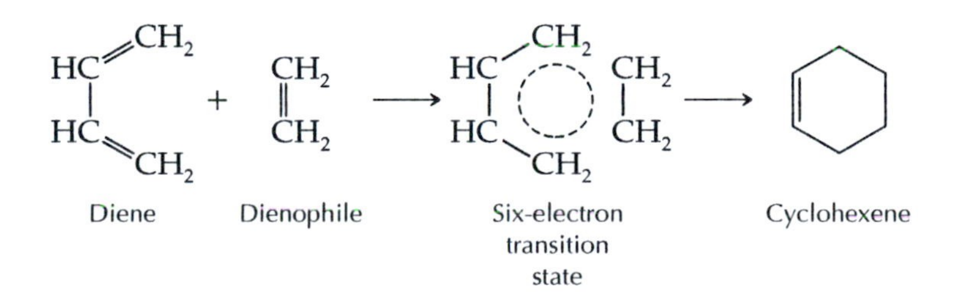

Diels-Alder reactions are stereospecific with respect to the geometry of the dienophile, and the configuration at the dienophile's C=C is retained in the product. Thus, a cis dienophile will produce a cis Diels-Alder product and a trans dienophile will produce a trans product.

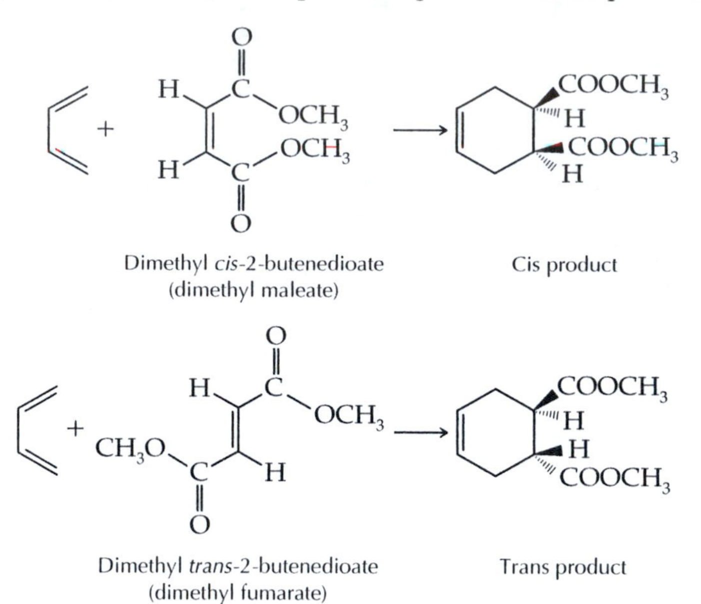

Dimethyl *cis*-2-butenedioate (dimethyl maleate) Cis product

Dimethyl *trans*-2-butenedioate (dimethyl fumarate) Trans product

Because a carbon-carbon double bond in a cyclohexene ring must have a cis configuration (*trans*-cyclohexene is unknown), the trans-like conformation of butadiene must rearrange to the cis-like conformation before reaction with the dienophile can proceed.

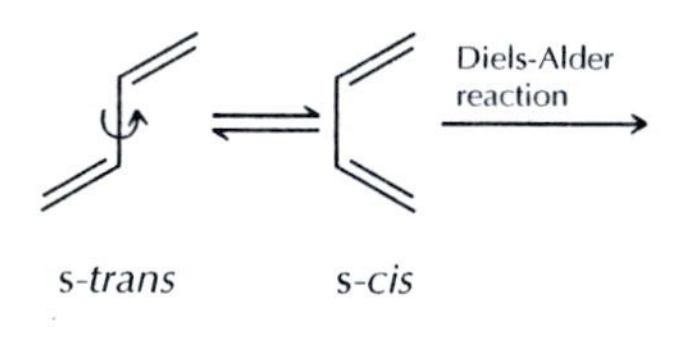

The Diels-Alder reaction is also stereoselective with respect to the orientation of the reactants in the transition state. When a cyclic diene reacts with a cyclic dienophile, a bridged tricyclic product forms. The new cis five-membered ring in the product has an *endo* configuration, where the ring is trans to the saturated carbon bridge. *Endo* products are frequently formed when Diels-Alder reactions produce bridged products.

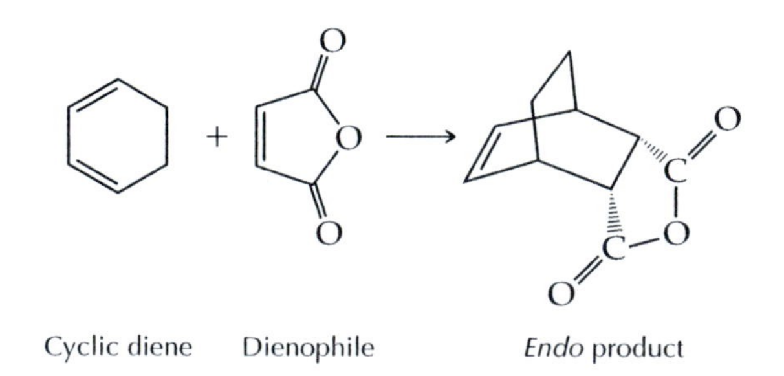

Cyclic diene Dienophile *Endo* product

Although simple in concept, the Diels-Alder reaction of 1,3-butadiene and ethene proceeds slowly and with a modest yield of cyclohexene. The reaction has a much faster rate when the C=C of the dienophile is electron-deficient and the diene is electron-rich, because these electronic factors cause more favorable orbital overlap in the Diels-Alder transition state. For example, in Project 10.1, the two electron-withdrawing carbonyl groups in maleic anhydride and the two methyl groups in 2,3-dimethyl-1,3-butadiene significantly speed up the Diels-Alder reaction.

Relative Reaction Rates

Perhaps the simplest way to measure relative reaction rates of two substrates is to run a competition experiment. The competition experiments for Project 10.1 need to have both dienes, 2,3-dimethyl-1,3-butadiene and 1,3-cyclohexadiene, as well as the dienophile, present in the same reaction mixture. Thus, both dienes compete directly for reaction with the dienophile.

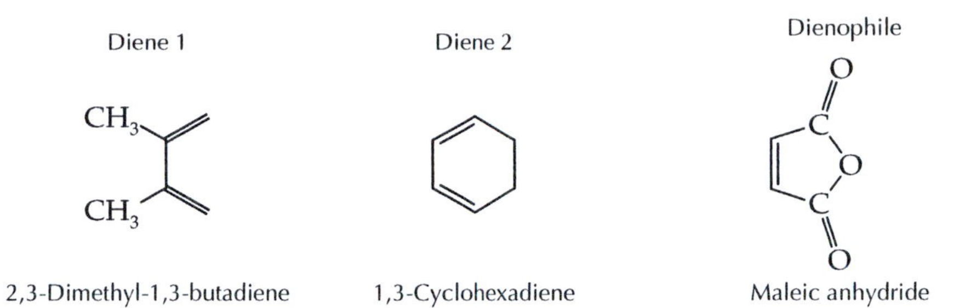

In general terms, the two competitive reactions are

$$\text{Diene 1} + \text{Dienophile} \xrightarrow{k_1} \text{Product 1} \tag{1}$$

$$\text{Diene 2} + \text{Dienophile} \xrightarrow{k_2} \text{Product 2} \tag{2}$$

The rate laws for the two reactions are

$$\text{Rate}_1 = k_1\,[\text{Diene 1}]\,[\text{Dienophile}] \tag{3}$$

$$\text{Rate}_2 = k_2\,[\text{Diene 2}]\,[\text{Dienophile}] \tag{4}$$

In a competition experiment, the concentration (moles/liter) of the dienophile in equations (1) and (2) is the same. Division of equation 3 by equation 4 gives the relative rate ratio

$$\frac{\text{Rate}_1}{\text{Rate}_2} = \frac{k_1\,[\text{Diene 1}]}{k_2\,[\text{Diene 2}]} \tag{5}$$

If the molar concentrations of the two dienes are much greater than the concentration of the dienophile or if the rate data are obtained in the very early part of the reaction, say, less than 5–10% reaction, the concentrations of the two dienes do not change much.

Under these conditions and when a competition experiment is run so that the initial molar concentrations of the two dienes are equal,

$$\frac{\text{Rate}_1}{\text{Rate}_2} \cong \frac{k_1}{k_2} \tag{6}$$

The rate-constant ratio k_1/k_2 is approximately equal to the relative rate ratio for the two dienes. $\text{Rate}_1/\text{Rate}_2$ can most easily be determined by the relative amounts of the two Diels-Alder products that are produced.

In Projects 10.2 and 10.3, the same experimental approach is used to compare the reaction rates of the diene 2,3-dimethyl-1,3-butadiene with two dienophiles that differ only in the geometry of the carbon-carbon double bond. In Project 10.2, the relative reaction rates of the cis and trans isomers of diethyl 2-butenedioate are compared.

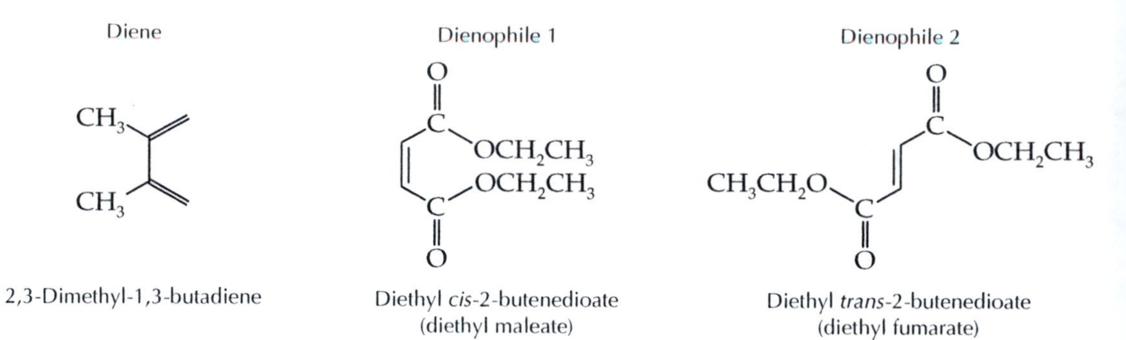

2,3-Dimethyl-1,3-butadiene

Diethyl *cis*-2-butenedioate (diethyl maleate)

Diethyl *trans*-2-butenedioate (diethyl fumarate)

In Project 10.3 one of the dienophiles is an alkene, dimethyl *cis*-2-butene-dioate, and the other is its analogous alkyne, dimethyl 2-butynedioate.

Diene

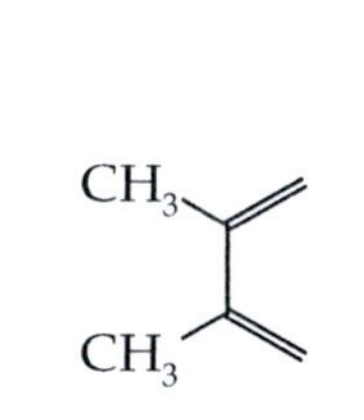

2,3-Dimethyl-1,3-butadiene

Dienophile 1

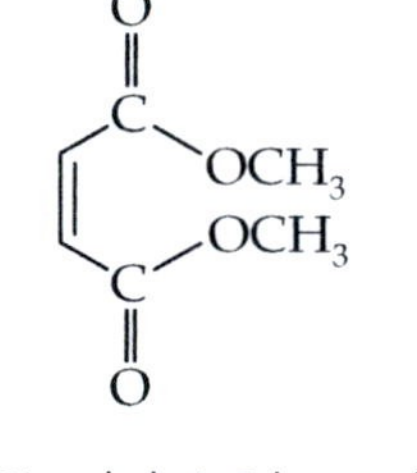

Dimethyl *cis*-2-butenedioate (dimethyl maleate)

Dienophile 2

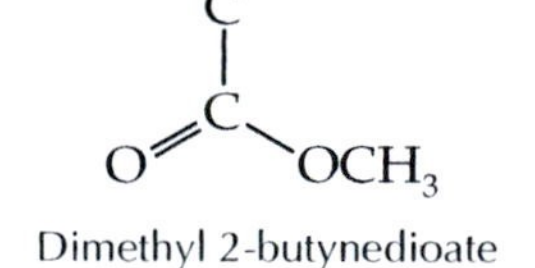

Dimethyl 2-butynedioate (dimethyl acetylenedicarboxylate)

In general terms, the two competitive reactions are

$$\text{Diene} + \text{Dienophile 1} \xrightarrow{k_1} \text{Product 1} \quad (7)$$

$$\text{Diene} + \text{Dienophile 2} \xrightarrow{k_2} \text{Product 2} \quad (8)$$

The rate laws for the two reactions are

$$\text{Rate}_1 = k_1 \,[\text{Diene}]\,[\text{Dienophile 1}] \quad (9)$$

$$\text{Rate}_2 = k_2 \,[\text{Diene}]\,[\text{Dienophile 2}] \quad (10)$$

In the competition experiment, the concentration (moles/liter) of diene in equations (9) and (10) is the same. Division of the two equations gives

$$\frac{\text{Rate}_1}{\text{Rate}_2} = \frac{k_1\,[\text{Dienophile 1}]}{k_2\,[\text{Dienophile 2}]} \quad (11)$$

Again, if the molar concentrations of the dienophiles are equal and if the rate data are obtained from the very early part of the reaction,

$$\frac{\text{Rate}_1}{\text{Rate}_2} \cong \frac{k_1}{k_2} \tag{12}$$

Thus, k_1/k_2 equals the relative rate ratio for the two dienophiles. $\text{Rate}_1/\text{Rate}_2$ can be determined most directly by the relative concentrations of the two Diels-Alder products that are produced.

Analysis of Competition Reaction Mixtures

Gas chromatography is the best way to measure the relative concentrations of Diels-Alder products that you need to calculate the relative reaction-rates of these dienes and dienophiles [see Technique 16, especially 16.6 and 16.7], although NMR spectroscopy could also be useful. Because the different dienes in Project 10.1 or dienophiles in Projects 10.2 and 10.3 have very similar structures, you can assume that in a given competition experiment the GC molar response factors of the two products are the same.

If you begin a competition experiment with equimolar concentrations of the reactants you are comparing, you can simply determine the relative amounts of the two products after 5–10% of the reaction has taken place. This can be done by first running each Diels-Alder reaction separately to determine the GC retention times of their products. At the same time, you should obtain an NMR spectrum of the reaction mixtures to determine if any NMR peaks for your reaction products can be useful in corroborating their relative concentrations [see Technique 19, especially 19.9].

It is also possible to measure the relative rates in a competition reaction by determining the relative amounts of the two competing substrates that remain unreacted. This approach, however, has two pitfalls. First, you would be determining a small extent of reaction by comparing differences in large numbers. Second, you would need to use an internal GC standard to determine the relative amounts of the two competing substrates that have reacted. Determining the relative amounts of the two Diels-Alder products themselves is much more straightforward.

Designing the Synthesis of an Aromatic Ring by Dehydrogenation

In Project 10.4 you have the opportunity to design and carry out a Diels-Alder reaction using a dienophile with a carbon-carbon triple bond:

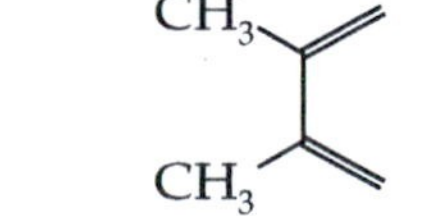
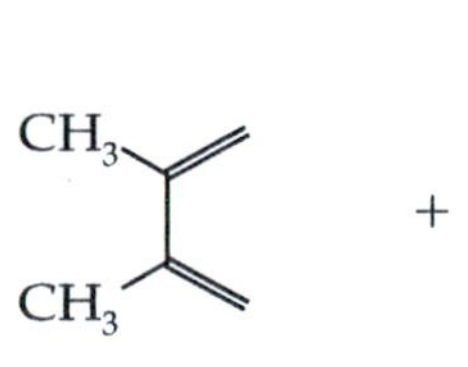

2,3-Dimethyl-1,3-butadiene

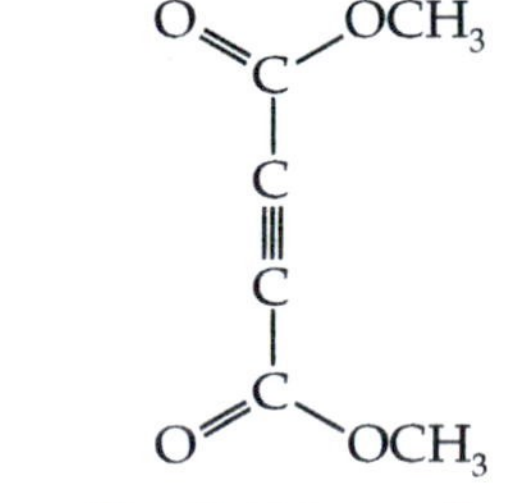

Dimethyl 2-butynedioate (dimethyl acetylenedicarboxylate)

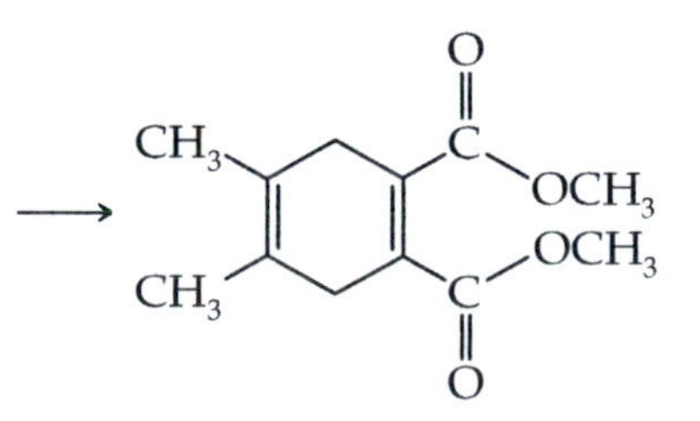

Dimethyl 4,5-dimethylcyclohexa-1,4-diene-1,2-dicarboxylate

The Diels-Alder reaction with an alkyne produces a cyclohexadiene, which can be converted to the highly stable benzene ring by dehydrogenation.

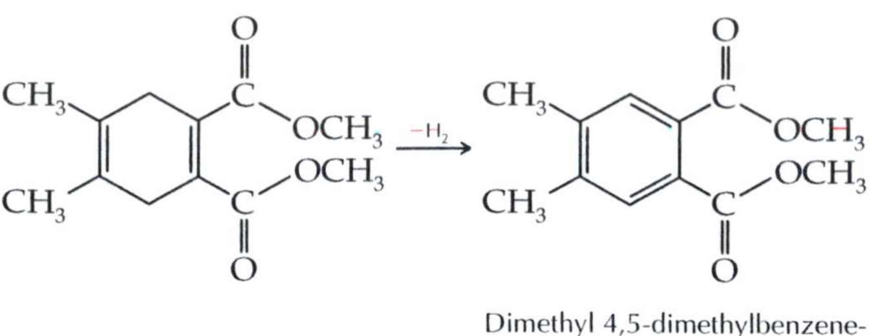

Dimethyl 4,5-dimethylbenzene-1,2-dicarboxylate

This dehydrogenation reaction can take place under exceptionally mild conditions due to the driving force of the stability of the benzene ring. Even in the absence of a catalyst it is possible to produce the aromatic compound by heating the Diels-Alder product in the presence of air. However, a better way to carry out the reaction is to use a metal catalyst like that used for the addition of hydrogen gas to an alkene, except that in this case the metal catalyst is catalyzing an oxidation rather than a reduction reaction. In the absence of a H_2 atmosphere, Pd/C catalyzes the loss of H_2 from the cyclohexadiene. Another way to carry out the dehydrogenation is to react the cyclohexadiene with a mild oxidizing agent, such as elemental sulfur or tetrachloro-*p*-benzoquinone (chloranil). When sulfur is the hydrogen acceptor, H_2S gas forms, and when the quinone is the hydrogen acceptor, tetrachlorohydroquinone forms.

Perhaps the most efficient way to design your Diels-Alder reaction of 2,3-dimethyl-1,3-butadiene and dimethyl 2-butynedioate is by modeling it on other Diels-Alder syntheses in Project 10. In the design of the dehydrogenation reaction, however, you will need to use the literature of organic chemistry.

If the SciFinder Scholar database is available on your campus, this excellent search engine is a good place to begin the design of your synthesis of dimethyl 4,5-dimethylbenzene-1,2-dicarboxylate. You should be able to find specific references to primary chemical journal articles for the synthesis of this compound. In addition, the following reference books, which contain methods, procedures, and reactions used in organic synthesis, may be helpful.

1. Larock, R. C. *Comprehensive Organic Transformations: A Guide to Functional Group Preparations;* 2nd ed.; Wiley: New York, 1999.
2. Fieser, L. F.; Fieser, M. *Reagents for Organic Synthesis;* 19 vols.; Wiley: New York, 1967–2000.
3. *Handbook of Reagents for Organic Synthesis;* 4 vols.; Wiley: Chichester, UK, 1999.
4. Harrison, I. T.; Wade, Jr., L. G.; Smith, M. B. (Eds.) *Compendium of Organic Synthetic Methods;* 8 Vols.; Wiley: New York, 1971–1995.
5. Mackie, R. D.; Smith, D. M.; Aitken, R. A. *Guidebook to Organic Synthesis;* 2nd ed.; Halsted: New York, 1990.

10.1 Competition Reactions of 2,3-Dimethyl-1,3-butadiene and 1,3-Cyclohexadiene with Maleic Anhydride

QUESTION: Which diene, 2,3-dimethyl-1,3-butadiene or 1,3-cyclohexadiene, reacts faster with maleic anhydride and what are their relative rates?

TEAMWORK: Work in teams of two students for this project.

PRELABORATORY ASSIGNMENT: Predict the chemical shifts and relative integrations of the various protons in the NMR spectrum of 4,5-dimethylcyclohex-4-ene-*cis*-1,2-dicarboxylic anhydride.

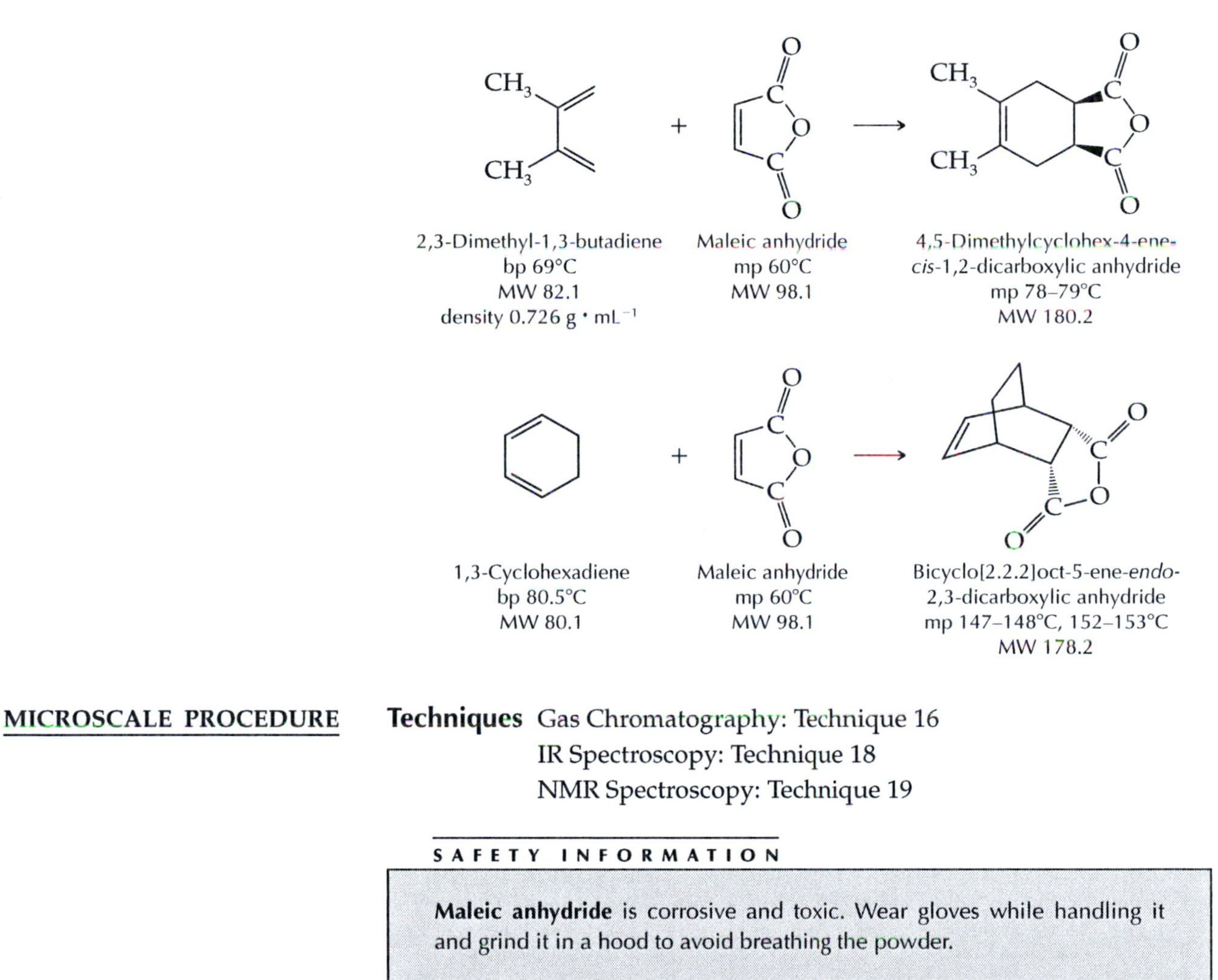

MICROSCALE PROCEDURE

Techniques Gas Chromatography: Technique 16
IR Spectroscopy: Technique 18
NMR Spectroscopy: Technique 19

SAFETY INFORMATION

Maleic anhydride is corrosive and toxic. Wear gloves while handling it and grind it in a hood to avoid breathing the powder.

2,3-Dimethyl-1,3-butadiene is extremely volatile and flammable.

Hexane is extremely volatile and flammable. Heat it only on a steam bath or in a hot-water bath.

Synthesis of 4,5-Dimethylcyclohex-4-ene-cis-1,2-dicarboxylic Anhydride

Weigh 4.00 mmol of powdered maleic anhydride; grind the maleic anhydride with a mortar and pestle before weighing it if it has not been ground previously. Measure 4.00 mmol of 2,3-dimethyl-1,3-butadiene with a graduated pipet and transfer it to a 10-mL round-bottomed flask. Add the powdered maleic anhydride to the diene and fit the flask with a water-cooled reflux condenser. Within a few minutes, the reaction should begin spontaneously. Heat will be evolved as the reaction takes place, and the reaction mixture will boil vigorously.

When the reaction has ceased, allow the flask to cool to room temperature and the reaction mixture to solidify. It may take a few minutes before crystals appear. Remove the condenser. Add 2 mL of cold water and break up any large chunks of product with a microspatula until the product consists of fine white crystals.

Allow the crystals to settle to the bottom of the flask. Using a Pasteur pipet fitted with a rubber bulb, expel the air from the rubber bulb and place the tip of the Pasteur pipet firmly against the bottom of the flask as shown in Figure 10.1. The pipet must be in a vertical position. Gently release the pressure on the rubber bulb. The supernatant liquid will be drawn into the pipet with few, if any, crystals. Remove the pipet and transfer the liquid to a 50-mL beaker. Continue this process until all the liquid is removed from the crystals. Repeat the washing process with 2-mL portions of water until a drop of the wash liquid placed on Congo Red test paper no longer gives an acid reaction. Three or four washes are usually sufficient. Then collect the solid product on a small Buchner funnel using vacuum filtration. Use the vacuum system to draw air over the crystals for 20 min.

Congo Red paper turns blue in acidic solution.

Transfer the dry crystals to a 50-mL Erlenmeyer flask and recrystallize the product from hexane, using a steam bath or a beaker of water at 70–75°C as the heat source. Calculate the percent yield and obtain the product's melting point and infrared spectrum. The ^{1}H NMR spectrum of the product is also quite revealing. Obtain the ^{1}H NMR spectrum of your 4,5-dimethylcyclohex-4-ene-*cis*-1,2-dicarboxylic anhydride product in $CDCl_3$, as directed by your instructor. Store your product in a tightly capped vial for use as a reference compound throughout the project.

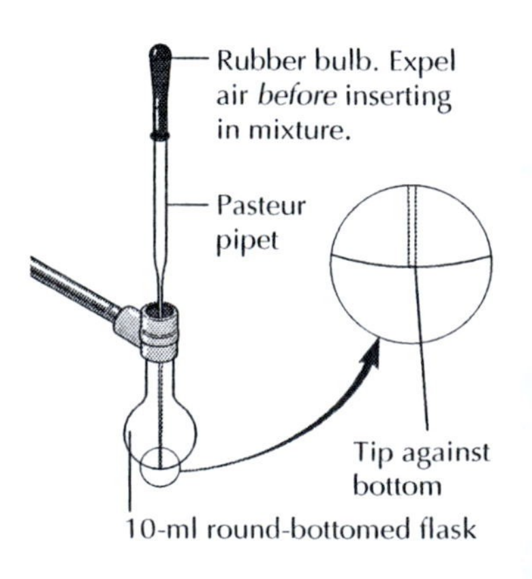

FIGURE 10.1 Position of Pasteur pipet for removing supernatant liquid from crystals.

CLEANUP: Combine the aqueous washes and the aqueous filtrate and neutralize the solution with sodium carbonate **(Caution: Foaming.)** before washing it down the sink or pouring it into the container for aqueous inorganic waste. Pour the hexane filtrate from the recrystallization into the container for flammable (organic) waste.

Synthesis of Bicyclo[2.2.2]oct-5-ene-endo-2,3-dicarboxylic Anhydride

As preparation for the relative-rate study, design the synthesis of bicyclo[2.2.2]oct-5-ene-*endo*-2,3-dicarboxylic anhydride by reacting 4.00 mmol of maleic anhydride and 4.00 mmol of 1,3-cyclohexadiene. Model the protocol on the previous reaction of 2,3-dimethyl-1,3-butadiene and maleic anhydride, but use more heat to get the reaction going; a warm-water bath is a useful heating source. No workup is necessary. Your product mixture can be assayed directly by GC analysis of the reaction mixture dissolved in hexane or ether (5–10% solution) or by NMR analysis ($CDCl_3$). Store your product in a tightly capped vial for use as a reference compound throughout the project.

Determination of GC Parameters for Analysis of Product Mixture

Gas chromatography is an excellent technique for determining the relative rates. Use your individual Diels-Alder products to determine retention times on a nonpolar GC column. Temperature programming on a capillary GC instrument works especially well. If your gas chromatograph does not have temperature programming, begin with a column temperature of 140°C and determine the GC conditions that will analyze a mixture of the two products of the diene present in the reaction. You need to determine the chromatographic conditions that will work best on your particular instrument [see Techniques 16.6 and 16.7]. Alternatively, NMR spectroscopy can be used to analyze the product mixture.

Competition Experiments

In a competition reaction, use 2.00 mmol of maleic anhydride and 1.00 mmol each of 2,3-dimethyl-1,3-butadiene and 1,3-cyclohexadiene so that you have equimolar concentrations. Carrying out the competition reaction in refluxing diethyl ether for 5–10 min gives good GC data. When running a competition experiment, it is not necessary to isolate and purify the products. Take samples from the early part of the reaction with no more than 5–10% conversion to the Diels-Alder products. You need use only a GC or NMR analysis method that can determine the relative concentrations of the products.

CLEANUP: Your instructor must approve your cleanup procedures before you begin your experiments. Prepare a list of all by-products from each reaction. Consult the cleanup procedures in experiments you have done previously for ways of handling similar materials. Your instructor may also give you specific guidelines that pertain to disposal regulations in your institution and community.

Interpretation of Experimental Data

Analyze your gas chromatograms or NMR spectra from the competition experiments to determine the relative concentrations of the two products in the reaction mixtures. Calculate the relative rate ratio for the two dienes. Explain your results.

10.2 Competition Reactions of 2,3-Dimethyl-1,3-butadiene with Diethyl *trans*-2-Butenedioate and Diethyl *cis*-2-Butenedioate

QUESTION: Which dienophile, diethyl *trans*-2-butenedioate or diethyl *cis*-2-butenedioate, reacts faster with 2,3-dimethyl-1,3-butadiene and what are their relative rates?

TEAMWORK: Work in teams of two students for this project.

PRELABORATORY ASSIGNMENT: Design reaction conditions for determining the relative rates of diethyl *trans*-2-butenedioate and diethyl *cis*-2-butenedioate with 2,3-dimethyl-1,3-butadiene. Your plan should include quantities of reagents, reaction conditions and equipment, assay methods, balanced chemical equations, and safety considerations.

MICROSCALE PROCEDURE

Techniques Gas Chromatography: Technique 16
IR Spectroscopy: Technique 18
NMR Spectroscopy: Technique 19

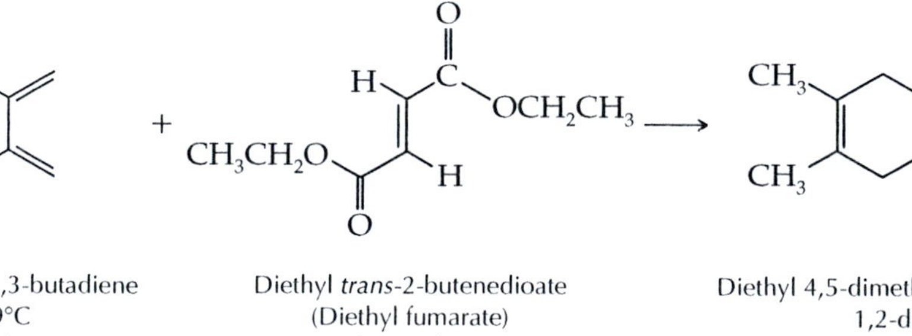

2,3-Dimethyl-1,3-butadiene
bp 69°C
MW 82.1
density 0.726 g · mL^{-1}

Diethyl *trans*-2-butenedioate
(Diethyl fumarate)
bp 214°C
MW 172.2
density 1.045 g · mL^{-1}

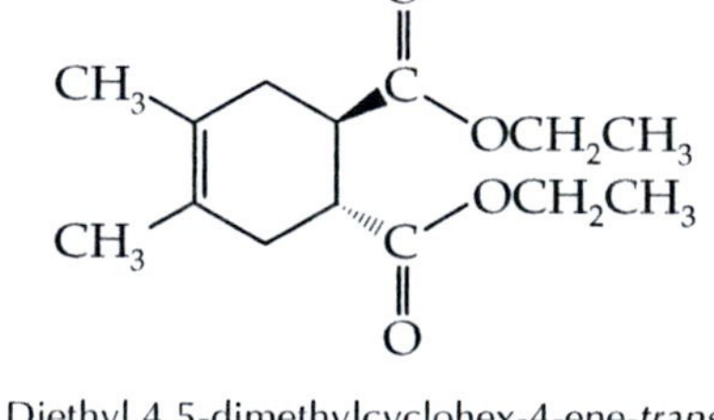

Diethyl 4,5-dimethylcyclohex-4-ene-*trans*-1,2-dicarboxylate
bp 104°C/5 Torr
MW 254.3

2,3-Dimethyl-1,3-butadiene
bp 69°C
MW 82.1
density 0.726 g · mL^{-1}

Diethyl *cis*-2-butenedioate
(Diethyl maleate)
bp 223°C
MW 172.2
density 1.006 g · mL^{-1}

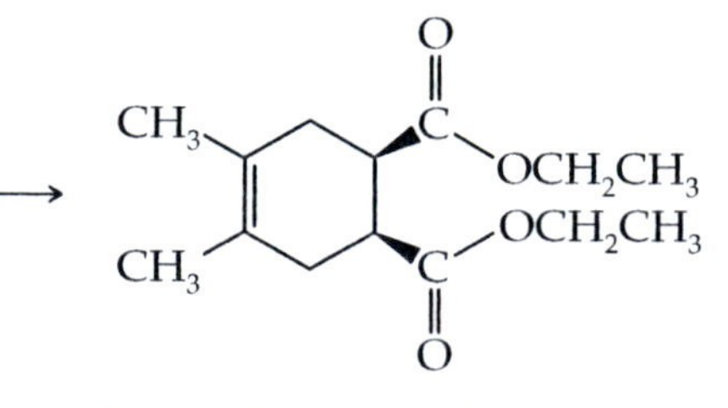

Diethyl 4,5-dimethylcyclohex-4-ene-*cis*-1,2-dicarboxylate
bp 153°C/7 Torr
MW 254.3

SAFETY INFORMATION

2,3-Dimethyl-1,3-butadiene is extremely volatile and flammable.

Diethyl *trans*-2-butenedioate (diethyl fumarate) and **diethyl *cis*-2-butenedioate (diethyl maleate)** require no hazardous-chemical handling procedures.

Synthesis of Diels-Alder Products

You will need to run each Diels-Alder reaction separately using 4.00 mmol of each reactant before you start any competition reactions. This will allow you to characterize the reactants and products by GC or NMR spectroscopy. The diethyl *cis*- and *trans*-2-butenedioates react more slowly than maleic anhydride (Project 10.1) in Diels-Alder reactions. Therefore, you will have to heat the reaction mixtures at reflux with no solvent for 1–2 h for them to go to completion. Efficient cooling of the condenser above the reaction flask is essential to prevent loss of volatile 2,3-dimethyl-1,3-butadiene during the reflux period. Remember that these Diels-Alder products are liquids, not solids. Reaction mixtures can be used without purification for GC analysis. Store your products in tightly capped vials to use as references throughout the project.

GC or NMR Analysis of Product Mixtures

Gas chromatography is an excellent technique for determining the relative rates. Use your individual Diels-Alder products to determine retention times on a nonpolar GC column. Temperature programming on a capillary GC instrument works especially well. If your gas chromatograph does not have temperature programming, begin with a column temperature of 140°C and determine the GC conditions that will analyze a mixture of the two products. You need to determine the chromatographic conditions that will work best on your particular instrument [see Techniques 16.6 and 16.7].

Alternatively, NMR can be used to analyze the reaction mixtures. If you use NMR, remember that you are analyzing a mixture of reactants and products. You will need to have the NMR spectra of the reactants at hand as you analyze the spectra of your reaction mixtures. Pay special attention to the predicted chemical shifts of the compounds you are working with.

Competition Experiments

For competition experiments, use 2.00 mmol of 2,3-dimethyl-1,3-butadiene and 1.00 mmol of each dienophile so that you have equimolar concentrations in the reaction mixtures. Use gentle heating under water-cooled reflux for a short time. Capillary GC is an excellent technique for obtaining relative rates from the first part of competition reactions. When running a competition experiment, it is not necessary to isolate and purify the products. Take samples from the early part of the reaction with no more than

5–10% conversion to the Diels-Alder products. You need to use only a GC or NMR method that can determine the relative concentrations of the products. Remember that capillary GC analysis must be done on dilute solutions (5–10% solutions), while NMR samples should be prepared according to your instructor's directions in $CDCl_3$ and may need to contain as much as 10–25 mg of the compounds you are assaying.

CLEANUP: Your instructor must approve your cleanup procedures before you begin your experiments. Prepare a list of all by-products from each reaction. Consult the cleanup procedures in experiments you have done previously for ways of handling similar materials. Your instructor may also give you specific guidelines that pertain to disposal regulations in your institution and community.

Interpretation of Experimental Data

Analyze your gas chromatograms or NMR spectra from the competition experiments to determine the relative concentrations of the two products in the reaction mixtures. Calculate the relative rate ratio for the two dienophiles. Explain your results.

10.3 Competition Reactions of 2,3-Dimethyl-1,3-butadiene with Dimethyl *cis*-2-Butenedioate and Dimethyl 2-Butynedioate

QUESTION: Which dienophile, dimethyl *cis*-2-butenedioate or dimethyl 2-butynedioate, reacts faster with 2,3-dimethyl-1,3-butadiene and what are their relative rates?

TEAMWORK: Work in teams of two students for this project.

PRELABORATORY ASSIGNMENT: Design reaction conditions for determining the relative rates of dimethyl *cis*-2-butenedioate and dimethyl 2-butynedioate with 2,3-dimethyl-1,3-butadiene. Your plan should include quantities of reagents, reaction conditions and equipment, assay methods, balanced chemical equations, and safety considerations.

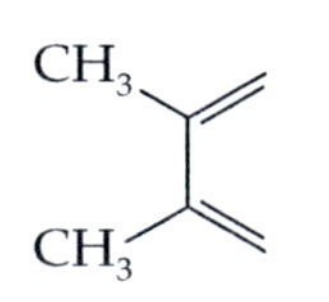

2,3-Dimethyl-1,3-butadiene
bp 69°C
MW 82.1
density 0.726 g · mL^{-1}

+

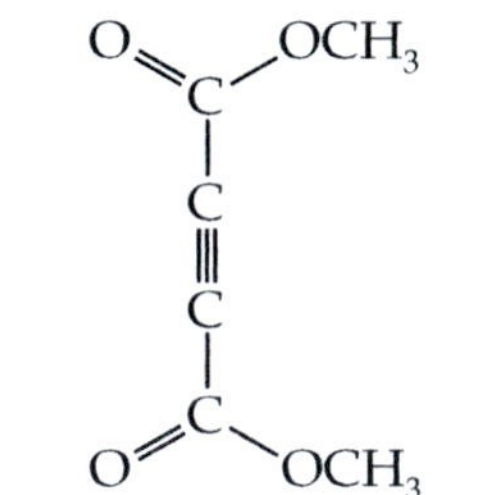

Dimethyl 2-butynedioate
(Dimethyl acetylenedicarboxylate)
bp 95–98°C/19 Torr
MW 142.1
density 1.156 g · mL^{-1}

⟶

Dimethyl 4,5-dimethylcyclohexa-1,4-diene-1,2-dicarboxylate
mp 65–66°C, 71–72°C, 75–76°C
MW 224.2

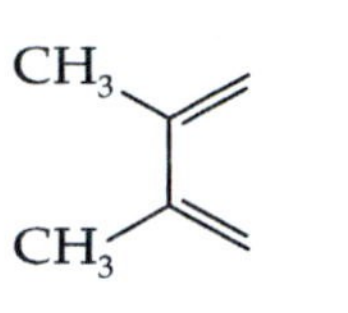

+

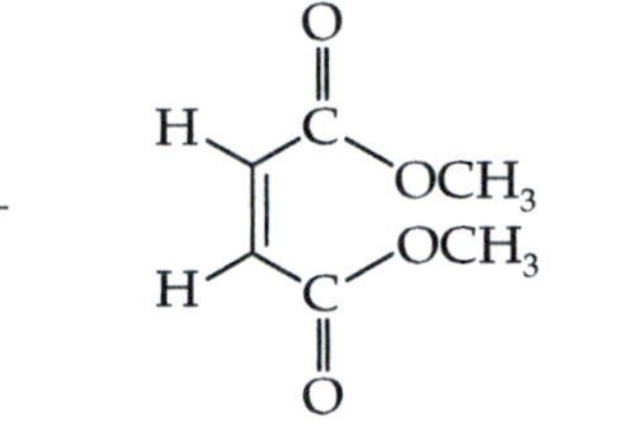

→

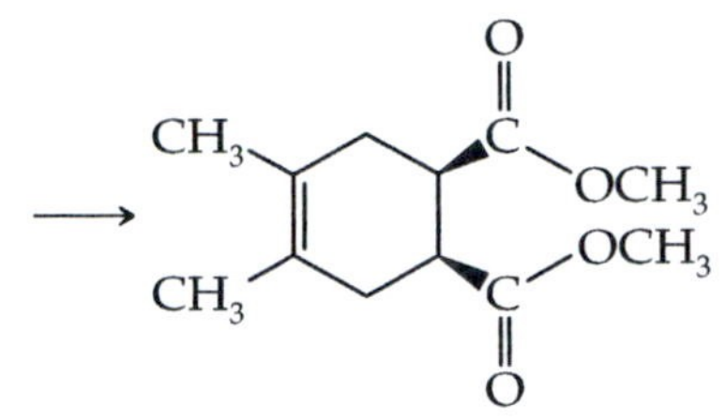

2,3-Dimethyl-1,3-butadiene
bp 69°C
MW 82.1
density 0.726 g • mL^{-1}

Dimethyl *cis*-2-butenedioate
(Dimethyl maleate)
bp 202°C
MW 144.1
density 1.161 g • mL^{-1}

Dimethyl 4,5-dimethylcyclohex-4-ene-*cis*-1,2-dicarboxylate
bp 157°C/20 Torr
MW 226.2

MICROSCALE PROCEDURE

Techniques Gas Chromatography: Technique 16
IR Spectroscopy: Technique 18
NMR Spectroscopy: Technique 19

SAFETY INFORMATION

Wear gloves while conducting all experimental work for this project.

2,3-Dimethyl-1,3-butadiene is extremely volatile and flammable.

Dimethyl *cis*-2-butenedioate (dimethyl maleate) is toxic and an irritant. Avoid contact with skin, eyes, and clothing.

Dimethyl 2-butynedioate (dimethyl acetylenedicarboxylate) is corrosive and a lachrymator (it causes tears). Wear gloves and avoid contact with skin, eyes, and clothing. Conduct all experiments with this reagent in a hood.

Synthesis of Diels-Alder Adducts

You should run each Diels-Alder reaction separately using 4.00 mmol of each reactant before you start any competition reactions. This will allow you to characterize the reactants and products by GC or NMR spectroscopy. Dimethyl *cis*-2-butenedioate and dimethyl 2-butynedioate **(Caution: The latter compound is a lachrymator; carry out the reaction in a hood.)** react more slowly than maleic anhydride (Project 10.1) in Diels-Alder reactions. Therefore, you will have to heat the reaction mixtures at reflux with no solvent for 1–2 h for them to go to completion. Efficient water cooling of the condenser above the reaction flask is essential to prevent loss of volatile 2,3-dimethyl-1,3-butadiene during the reflux period. Remember that one of these Diels-Alder products is a liquid rather than a solid. Reaction mixtures can be used without purification for GC analysis. Store your products in tightly capped vials to use for reference throughout the project.

GC or NMR Analysis of Product Mixtures

Determine the retention times of your Diels-Alder products on a nonpolar GC column. Gas chromatography is an excellent technique for determining the relative rates. Temperature programming on a capillary GC instrument works especially well. If your gas chromatograph does not have temperature programming, begin with a column temperature of 140°C and determine the GC conditions that will analyze a mixture of the two products. You need to determine the chromatographic conditions that will work best [see Techniques 16.6 and 16.7].

Alternatively, NMR spectroscopy can be used to analyze the reaction mixtures ($CDCl_3$ as the solvent). If you use NMR, remember that you are analyzing a mixture of reactants and products. You will need to have the NMR spectra of the reactants at hand as you analyze the spectra of your reaction mixtures. Pay special attention to the predicted chemical shifts of the compounds you are working with.

Competition Experiments

Use gentle heating of the reaction mixture for 10–20 min, along with GC analysis, to obtain relative rates from the first part of these competition reactions. Use 1.00 mmol of each dienophile so that you have equimolar concentrations of the dienophiles present in the reaction, and 2.00 mmol of 2,3-dimethyl-1,3-butadiene. Take samples from the early part of the reaction with no more than 5–10% conversion to the Diels-Alder products. When running a competition experiment, it is not necessary to isolate and purify the products. You need use only a GC or NMR method that can determine the relative concentrations of the reactants and products.

CLEANUP: Your instructor must approve your cleanup procedures before you begin your experiments. Prepare a list of all by-products from each reaction. Consult the cleanup procedures in experiments you have done previously for ways of handling similar materials. Your instructor may also give you specific guidelines that pertain to disposal regulations in your institution and community.

Interpretation of Experimental Data

Analyze your gas chromatograms or NMR spectra from the competition experiments to determine the relative concentrations of the two products in the reaction mixtures. Calculate the relative rate ratio for the two dienophiles. Explain your results.

10.4 Designing and Carrying Out a Synthesis of Dimethyl 4,5-Dimethylbenzene-1,2-dicarboxylate by Diels-Alder and Dehydrogenation Chemistry

PURPOSE: To design a Diels-Alder reaction of 2,3-dimethyl-1,3-butadiene with dimethyl butynedioate and the dehydrogenation of the product to an aromatic compound.

TEAMWORK: Work in teams of two students for this project.

PRELABORATORY ASSIGNMENT: Design reaction conditions for the synthesis of dimethyl 4,5-dimethylcyclohexa-1,4-diene-1,2-dicarboxylate by a Diels-Alder reaction. Your plan should include quantities of reagents, reaction conditions and equipment, assay methods, balanced chemical equations, and safety considerations.

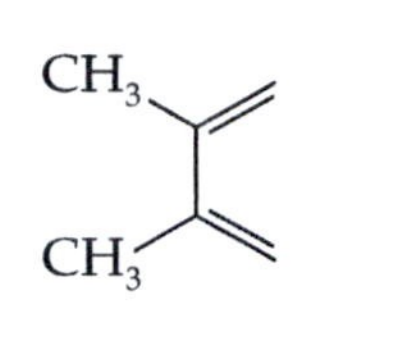

2,3-Dimethyl-1,3-butadiene
bp 69°C
MW 82.1
density 0.726 g · mL^{-1}

+

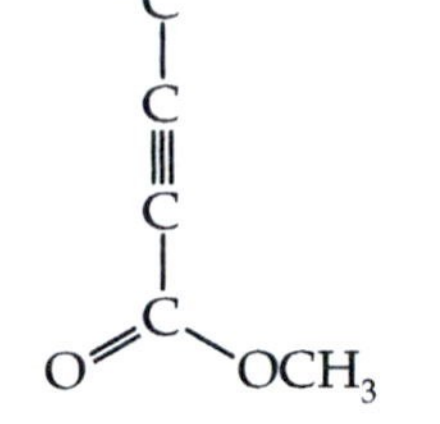

Dimethyl 2-butynedioate
(Dimethyl acetylenedicarboxylate)
bp 95–98°C/19 Torr
MW 142.1
density 1.156 g · mL^{-1}

⟶

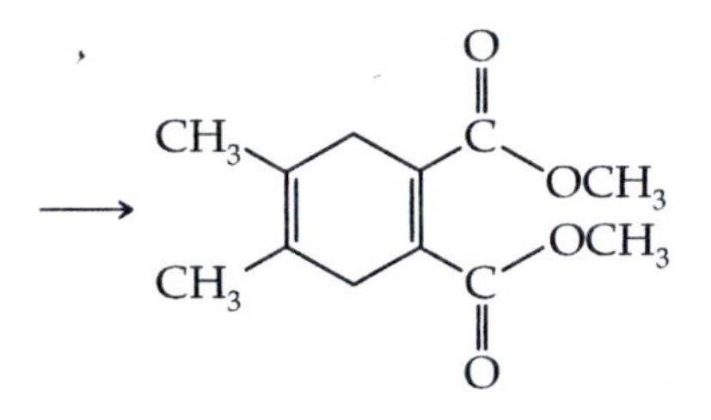

Dimethyl 4,5-dimethylcyclohexa-1,4-diene-1,2-dicarboxylate
mp 65–66°C, 71–72°C, 75–76°C
MW 224.2

↙

+ H_2

Dimethyl 4,5-dimethylbenzene-1,2-dicarboxylate
mp 54–55°C
MW 222.2

MINISCALE PROCEDURE

Techniques Gas Chromatography: Technique 16
IR Spectroscopy: Technique 18
NMR Spectroscopy: Technique 19

SAFETY INFORMATION

Wear gloves while conducting all experimental work for this project.

2,3-Dimethyl-1,3-butadiene is extremely volatile and flammable.

Dimethyl 2-butynedioate (dimethyl acetylenedicarboxylate) is corrosive and a lachrymator (it causes tears). Wear gloves and avoid contact with skin, eyes, and clothing. Conduct all experiments with this reagent in a hood.

Synthesis of Dimethyl 4,5-Dimethylcyclohexa-1,4-diene-1,2-dicarboxylate.

Carry out the synthesis on a 10.0-mmol scale. **(Caution: Dimethyl 2-butynedioate is a lachrymator; carry out the reaction in a hood.)** Either GC or NMR can serve to assay the progress of the reaction. If you carry out the reaction at reflux, efficient water cooling of the condenser above the reaction flask is essential to prevent loss of volatile 2,3-dimethyl-1,3-butadiene. Recrystallize the crude crystalline product from hexane, using a steam bath or a beaker of water at 70–75°C as the heat source. After recrystallization, your reaction product can be characterized by one or more of the following methods, as directed by your instructor: melting point, ^{1}H NMR, IR, and GC. It is vital that you characterize an intermediate product before trying to use it as a substrate for the next reaction.

Synthesis of Dimethyl 4,5-Dimethylbenzene-1,2-dicarboxylate

Your procedure must be approved by your instructor before you begin the synthesis. Either GC or NMR can serve to assay the progress of the dehydrogenation reaction. If you use Pd/C to catalyze the reaction, GC samples must be diluted to a 5–10% solution in ether or hexane and filtered through a filter pipet [see Technique 8.9, Figure 8.16] before injection into the GC instrument. Do not use GC analysis on unpurified reaction mixtures if you used sulfur or chloranil as a hydrogen receptor. NMR spectroscopy can be used for assaying the progress of these dehydrogenation reactions. See Technique 19, Tables 19.3 and 19.6, for help in predicting the chemical shifts of the protons in your anticipated product.

CLEANUP: Your instructor must approve your cleanup procedures before you begin your experiments. Prepare a list of all by-products from each reaction. Consult the cleanup procedures in experiments you have done previously for ways of handling similar materials. Your instructor may also give you specific guidelines that pertain to disposal of reaction by-products in your institution and community.

Interpretation of Experimental Data

Analyze all melting-point data, IR and NMR spectra, and gas chromatograms that you obtained for the products of the two reactions that you carried out. Did you have a pure compound to use as the substrate in the dehydrogenation reaction? Does the NMR spectrum of your final product indicate that you did indeed form an aromatic ring in the dehydrogenation reaction?

References

1. Mohrig, J. R.; Hammond, C. N.; Schatz, P. F.; Morrill, T. C. *Techniques in Organic Chemistry*; W. H. Freeman and Company: New York, 2003, Appendix, "The Literature of Organic Chemistry."
2. Wasserman, A. *Diels-Alder Reactions*; Elsevier: New York, 1965.

Questions

1. Predict the relative rates for the reaction of butadiene and (a) ethene, (b) methyl propenoate, and (c) maleic anhydride.
2. What product is expected from the reaction of fumaric acid (*trans*-2-butenedioic acid) with 1,3-butadiene?
3. One of the following compounds reacts more readily as the diene in a Diels-Alder reaction. Which one and why?

4. What purpose did the water washes serve in Project 10.1?
5. Propose a Diels-Alder reaction that would lead to each of the following.

a.

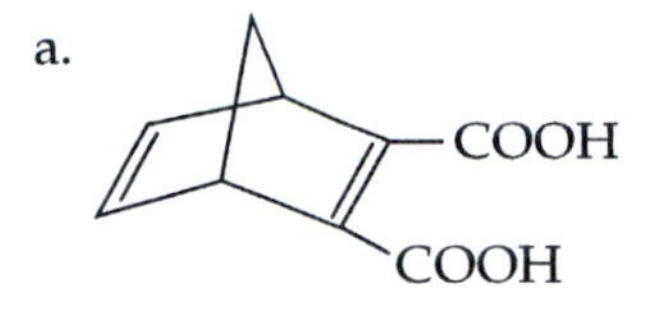

b.

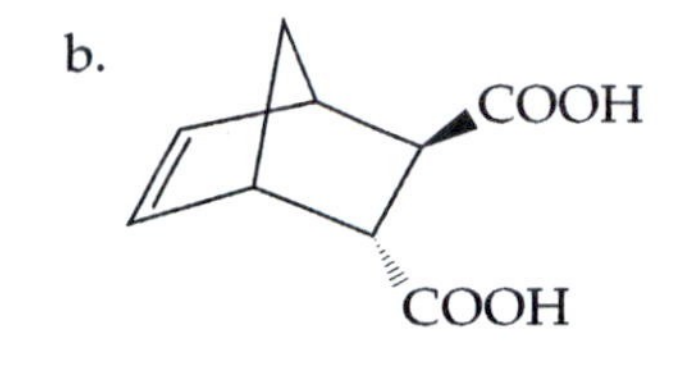

c.

CO_2H

CO_2H

PROJECT

11 ALDOL-DEHYDRATION CHEMISTRY USING UNKNOWN ALDEHYDES AND KETONES

QUESTION: What are the structures of an "unknown" aldehyde and ketone that react with NaOH in ethanol to form an aldol condensation product followed by its dehydration?

In this two-week project you will carry out an aldol condensation between an unknown aldehyde and an unknown ketone. Your task is to identify the aldehyde and ketone by determining the melting point and analyzing the NMR spectrum of the dehydration product of their aldol condensation. You will also prepare solid derivatives of the aldehyde and ketone and use their melting points as additional data in determining their identities.

Aldol Condensation-Dehydration Chemistry

The formation of carbon-carbon bonds is of fundamental importance in synthetic organic chemistry, and the aldol condensation has a long and successful history as a method of carbon-carbon bond formation. The base-promoted condensation of a molecule of benzaldehyde with a molecule of acetophenone (1-phenylethanone) represents a typical aldol condensation between two different carbonyl compounds. The aldol product can subsequently lose a molecule of water (dehydrate) to form a conjugated ketone:

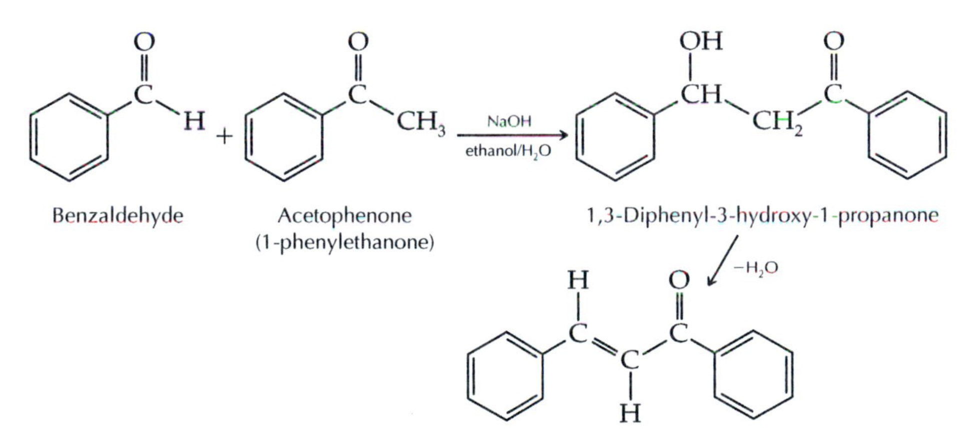

Whereas the ketone has α-protons and can react with OH^- to form an enolate anion, benzaldehyde cannot do so because it has no α-protons. The aldehyde, however, is more susceptible to nucleophilic attack by an enolate anion than is the ketone. Therefore, this crossed aldol condensation can produce a relatively pure product. The reaction mechanism involves base-catalyzed abstraction of an α-proton of acetophenone, followed by condensation of the resonance-stabilized enolate anion with a molecule of benzaldehyde:

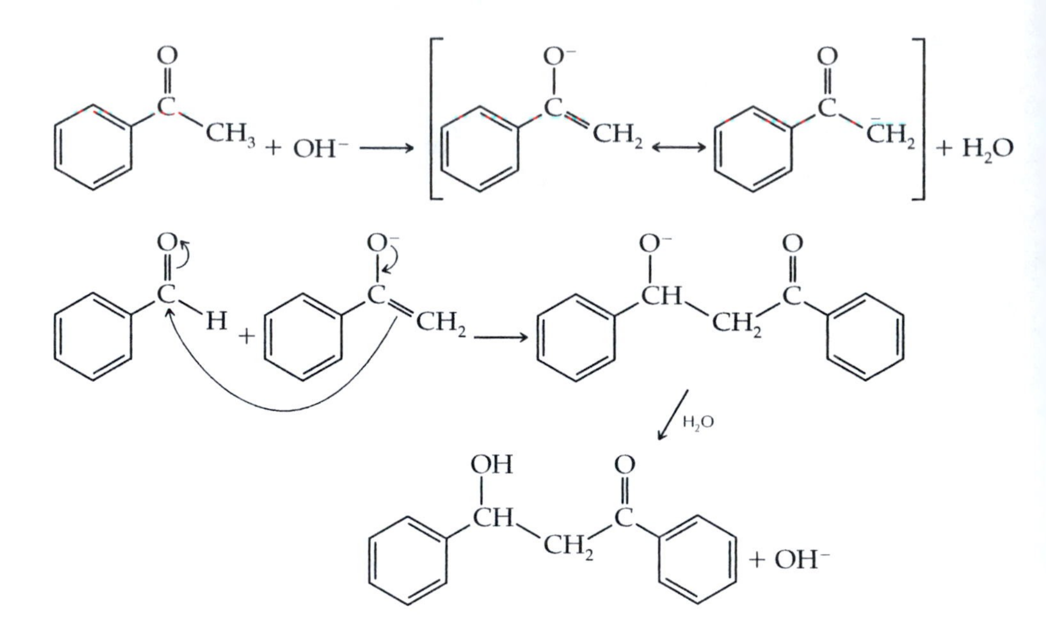

The elimination of water from the aldol product occurs in two steps. The first step is the formation of an enolate anion by base-catalyzed abstraction of an α-proton from the aldol condensation product. The second is the expulsion of the hydroxi1de anion to form the conjugated dehydration product:

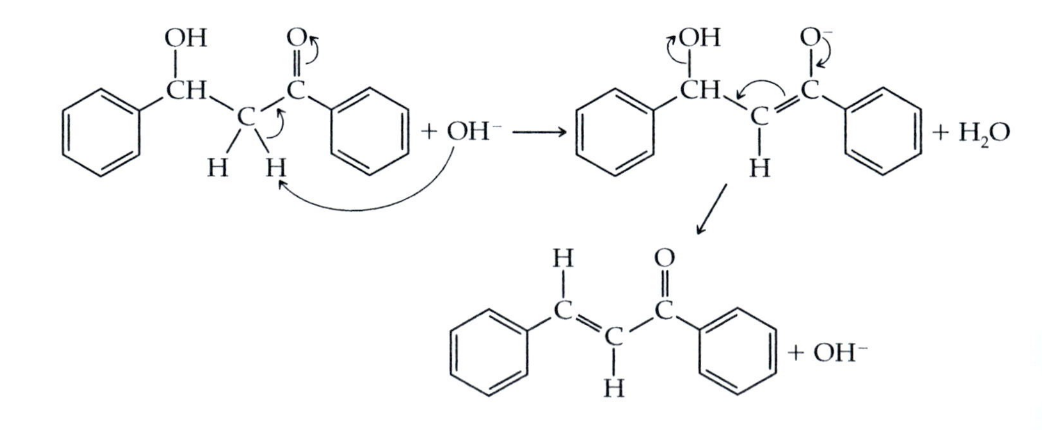

Often the dehydration of the initially formed aldol product does not occur spontaneously. However, dehydration is strongly favored in the reaction that you will be carrying out. The rate of dehydration is more favorable because loss of hydroxide produces a product with extended conjugation. This conjugation also influences the equilibrium constant for the reaction, and even more important is the fact that the dehydration product precipitates from the reaction mixture.

Symmetrical ketones are used for all the aldol condensation reactions in Project 11. Not only does dehydration occur under the reaction conditions, but a "double condensation" occurs. Two molecules of an aromatic aldehyde condense with one molecule of a symmetrical ketone to form, after dehydration, an extensively conjugated product. Using an excess of the aldehyde ensures that a double aldol-dehydration cycle occurs in the reaction. The double aldol-dehydration cycle that would occur with 3-pentanone and benzaldehyde follows:

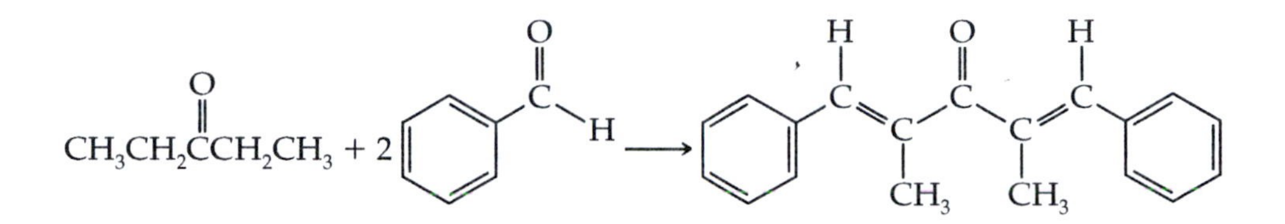

NMR Spectroscopy of the Aldol-Dehydration Product

To characterize your unknown compounds, you need to obtain an NMR spectrum of your aldol-dehydration product as well as synthesize solid derivatives of the aldehyde and the ketone. Analysis of the ^{1}H NMR spectrum will allow you to draw conclusions about the structure of your aldol-dehydration product and, by deduction, the structures of the aldehyde and ketone you used as starting materials for the reaction.

The β-proton of a conjugated ketone is more strongly deshielded than the α-proton, even though the α-proton is closer to the electron-withdrawing carbonyl group. This phenomenon can be understood when the resonance in a conjugated ketone is considered:

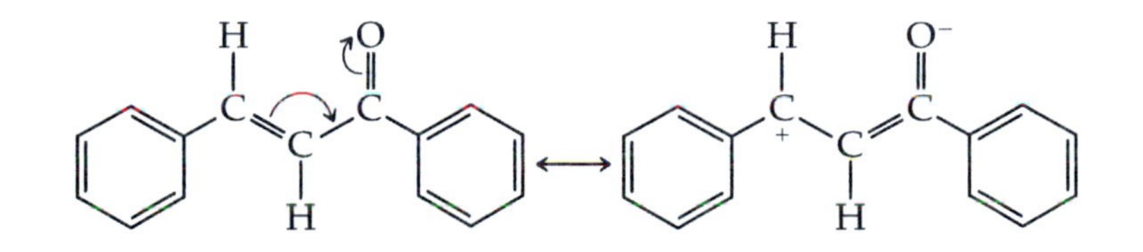

The substantial amount of positive charge at the β-carbon atom strongly deshields the β-proton. The phenyl group also deshields the proton that is β to the carbonyl group. As a consequence, the β-proton (and sometimes even the α-proton) is superimposed on the aromatic protons, and NMR peaks in the aromatic region can be difficult to assign with certainty. However, it is not difficult to distinguish a *para*-disubstituted benzene ring from a monosubstituted benzene ring by NMR analysis.

In addition, the structure of the ketone portion of the aldol-dehydration product can readily be ascertained by its NMR spectrum. Two of the ketones that are possible unknowns contain cyclohexane rings, which have axial and equatorial hydrogen atoms that usually have different chemical shifts. An equatorial hydrogen on a cyclohexane-carbon atom generally has a chemical shift approximately 0.5 ppm greater than an axial hydrogen on the same carbon atom. Although the NMR spectrum of your aldol-dehydration product may be complex, it will contain valuable information about the structure of the product and thus indirectly about the structures of your unknown aldehyde and ketone.

You will be assigned an aldehyde and a ketone from the following list as the starting reagents for your aldol condensation.

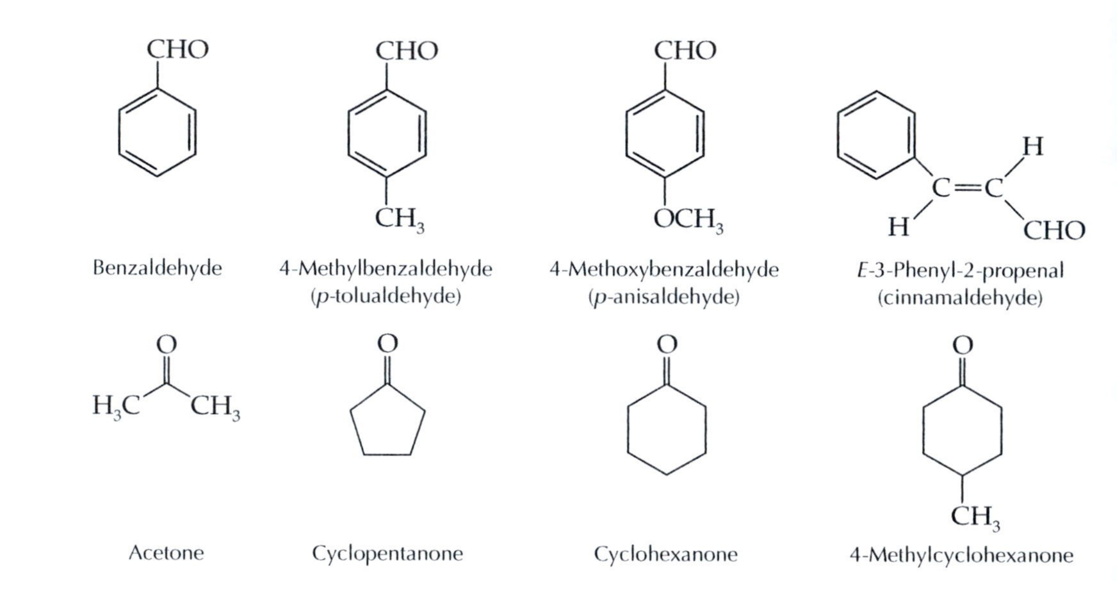

Table 11.1 lists the melting points of the aldol dehydration products formed from these aldehydes and ketones. The identification of the aldol-dehydration product deduced from these melting points and from the ^1H NMR data must, of course, be the same.

Arylhydrazone Derivatives of Aldehydes and Ketones

Both aldehydes and ketones usually undergo rapid reaction with arylhydrazines such as phenylhydrazine or 2,4-dinitrophenylhydrazine:

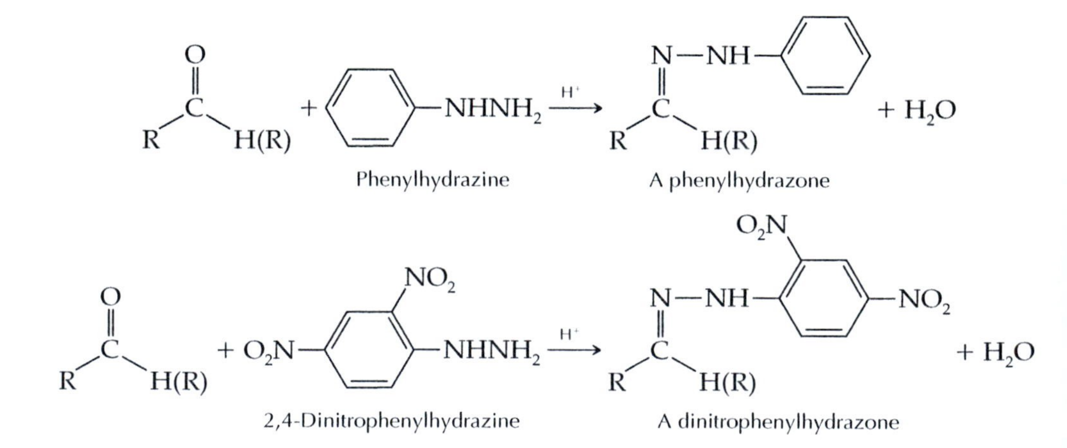

TABLE 11.1 Melting points of aldol-dehydration products

	Ketones			
Aldehyde	Acetone	Cyclopentanone	Cyclohexanone	4-Methylcyclohexanone
Benzaldehyde	113°C	189°C	118°C	98–99°C
4-Methylbenzaldehyde	175°C	235–236°C	170°C	133–135°C
4-Methoxybenzaldehyde	129–130°C	212°C	159°C	141–142°C
E-3-Phenyl-2-propenal	144°C	225°C	180°C	163–164°C

These hydrazones are normally high melting-point solids, especially in the case of the deeply colored 2,4-dinitrophenylhydrazones. Their melting points can serve as aids in identifying your starting carbonyl compounds. Before the advent of modern infrared and NMR spectroscopy, the use of solid derivatives of liquid compounds was widespread in organic chemistry. You will prepare a 2,4-dinitrophenylhydrazone derivative of your ketone, and you have the choice of preparing the phenylhydrazone or the 2,4-dinitrophenylhydrazone derivative of your aldehyde.

In summary, the following strategy will be useful in your study.

1. Carry out the aldol condensation and dehydration between an "unknown" aldehyde and an "unknown" ketone.
2. Obtain the melting point and ^{1}H NMR spectrum for your product.
3. Prepare the phenylhydrazone or 2,4-dinitrophenylhydrazone of the aldehyde and the 2,4-dinitrophenylhydrazone of the ketone and determine their melting points.
4. Deduce the structure of your aldol-dehydration product and your starting materials by evaluating the data collected in steps 2 and 3.

11.1 Preparation and Characterization of the Aldol-Dehydration Product

QUESTION: What is the structure of an aldol-dehydration product obtained by reaction of NaOH in ethanol with an "unknown" aldehyde and ketone?

PRELABORATORY ASSIGNMENT: The choice of an appropriate solvent is crucial for a successful recrystallization. Describe how you will determine the proper recrystallization solvent for your product in Project 11.1.

MINISCALE PROCEDURE

Techniques Selecting a Recrystallization Solvent: Technique 9.2
Miniscale Recrystallization: Technique 9.5

SAFETY INFORMATION

Wear gloves while performing the experiment and work in a hood, if possible.

The **aldehydes** and **ketones** used in this experiment are skin and eye irritants.

Aqueous **sodium hydroxide** solutions are corrosive and cause burns.

Select (or your instructor will assign) a set of unknowns. The vials contain enough of each compound for the aldol-dehydration reaction and the preparation of a derivative.

Having an excess of aldehyde ensures that a double condensation will occur.

With a 1-mL graduated pipet, measure 0.80 mL of the unknown aldehyde and place it in a 25-mL Erlenmeyer flask; note its number in your lab notebook. Using another 1-mL graduated pipet, measure 0.20 mL of the unknown ketone (note its number in your lab notebook) and add that compound to the same Erlenmeyer flask. Add 4.0 mL of 95% ethanol and 3.0 mL of 2 M sodium hydroxide solution to the flask. Stir the solution with a magnetic stirrer for 15 min—or longer, if precipitate is still forming. If the solution is only cloudy or very little precipitate has formed after 15 min, heat the reaction mixture on a steam bath or in a boiling-water bath for 10–15 min. Cool the flask to room temperature.

The various aldol-dehydration products form at different rates.

When precipitation is complete, cool the flask in an ice-water bath for 10 min. While the flask is cooling, place 8 mL of 95% ethanol in a test tube; and 4 mL of 4% acetic acid in 95% ethanol (v/v) in another test tube. Chill the test tubes containing these solutions in the ice-water bath.

Collect the product by vacuum filtration on a Buchner funnel. Disconnect the vacuum and rinse the product with 4 mL of the ice-cold ethanol. Reconnect the vacuum and draw the liquid from the product. Repeat this washing procedure with the 4 mL of acetic acid/ethanol solution, and finally wash the crude product with the remaining 4 mL of ethanol.

Before recrystallizing the crude product, you need to find a suitable solvent. Test 95% ethanol and toluene according to the procedure described in Technique 9.2. The addition of a few drops of hexane may be necessary to promote crystallization from a toluene solution. If neither of these recrystallization solvents is satisfactory for your aldol-dehydration product, test a 9:1 (v/v) mixture of 95% ethanol/acetone. Once you have selected a suitable solvent, carry out the recrystallization. Allow the recrystallized product to dry until the next laboratory period.

Weigh your aldol-dehydration product and determine its melting point. Obtain an NMR spectrum of your product using $CDCl_3$ as the solvent, as directed by your instructor [see Technique 19.2].

Cleanup: Place all filtrates in the container for flammable (organic) waste.

11.2 Preparation and Characterization of Solid Derivatives of the Unknown Aldehyde and Ketone

QUESTION: How can you use solid derivatives to confirm the structures of your "unknown" aldehyde and ketone?

PRELABORATORY ASSIGNMENT: Describe, using chemical reactions, how the arylhydrazone formation can be used to characterize your aldol-dehydration product.

MICROSCALE PROCEDURE

Techniques Mixed Solvent Recrystallization: Technique 9.2
Microscale Recrystallization: Technique 9.7a

SAFETY INFORMATION

Wear gloves while performing the following procedures and work in a hood, if possible.

The **aldehydes** and **ketones** used in this experiment are skin and eye irritants.

2,4-Dinitrophenylhydrazine is toxic and an irritant. **Phenylhydrazine** is toxic and a suspected cancer agent.

Your instructor will advise you about whether you will prepare the phenylhydrazone or the 2,4-dinitrophenylhydrazone of your unknown aldehyde. If you are preparing the 2,4-dinitrophenylhydrazone, use the same procedure as the one used for the ketone derivative.

Compare the melting point of the phenylhydrazone or the 2,4-dinitrophenylhydrazone of your aldehyde, whichever you prepared, with those in Table 11.2. You will need to use your interpretation of the NMR spectrum of your aldol-dehydration product to distinguish between the two aldehydes that have similar melting points for the 2,4-dinitrophenylhydrazone derivative.

Phenylhydrazone of Unknown Aldehyde

Place 4 Pasteur pipet drops of the unknown aldehyde and 0.5 mL of ethanol in a 10 × 75 mm test tube. Add water dropwise, shaking after

TABLE 11.2 Melting points of phenylhydrazones and 2,4-dinitrophenylhydrazones of aldehydes

Aldehyde	Phenylhydrazone melting point, °C	2,4-Dinitrophenylhydrazone melting point, °C
Benzaldehyde	158	237
4-Methylbenzaldehyde	114	232
4-Methoxybenzaldehyde	120	254
E-3-Phenyl-2-propenal	168	255 (decomposition)

each drop, until the cloudiness produced just disappears (an indication of saturation). If too much water is used, add 1 or 2 drops of ethanol to give a clear solution. Add 4 drops of phenylhydrazine and shake the test tube to mix the contents. A precipitate will form within a few minutes. Chill the test tube in an ice-water bath and collect the phenylhydrazone on a Hirsch funnel by vacuum filtration. Wash the precipitate three times with 0.5 mL of ice-cold ethanol.

Recrystallize the phenylhydrazone from an ethanol/water mixed solvent by dissolving the product in ethanol, then adding water to the hot solution until it is just cloudy [see Technique 9.2]. Cool the solution to room temperature, then chill the recrystallized mixture in an ice-water bath before collecting the crystals on a Hirsch funnel by vacuum filtration. Allow the phenylhydrazone to dry and then determine its melting point. Compare the melting point of your derivative with those given in Table 11.2.

CLEANUP: Pour the filtrate from the reaction mixture and from the recrystallization into the container for flammable (organic) waste.

2,4-Dinitrophenylhydrazone of Unknown Aldehyde and Ketone

Dissolve 4 Pasteur pipet drops of the unknown ketone in 0.5 mL ethanol in a 10 × 75 mm test tube. Add 1.5 mL of the 2,4-dinitrophenylhydrazine reagent solution.* Shake the tube to mix the contents. A precipitate usually forms immediately. Let the mixture stand at room temperature for 15 min before collecting the solid product on a Hirsch funnel by vacuum filtration. Wash the crystals with three 1-mL portions of cold ethanol. To do this, remove the suction, add the ethanol, and stir the crystals gently to wash them completely. Then apply the suction again. Recrystallize the 2,4-dinitrophenylhydrazone from ethanol [see Technique 9.7a]. Allow it to dry and determine its melting point. Compare the melting point of your derivative with those given in Table 11.3.

TABLE 11.3 Melting points of 2,4-dinitrophenylhydrazones of ketones

Ketone	2,4-Dinitrophenylhydrazone melting point, °C
Acetone	126
Cyclopentanone	142
Cyclohexanone	162
4-Methylcyclohexanone	134

*Note to instructor: Prepare the 2,4-dinitrophenylhydrazine reagent by carefully dissolving 8.0 g of 2,4-dinitrophenylhydrazine in 40 mL of concentrated sulfuric acid, then adding 60 mL of water slowly while stirring the mixture. Add 200 mL of reagent-grade ethanol to this warm solution and filter the mixture if any solid precipitates.

CLEANUP: Pour the filtrate from the reaction mixture and the recrystallization into the container for flammable (organic) waste.

Interpretation of Experimental Data

Analyze the NMR spectrum of your aldol-dehydration product and make structural assignments for as many signals as you can [see Technique 19, especially Tables 19.2, 19.3, and 19.5]. Use the melting points of your aldol-dehydration product and the derivatives to determine the identity of your unknown aldehyde and ketone. Does the NMR spectrum of your aldol-dehydration product support these identifications? Explain your reasoning.

Draw the structure of your aldol-dehydration product and write a balanced equation for its formation. Calculate the theoretical yield and percent yield. Write equations for the formation of your hydrazones using the structures of your identified aldehyde and ketone.

References

1. Hathaway, B. *J. Chem. Educ.* **1987,** *64,* 367–368.
2. Pickering, M. *J. Chem. Educ.* **1991,** *68,* 232–234.
3. Shriner, R. L.; Hermann, C. K. F.; Morrill, T. C.; Curtin, D. Y.; Fuson, R. C. *The Systematic Identification of Organic Compounds;* 7th ed.; Wiley: New York, 1998.
4. Rappoport, Z. *Handbook of Tables for Organic Compound Identification;* 3rd ed.; CRC Press: Boca Raton, 1967.

Questions

1. To undergo an aldol condensation with itself, an aldehyde or ketone must contain at least one α-proton. Which of the following compounds have no α-hydrogen atoms? (a) acetone; (b) 2-butanone; (c) diphenylketone; (d) 2,2-dimethyl-1-phenyl-1-propanone; (e) cyclobutanone.
2. Write the mechanism for the aldol condensation of two molecules of propanal in a $NaOH/H_2O$ solution.
3. Although acid is not needed to dehydrate the aldol products in the study you did, it will enhance the rate of the elimination reaction of the aldol product obtained in Question 2. Write a mechanism for this acid-catalyzed dehydration reaction.
4. Why is a solution of 4% acetic acid in 95% ethanol used to wash the crude aldol-dehydration product?
5. The crossed aldol condensation of acetaldehyde with benzaldehyde is aided by the fact that benzaldehyde bears no α-protons. However, a possible competitive reaction is the "self-condensation" of two molecules of acetaldehyde. Slow addition of the acetaldehyde to a benzaldehyde solution containing an NaOH/ethanol/water mixture will often result in less self-condensation. Explain.

PROJECT

12

BIOCHEMICAL CATALYSIS AND THE STEREOCHEMISTRY OF BOROHYDRIDE REDUCTION

QUESTION: What is the stereoselectivity in the sodium borohydride reduction of benzoin to 1,2-diphenyl-1,2-ethanediol?

In this three-week project you will use thiamine hydrochloride (vitamin B_1) to catalyze the benzoin condensation of two molecules of benzaldehyde. Then you will reduce benzoin, an α-hydroxyketone, to 1,2-diphenyl-1,2-ethanediol with sodium borohydride and determine the stereoselectivity of borohydride addition to the acyclic ketone. To distinguish between the two possible diastereomers, you will synthesize a cyclic acetal of the 1,2-diol, whose symmetry elements will allow a structure proof by NMR spectroscopy.

Benzoin Condensation

One of the greatest challenges in organic synthesis is the controlled formation of carbon-carbon bonds. The synthesis of a complex molecule from simple precursors involves the formation of many such bonds, and many organic chemists have devoted their careers to the discovery of more efficient and specific synthetic methods. The benzoin condensation is a classic example of the specific catalysis of carbon-carbon bond formation.

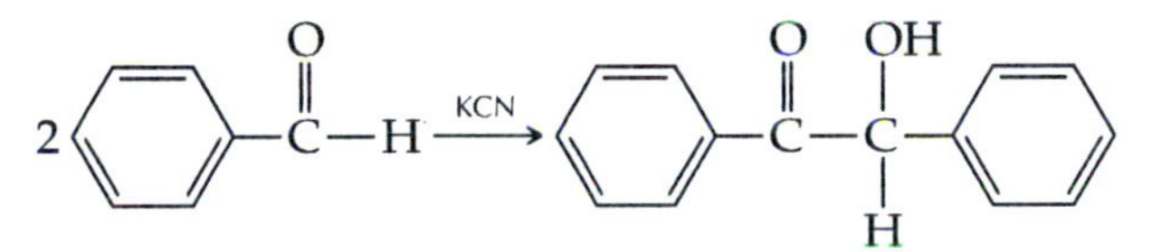

The remarkably specific catalysis by cyanide ion in this reaction was first studied in the early years of the twentieth century. The cyanide anion is a highly specific catalyst because it can perform three functions. Cyanide can act as a nucleophile, increase the acidity of the aldehydic proton, and serve as a leaving group. Thus, cyanide activates the substrate and then leaves after its work is done.

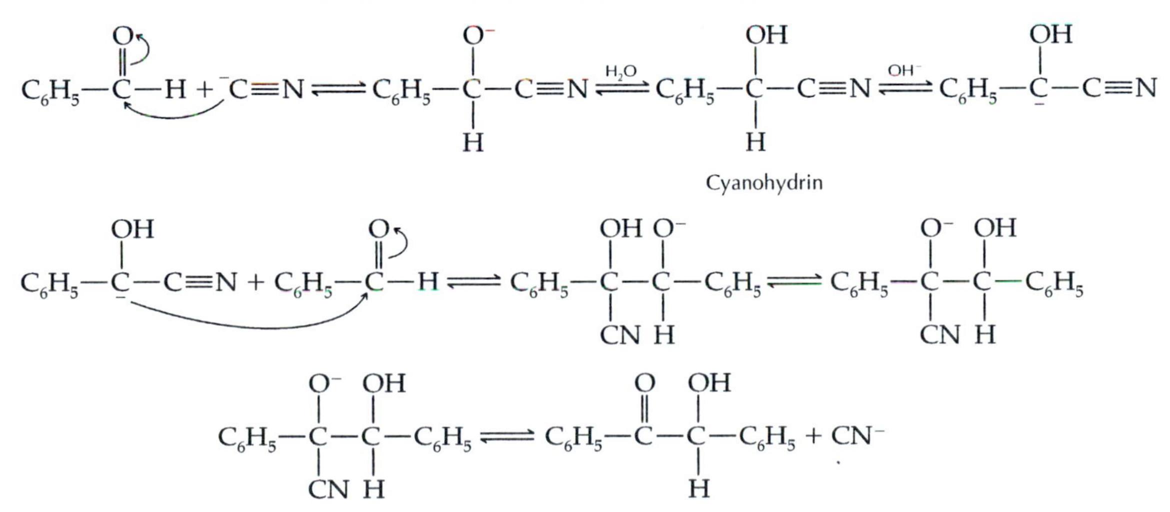

The electron-withdrawing cyano group gives a tremendous boost to the acidity of the α-proton in the cyanohydrin; removal of the proton by base then produces a carbon nucleophile. The carbon nucleophile reacts at the carbonyl-carbon atom of a second molecule of benzaldehyde, making a carbon-carbon bond. Loss of cyanide ion gives the product, benzoin, and the cyanide ion recycles as a catalyst.

Although the cyanide-catalyzed benzoin condensation is an important reaction, it is not without danger. Potassium cyanide and hydrogen cyanide are lethal poisons that can kill with little warning. Death may result from a few minutes' exposure to 300 ppm of hydrogen cyanide in the atmosphere. Vitamin B_1 provides a far safer catalytic agent for the benzoin condensation.

Thiamine (Vitamin B_1)

Vitamins are organic compounds that are necessary for good health and growth, but that cannot be synthesized efficiently in our bodies. A balanced diet contains sufficient quantities of the essential vitamins, but when people choose or are forced to restrict their diets, they may not ingest all the vitamins necessary for sound health. A deficiency of thiamine in the human diet can result in fatigue and depression. People who eat insufficient thiamine may get a disease of the nervous system called beriberi, which is characterized by partial paralysis of the extremities, emaciation, and anemia. When a dietary growth factor from rice polishings was isolated in 1911 and shown to cure beriberi, the term "vitamine," or vital amine, was proposed for the trace nutrient. Later, when it was discovered that not all of these growth factors are amines, the final letter was dropped, giving us the word "vitamin."

Many of the enzymes that catalyze and control our metabolic chemistry require additional small organic molecules as cocatalysts. These molecules are called ***coenzymes.*** In many biochemical reactions, much of the bond making and breaking takes place in the coenzymes and the substrates, although enzymes are necessary to assemble the reactants in the necessary spatial relationships as well as to provide additional catalysis. Many vital coenzymes are synthesized in our bodies from the vitamins we eat.

Thiamine pyrophosphate (TPP) is a coenzyme of universal occurrence. It is easily formed in human beings and other animals from vitamin B_1. The mechanism of the biochemical action of TPP has been shown to involve the removal of a relatively acidic proton on the five-membered thiazolium ring of thiamine by a basic group at the active site of the enzyme. This reaction produces a carbanion that is nucleophilic and can attack the carbonyl carbon of an aldehyde in the same way a cyanide ion does. In this respect, thiamine is the biochemical analogue of cyanide.

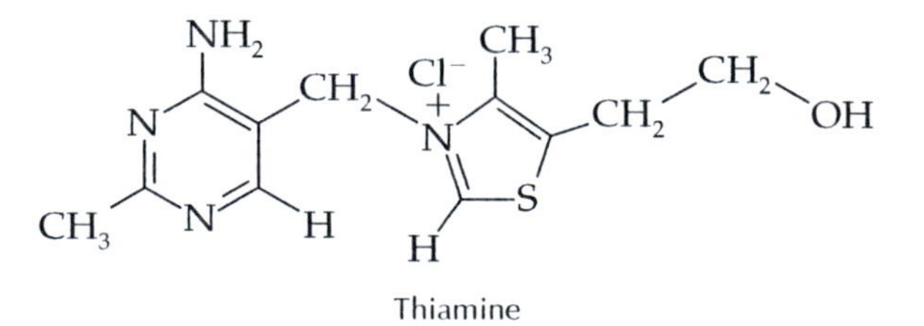

Thiamine

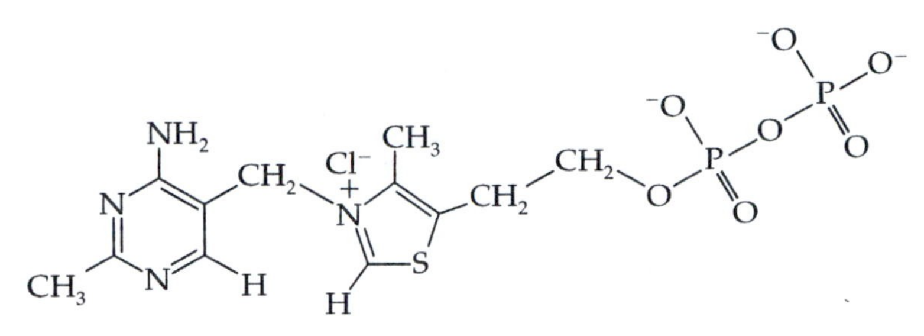

Thiamine pyrophosphate (TPP)

The mechanism by which thiamine catalyzes the benzoin condensation is very similar to the action of TTP under physiological conditions, except that no enzyme is present. Although a complete mechanism for the thiamine catalysis of the benzoin condensation takes quite a bit of space because the molecules involved are reasonably complex, no new kinds of catalysis are present. The same laws of chemistry and physics are at work.

The most important steps in the mechanism are shown in Figure 12.1. Notice how similar the role of thiamine is compared to cyanide. Thiamine is able to serve the same three critical functions. It acts as a nucleophile, boosts the acidity of the aldehydic proton of benzaldehyde, and then leaves.

Reduction Using Sodium Borohydride

Oxidation-reduction reactions are extremely diverse and proceed by a variety of mechanisms. The unifying factor is that one substrate is oxidized and another is reduced. In the inorganic chemistry of metal cations, the process can be thought of most easily as an electron transfer phenomenon; that is, oxidation is a loss of electrons and reduction is a gain of electrons. While this approach also applies to covalent organic compounds, it can be awkward to use. It is often easier to think of reduction as the gain of hydrogen, the equivalent of H_2. When reduction of an organic substrate is necessary, the reducing agent is often an inorganic compound. Sodium borohydride has become a popular mild reducing agent because of its functional group selectivity and ease of handling.

The probable reaction mechanism for the reduction of benzoin with borohydride involves a nucleophilic hydride ion transfer from boron to the electropositive carbonyl carbon of the ketone:

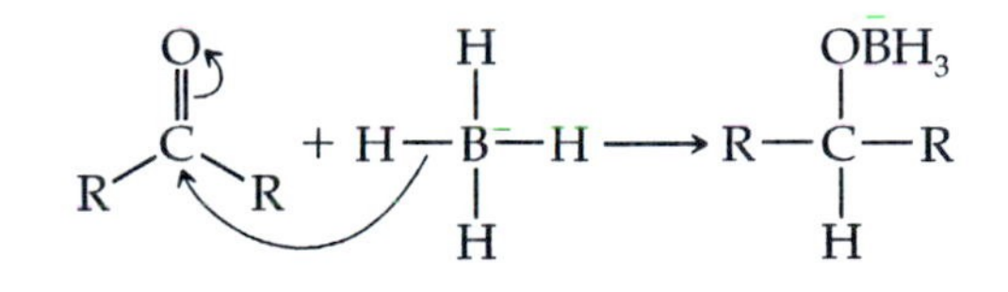

Substitution of an alkoxy group for a hydrogen atom on boron does not seriously alter the reducing ability of the borohydride. Thus, when BH_4^- reduces an organic compound to an alcohol (ROH), it is oxidized to $B(OR)_4^-$. The overall stoichiometry is

$$4R{-}\overset{\overset{\displaystyle O}{\|}}{C}{-}R + Na^+BH_4^- \longrightarrow (R_2CHO)_4B^-Na^+$$

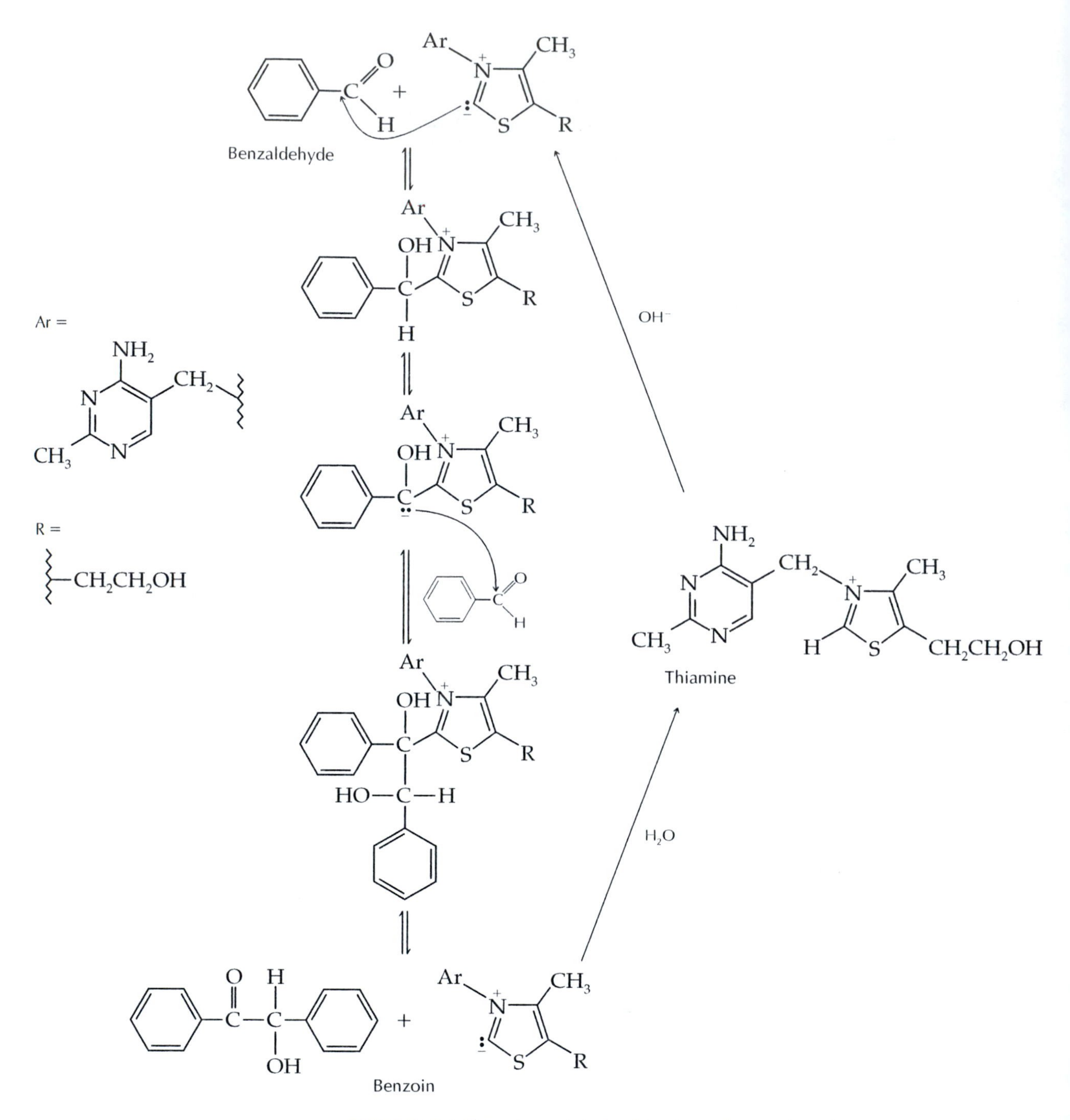

FIGURE 12.1 Thiamine catalysis of the benzoin condensation.

After the reaction is finished and water is added, $B(OR)_4^-$ is hydrolyzed, producing salts of boric acid [$B(OH)_3$].

$$(R_2CHO)_4B^-Na^+ + 2H_2O \longrightarrow 4R_2CHOH + NaBO_2$$

Hydride reducing agents are also bases, which can react with Brønsted acids such as water to produce H_2 gas. Sodium borohydride is a relatively safe reducing reagent. In fact, water and alcohols can be used as solvents for sodium borohydride reductions.

The reaction of sodium borohydride and benzoin produces 1,2-diphenyl-1,2-ethanediol.

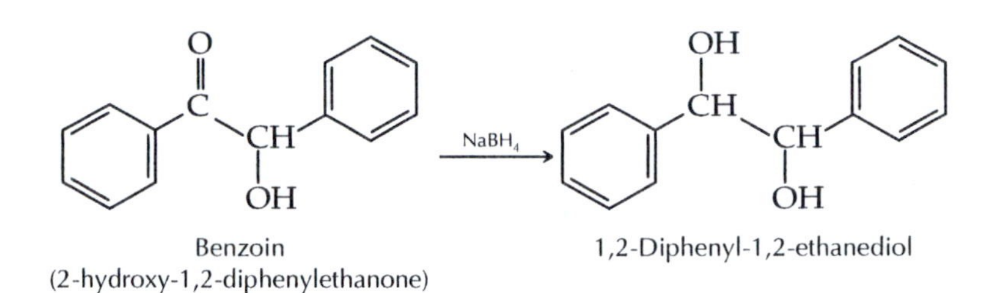

Benzoin (2-hydroxy-1,2-diphenylethanone) 1,2-Diphenyl-1,2-ethanediol

There are two diastereomeric products, meso 1,2-diphenyl-1,2-ethanediol and (±)-1,2-diphenyl-1,2-ethanediol. The achiral meso diol has a plane of symmetry. The (±)-diol, however, is chiral and forms as a racemic mixture.

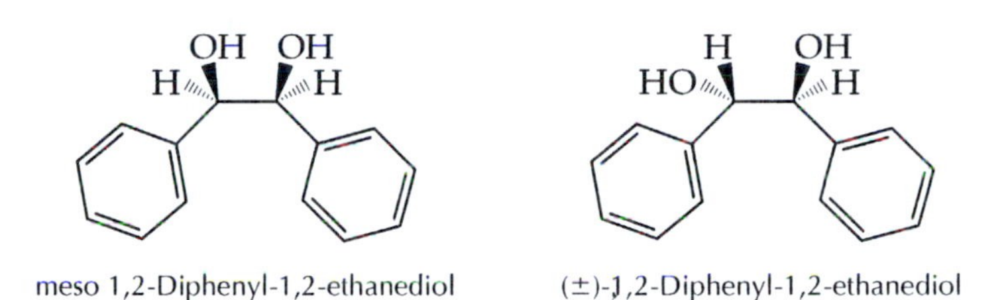

meso 1,2-Diphenyl-1,2-ethanediol (±)-1,2-Diphenyl-1,2-ethanediol

In the borohydride reduction of benzoin, one of these diastereomers is produced to a much larger extent than the other. You might think that the preferential attack of the borohydride transfers hydride to the least hindered face of the carbonyl group, which would produce the meso diastereomer. On the other hand, the (±)-diol is more stable and could be the preferred reduction product.

Synthesis of the Cyclic Acetal of 1,2-Diphenyl-1,2-ethanediol

Preparation of the cyclic acetal of your 1,2-diphenyl-1,2-ethanediol will allow you to determine the structure of the major stereoisomer formed by the $NaBH_4$ reduction of benzoin as well as the ratio of the major to the minor diastereomer formed in the reaction. Rather than using acetone itself in the synthesis of the cyclic acetal, the dimethyl acetal of acetone, 2,2-dimethoxypropane, is used.

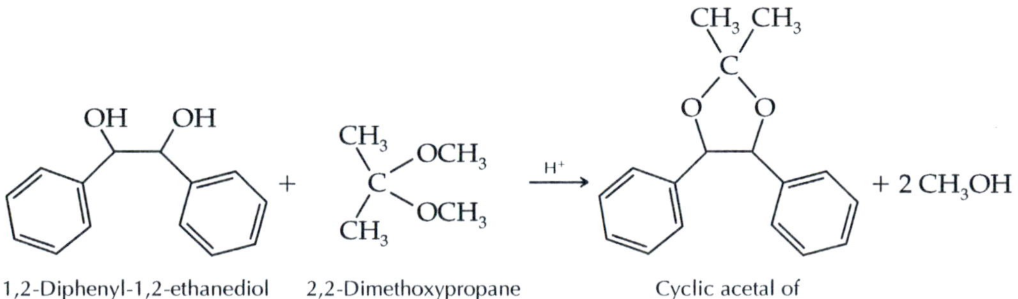

1,2-Diphenyl-1,2-ethanediol 2,2-Dimethoxypropane Cyclic acetal of 1,2-diphenyl-1,2-ethanediol

The 1,2-diphenyl-1,2-ethanediol is converted into the cyclic acetal and methanol. Formation of an acetal has an enthalpy of almost zero, but the liberation of two molecules of methanol gives the reaction a positive entropy. The favorable entropy, plus a large excess of 2,2-dimethoxypropane, work together to produce a high yield of the cyclic acetal.

Formation of the cyclic acetal is an acid-catalyzed S_N1 reaction. After protonation of 2,2-dimethoxypropane, a molecule of methanol leaves, forming a resonance-stabilized carbocation.

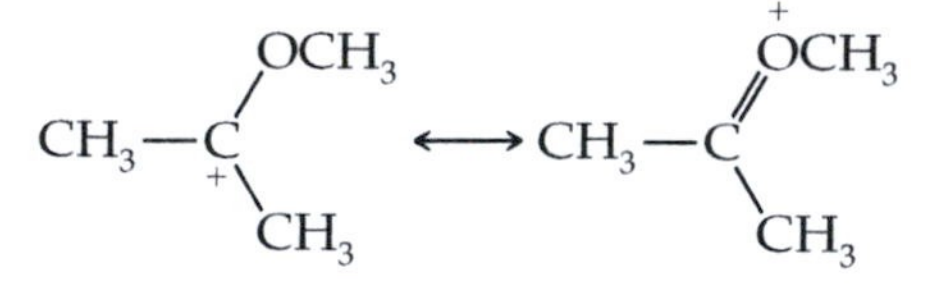

A hydroxyl group of the diol then reacts with the carbocation, and loss of a proton leads to the hemiacetal. Subsequent protonation of the remaining methoxy group allows it to leave, forming another resonance-stabilized carbocation. The remaining hydroxyl group then bonds to the carbocation center. Loss of a proton finally produces the cyclic acetal. Without the presence of the *p*-toluenesulfonic acid catalyst, which has a pK_a of -2, there is no leaving group that is good enough to produce the carbocation intermediates, even though they are resonance stabilized. Thus, the presence of a strong acid is required to catalyze the formation of the cyclic acetal.

NMR Analysis of the Cyclic Acetal of 1,2-Diphenyl-1,2-ethanediol

NMR analysis can easily differentiate the cyclic acetals of the meso and the (±) isomers of 1,2-diphenyl-1,2-ethanediol and allow you to determine the ratio of the major to the minor diastereomer formed in the borohydride reaction.

meso Diastereomer (±)-Diastereomer

The two methyl groups of the (±)-1,2-diphenyl-1,2-ethanediol acetal have identical environments. Each methyl group is cis to a hydrogen atom and a phenyl group, so the environment of each is the same, and a single methyl peak is present at 1.7 ppm in the NMR spectrum. However, there are two different environments for the methyl groups of the meso isomer. One methyl group is cis to two phenyl groups and the other is cis to two hydrogen atoms. In the NMR spectrum of meso 1,2-diphenyl-1,2-ethanediol acetal, one methyl group appears at 1.6 ppm and the other at 1.8 ppm. Thus, the NMR peak pattern in the 1.5–2.0 ppm region should allow you to determine which stereoisomer of 1,2-diphenyl-1,2-ethanediol is the major product of $NaBH_4$ reduction. In addition, the methine protons of the (±)-cyclic acetal appear at 4.75 ppm, whereas those of the meso diastereomer appear at 5.52 ppm. Integration of these two peaks will give the ratio of the two diastereomers that form in the $NaBH_4$ reduction.

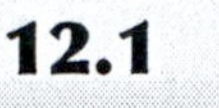

12.1 Thiamine-Catalyzed Benzoin Condensation

PURPOSE: To synthesize benzoin using vitamin B_1 as the catalyst.

PRELABORATORY ASSIGNMENT: Predict the major functional group absorptions that will appear in the IR spectrum of benzoin. At what frequencies would these absorptions appear?

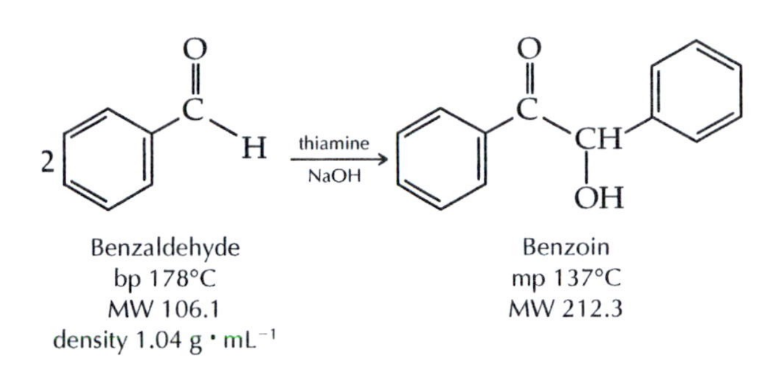

MINISCALE PROCEDURE

Techniques Recrystallization: Technique 9.5
Thin-Layer Chromatography: Technique 15
IR Spectroscopy: Technique 18

SAFETY INFORMATION

Benzaldehyde is toxic and an irritant. Wear gloves and avoid contact with skin, eyes, and clothing.

Aqueous **sodium hydroxide** solutions are corrosive and cause burns. Solutions as dilute as 9% (2.5 M) can cause severe eye injury.

Weigh 0.75 g of recently purchased thiamine hydrochloride (vitamin B_1)* and dissolve it in 1.7 mL of distilled water in a 25-mL round-bottomed flask. Add 7 mL of 95% ethanol and cool the resulting solution in an ice-water bath. While continuing to cool the thiamine hydrochloride solution in the ice-water bath, slowly add 1.5 mL of cold 3 M sodium hydroxide solution over a 5-min period. Gently swirl the thiamine solution during the entire addition to ensure thorough mixing. The solution will become yellow.

Measure 4.0 mL of benzaldehyde (freshly distilled or from a newly opened bottle) and add it to the reaction mixture. Fit the flask with a

*Thiamine hydrochloride is sensitive to heat and highly alkaline pH. It should be stored in the refrigerator when not being used. It is important not to heat the reaction mixture any more than necessary and not to add any more sodium hydroxide than needed.

water-cooled condenser. Heat the reaction at 60°C for 90 min in a water bath or stopper the reaction and allow it to stand in your laboratory drawer until your next laboratory period (48 h or more).

If the reaction mixture was heated, allow it to cool to room temperature. Precipitate the benzoin by cooling the flask in an ice-water bath until the temperature is about 10°C. Benzoin appears as white crystals. If the benzaldehyde was not completely pure, some oil may also be floating on the surface of the reaction mixture. This oil is probably a mixture of benzaldehyde and some benzoin. It may be analyzed by thin-layer chromatography on silica gel plates, using a polar developing solvent [see Technique 15].

Collect the crude solid product by vacuum filtration on a Buchner funnel. Wash it with 25 mL of cold water while it is still on the funnel. Set aside 10–15 mg of crude benzoin for a melting-point determination before you recrystallize the rest from 95% ethanol [see Technique 9.3]. Approximately 8 mL of 95% ethanol are required per gram of crude benzoin. Be careful not to add more than the minimum amount of ethanol needed to dissolve the product.

Determine the melting points of both the dried crude and recrystallized benzoin samples. In some sources benzoin is reported to melt at 137°C, in others at 133–134°C. Calculate your percent yield. Typical yields are 40–50% if the recrystallization is done carefully. Obtain and analyze an IR spectrum of your product. Your instructor may also ask you to obtain and analyze the NMR spectrum of benzoin.

CLEANUP: If the filtrate from the reaction mixture has an upper layer of oil, transfer it to a separatory funnel. Drain the lower aqueous layer into a beaker and neutralize the solution to pH 7, using 10% hydrochloric acid before washing it down the sink or pouring it into the container for aqueous inorganic waste. Pour the oil remaining in the separatory funnel and the filtrate remaining from the recrystallization into the container for flammable (organic) waste.

12.2 Stereoselectivity of the Borohydride Reduction of Benzoin

QUESTION: What is the stereoselectivity of the $NaBH_4$ reduction of benzoin to 1,2-diphenyl-1,2-ethanediol?

PRELABORATORY ASSIGNMENT: Propose an appropriate developing solvent for a TLC analysis to determine if your $NaBH_4$ reduction is complete. Describe how the R_f values for benzoin and 1,2-diphenyl-1,2-ethanediol will differ.

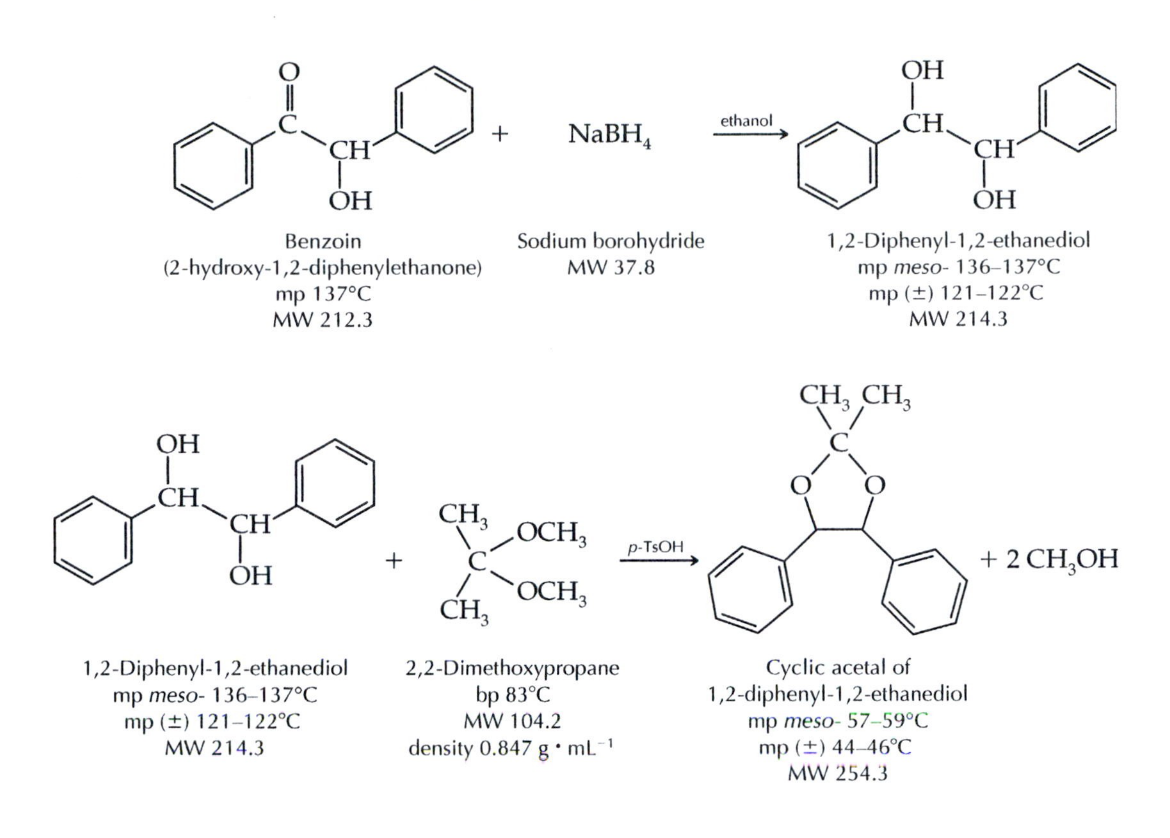

MINISCALE PROCEDURE

Techniques Thin-Layer Chromatography: Technique 15
Mixed Solvent Recrystallization: Technique 9.2
Extraction: Technique 8.2
NMR Spectroscopy: Technique 19

$NaBH_4$ is moisture sensitive; keep the bottle tightly closed when not in use. Store the reagent in a desiccator.

SAFETY INFORMATION

Sodium borohydride is harmful if swallowed, inhaled, or absorbed through the skin. Avoid breathing the dust. Avoid contact with skin, eyes, and clothing. It decomposes to flammable, explosive hydrogen gas.

Hydrochloric acid (6 M) is corrosive and a skin irritant. Avoid contact with skin, eyes, and clothing.

2,2-Dimethoxypropane is an irritant. Wear gloves while handling it and avoid contact with skin, eyes, and clothing.

***p*-Toluenesulfonic acid monohydrate** is toxic and corrosive. Wear gloves while handling it and avoid contact with skin, eyes, and clothing.

Dichloromethane is toxic, an irritant, absorbed through the skin, and harmful if swallowed or inhaled. Use it only in a hood and wear neoprene gloves. Wash your hands thoroughly after handling it.

Acetone, hexane, and **pentane** are volatile and extremely flammable.

$NaBH_4$ Reduction

Add 250 mg of sodium borohydride to a magnetically stirred suspension of 1.0 g of benzoin in 15 mL of 100% ethanol in a 125-mL Erlenmeyer flask. Any sodium borohydride clinging to the side of the flask can be rinsed into the reaction mixture with a small additional amount of absolute ethanol. Stir the reaction for 15 min and check to see if the reaction is complete by TLC. When it is finished, place the flask in a beaker that is positioned on the magnetic stirrer, and add crushed ice to the beaker. Dilute the stirred solution with 15 mL of water and carefully add 2 mL of 6 M hydrochloric acid to the solution. **(Caution: Foaming.)** Add another 5 mL of water and continue the stirring for a further 15 min.

Collect the white precipitate by vacuum filtration and wash it with approximately 25 mL of water. Determine the melting point of the dried crude product. After successfully completing the synthesis of the 1,2-diphenyl-1,2-ethanediol acetal, the remaining dried crude 1,2-diphenyl-1,2-ethanediol can be recrystallized from a mixture of acetone and hexane [see Technique 9.3]; 8 mL of reagent-grade acetone per gram of product works well.

Synthesis of the Cyclic Acetal

Place 0.40 g of your dried crude 1,2-diphenyl-1,2-ethanediol, 5 mL of 2,2-dimethoxypropane, and 10 mg of *p*-toluenesulfonic acid monohydrate in a 10-mL Erlenmeyer flask. Cork the flask and stir the mixture at room temperature for 1 h and then pour it into 50 mL of dichloromethane. Extract the resulting mixture successively with 25-mL portions of water, saturated sodium bicarbonate solution, and water. Dry the organic phase with anhydrous magnesium sulfate for at least 10 min. Filter the solution into a dry flask. Recover the crude product by evaporation of the solvent using a stream of nitrogen in a hood or a rotary evaporator [see Technique 8.9].

Evaporation of the solvent yields the cyclic acetal of 1,2-diphenyl-1,2-ethanediol as a residue. Prepare a sample of the crude product for NMR analysis.

If you wish to recrystallize the cyclic acetal so that only the major diastereomer is present, it can be done from a small amount of pentane. Dissolve the crude product at room temperature and cool the solution to −5°C to 0°C in an ice-salt mixture. Crystallization can be induced by scratching the inside of the flask with a glass rod. Collect the recrystallized product by vacuum filtration. Wash the chunky white crystals with a small amount of chilled pentane.

CLEANUP: Neutralize the filtrate from the reduction reaction with solid sodium carbonate before washing it down the sink or pouring it into the container for aqueous inorganic waste. Wash the aqueous solutions remaining from the extraction of the cyclic acetal down the sink or pour them into the container for aqueous inorganic waste. Pour the filtrate from the recrystallization of the reduction product and of the cyclic acetal into the container for flammable (organic) waste.

Interpretation of Experimental Data

Analyze your NMR spectrum of the cyclic acetal of your 1,2-diphenyl-1,2-ethanediol product mixture. Which diastereomer is the major product of the borohydride reduction? What is the ratio of the major product to the minor one? Explain the stereoselectivity of the borohydride reduction.

Reference

1. Rowland, A. T. *J. Chem. Educ.* **1983,** *60,* 1084–85.

Questions

1. How does a base, such as NaOH, help to catalyze the benzoin condensation when thiamine is present?
2. Suggest reasons why benzoic acid impurity in the benzaldehyde may drastically reduce the yield of the thiamine-promoted condensation of benzaldehyde.
3. More NaOH solution could both catalyze the condensation and neutralize the benzoic acid present if old benzaldehyde were used in the synthesis of benzoin. Unfortunately, too much base catalyzes the formation of noxious hydrogen sulfide. Where does the hydrogen sulfide come from?
4. What modifications in the reaction conditions would be necessary if an enzyme were used to catalyze the reaction?
5. Why is hydrochloric acid added to the reaction mixture for reduction of benzoin with $NaBH_4$ after the reduction is complete?
6. Write out a complete reaction mechanism for the synthesis of the 1,2-diphenyl-1,2-ethanediol acetal from 1,2-diphenyl-1,2-ethanediol and 2,2-dimethoxypropane in the presence of acid.

PROJECT

13 DESIGNING A MULTISTEP SYNTHESIS

PURPOSE: To design and carry out a two- to three-step organic synthesis.

In this three- to four-week project you and your partner will design and carry out a multistep synthesis or three separate one-step syntheses. You can decide what direction the experimental work will take over the several weeks of the project. The instructor will either assign to each team an alkyl halide or alcohol as the starting material or allow the team to select one from a list distributed in the laboratory. Before you begin experimental work, you must submit a plan for each project to the instructor.

Guidelines for the Project

Your challenge is to carry out a multistep synthesis starting with 15 g of either an alkyl halide or a primary/secondary alcohol. Alternatively, you may be asked to carry out three separate syntheses from the 15 g of starting material. Whatever the option, the product of each synthetic step must be characterized by one or more of the following methods: melting point/boiling point, gas chromatography, or ^{1}H NMR/IR spectroscopy. Characterization is particularly important in a sequential multistep synthesis. It is senseless to perform the next step until you know the structure and purity of the starting material. If you are carrying out three separate syntheses, you should aim to hand in at least a 1-g sample of each product to your instructor at the end of the project.

You may not find a procedure for the specific compound that you wish to make, but you should be able to find a preparation for an analogous compound. For each reaction you will have to adapt the procedure to your compound and the amount of starting material that you are using. Note that it is usually easier to work with small amounts of solids than of liquids. If your final product is a liquid, it should have no more than eight or nine carbon atoms to keep its boiling point below 220°C so that a vacuum distillation is unnecessary.

You can model your synthetic reactions on other experiments or projects in this or another lab text or from procedures in the chemical literature (Ref. 2). You also need to prepare a list of the chemicals and equipment required for each step. Your instructor will provide specific deadlines for submitting this information.

Starting with 15 g of your substrate may seem excessive, but it is important to look at the kinds of losses that can occur in a multistep synthesis. If we assume that each step of a three-step synthesis has a 70% yield and that the molecular weights stay reasonably constant, the recovered yield after three steps is

$$15\text{ g} \times 0.70 \times 0.70 \times 0.70 = 5.1\text{ g}$$

If each step has a 60% yield, the recovered yield is

$$15\text{ g} \times 0.60 \times 0.60 \times 0.60 = 3.2\text{ g}$$

If one of the steps has a 70% yield and the other two have 50% yields, the calculation shows that 15 g of starting material would produce a final yield of 2.6 g. It is important to carry out chemical reactions that can be expected to produce high yields.

After you and your instructor have settled on a suitable starting material and before you begin experimental work, you need to submit to your instructor a plan for your project that outlines the syntheses you intend to undertake. Your plan should include a reference for each synthesis, balanced equations, the modified quantities of reagents scaled to your actual starting amount of alkyl halide or alcohol, the necessary safety considerations, and the by-products you will have left at the end of each synthesis. You need to find the physical constants for your products. Sources for this information include the *CRC Handbook of Chemistry and Physics*, *Aldrich Catalog Handbook of Fine Chemicals*, and *The Dictionary of Organic Compounds*.

Synthesis Options

Option 1: You can choose a three-step synthesis in which the second and third reactions start with the product of the previous reaction. You will probably want to use all the alkyl halide or alcohol starting material in the first step. You might need to use microscale techniques for the third reaction if your yields have been small in the two previous steps. Microscale distillation techniques should be used for any product with a volume of less than 3 mL.

Option 2: You can select three single-step reactions, in each case starting with your alkyl halide or alcohol. A particular type of reaction may not be repeated. Using 5 g of substrate for each synthesis is often appropriate, but you may use more for one synthesis and then use a microscale reaction for another synthesis. This may be a good strategy if you carry out a miniscale synthesis that involves a final distillation.

Option 3: You can select a two-step synthesis in which the starting material for the second reaction is the product of the first. Here we suggest that you use 10 g of your starting material for the first step. The third reaction would be a one-step synthesis, again starting with the alkyl halide or alcohol.

SAFETY INFORMATION

You need to find safety information pertaining to the reagents you will be using. Sources of safety information about chemicals are discussed in Technique 1.6, "Safety Information."

CLEANUP: Your instructor must approve your cleanup procedures before you begin the project. Prepare a list of all by-products remaining after each synthesis. Some examples of by-products include the aqueous layer from an extraction, used drying agent, solvent distilled from a product, and solutions prepared for GC analysis. Consult the cleanup procedures in experiments that you have done previously for ways of handling similar materials. Your instructor may also give you specific guidelines that pertain to disposal of by-products in your institution and community.

Suggestions for Syntheses

Suggestions for possible synthetic routes follow. You will find additional reactions of alkyl halides and alcohols described in your textbook. However, it is important to remember that reactions described in your textbook may be very complex to carry out in practice. Keep a healthy sense of perspective about what you can accomplish in a few hours of laboratory time. If there is a particular reaction that you would like to try, discuss it with your instructor beforehand.

Reactions of Primary Alkyl Halides

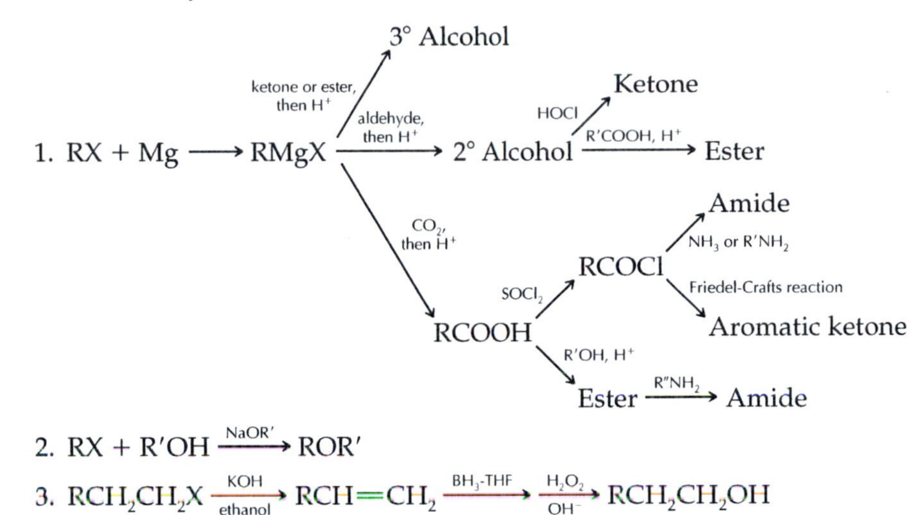

Reactions of Secondary and Tertiary Alkyl Halides

1. Any of the Grignard reactions and subsequent reactions listed for primary alkyl halides
2. A Friedel-Crafts synthesis with a tertiary halide or any secondary halide that cannot undergo rearrangement

Reactions of Primary and Secondary Alcohols

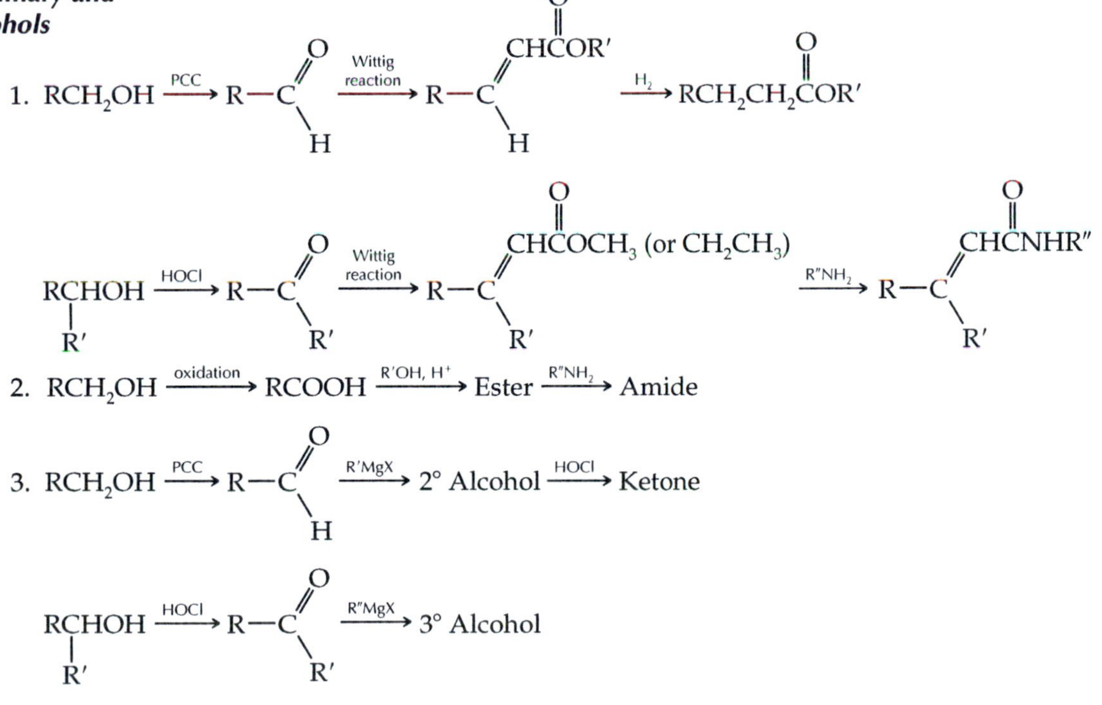

Report

Your instructor will specify the report format for your project. You may also be asked to give an oral presentation or prepare a poster about your project.

References

1. Potter, N. H.; McGrath, T. F. *J. Chem. Educ.* **1989**, *66*, 666–667.
2. Mohrig, J. R.; Hammond, C. N.; Schatz, P. F.; Morrill, T. C. *Techniques in Organic Chemistry*; W. H. Freeman and Company: New York, 2003, Appendix, "The Literature of Organic Chemistry."

PROJECT 14

SUGARS: GLUCOSE PENTAACETATES

QUESTION: What is the relative stability of α- and β-D-glucose pentaacetate? How can you distinguish between kinetic and equilibrium control in the synthesis of the glucose pentaacetates?

> In this three- to four-week project you and a partner will study the reactivity of the D-glucose pentaacetates under conditions of kinetic and equilibrium control. You will synthesize the α- and β-isomers of glucose pentaacetate under acidic and basic conditions and use NMR spectroscopy to study their interconversion, determining their relative thermodynamic stability in the process.

Structure and Reactivity of Glucose

Through the process of photosynthesis, green plants synthesize the carbohydrates that are vital foods for animals. The simplest carbohydrates, such as glucose and sucrose, are called sugars because of their sweetness. Cellulose is the chief structural material of plants, and amylose is one of the major components of starch, our main source of energy. Both cellulose and amylose are linear polymers of glucose, the most abundant organic monomer unit on earth. Enzymes in our bodies catalyze formation of glucose by the hydrolysis of starch. Glucose, a six-carbon sugar, is an excellent fast-energy source. Because of their great importance in nature, carbohydrates have been studied extensively by organic chemists and biochemists.

D-(+)-Glucose, the enantiomer found commonly in nature, is a chiral molecule with four stereocenters. In water solution, glucose exists almost entirely as the cyclic hemiacetal, formed by the addition of the hydroxyl group at C-5 to the carbonyl double bond of the aldehyde. This reaction produces a new stereocenter at C-1, so there are two cyclic diastereomers of D-glucose:

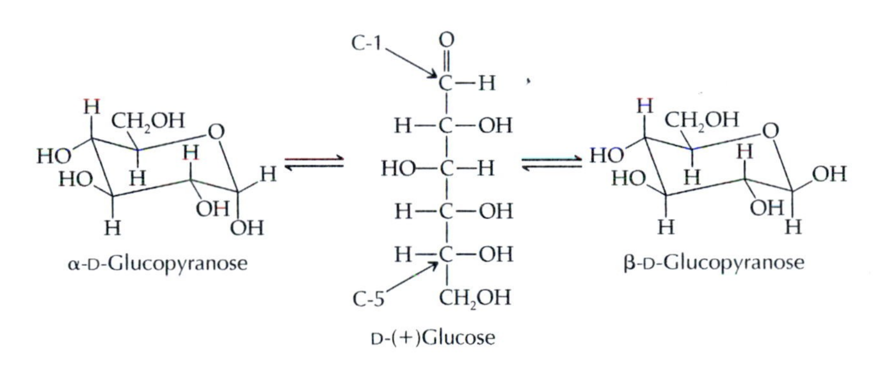

α-D-Glucopyranose

D-(+)Glucose

β-D-Glucopyranose

Cyclic diastereomers of sugars that differ from each other in their configurations only at C-1 are called ***anomers.*** C-1 is at the far right in each of the rings shown. The cyclic diastereomers of glucose are called glucopyranoses to indicate that they have six atoms in the ring (five carbons and one oxygen). The two cyclic diastereomers are differentiated by the Greek letters α and β.

β-D-Glucose has an optical rotation $[\alpha]_D^{20} = +18.7°$; the α-anomer has $[\alpha]_D^{20} = +112°$. If either of them is dissolved in water, the optical rotation gradually changes until the equilibrium value of $[\alpha]_D^{20} = +52.7°$ is reached, a process called ***mutarotation.*** Both acids and bases catalyze this equilibration. In water solution, the equilibrium mixture contains 63.6% of the β-anomer and 36.4% of the α-anomer. It is worth noting that cellulose, which humans cannot metabolize, and amylose, which we metabolize readily, differ only in their anomeric configurations. Whereas amylose has α-linkages between glucose units, cellulose has β-linkages.

In the presence of acetic anhydride, the esterification of D-(+)-glucose produces glucose pentaacetates, in which each of the five hydroxyl groups of the cyclic hemiacetal is transformed into an acetate group. Either acidic or basic conditions may be used to catalyze the esterification, but different anomers are produced with different catalysts.

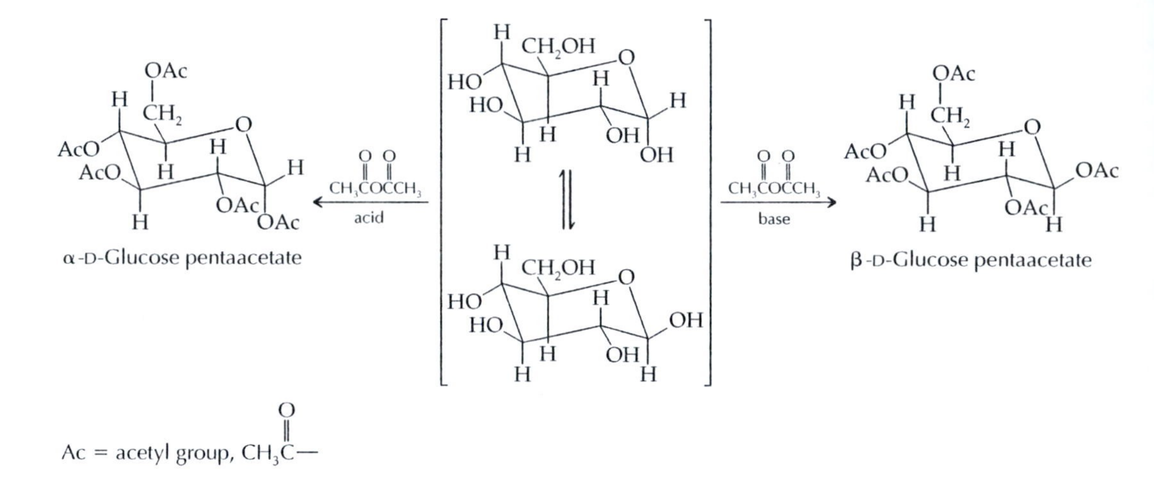

Ac = acetyl group, $CH_3\overset{O}{\overset{\|}{C}}—$

You will use molecular iodine (I_2) as the acidic catalyst in Project 14.1. Reaction of D-(+)-glucose with acetic anhydride in the presence of iodine, a mild Lewis acid, leads primarily to the α-D-glucose pentaacetate. The Lewis acid catalyzes the formation of an acetate ester by coordinating with the carbonyl oxygen of acetic anhydride, thereby withdrawing electron density from the carbonyl carbon. This makes it easier for a hydroxyl group of glucose to attack the carbonyl group, initiating the transfer of an acetyl group from acetic anhydride to the hydroxyl group.

Anhydrous sodium acetate is the basic catalyst in Project 14.1. Using basic catalysis, the β-anomer is the favored product. The base catalyzes ester formation by converting a hydroxyl group into an alkoxide anion, a much better nucleophile for the attack on acetic anhydride.

Kinetic and Equilibrium Control in the Glucose Pentaacetate System

When I_2 is used to catalyze the reaction of glucose and acetic anhydride, α-glucose pentaacetate is the major product. However, when sodium acetate is the catalyst, the major product is β-glucose pentaacetate.

It is not possible that both products are more stable. One of the glucose pentaacetates must be less stable than the other. The question is which one.

When two different products are possible in a chemical reaction and the isolated product is the one that forms faster, the reaction is said to be under ***kinetic control.*** Often the more stable product forms faster, but this is not always the case. Sometimes, a less stable product forms quickly and then gets trapped.

If the two products are in equilibrium with each other and the most stable product mixture results, even if it is not the mixture formed initially, the reaction is said to be under ***thermodynamic or equilibrium control.*** The energy diagram shown in Figure 14.1 may help you visualize the concepts of equilibrium and kinetic control, where the products of the two processes are different. The energy barrier to the product on the left is lower and it is formed faster (kinetic control). The product on the right is formed more slowly, since it has a higher energy transition state, but it has lower energy than the product on the left. If the product on the right is the one isolated, the reaction is under equilibrium control.

In a process that behaves as shown in Figure 14.1, the lower energy product is not formed first. It takes reaction conditions that allow an equilibrium to develop so that the product of kinetic control can return to the reactants many times. Every so often the thermodynamic product forms, and once it does, it is much harder for it to get over the high-energy barrier of the reverse reaction. Reactions run at low temperature and for short times favor kinetically controlled products. Reactions that have inherently slow reverse rates also favor kinetically controlled products.

At equilibrium, when there is sufficient energy to overcome all transition-state barriers, the outcome of a reaction is determined by the relative energies of the products. If the energy difference between two

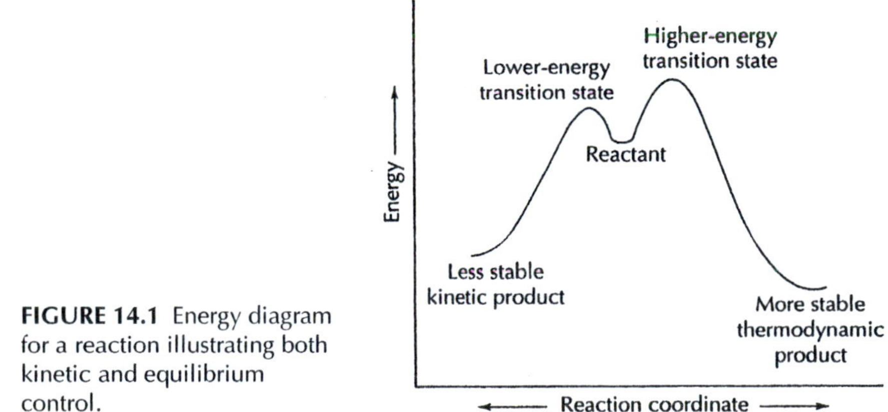

FIGURE 14.1 Energy diagram for a reaction illustrating both kinetic and equilibrium control.

products is small enough, equilibrium control may produce not a pure product but a mixture of products. For example, if K_{eq} is 9, there will be a 90/10 ratio of the two products at equilibrium.

^{1}H NMR Analysis of α- and β-Glucose Pentaacetates

Historically, the assignment of the configuration at C-1 of carbohydrates presented a difficult challenge. Today, however, ^{1}H NMR spectroscopy provides a convenient way to assign the configurations. Overall, the α- and β-glucose pentaacetates give complex ^{1}H NMR spectra as a result of the large number of protons with nearly identical chemical shifts. However, the proton attached to C-1 of each compound, which appears farthest downfield because of the strong deshielding of the two nearby electronegative oxygen atoms, is very easy to pick out. C-1 is the only carbon atom in the pentaacetates that is bonded to two oxygen atoms. The anomeric proton at C-1 shows a simple spin-spin splitting pattern. The protons at C-1 in both the α- and β-D-glucose pentaacetates have only one nearby proton neighbor, the one at C-2. Therefore, in each anomer the signal for the proton at C-1 appears as a doublet (Figure 14.2).

As you can see in the NMR spectra, there are differences in the peaks due to the α- and β-anomeric protons of the glucose pentaacetates. They differ both in chemical shift and in the magnitude of the coupling constant. ^{1}H NMR studies on a large number of cyclohexane derivatives have shown that equatorial protons generally appear about 0.5 ppm farther downfield than do axial protons. This pattern also holds true for

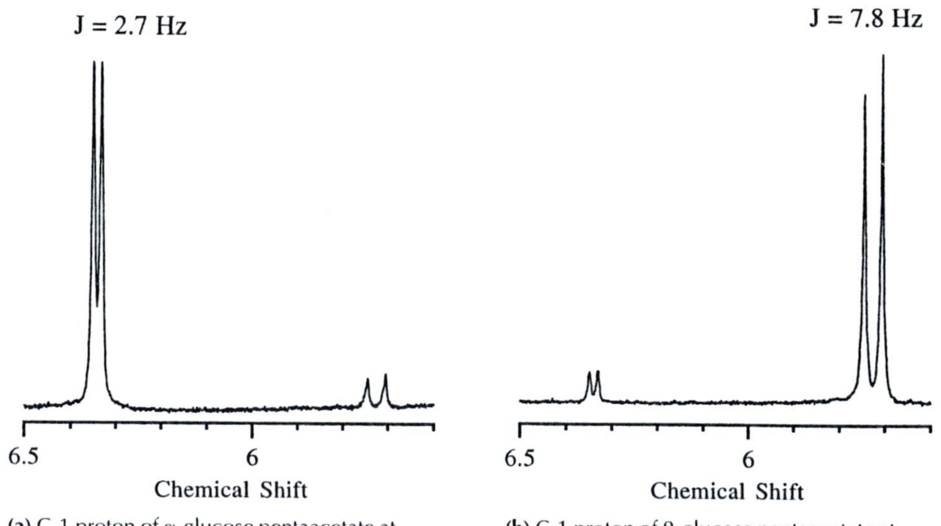

(a) C-1 proton of α-glucose pentaacetate at 6.33 ppm

(b) C-1 proton of β-glucose pentaacetate at 5.72 ppm

FIGURE 14.2 Expansions of the 5.6–6.5 ppm region of the ^{1}H NMR spectra of α- and β-glucose pentaacetate showing the signals for the C-1 anomeric protons. Each spectrum indicates that a small amount of the other isomer is also present.

sugars in the pyranose-ring chair conformations. Axial protons are shielded a small amount by diamagnetic anisotropy, which shifts their peaks upfield.

The coupling constants also provide useful information about the configurations of the α- and β-glucose pentaacetates. Analysis of spin-spin splitting from the ^{1}H NMR spectra of a large number of compounds shows that the efficiency of the coupling interaction of protons on adjacent carbon atoms depends on their conformational relationships. When the neighboring protons are eclipsed or anti to each other, their coupling constants (J) are largest. When they are gauche, the coupling constants are far smaller [see Technique 19, Figures 19.16 and 19.17].

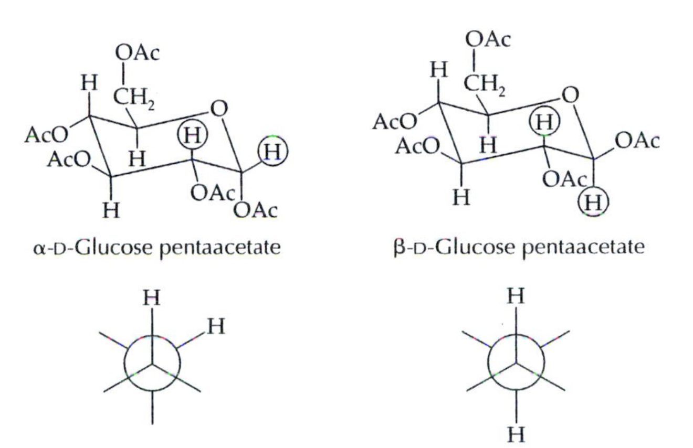

The chair conformation of a glucose pentaacetate, which has the majority of the large acetoxy groups and the acetoxymethyl group in equatorial positions, is more stable than the other chair conformation where these large groups are axial. In effect, α- and β-glucose pentaacetate are each frozen in a single chair conformation. The two axial protons at C-1 and C-2 of β-glucose pentaacetate have a larger coupling constant (7.8 Hz) than do the axial (C-2) and equatorial (C-1) protons on the α-anomer (2.7 Hz).

14.1 Synthesis of α-D-Glucose Pentaacetate and β-D-Glucose Pentaacetate

PURPOSE: To synthesize α- and β-D-glucose pentaacetate from glucose and acetic anhydride for the study of kinetic and equilibrium control in their formation.

PRELABORATORY ASSIGNMENT: Write a reasonable reaction mechanism for the formation of glucose pentaacetate from glucose and acetic anhydride, catalyzed by sodium acetate.

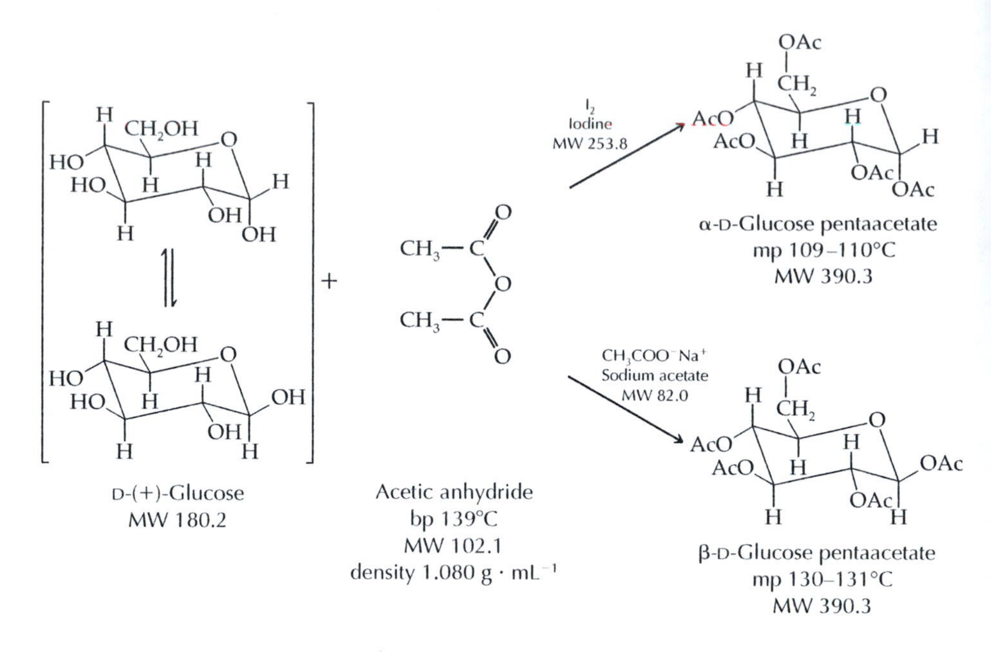

MINISCALE PROCEDURE

SAFETY INFORMATION

> **Acetic anhydride** is a strong irritant. Avoid contact with skin, eyes, and clothing. Pour acetic anhydride in the hood.
>
> **Iodine** is toxic and corrosive and emits harmful vapor by sublimation. Wear gloves and weigh it in a hood.

Synthesis and Purification of α-D-Glucose Pentaacetate

Place 10 mL of acetic anhydride, 2.0 g of powdered D-(+)-glucose, and a small magnetic stirring bar in a 25-mL round-bottomed flask equipped with a water-cooled condenser. Then add 0.50 g of iodine to the stirred mixture, which turns dark brown almost immediately. Stir the reaction in a hot-water bath at 85–95°C for 60 min.

Cool the reaction mixture and begin the workup by adding 10% sodium bisulfite solution to the stirred mixture until the brown color of iodine disappears (approximately 3–4 mL should be needed). The solution may continue to have a light yellow appearance. Pour the reaction mixture onto approximately 60 mL of ice in a 100-mL beaker. Rinse any residual material in the flask into the beaker with a small amount of water, and stir the mixture in the beaker using a glass stirring rod. At first a gummy material forms. On continued stirring, the ice melts and the gummy material changes into a granular or powdery solid. Collect the product by vacuum filtration. Set aside a small portion of the crude reaction product to dry before preparing a sample for NMR analysis in $CDCl_3$ solution [see Technique 19.2].

Recrystallize the remaining crude product from 1:2 (v/v) methanol/water solution. Approximately 10 mL of this solvent mixture is required for each gram of glucose pentaacetate. Weigh your dried, recrystallized product and determine its melting point. Calculate the percent yield. Prepare a sample of the recrystallized product for NMR analysis. Save your recrystallized product for the study of kinetic and equilibrium control in 14.2.

CLEANUP: Neutralize the filtrate from the reaction mixture with sodium carbonate **(Caution: Foaming.)** before washing it down the sink or pouring it into the container for aqueous inorganic waste. Place the filtrate from the recrystallization in the container for flammable (organic) waste.

Synthesis and Purification of β-D-Glucose Pentaacetate

Place 10 mL of acetic anhydride, 2.0 g of powdered D-(+)-glucose and 1.6 g of powdered, anhydrous sodium acetate in a 25-mL round-bottomed flask equipped with a water-cooled condenser. Use a mortar and pestle to powder the reagents if necessary. Add a small magnetic stirring bar to the flask. Heat the stirred reaction mixture in a boiling water bath for 1.5 h.

As you stir approximately 100 mL of ice in a beaker, slowly pour in the reaction mixture. At first a gummy material forms. On continued stirring, the ice melts and the gummy material changes into a granular or powdery solid. Collect the crude product by vacuum filtration. Set aside a small portion of the crude reaction product to dry before preparing a sample for NMR analysis in $CDCl_3$ solution [see Technique 19.2].

Recrystallize the remaining crude product from 1:2 (v/v) methanol/water solution. Approximately 10 mL of this solvent mixture is required for each gram of β-D-glucose pentaacetate. The crystals often have a beautiful, lustrous appearance.

Weigh your dried product, determine its melting point, and calculate the percent yield in the reaction. Analyze your recrystallized product by ^1H NMR spectroscopy. Save the recrystallized product for use in 14.2.

CLEANUP: Neutralize the filtrate from the reaction mixture with solid sodium carbonate **(Caution: Foaming.)** before washing it down the sink or pouring it into the container for aqueous inorganic waste. Place the filtrate from the recrystallization in the container for flammable (organic) waste.

14.2 Investigation of Kinetic and Equilibrium Control in the Glucose Pentaacetate System

QUESTION: What is the relative stability of α- and β-D-glucose pentaacetate? How can you account for the product mixtures that form under conditions of acidic and basic catalysis in the synthesis of α- and β-D-glucose pentaacetate?

PRELABORATORY ASSIGNMENT: Analyze the NMR spectra of the crude and recrystallized α- and β-D-glucose pentaacetate samples prepared in Project 14.1 by integrating the peaks in the 5.5–6.5 ppm region. Use an expansion of this region for careful integrations. Determine the ratio of α- to β-D-glucose pentaacetate in each of the four samples.

MICROSCALE PROCEDURE

You and your partner will design a set of experiments that will answer the questions posed. First you need to analyze the NMR spectra of the two crude reaction products that you isolated. What is the ratio of α- to β-D-glucose pentaacetate from the acidic and basic reaction mixtures? One set of the products must result from kinetic control. The other set of products may result from equilibrium control, or it may be partially under kinetic control if there has been insufficient time to set up complete equilibration.

A set of experiments designed to make sense of your data involves testing the equilibration of α-D-glucose pentaacetate and β-D-glucose pentaacetate in separate experiments. It is crucial that you have almost pure samples of the two anomers available, which is why you recrystallized your crude reaction products. If each recrystallized sample is not at least 95% pure, you may have to carry out another recrystallization.

Isomerization Method 1. Probably the simplest method for discovering which isomer isomerizes to the other is to use a catalytic amount of concentrated H_2SO_4 in acetic anhydride. Good initial conditions for attempted isomerizations are 150 mg of the glucose pentaacetate with 4 drops of concentrated H_2SO_4 in 0.75 mL acetic anhydride in a small test tube at 75–80°C for 20–30 min. The product can be recovered from 5–10 mL ice water by the method used in Project 14.1. After the product is dried, its composition can be determined by integration of its NMR spectrum.

Isomerization Method 2. An alternative method for determining which isomer isomerizes to the other uses the same reaction conditions as those given in isomerization method 1 but with a different workup. Rather than recovering the glucose pentaacetate from ice water, a few drops of the reaction mixture are added directly to 0.5 mL $CDCl_3$ in a small test tube. You must also have in the test tube an amount of solid anhydrous sodium acetate adequate to neutralize all the H_2SO_4 present. After the mixture has been well shaken, the organic solution is filtered directly into an NMR tube using a Pasteur pipet with cotton packed tightly at the top of the tip [see Technique 8.9]. Add additional $CDCl_3$ to the NMR tube to bring the sample to the necessary height and thoroughly mix the sample.

This workup is easier than the ice-water method but requires an accurate calculation on the amount of sodium acetate needed to neutralize the sulfuric acid. In addition, the methyl group of acetic anhydride is by far the largest peak in the NMR spectrum, but the C-1 protons of the glucose pentaacetates should be readily apparent in the 5.5–6.5 ppm region, with an adequate signal-to-noise ratio for NMR integrations.

14.3 Computational Chemistry Experiment

For aqueous solutions of glucose, the β-anomer, which has the C-1 hydroxyl group in the equatorial position, is about 0.33 kcal/mol lower in energy than the α-anomer, which has the C-1 hydroxyl group in the axial position. At equilibrium, the mixture is 63.6% β-anomer and 36.4% α-anomer. This equilibrium is consistent with the observation that the lowest energy conformers of substituted cyclohexane rings have the large substituents in the equatorial position. However, in some derivatives of glucose, the substituent attached to the anomeric carbon prefers to be in the axial position, an example of the anomeric effect.

Build molecules of *axial*-cyclohexyl acetate and *equatorial*-cyclohexyl acetate. Optimize their geometries using AM1 semiempirical calculations. Record the heats of formation. Which conformer is lower in energy? Then build molecules of *axial*-2-acetoxytetrahydropyran and *equatorial*-2-acetoxytetrahydropyran. Optimize their geometries using AM1 semiempirical calculations. Record the heats of formation. Which conformer is lower in energy?

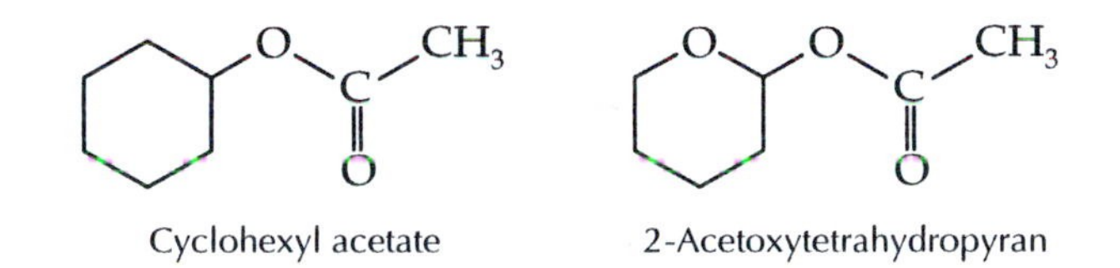

Using the difference in the heats of formation of the 2-acetoxytetrahydropyran conformers, calculate an equilibrium constant (K_{eq}) for the interconversion of α- and β-D-glucose pentaacetate. Assume that the entropy change is close to zero, so that $\Delta H° \cong -RT \ln K_{eq}$. Are your calculations consistent with the results from your investigations of kinetic and equilibrium control in the glucose pentaacetate system? Explain.

References

1. Pearson, W. A.; Spessard, G. O. *J. Chem. Educ.* **1975,** *52,* 814–15.
2. Schatz, P. *J. Chem. Educ,* **2001,** *78,* 1378.

Questions

1. What cyclic structures other than glucopyranoses might D-(+)-glucose form?
2. An alternative method for synthesizing glucose pentaacetates substitutes acetyl chloride (CH_3COCl) for acetic anhydride. What advantages or disadvantages would this method have over the acetic anhydride method?
3. In isomerization method 2 of Project 14.2, what undesirable result might occur if the H_2SO_4 is not neutralized with sodium acetate before the final NMR sample is prepared?
4. Sketch an energy-reaction progress diagram for the conversion of D-(+)-glucose to α- and β-D-glucose pentaacetate.

PROJECT

15 ENANTIOSELECTIVE SYNTHESIS BY ENZYMATIC HYDROLYSIS USING PIG-LIVER ESTERASE

QUESTION: How can one use NMR to determine the enantioselectivity of enzymatic hydrolysis of a diester?

> In this three-week project you and your partner will carry out four synthetic reactions, culminating with an enantioselective hydrolysis using the enzyme pig-liver esterase. You will be able to determine the enantiomeric excess in the enzymatic hydrolysis by NMR spectroscopy of the salt formed by the monocarboxylic acid product and an optically pure amine.

Diels-Alder Reaction and Esterification of the Product

In Project 15.1 you will carry out a Diels-Alder reaction to produce cyclohex-4-ene-*cis*-1,2-dicarboxylic anhydride, followed by its conversion to the dimethyl ester.

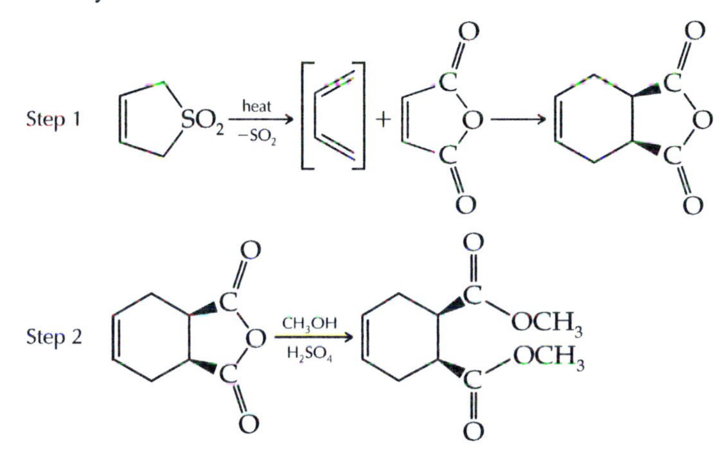

The cis configuration of the dienophile maleic anhydride and the stereospecificity of the Diels-Alder reaction determine the cis geometry of the Diels-Alder product. Because the conjugated diene 1,3-butadiene is a gas at room temperature, it is awkward to handle. In Project 15.1, butadiene is formed in situ, within the reaction mixture, by the extrusion of sulfur dioxide from the solid reactant, butadiene sulfone. As butadiene forms, it reacts quickly with the dienophile to form the Diels-Alder product. Since a carbon-carbon double bond in a cyclohexene ring must have a cis configuration (*trans*-cyclohexene is unknown), the trans-like conformation of butadiene must rearrange to the cis-like conformation before reaction with maleic anhydride can proceed. The reaction is carried out at the boiling point of dimethylbenzene (xylene).

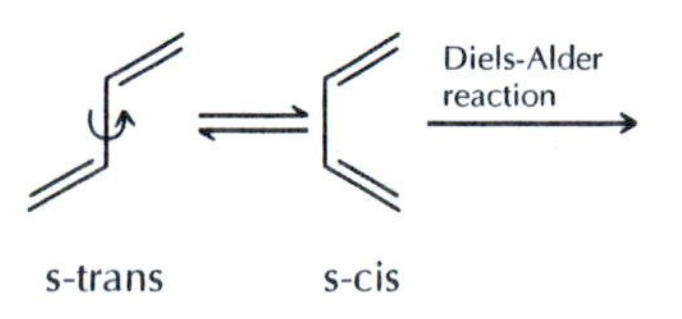

The Diels-Alder reaction has great utility in the chemical synthesis of six-membered ring compounds, from simple cyclohexanes to complex steroids. Otto Diels and Kurt Alder received the Nobel prize in 1950 for discovering the reaction that bears their names. It is a [4 + 2] cycloaddition reaction in which two molecules add together to form a cyclic compound in one step. The conjugated diene provides four π-electrons, and the dienophile provides the other two π-electrons. This allows a concerted reaction with a stabilized aromatic-like transition state having six π-electrons. In the transition state, the *p*-orbitals of all six carbon atoms overlap simultaneously.

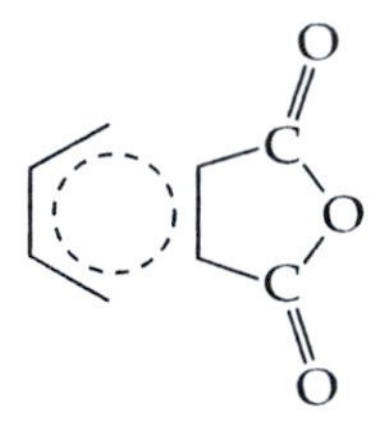

The Diels-Alder reaction has a faster rate when the carbon-carbon double bond of the dienophile is electron-deficient. The two electron-withdrawing groups on maleic anhydride speed up the reaction significantly.

The second step of Project 15.1 is the reaction of the bicyclic anhydride produced in the Diels-Alder reaction with methanol in the presence of sulfuric acid. The esterification reaction involves the addition of methanol to the carbonyl group and breaking of the anhydride linkage, giving the monomethyl ester. Further reaction with a second molecule of methanol produces dimethyl cyclohex-4-ene-*cis*-1,2-dicarboxylate.

The mechanism of acid-catalyzed ester formation from a carboxylic acid is well understood. The first step is protonation by the sulfuric acid catalyst to produce the conjugate acid of the carboxylic acid:

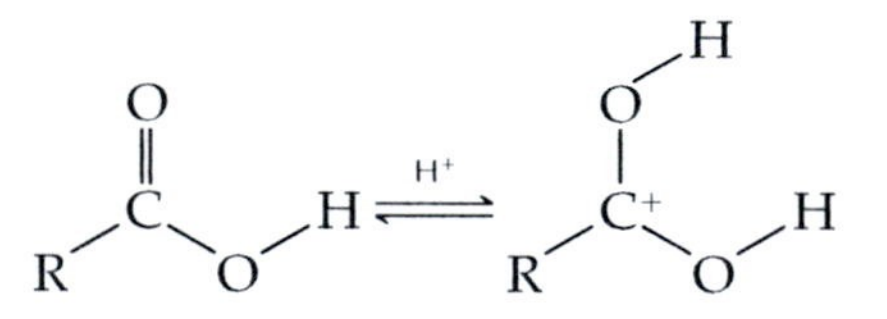

This protonated intermediate is very reactive and highly vulnerable to attack by an alcohol molecule:

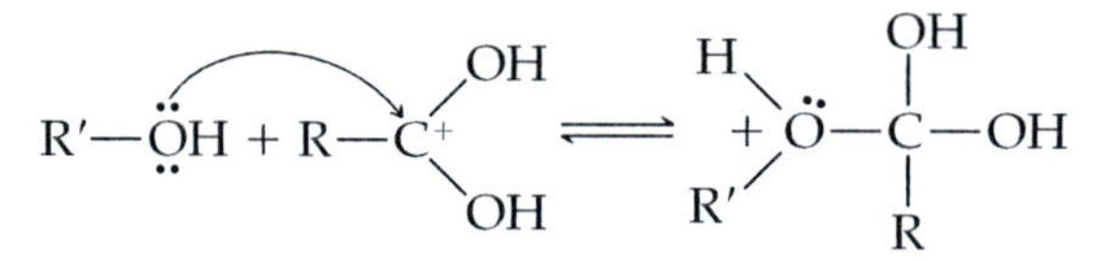

A proton shift then produces a new intermediate from which water can be lost:

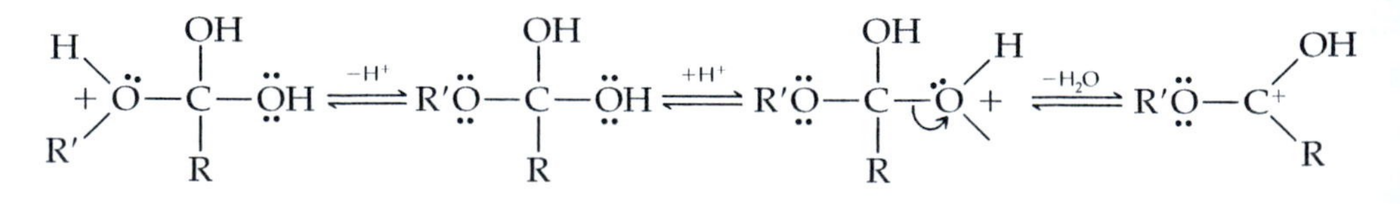

After water is lost, the carbocation loses a proton to form the ester:

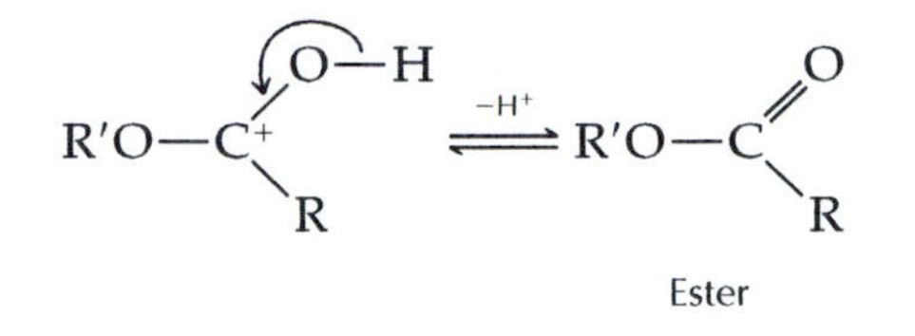

Synthesis of the Monomethyl Ester

In Project 15.2 you will also react cyclohex-4-ene-*cis*-1,2-dicarboxylic anhydride with methanol, but in this instance you will use methanol alone to synthesize the monomethyl ester methyl hydrogen cyclohex-4-ene-*cis*-1,2-dicarboxylate, which is necessary as a standard for evaluating the enantioselective success of the enzymatic hydrolysis.

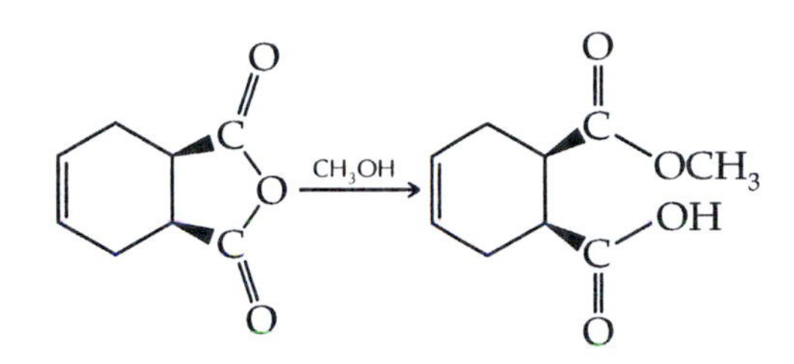

Reaction of cyclohex-4-ene-*cis*-1,2-dicarboxylic anhydride with methanol at a gentle reflux in the absence of a strong acid catalyst breaks the anhydride linkage. The anhydride is more reactive than a carboxylic acid to nucleophilic attack by methanol at the carbonyl group. Without catalysis by a strong acid, the monomethyl ester is the sole product.

Enzymatic Hydrolysis of Dimethyl Cyclohex-4-ene-cis-1,2-dicarboxylate

In Project 15.3 you will use an enzyme isolated from pig livers, called an esterase because in neutral water solution it catalyzes the cleavage of an ester into a carboxylic acid and an alcohol. X-ray diffraction studies have shown that the major hydrolysis product in Project 15.3 is 1-methyl hydrogen (1*S*,2*R*)-(+)-cyclohex-4-ene-*cis*-1,2-dicarboxylate.

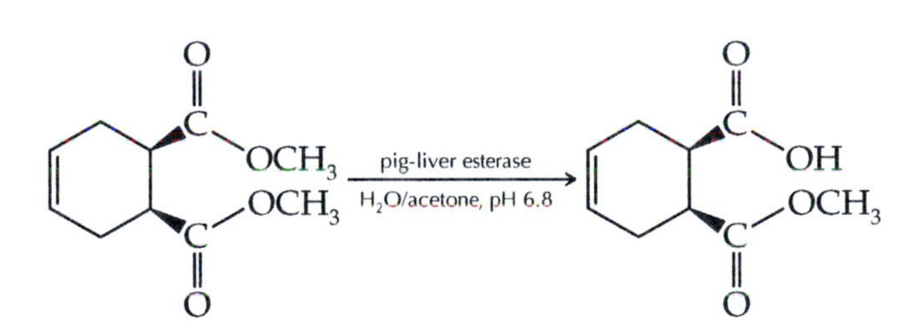

Virtually all chiral molecules in nature exist as pure enantiomers. These optically active compounds are vital for the efficient operation of the metabolic processes on which life depends. The enzymes that catalyze our metabolic reactions are proteins. Their active sites provide a chiral environment that can catalyze stereospecific reactions. Enzymes have the ability to produce stereospecific transformations on both chiral and nonchiral molecules. This property allows chemists to use enzymes to synthesize one enantiomer of a chiral product from a nonchiral substrate. The synthesis of pharmaceuticals, so important to the success of modern medicine, places great emphasis on the production of optically active drugs, which can be more effective and have fewer side effects then racemic drugs. Enzymes are useful in making optically active chiral precursors from which drugs can be synthesized.

An esterase enzyme has the ability to catalyze the hydrolysis of esters. Some esterases can recognize and act on many different esters. An ideal esterase for use in organic synthesis has low substrate selectivity so that it can catalyze reactions on a variety of substrates, while at the same time it is capable of producing high stereoselectivity, especially enantioselectivity, in the reactions it catalyzes. The substrate in Project 15.3 is a nonchiral *meso* diester, while the product of the hydrolysis is an optically active chiral monoester.

NMR Analysis of Enantiomeric Composition

The stereoselectivity of a reaction that produces an optically active compound is given by the ***enantiomeric excess (% ee)*** of the product. To calculate the % ee, you divide the difference between the amounts of the two enantiomers by the sum of the two enantiomers. Let us assume that a mixture of enantiomers contains 96% of the (+)-isomer and 4% of the (−)-isomer. The enantiomeric excess is then 92% ee.

$$\frac{96 - 4}{96 + 4} \times 100 = 92\% \text{ ee}$$

You will measure the enantiomeric composition of methyl hydrogen cyclohex-4-ene-*cis*-1,2-dicarboxylate by adding an equimolar quantity of an optically pure amine, (*S*)-(−)-α-methylbenzylamine. When the optically pure amine is added to a mixture of the two enantiomers of the monoester, a mixture of diastereomeric salts is formed. The NMR spectrum of the mixture displays two methyl signals for the monoester, one at 3.54 ppm and the other at 3.53 ppm. The ratio of these two signals allows you to calculate the enantiomeric excess obtained in the enzymatic hydrolysis.

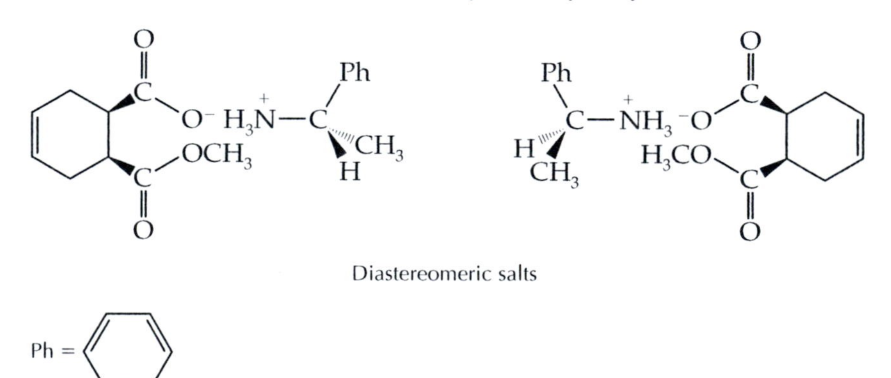

Diastereomeric salts

15.1 Synthesis of Dimethyl Cyclohex-4-ene-*cis*-1,2-dicarboxylate by Diels-Alder Cycloaddition and Esterification

PURPOSE: To synthesize a cyclic diester for use as a substrate in an enzymatic hydrolysis.

PRELABORATORY ASSIGNMENT: Predict the chemical shifts and relative integrations of the various protons in the NMR spectrum of cyclohex-4-ene-*cis*-1,2-dicarboxylic anhydride.

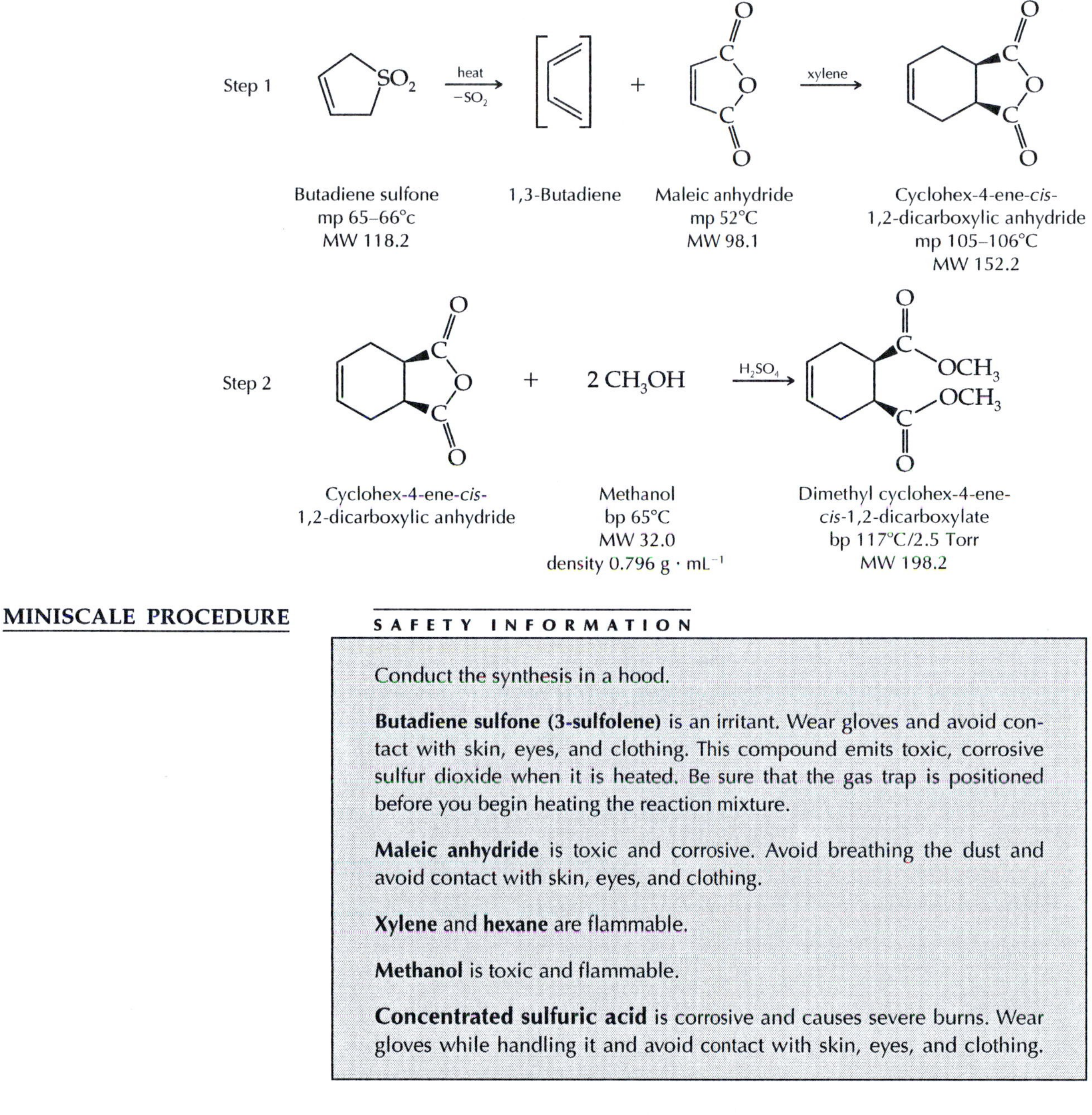

MINISCALE PROCEDURE

SAFETY INFORMATION

Conduct the synthesis in a hood.

Butadiene sulfone (3-sulfolene) is an irritant. Wear gloves and avoid contact with skin, eyes, and clothing. This compound emits toxic, corrosive sulfur dioxide when it is heated. Be sure that the gas trap is positioned before you begin heating the reaction mixture.

Maleic anhydride is toxic and corrosive. Avoid breathing the dust and avoid contact with skin, eyes, and clothing.

Xylene and **hexane** are flammable.

Methanol is toxic and flammable.

Concentrated sulfuric acid is corrosive and causes severe burns. Wear gloves while handling it and avoid contact with skin, eyes, and clothing.

Cyclohex-4-ene-cis-1,2-dicarboxylic anhydride

Place 5.0 g of 3-sulfolene, 3.0 g of maleic anhydride, and 2.5 mL of xylenes in a 25-mL round-bottomed flask equipped with a reflux condenser and add a magnetic stirring bar. Attach a gas trap to the top of the flask, as shown in Technique 7.4, Figure 7.5. Stir the mixture and gently heat it until it becomes homogeneous. Then heat the mixture at reflux for 30 min. Remove the gas trap.

Add a mixture of 2.5 mL of hexane and 2.5 mL of xylene dropwise down the condenser, and discontinue the heating and stirring. On cooling, crystals form in the reaction flask. Cool the flask in an ice-water bath to encourage further crystallization.

Collect the crystals on a Buchner funnel by vacuum filtration, wash them with a small amount of cold hexane, and then allow them to air dry. Weigh the collected crystals, determine the melting point, and prepare a sample for NMR analysis in $CDCl_3$ solution [see Technique 19.2]. Concentration of the filtrate followed by cooling can yield an additional crop of crystals. Usually, further purification is not necessary for the next reaction. However, if it is required, the crystals can be recrystallized from hexane.

CLEANUP: Pour the filtrate into the container for flammable organic waste. Neutralize the water in the gas trap with solid sodium carbonate before washing the solution down the sink or pouring it into the container for inorganic waste.

Dimethyl Cyclohex-4-ene-cis-1,2-dicarboxylate

Place 2.0 g of cyclohex-4-ene-*cis*-1,2-dicarboxylic anhydride, 20 mL of anhydrous methanol, and 10 drops of concentrated sulfuric acid in a 50-mL round-bottomed flask equipped with a water-cooled reflux condenser. Gently heat the mixture at reflux for 1 hr in a 75–80°C water bath. After cooling the reaction mixture, pour it into 80 mL of saturated sodium bicarbonate solution. **(Caution: Foaming.)** Extract the mixture three times with 40-mL portions of diethyl ether and dry the combined ether extracts over anhydrous magnesium sulfate. Remove the ether by evaporation under a stream of nitrogen in a hood or with a rotary evaporator. Prepare a sample for NMR analysis in $CDCl_3$ solution [see Technique 19.2]. Use the NMR spectrum to assess the purity of your product.

CLEANUP: Pour the aqueous extract remaining from the extractions into the container for aqueous inorganic waste. Place the spent drying agent in the container for solid inorganic waste.

15.2 Synthesis of (±)-Methyl Hydrogen Cyclohex-4-ene-*cis*-1,2-dicarboxylate

PURPOSE: To synthesize the racemic monoester that you will use as a reference for your enzymatic hydrolysis of dimethyl cyclohex-4-ene-*cis*-1,2-dicarboxylate.

PRELABORATORY ASSIGNMENT: You need equimolar quantities of the optically pure amine (*S*)-(−)-α-methylbenzylamine and of methyl hydrogen cyclohex-4-ene-*cis*-1,2-dicarboxylate to prepare the ammonium salt of the carboxylic acid for NMR determination of the enantiomeric composition of the monoester. Calculate how many milligrams of the amine will be necessary to make this ammonium salt if you prepare the NMR sample with 20 mg of the monoester.

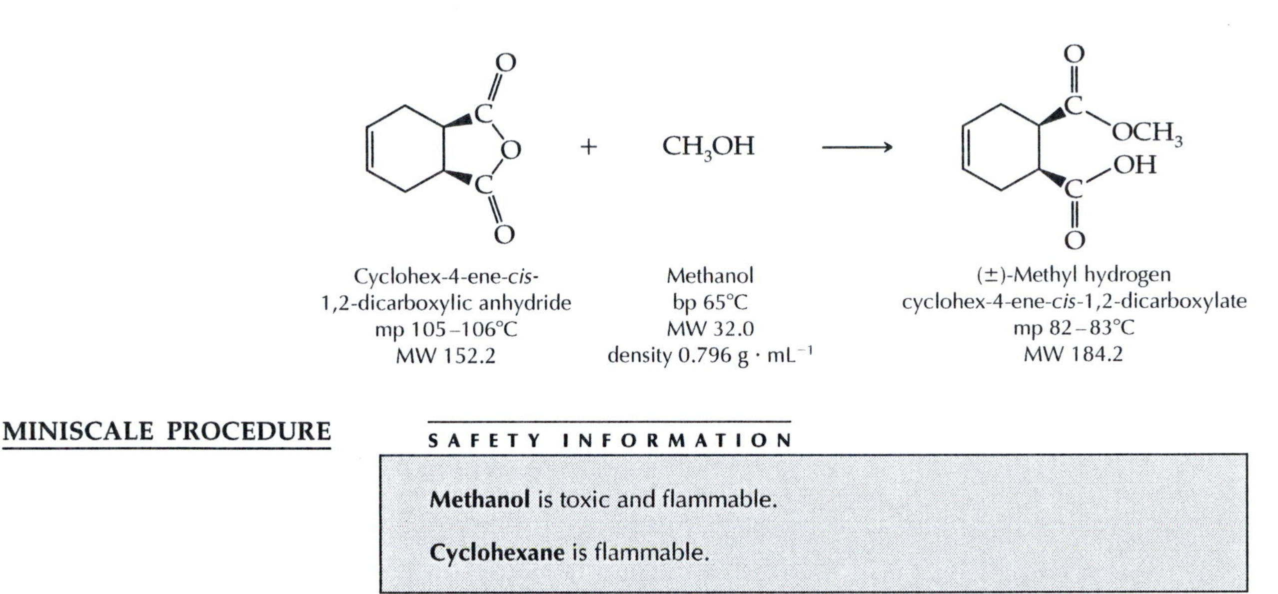

MINISCALE PROCEDURE

SAFETY INFORMATION

Methanol is toxic and flammable.

Cyclohexane is flammable.

Place 1.0 g of cyclohex-4-ene-*cis*-1,2-dicarboxylic anhydride and 10 mL of anhydrous methanol in a 25-mL round-bottomed flask equipped with a reflux condenser. Heat the mixture at a gentle reflux for 1 h. Remove the methanol by rotary evaporation or by evaporation with a stream of nitrogen. On standing overnight, the residue solidifies. It can be crystallized from cyclohexane to afford methyl hydrogen cyclohex-4-ene-*cis*-1,2-dicarboxylate as colorless prisms.

Prepare a sample of the product for NMR analysis in $CDCl_3$ solution. In your spectral analysis, pay particular attention to the expanded portion of the spectrum near 3.5 ppm, which is due to the methyl group on the oxygen atom of the ester. The enantiomeric composition of methyl hydrogen cyclohex-4-ene-*cis*-1,2-dicarboxylate is measured by adding an equimolar quantity of an optically pure amine, (S)-(−)-α-methylbenzylamine. Obtain an NMR spectrum of the resulting ammonium salt in $CDCl_3$.

15.3 Enantioselective Enzymatic Hydrolysis of Dimethyl Cyclohex-4-ene-*cis*-1,2-dicarboxylate

QUESTION: What is the enantioselectivity of the enzymatic hydrolysis of dimethyl cyclohex-4-ene-*cis*-1,2-dicarboxylate with pig-liver esterase?

PRELABORATORY ASSIGNMENT: Analyze the NMR spectrum of the (S)-(−)-α-methylbenzylammonium methyl cyclohex-4-ene-*cis*-1,2-dicarboxylate diastereomeric salt that you obtained in Project 15.2.

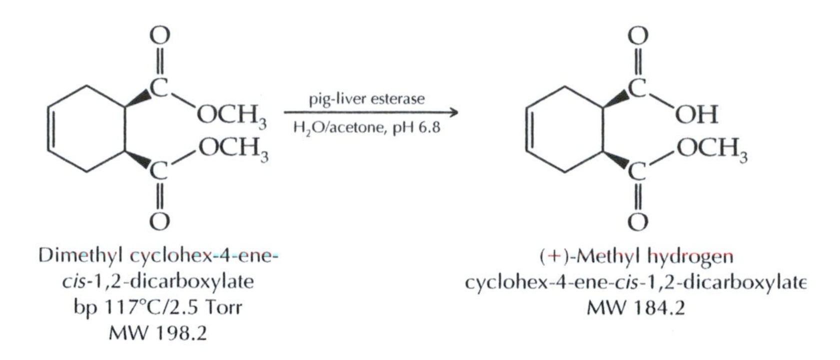

Dimethyl cyclohex-4-ene-*cis*-1,2-dicarboxylate
bp 117°C/2.5 Torr
MW 198.2

(+)-Methyl hydrogen cyclohex-4-ene-*cis*-1,2-dicarboxylate
MW 184.2

MINISCALE PROCEDURE

SAFETY INFORMATION

Acetone and **ethyl acetate** are flammable and irritants. Avoid contact with skin, eyes, and clothing.

Hydrochloric acid (2 M) is an irritant. Avoid contact with skin, eyes, and clothing.

Place 0.24g of dimethyl cyclohex-4-ene-*cis*-1,2-dicarboxylate and 5 mL of reagent-grade acetone into a 125-mL Erlenmeyer flask. Add 45 mL of pH 6.8 phosphate buffer to this mixture (Note 1). Gently stir the mixture and add 0.020 g of pig-liver esterase (Note 2) to it. Stir the mixture gently at room temperature for one week.

Work up the aqueous reaction mixture by acidifying it to pH 3 by *careful* addition of 2 M hydrochloric acid. Then extract the mixture five times with 20-mL portions of ethyl acetate (Note 3). Wash the combined ethyl acetate extract with 25 mL of water and then with 25 mL of saturated sodium chloride solution. Dry the solution over anhydrous magnesium sulfate.

Evaporation of the solvent with a rotary evaporator or a stream of nitrogen in a hood leaves a residue of methyl hydrogen cyclohex-4-ene-*cis*-1,2-dicarboxylate. Prepare the product for NMR analysis by adding an equimolar quantity of (*S*)-(−)-α-methylbenzylamine. Use the technique you used in Project 15.2 to determine the enantiomeric composition of methyl hydrogen cyclohex-4-ene-*cis*-1,2-dicarboxylate.

CLEANUP: Combine the aqueous extracts from the extractions and neutralize the solution with solid sodium carbonate before washing it down the sink or pouring it into the container for inorganic waste. Place the spent drying agent in the container for solid inorganic waste.

Interpretation of Experimental Data

In your spectral analysis, pay particular attention to the expanded portion of the spectrum near 3.5 ppm, which is where the methyl groups on the ester appear. This NMR region will be used for comparison with the

NMR spectrum of the (*S*)-(−)-α-methylbenzylamine salt of the racemic product from Project 15.2. If your enzymatic hydrolysis is stereospecific, only one diastereomeric salt will be formed and only one signal for the methyl ester will appear in the 3.5 ppm region. If two signals appear in this region, their ratio will allow you to calculate the enantiomeric excess obtained in the enzymatic hydrolysis.

Notes About Experimental Procedure

1. The phosphate buffer solution is prepared from 5.44 g of KH_2PO_4, 1.47 g of NaOH, and 400 mL of distilled water.
2. Pig-liver esterase used in this procedure is the crude lyophilized powder from Sigma-Aldrich; approximate activity = 20 units/mg.
3. Some problems with the formation of an emulsion during the extraction may be encountered. If an emulsion forms, separate it by draining most of the aqueous phase and then use saturated sodium chloride solution to break up the emulsion and complete the separation.

References

1. Mohr, P.; Waespe-Sarcevic, N.; Tamm, C.; Gawronska, K.; Gawronski, J. K. *Helv. Chim. Acta* **1983,** *66,* 2501–2511.
2. Kobayashi, S.; Kamiyama, K.; Iimori, T.; Ohno, M. *Tetrahedron Lett.* **1984,** *25,* 2557–2560.

Questions

1. What product is expected from the reaction of fumaric acid (*trans*-2-butenedioic acid) with 1,3-butadiene?
2. One of the following compounds reacts much more readily as the diene in a Diels-Alder reaction. Which one and why?

3. How does using methanol as the reaction solvent in Project 15.2 ensure that the reaction goes to completion?
4. Write out a mechanism for the reaction of methanol and the dicarboxylic anhydride in Project 15.2.
5. Why is acetone added as a cosolvent in the enzymatic hydrolysis reaction?
6. In Project 15.3, why is the aqueous reaction mixture brought to pH 3 with hydrochloric acid before the extractions with ethyl acetate?
7. What allows the diastereomeric (*S*)-α-methylbenzylammonium salt of the racemic methyl (1*S*,2*R*), (1*R*,2*S*)-cyclohex-4-ene-*cis*-1,2-dicarboxylate to have different NMR chemical shifts for the methyl protons of the ester?

Index

Note: Page numbers followed by f indicate figures; those followed by t indicate tables.

D

H

I

R

S

T